COURSE OF THEORETICAL PHYSICS

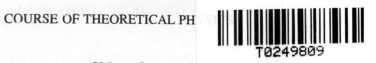

Volume 2

THE CLASSICAL THEORY OF FIELDS

Other Titles in-the COURSE OF THEORETICAL PHYSICS by
L. D. Landau and E. M. Lifshitz

THE CLASSICAL THEORY
OF
FIELDS

Fourth Revised English Edition

L. D. LANDAU AND E. M. LIFSHITZ
Institute for Physical Problems, Academy of Sciences of the U.S.S.R.

Translated from the Russian
by
MORTON HAMERMESH
University of Minnesota

ELSEVIER
BUTTERWORTH
HEINEMANN

AMSTERDAM • BOSTON • HEIDELBERG • LONDON • NEW YORK • OXFORD
PARIS • SAN DIEGO • SAN FRANCISCO • SINGAPORE • SYDNEY • TOKYO

Butterworth-Heinemann is an imprint of Elsevier
Linacre House, Jordan Hill, Oxford OX2 8DP, UK
30 Corporate Drive, Suite 400, Burlington, MA 01803, USA

Translated from the Sixth Revised Edition of Teoriya Pola,
Nauka, Moscow 1973

First published by Pergamon Press Ltd 1951

British Library Cataloguing in Publication Data
A catalogue record for this book is available from the British Library

Library of Congress Cataloging-in-Publication Data
Landau, Lev Davidovich, *1908–1968*
  The classical theory of fields (course of theoretical physics; v.2)
  Includes bibliographical references
  1. Electromagnetic fields   2. Field theory (physics)
  I. Lifshitz, Evgenii Mikhailovich, joint author   II. Title
  QXC665.E4L3713   1975 530.1′4   75–4737

ISBN: 978-0-7506-2768-9

For information on all Butterworth-Heinemann publications
visit our website at books.elsevier.com

Working together to grow
libraries in developing countries

www.elsevier.com | www.bookaid.org | www.sabre.org

ELSEVIER   BOOK AID
International   Sabre Foundation

Transferred to Digital Printing 2010

CONTENTS

CONTENTS

EXCERPTS FROM THE PREFACES TO THE FIRST AND SECOND EDITIONS

THIS book is devoted to the presentation of the theory of the electromagnetic and gravitational fields, i.e. electrodynamics and general relativity. A complete, logically connected theory of the electromagnetic field includes the special theory of relativity, so the latter has been taken as the basis of the presentation. As the starting point of the derivation of the fundamental relations we take the variational principles, which make possible the attainment of maximum generality, unity and simplicity of presentation.

In accordance with the overall plan of our Course of Theoretical Physics (of which this book is a part), we have not considered questions concerning the electrodynamics of continuous media, but restricted the discussion to "microscopic electrodynamics"—the electrodynamics of point charges *in vacuo*.

The reader is assumed to be familiar with electromagnetic phenomena as discussed in general physics courses. A knowledge of vector analysis is also necessary. The reader is not assumed to have any previous knowledge of tensor analysis, which is presented in parallel with the development of the theory of gravitational fields.

Moscow, December 1939
Moscow, June 1947

L. LANDAU, E. LIFSHITZ

PREFACE TO THE FOURTH ENGLISH EDITION

THE first edition of this book appeared more than thirty years ago. In the course of reissues over these decades the book has been revised and expanded; its volume has almost doubled since the first edition. But at no time has there been any need to change the method proposed by Landau for developing the theory, or his style of presentation, whose main feature was a striving for clarity and simplicity. I have made every effort to preserve this style in the revisions that I have had to make on my own.

As compared with the preceding edition, the first nine chapters, devoted to electrodynamics, have remained almost without changes. The chapters concerning the theory of the gravitational field have been revised and expanded. The material in these chapters has increased from edition to edition, and it was finally necessary to redistribute and rearrange it.

I should like to express here my deep gratitude to all of my helpers in this work—too many to be enumerated—who, by their comments and advice, helped me to eliminate errors and introduce improvements. Without their advice, without the willingness to help which has met all my requests, the work to continue the editions of this course would have been much more difficult. A special debt of gratitude is due to L. P. Pitaevskii, with whom I have constantly discussed all the vexing questions.

The English translation of the book was done from the last Russian edition, which appeared in 1973. No further changes in the book have been made. The 1994 corrected reprint includes the changes made by E. M. Lifshitz in the Seventh Russian Edition published in 1987.

I should also like to use this occasion to sincerely thank Prof. Hamermesh, who has translated this book in all its editions, starting with the first English edition in 1951. The success of this book among English-speaking readers is to a large extent the result of his labour and careful attention.

E. M. LIFSHITZ

PUBLISHER'S NOTE

As with the other volumes in the *Course of Theoretical Physics,* the authors do not, as a rule, give references to original papers, but simply name their authors (with dates). Full bibliographic references are only given to works which contain matters not fully expounded in the text.

EDITOR'S PREFACE TO THE
SEVENTH RUSSIAN EDITION

E. M. Lifshitz began to prepare a new edition of *Teoria Polia* in 1985 and continued his work on it even in hospital during the period of his last illness. The changes that he proposed are made in the present edition. Of these we should mention some revision of the proof of the law of conservation of angular momentum in relativistic mechanics, and also a more detailed discussion of the question of symmetry of the Christoffel symbols in the theory of gravitation. The sign has been changed in the definition of the electromagnetic field stress tensor. (In the present edition this tensor was defined differently than in the other volumes of the Course.)

June 1987
L. P. PITAEVSKII

NOTATION

Three-dimensional quantities

Three-dimensional tensor indices are denoted by Greek letters
Element of volume, area and length: dV, $d\mathbf{f}$, $d\mathbf{l}$
Momentum and energy of a particle: $\mathbf{p}$ and $\mathscr{E}$
Hamiltonian function: $\mathscr{H}$
Scalar and vector potentials of the electromagnetic field: ϕ and $\mathbf{A}$
Electric and magnetic field intensities: $\mathbf{E}$ and $\mathbf{H}$
Charge and current density: ρ and $\mathbf{j}$
Electric dipole moment: $\mathbf{d}$
Magnetic dipole moment: m

Four-dimensional quantities

Four-dimensional tensor indices are denoted by Latin letters i, k, l, ... and take on the values
  0, 1, 2, 3
We use the metric with signature $(+ - - -)$
Rule for raising and lowering indices—see p. 14
Components of four-vectors are enumerated in the form $A^i = (A^0, \mathbf{A})$
Antisymmetric unit tensor of rank four is e^{iklm}, where $e^{0123} = 1$ (for the definition, see p. 17)
Element of four-volume $d\Omega = dx^0 dx^1 dx^2 dx^3$
Element of hypersurface dS^i (defined on pp. 20–21)
Radius four-vector: $x^i = (ct, \mathbf{r})$
Velocity four-vector: $u^i = dx^i/ds$
Momentum four-vector: $p = (\mathscr{E}/c, \mathbf{p})$
Current four-vector: $j^i = (c\rho, \rho\mathbf{v})$
Four-potential of the electromagnetic field: $A^i = (\phi, \mathbf{A})$

Electromagnetic field four-tensor $F_{ik} = \dfrac{\partial A_k}{\partial x^i} - \dfrac{\partial A_i}{\partial x^k}$ (for the relation of the components of
  F_{ik} to the components of $\mathbf{E}$ and $\mathbf{H}$, see p. 65)
Energy-momentum four-tensor T^{ik} (for the definition of its components, see p. 83)

CHAPTER 1

THE PRINCIPLE OF RELATIVITY

§ 1. Velocity of propagation of interaction

For the description of processes taking place in nature, one must have a *system of reference*. By a system of reference we understand a system of coordinates serving to indicate the position of a particle in space, as well as clocks fixed in this system serving to indicate the time.

There exist systems of reference in which a freely moving body, i.e. a moving body which is not acted upon by external forces, proceeds with constant velocity. Such reference systems are said to be *inertial*.

If two reference systems move uniformly relative to each other, and if one of them is an inertial system, then clearly the other is also inertial (in this system too every free motion will be linear and uniform). In this way one can obtain arbitrarily many inertial systems of reference, moving uniformly relative to one another.

Experiment shows that the so-called *principle of relativity* is valid. According to this principle all the laws of nature are identical in all inertial systems of reference. In other words, the equations expressing the laws of nature are invariant with respect to transformations of coordinates and time from one inertial system to another. This means that the equation describing any law of nature, when written in terms of coordinates and time in different inertial reference systems, has one and the same form.

The interaction of material particles is described in ordinary mechanics by means of a potential energy of interaction, which appears as a function of the coordinates of the interacting particles. It is easy to see that this manner of describing interactions contains the assumption of instantaneous propagation of interactions. For the forces exerted on each of the particles by the other particles at a particular instant of time depend, according to this description, only on the positions of the particles at this one instant. A change in the position of any of the interacting particles influences the other particles immediately.

However, experiment shows that instantaneous interactions do not exist in nature. Thus a mechanics based on the assumption of instantaneous propagation of interactions contains within itself a certain inaccuracy. In actuality, if any change takes place in one of the interacting bodies, it will influence the other bodies only after the lapse of a certain interval of time. It is only after this time interval that processes caused by the initial change begin to take place in the second body. Dividing the distance between the two bodies by this time interval, we obtain the *velocity of propagation of the interaction*.

We note that this velocity should, strictly speaking, be called the *maximum* velocity of propagation of interaction. It determines only that interval of time after which a change occurring in one body *begins* to manifest itself in another. It is clear that the existence of a

1

maximum velocity of propagation of interactions implies, at the same time, that motions of bodies with greater velocity than this are in general impossible in nature. For if such a motion could occur, then by means of it one could realize an interaction with a velocity exceeding the maximum possible velocity of propagation of interactions.

Interactions propagating from one particle to another are frequently called "signals", sent out from the first particle and "informing" the second particle of changes which the first has experienced. The velocity of propagation of interaction is then referred to as the *signal velocity*.

From the principle of relativity it follows in particular that the velocity of propagation of interactions is the *same* in *all* inertial systems of reference. Thus the velocity of propagation of interactions is a universal constant. This constant velocity (as we shall show later) is also the velocity of light in empty space. The velocity of light is usually designated by the letter c, and its numerical value is

$$c = 2.998 \times 10^{10} \text{ cm/sec.} \tag{1.1}$$

The large value of this velocity explains the fact that in practice classical mechanics appears to be sufficiently accurate in most cases. The velocities with which we have occasion to deal are usually so small compared with the velocity of light that the assumption that the latter is infinite does not materially affect the accuracy of the results.

The combination of the principle of relativity with the finiteness of the velocity of propagation of interactions is called the *principle of relativity of Einstein* (it was formulated by Einstein in 1905) in contrast to the principle of relativity of Galileo, which was based on an infinite velocity of propagation of interactions.

The mechanics based on the Einsteinian principle of relativity (we shall usually refer to it simply as the principle of relativity) is called *relativistic*. In the limiting case when the velocities of the moving bodies are small compared with the velocity of light we can neglect the effect on the motion of the finiteness of the velocity of propagation. Then relativistic mechanics goes over into the usual mechanics, based on the assumption of instantaneous propagation of interactions; this mechanics is called *Newtonian* or *classical*. The limiting transition from relativistic to classical mechanics can be produced formally by the transition to the limit $c \to \infty$ in the formulas of relativistic mechanics.

In classical mechanics distance is already relative, i.e. the spatial relations between different events depend on the system of reference in which they are described. The statement that two nonsimultaneous events occur at one and the same point in space or, in general, at a definite distance from each other, acquires a meaning only when we indicate the system of reference which is used.

On the other hand, time is absolute in classical mechanics; in other words, the properties of time are assumed to be independent of the system of reference; there is one time for all reference frames. This means that if any two phenomena occur simultaneously for any one observer, then they occur simultaneously also for all others. In general, the interval of time between two given events must be identical for all systems of reference.

It is easy to show, however, that the idea of an absolute time is in complete contradiction to the Einstein principle of relativity. For this it is suffcient to recall that in classical mechanics, based on the concept of an absolute time, a general law of combination of velocities is valid, according to which the velocity of a composite motion is simply equal to the (vector) sum of the velocities which constitute this motion. This law, being universal, should also be applicable to the propagation of interactions. From this it would follow that the velocity of propagation must be different in different inertial systems of reference, in contradiction to the principle of relativity. In this matter experiment completely confirms the principle of relativity. Measurements first performed by Michelson (1881) showed complete lack of dependence of the velocity of light on its direction of propagation; whereas according to classical mechanics the velocity of light should be smaller in the direction of the earth's motion than in the opposite direction.

Thus the principle of relativity leads to the result that time is not absolute. Time elapses differently in different systems of reference. Consequently the statement that a definite time interval has elapsed between two given events acquires meaning only when the reference frame to which this statement applies is indicated. In particular, events which are simultaneous in one reference frame will not be simultaneous in other frames.

To clarify this, it is instructive to consider the following simple example.

Let us look at two inertial reference systems K and K' with coordinate axes XYZ and $X'\,Y'\,Z'$ respectively, where the system K' moves relative to K along the $X(X')$ axis (Fig. 1).

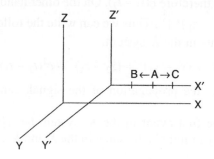

Fig. 1.

Suppose signals start out from some point A on the X' axis in two opposite directions. Since the velocity of propagation of a signal in the K' system, as in all inertial systems, is equal (for both directions) to c, the signals will reach points B and C, equidistant from A, at one and the same time (in the K' system)

But it is easy to see that the same two events (arrival of the signal at B and C) can by no means be simultaneous for an observer in the K system. In fact, the velocity of a signal relative to the K system has, according to the principle of relativity, the same value c, and since the point B moves (relative to the K system) toward the source of its signal, while the point C moves in the direction away from the signal (sent from A to C), in the K system the signal will reach point B earlier than point C.

Thus the principle of relativity of Einstein introduces very drastic and fundamental changes in basic physical concepts. The notions of space and time derived by us from our daily experiences are only approximations linked to the fact that in daily life we happen to deal only with velocities which are very small compared with the velocity of light.

§ 2. Intervals

In what follows we shall frequently use the concept of an *event*. An event is described by the place where it occurred and the time when it occurred. Thus an event occurring in a certain material particle is defined by the three coordinates of that particle and the time when the event occurs.

It is frequently useful for reasons of presentation to use a fictitious four-dimensional space, on the axes of which are marked three space coordinates and the time. In this space events are represented by points, called *world points*. In this fictitious four-dimensional space there corresponds to each particle a cetain line, called a *world line*. The points of this line determine the coordinates of the particle at all moments of time. It is easy to show that to a particle in uniform rectilinear motion there corresponds a straight world line.

We now express the principle of the invariance of the velocity of light in mathematical form. For this purpose we consider two reference systems K and K' moving relative to each other with constant velocity. We choose the coordinate axes so that the axes X and X' coincide, while the Y and Z axes are parallel to Y' and Z'; we designate the time in the systems K and K' by t and t'.

Let the first event consist of sending out a signal, propagating with light velocity, from a point having coordinates $x_1 y_1 z_1$ in the K system, at time t_1 in this system. We observe the propagation of this signal in the K system. Let the second event consist of the arrival of the signal at point $x_2 y_2 z_2$ at the moment of time t_2. The signal propagates with velocity c; the distance covered by it is therefore $c(t_1 - t_2)$. On the other hand, this same distance equals $[(x_2 - x_1)^2 + (y_2 - y_1)^2 + (z_2 - z_1)^2]^{\frac{1}{2}}$. Thus we can write the following relation between the coordinates of the two events in the K system:

$$(x_2 - x_1)^2 + (y_2 - y_1)^2 + (z_2 - z_1)^2 - c^2(t_2 - t_1)^2 = 0. \tag{2.1}$$

The same two events, i.e. the propagation of the signal, can be observed from the K' system:

Let the coordinates of the first event in the K' system be $x_1' y_1' z_1' t_1'$, and of the second: $x_2' y_2' z_2' t_2'$. Since the velocity of light is the same in the K and K' systems, we have, similarly to (2.1):

$$(x_2' - x_1')^2 + (y_2' - y_1')^2 + (z_2' - z_1')^2 - c^2(t_2' - t_1')^2 = 0. \tag{2.2}$$

If $x_1 y_1 z_1 t_1$ and $x_2 y_2 z_2 t_2$ are the coordinates of *any* two events, then the quantity

$$s_{12} = [c^2(t_2 - t_1)^2 - (x_2 - x_1)^2 - (y_2 - y_1)^2 - (z_2 - z_1)^2]^{\frac{1}{2}} \tag{2.3}$$

is called the *interval* between these two events.

Thus it follows from the principle of invariance of the velocity of light that if the interval between two events is zero in one coordinate system, then it is equal to zero in all other systems.

If two events are infinitely close to each other, then the interval ds between them is

$$ds^2 = c^2 dt^2 - dx^2 - dy^2 - dz^2. \tag{2.4}$$

The form of expressions (2.3) and (2.4) permits us to regard the interval, from the formal point of view, as the distance between two points in a fictitious four-dimensional space (whose axes are labelled by x, y, z, and the product ct). But there is a basic difference between the rule for forming this quantity and the rule in ordinary geometry: in forming the square of the interval, the squares of the coordinate differences along the different axes are summed, not with the same sign, but rather with varying signs.†

As already shown, if $ds = 0$ in one inertial system, then $ds' = 0$ in any other system. On

---

† The four-dimensional geometry described by the quadratic form (2.4) was introduced by H. Minkowski, in connection with the theory of relativity. This geometry is called *pseudo-euclidean,* in contrast to ordinary euclidean geometry.

the other hand, ds and ds' are infinitesimals of the same order. From these two conditions it follows that ds^2 and ds'^2 must be proportional to each other:

$$ds^2 = a\,ds'^2$$

where the coefficient a can depend only on the absolute value of the relative velocity of the two inertial systems. It cannot depend on the coordinates or the time, since then different points in space and different moments in time would not be equivalent, which would be in contradiction to the homogeneity of space and time. Similarly, it cannot depend on the direction of the relative velocity, since that would contradict the isotropy of space.

Let us consider three reference systems K, K_1, K_2, and let V_1 and V_2 be the velocities of systems K_1 and K_2 relative to K. We then have:

$$ds^2 = a(V_1)ds_1^2, \quad ds^2 = a(V_2)ds_2^2.$$

Similarly we can write

$$ds_1^2 = a(V_{12})ds_2^2,$$

where V_{12} is the absolute value of the velocity of K_2 relative to K_1. Comparing these relations with one another, we find that we must have

$$\frac{a(V_2)}{a(V_1)} = a(V_{12}). \tag{2.5}$$

But V_{12} depends not only on the absolute values of the vectors $\mathbf{V}_1$ and $\mathbf{V}_2$, but also on the angle between them. However, this angle does not appear on the left side of formula (2.5). It is therefore clear that this formula can be correct only if the function $a(V)$ reduces to a constant, which is equal to unity according to this same formula.

Thus,

$$ds^2 = ds'^2, \tag{2.6}$$

and from the equality of the infinitesimal intervals there follows the equality of finite intervals: $s = s'$.

Thus we arrive at a very important result: the interval between two events is the same in all inertial systems of reference, i.e. it is invariant under transformation from one inertial system to any other. This invariance is the mathematical expression of the constancy of the velocity of light.

Again let $x_1 y_1 z_1 t_1$ and $x_2 y_2 z_2 t_2$ be the coordinates of two events in a certain reference system K. Does there exist a coordinate system K', in which these two events occur at one and the same point in space?

We introduce the notation

$$t_2 - t_1 = t_{12}, \quad (x_2 - x_1)^2 + (y_2 - y_1)^2 + (z_2 - z_1)^2 = l_{12}^2.$$

Then the interval between events in the K system is:

$$s_{12}^2 = c^2 t_{12}^2 - l_{12}^2$$

and in the K' system

$$s_{12}'^2 = c^2 t_{12}'^2 - l_{12}'^2,$$

whereupon, because of the invariance of intervals,

$$c^2 t_{12}^2 - l_{12}^2 = c^2 t_{12}'^2 - l_{12}'^2 .$$

We want the two events to occur at the same point in the K' system, that is, we require $l_{12}' = 0$. Then

$$s_{12}^2 = c^2 t_{12}^2 - l_{12}^2 = c^2 t_{12}'^2 > 0.$$

Consequently a system of reference with the required property exists if $s_{12}^2 > 0$, that is, if the interval between the two events is a real number. Real intervals are said to be *timelike*.

Thus, if the interval between two events is timelike, then there exists a system of reference in which the two events occur at one and the same place. The time which elapses between the two events in this system is

$$t_{12}' = \frac{1}{c} \sqrt{c^2 t_{12}^2 - l_{12}^2} = \frac{s_{12}}{c}. \tag{2.7}$$

If two events occur in one and the same body, then the interval between them is always timelike, for the distance which the body moves between the two events cannot be greater than $c t_{12}$, since the velocity of the body cannot exceed c. So we have always

$$l_{12} < c t_{12}.$$

Let us now ask whether or not we can find a system of reference in which the two events occur at one and the same time. As before, we have for the K and K' systems $c^2 t_{12}^2 - l_{12}^2 = c^2 t_{12}'^2 - l_{12}'^2$. We want to have $t_{12}' = 0$, so that

$$s_{12}^2 = - l_{12}'^2 < 0.$$

Consequently the required system can be found only for the case when the interval s_{12} between the two events is an imaginary number. Imaginary intervals are said to be *spacelike*.

Thus if the interval between two events is spacelike, there exists a reference system in which the two events occur simultaneously. The distance between the points where the events occur in this system is

$$l_{12}' = \sqrt{l_{12}^2 - c^2 t_{12}^2} = i s_{12} . \tag{2.8}$$

The division of intervals into space- and timelike intervals is, because of their invariance, an absolute concept. This means that the timelike or spacelike character of an interval is independent of the reference system.

Let us take some event O as our origin of time and space coordinates. In other words, in the four-dimensional system of coordinates, the axes of which are marked x, y, z, t, the world point of the event O is the origin of coordinates. Let us now consider what relation other events bear to the given event O. For visualization, we shall consider only one space dimension and the time, marking them on two axes (Fig. 2). Uniform rectilinear motion of a particle, passing through $x = 0$ at $t = 0$, is represented by a straight line going through O and inclined to the t axis at an angle whose tangent is the velocity of the particle. Since the maximum possible velocity is c, there is a maximum angle which this line can subtend with the t axis. In Fig. 2 are shown the two lines representing the propagation of two signals (with the velocity of light) in opposite directions passing through the event O (i.e. going through $x = 0$ at $t = 0$). All lines representing the motion of particles can lie only in the regions aOc and dOb. On the lines ab and cd, $x = \pm ct$. First consider events whose world points lie within the region aOc. It is easy to show that for all the points of this region $c^2 t^2 - x^2 > 0$.

In other words, the interval between any event in this region and the event O is timelike. In this region $t > 0$, i.e. all the events in this region occur "after" the event O. But two events which are separated by a timelike interval cannot occur simultaneously in any reference system. Consequently it is impossible to find a reference system in which any of the events in region aOc occurred "before" the event O, i.e. at time $t < 0$. Thus all the events in region aOc are future events relative to O in *all* reference systems. Therefore this region can be called the *absolute future* relative to O.

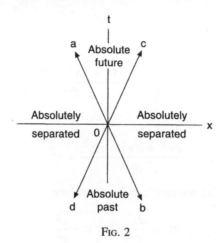

FIG. 2

In exactly the same way, all events in the region bOd are in the *absolute past* relative to O; i.e. events in this region occur before the event O in all systems of reference.

Next consider regions dOa and cOb. The interval between any event in this region and the event O is spacelike. These events occur at different points in space in every reference system. Therefore these regions can be said to be *absolutely remote* relative to O. However, the concepts "simultaneous", "earlier", and "later" are relative for these regions. For any event in these regions there exist systems of reference in which it occurs after the event O, systems in which it occurs earlier than O, and finally one reference system in which it occurs simultaneously with O.

Note that if we consider all three space coordinates instead of just one, then instead of the two intersecting lines of Fig. 2 we would have a "cone" $x^2 + y^2 + z^2 - c^2t^2 = 0$ in the four-dimensional coordinate system x, y, z, t, the axis of the cone coinciding with the t axis. (This cone is called the *light cone*.) The regions of absolute future and absolute past are then represented by the two interior portions of this cone.

Two events can be related causally to each other only if the interval between them is timelike; this follows immediately from the fact that no interaction can propagate with a velocity greater than the velocity of light. As we have just seen, it is precisely for these events that the concepts "earlier" and "later" have an absolute significance, which is a necessary condition for the concepts of cause and effect to have meaning.

§ 3. Proper time

Suppose that in a certain inertial reference system we observe clocks which are moving relative to us in an arbitrary manner. At each different moment of time this motion can be considered as uniform. Thus at each moment of time we can introduce a coordinate system rigidly linked to the moving clocks, which with the clocks constitutes an inertial reference system.

In the course of an infinitesimal time interval dt (as read by a clock in our rest frame) the moving clocks go a distance $\sqrt{dx^2 + dy^2 + dz^2}$. Let us ask what time interval dt' is indicated for this period by the moving clocks. In a system of coordinates linked to the moving clocks, the latter are at rest, i.e., $dx' = dy' = dz' = 0$. Because of the invariance of intervals

$$ds^2 = c^2 dt^2 - dx^2 - dy^2 - dz^2 = c^2 dt'^2,$$

from which

$$dt' = dt \sqrt{1 - \frac{dx^2 + dy^2 + dz^2}{c^2 dt^2}}.$$

But

$$\frac{dx^2 + dy^2 + dz^2}{dt^2} = v^2,$$

where v is the velocity of the moving clocks; therefore

$$dt' = \frac{ds}{c} = dt \sqrt{1 - \frac{v^2}{c^2}}. \tag{3.1}$$

Integrating this expression, we can obtain the time interval indicated by the moving clocks when the elapsed time according to a clock at rest is $t_2 - t_1$:

$$t_2' - t_1' = \int_{t_1}^{t_2} dt \sqrt{1 - \frac{v^2}{c^2}}. \tag{3.2}$$

The time read by a clock moving with a given object is called the *proper time* for this object. Formulas (3.1) and (3.2) express the proper time in terms of the time for a system of reference from which the motion is observed.

As we see from (3.1) or (3.2), the proper time of a moving object is always less than the corresponding interval in the rest system. In other words, moving clocks go more slowly than those at rest.

Suppose some clocks are moving in uniform rectilinear motion relative to an inertial system K. A reference frame K' linked to the latter is also inertial. Then from the point of view of an observer in the K system the clocks in the K' system fall behind. And conversely, from the point of view of the K' system, the clocks in K lag. To convince ourselves that there is no contradiction, let us note the following. In order to establish that the clocks in the K' system lag behind those in the K system, we must proceed in the following fashion. Suppose that at a certain moment the clock in K' passes by the clock in K, and at that moment the readings of the two clocks coincide. To compare the rates of the two clocks in K and K' we must once more compare the readings of the same moving clock in K' with the clocks in K. But now we compare this clock with *different* clocks in K—with those past which the clock in K' goes at ths new time. Then we find that the clock in K' lags behind the clocks in K with which it is being compared. We see that to compare the rates of clocks in two reference frames we require several clocks in one frame and one in the other, and that therefore this process is not symmetric with respect to the two systems. The clock that appears to lag is always the one which is being compared with different clocks in the other system.

If we have two clocks, one of which describes a closed path returning to the starting point (the position of the clock which remained at rest), then clearly the moving clock appears to lag relative to the one at rest. The converse reasoning, in which the moving clock would be considered to be at rest (and vice versa) is now impossible, since the clock describing a closed trajectory does not carry out a uniform rectilinear motion, so that a coordinate system linked to it will not be inertial.

Since the laws of nature are the same only for inertial reference frames, the frames linked to the clock at rest (inertial frame) and to the moving clock (non-inertial) have different properties, and the argument which leads to the result that the clock at rest must lag is not valid.

The time interval read by a clock is equal to the integral

$$\frac{1}{c}\int_a^b ds,$$

taken along the world line of the clock. If the clock is at rest then its world line is clearly a line parallel to the t axis; if the clock carries out a nonuniform motion in a closed path and returns to its starting point, then its world line will be a curve passing through the two points, on the straight world line of a clock at rest, corresponding to the beginning and end of the motion. On the other hand, we saw that the clock at rest always indicates a greater time interval than the moving one. Thus we arrive at the result that the integral

$$\int_a^b ds,$$

taken between a given pair of world points, has its maximum value if it is taken along the straight world line joining these two points.†

§ 4. The Lorentz transformation

Our purpose is now to obtain the formula of transformation from one inertial reference system to another, that is, a formula by means of which, knowing the coordinates x, y, z, t, of a certain event in the K system, we can find the coordinates x', y', z', t' of the same event in another inertial system K'.

In classical mechanics this question is resolved very simply. Because of the absolute nature of time we there have $t = t'$; if, furthermore, the coordinate axes are chosen as usual (axes X, X' coincident, Y, Z axes parallel to Y', Z', motion along X, X') then the coordinates y, z clearly are equal to y', z', while the coordinates x and x' differ by the distance traversed by one system relative to the other. If the time origin is chosen as the moment when the two coordinate systems coincide, and if the velocity of the K' system relative to K is V, then this distance is Vt. Thus

† It is assumed, of course, that the points a and b and the curves joining them are such that all elements ds along the curves are timelike.

This property of the integral is connected with the pseudo-euclidean character of the four-dimensional geometry. In euclidean space the integral would, of course, be a minimum along the straight line.

$$x = x' + Vt, \quad y = y', \quad z = z', \quad t = t'. \tag{4.1}$$

This formula is called the *Galileo transformation*. It is easy to verify that this transformation, as was to be expected, does not satisfy the requirements of the theory of relativity; it does not leave the interval between events invariant.

We shall obtain the relativistic transformation precisely as a consequence of the requirement that it leaves the interval between events invariant.

As we saw in § 2, the interval between events can be looked on as the distance between the corresponding pair of world points in a four-dimensional system of coordinates. Consequently we may say that the required transformation must leave unchanged all distances in the four-dimensional x, y, z, ct, space. But such transformations consist only of parallel displacements, and rotations of the coordinate system. Of these the displacement of the coordinate system parallel to itself is of no interest, since it leads only to a shift in the origin of the space coordinates and a change in the time reference point. Thus the required transformation must be expressible mathematically as a rotation of the four-dimensional x, y, z, ct, coordinate system.

Every rotation in the four-dimensional space can be resolved into six rotations, in the planes xy, zy, xz, tx, ty, tz (just as every rotation in ordinary space can be resolved into three rotations in the planes xy, zy and xz). The first three of these rotations transform only the space coordinates; they correspond to the usual space rotations.

Let us consider a rotation in the tx plane; under this, the y and z coordinates do not change. In particular, this transformation must leave unchanged the difference $(ct)^2 - x^2$, the square of the "distance" of the point (ct, x) from the origin. The relation between the old and the new coordinates is given in most general form by the formulas:

$$x = x' \cosh \psi + ct' \sinh \psi, \quad ct = x' \sinh \psi + ct' \cosh \psi, \tag{4.2}$$

where ψ is the "angle of rotation"; a simple check shows that in fact $c^2 t^2 - x^2 = c^2 t'^2 - x'^2$. Formula (4.2) differs from the usual formulas for transformation under rotation of the coordinate axes in having hyperbolic functions in place of trigonometric functions. This is the difference between pseudo-euclidean and euclidean geometry.

We try to find the formula of transformation from an inertial reference frame K to a system K' moving relative to K with velocity V along the x axis. In this case clearly only the coordinate x and the time t are subject to change. Therefore this transformation must have the form (4.2). Now it remains only to determine the angle ψ, which can depend only on the relative velocity V.†

Let us consider the motion, in the K system, of the origin of the K' system. Then $x' = 0$ and formulas (4.2) take the form:

$$x = ct' \sinh \psi, \quad ct = ct' \cosh \psi,$$

or dividing one by the other,

$$\frac{x}{ct} = \tanh \psi.$$

But x/t is clearly the velocity V of the K' system relative to K. So

---

† Note that to avoid confusion we shall always use V to signify the constant relative velocity of two inertial systems, and v for the velocity of a moving particle, not necessarily constant.

$$\tanh \psi = \frac{V}{c}.$$

From this

$$\sinh \psi = \frac{\dfrac{V}{c}}{\sqrt{1 - \dfrac{V^2}{c^2}}}, \qquad \cosh \psi = \frac{1}{\sqrt{1 - \dfrac{V^2}{c^2}}}.$$

Substituting in (4.2), we find:

$$x = \frac{x' + Vt'}{\sqrt{1 - \dfrac{V^2}{c^2}}}, \quad y = y', \quad z = z', \quad t = \frac{t' + \dfrac{V}{c^2}x'}{\sqrt{1 - \dfrac{V^2}{c^2}}}. \tag{4.3}$$

This is the required transformation formula. It is called the *Lorentz transformation,* and is of fundamental importance for what follows.

The inverse formulas, expressing x', y', z', t' in terms of x, y, z, t, are most easily obtained by changing V to $-V$ (since the K system moves with velocity $-V$ relative to the K' system). The same formulas can be obtained directly by solving equations (4.3) for x', y', z', t'.

It is easy to see from (4.3) that on making the transition to the limit $c \to \infty$ and classical mechanics, the formula for the Lorentz transformation actually goes over into the Galileo transformation.

For $V > c$ in formula (4.3) the coordinates x, t are imaginary; this corresponds to the fact that motion with a velocity greater than the velocity of light is impossible. Moreover, one cannot use a reference system moving with the velocity of light—in that case the denominators in (4.3) would go to zero.

For velocities V small compared with the velocity of light, we can use in place of (4.3) the approximate formulas:

$$x = x' + Vt', \quad y = y', \quad z = z', \quad t = t' + \frac{V}{c^2}x'. \tag{4.4}$$

Suppose there is a rod at rest in the K system, parallel to the X axis. Let its length, measured in this system, be $\Delta x = x_2 - x_1$ (x_2 and x_1 are the coordinates of the two ends of the rod in the K system). We now determine the length of this rod as measured in the K' system. To do this we must find the coordinates of the two ends of the rod (x_2' and x_1') in this system at one and the same time t'. From (4.3) we find:

$$x_1 = \frac{x_1' + Vt'}{\sqrt{1 - \dfrac{V^2}{c^2}}}, \qquad x_2 = \frac{x_2' + Vt'}{\sqrt{1 - \dfrac{V^2}{c^2}}}.$$

The length of the rod in the K' system is $\Delta x' = x_2' - x_1'$; subtracting x_1 from x_2, we find

$$\Delta x = \frac{\Delta x'}{\sqrt{1 - \dfrac{V^2}{c^2}}}.$$

The *proper length* of a rod is its length in a reference system in which it is at rest. Let us denote it by $l_0 = \Delta x$, and the length of the rod in any other reference frame K' by l. Then

$$l = l_0 \sqrt{1 - \frac{V^2}{c^2}}. \tag{4.5}$$

Thus a rod has its greatest length in the reference system in which it is at rest. Its length in a system in which it moves with velocity V is decreased by the factor $\sqrt{1 - V^2/c^2}$. This result of the theory of relativity is called the *Lorentz contraction*.

Since the transverse dimensions do not change because of its motion, the volume $\mathscr{V}$ of a body decreases according to the similar formula

$$\mathscr{V} = \mathscr{V}_0 \sqrt{1 - \frac{V^2}{c^2}}, \tag{4.6}$$

where $\mathscr{V}_0$ is the *proper volume* of the body.

From the Lorentz transformation we can obtain anew the results already known to us concerning the proper time (§ 3). Suppose a clock to be at rest in the K' system. We take two events occurring at one and the same point x', y', z' in space in the K' system. The time between these events in the K' system is $\Delta t' = t'_2 - t'_1$. Now we find the time Δt which elapses between these two events in the K system. From (4.3), we have

$$t_1 = \frac{t'_1 + \frac{V}{c^2}x'}{\sqrt{1 - \frac{V^2}{c^2}}}, \quad t_2 = \frac{t'_2 + \frac{V}{c^2}x'}{\sqrt{1 - \frac{V^2}{c^2}}},$$

or, subtracting one from the other,

$$t_2 - t_1 = \Delta t = \frac{\Delta t'}{\sqrt{1 - \frac{V^2}{c^2}}},$$

in complete agreement with (3.1).

Finally we mention another general property of Lorentz transformations which distinguishes them from Galilean transformations. The latter have the general property of commutativity, i.e. the combined result of two successive Galilean transformations (with different velocities V_1 and V_2) does not depend on the order in which the transformations are performed. On the other hand, the result of two successive Lorentz transformations does depend, in general, on their order. This is already apparent purely mathematically from our formal description of these transformations as rotations of the four-dimensional coordinate system: we know that the result of two rotations (about different axes) depends on the order in which they are carried out. The sole exception is the case of transformations with parallel vectors V_1 and V_2 (which are equivalent to two rotations of the four-dimensional coordinate system about the same axis).

§ 5. Transformation of velocities

In the preceding section we obtained formulas which enable us to find from the coordinates of an event in one reference frame, the coordinates of the same event in a second reference frame. Now we find formulas relating the velocity of a material particle in one reference system to its velocity in a second reference system.

Let us suppose once again that the K' system moves relative to the K system with velocity V along the x axis. Let $v_x = dx/dt$ be the component of the particle velocity in the K system and $v_x' = dx'/dt'$ the velocity component of the same particle in the K' system. From (4.3), we have

$$dx = \frac{dx' + V dt'}{\sqrt{1 - \frac{V^2}{c^2}}}, \quad dy = dy', \quad dz = dz', \quad dt = \frac{dt' + \frac{V}{c^2} dx'}{\sqrt{1 - \frac{V^2}{c^2}}}.$$

Dividing the first three equations by the fourth and introducing the velocities

$$\mathbf{v} = \frac{d\mathbf{r}}{dt}, \quad \mathbf{v}' = \frac{d\mathbf{r}'}{dt'},$$

we find

$$v_x = \frac{v_x' + V}{1 + v_x' \frac{V}{c^2}}, \quad v_y = \frac{v_y' \sqrt{1 - \frac{V^2}{c^2}}}{1 + v_x' \frac{V}{c^2}}, \quad v_z = \frac{v_z' \sqrt{1 - \frac{V^2}{c^2}}}{1 + v_x' \frac{V}{c^2}}. \tag{5.1}$$

These formulas determine the transformation of velocities. They describe the law of composition of velocities in the theory of relativity. In the limiting case of $c \to \infty$, they go over into the formulas $v_x = v_x' + V$, $v_y, = v_y'$, $v_z = v_z'$ of classical mechanics.

In the special case of motion of a particle parallel to the X axis, $v_x = v$, $v_y = v_z = 0$. Then $v_y' = v_z' = 0$, $v_x' = v'$, so that

$$v = \frac{v' + V}{1 + v' \frac{V}{c^2}}. \tag{5.2}$$

It is easy to convince oneself that the sum of two velocities each smaller than the velocity of light is again not greater than the light velocity.

For a velocity V significantly smaller than the velocity of light (the velocity v can be arbitrary), we have approximately, to terms of order V/c:

$$v_x = v_x' + V\left(1 - \frac{v_x'^2}{c^2}\right), \quad v_y = v_y' - v_x' v_y' \frac{V}{c^2}, \quad v_z = v_z' - v_x' v_z' \frac{V}{c^2}.$$

These three formulas can be written as a single vector formula

$$\mathbf{v} = \mathbf{v}' + \mathbf{V} - \frac{1}{c^2}(\mathbf{V} \cdot \mathbf{v}')\mathbf{v}'. \tag{5.3}$$

We may point out that in the relativistic-law of addition of velocities (5.1) the two velocities $\mathbf{v}'$ and $\mathbf{V}$ which are combined enter unsymmetrically (provided they are not both directed along the x axis). This fact is related to the noncommutativity of Lorentz transformations which we mentioned in the preceding section.

Let us choose our coordinate axes so that the velocity of the particle at the given moment lies in the XY plane. Then the velocity of the particle in the K system has components $v_x = v \cos \theta$, $v_y = v \sin \theta$, and in the K' system $v'_x = v' \cos \theta'$, $v'_y = v' \sin \theta'$ (v, v', θ, θ' are the absolute values and the angles subtended with the X, X' axes respectively in the K, K' systems), With the help of formula (5.1), we then find

$$\tan \theta = \frac{v' \sqrt{1 - \frac{v^2}{c^2}} \sin \theta'}{v' \cos \theta' + V}. \tag{5.4}$$

This formula describes the change in the direction of the velocity on transforming from one reference system to another.

Let us consider a very important special case of this formula, namely, the deviation of light in transforming to a new reference system—a phenomenon known as the *aberration of light*. In this case $v = v' = c$, so that the preceding formula goes over into

$$\tan \theta = \frac{\sqrt{1 - \frac{V^2}{c^2}}}{\frac{V}{c} + \cos \theta'} \sin \theta'. \tag{5.5}$$

From the same transformation formulas (5.1) it is easy to obtain for $\sin \theta$ and $\cos \theta$:

$$\sin \theta = \frac{\sqrt{1 - \frac{V^2}{c^2}}}{1 + \frac{V}{c} \cos \theta'} \sin \theta', \qquad \cos \theta = \frac{\cos \theta' + \frac{V}{c}}{1 + \frac{V}{c} \cos \theta'}. \tag{5.6}$$

In case $V \ll c$, we find from this formula, correct to terms of order V/c:

$$\sin \theta - \sin \theta' = -\frac{V}{c} \sin \theta' \cos \theta'.$$

Introducing the angle $\Delta \theta = \theta' - \theta$ (the aberration angle), we find to the same order of accuracy

$$\Delta \theta = \frac{V}{c} \sin \theta', \tag{5.7}$$

which is the well-known elementary formula for the aberration of light.

§ 6. Four-vectors

The coordinates of an event (ct, x, y, z) can be considered as the components of a four-dimensional radius vector (or, for short, a four-radius vector) in a four-dimensional space. We shall denote its components by x^i, where the index i takes on the values 0, 1, 2, 3, and

$$x^0 = ct, \qquad x^1 = x, \qquad x^2 = y, \qquad x^3 = z.$$

The square of the "length" of the radius four-vector is given by

$$(x^0)^2 - (x^1)^2 - (x^2)^2 - (x^3)^2.$$

It does not change under any rotations of the four-dimensional coordinate system, in particular under Lorentz transformations.

In general a set of four quantities A^0, A^1, A^2, A^3 which transform like the components of the radius four-vector x^i under transformations of the four-dimensional coordinate system is called a *four-dimensional vector* (*four-vector*) A^i. Under Lorentz transformations,

$$A^0 = \frac{A'^0 + \frac{V}{c}A'^1}{\sqrt{1 - \frac{V^2}{c^2}}}, \quad A^1 = \frac{A'^1 + \frac{V}{c}A'^0}{\sqrt{1 - \frac{V^2}{c^2}}}, \quad A^2 = A'^2, \quad A^3 = A'^3. \tag{6.1}$$

The square magnitude of any four-vector is defined analogously to the square of the radius four-vector:

$$(A^0)^2 - (A^1)^2 - (A^2)^2 - (A^3)^2.$$

For convenience of notation, we introduce two "types" of components of four-vectors, denoting them by the symbols A^i and A_i, with superscripts and subscripts. These are related by

$$A_0 = A^0, \quad A_1 = -A^1, \quad A_2 = -A^2, \quad A_3 = -A^3. \tag{6.2}$$

The quantities A^i are called the *contravariant*, and the A_i the *covariant* components of the four-vector. The square of the four-vector then appears in the form

$$\sum_{i=0}^{3} A^i A_i = A^0 A_0 + A^1 A_1 + A^2 A_2 + A^3 A_3.$$

Such sums are customarily written simply as $A^i A_i$, omitting the summation sign. One agrees that one sums over any repeated index, and omits the summation sign. Of the pair of indices, one must be a superscript and the other a subscript. This convention for summation over "dummy" indices is very convenient and considerably simplifies the writing of formulas.

We shall use Latin letters i, k, l, ... , for four-dimensional indices, taking on the values 0, 1, 2, 3.

In analogy to the square of a four-vector, one forms the *scalar product* of two different four-vectors:

$$A^i B_i = A^0 B_0 + A^1 B_1 + A^2 B_2 + A^3 B_3.$$

It is clear that this can be written either as $A^i B_i$ or $A_i B^i$—the result is the same. In general one can switch upper and lower indices in any pair of dummy indices.†

The product $A^i B_i$ is a *four-scalar*—it is invariant under rotations of the four-dimensional coordinate system. This is easily verified directly,‡ but it is also apparent beforehand (from the analogy with the square $A^i A_i$) from the fact that all four-vectors transform according to the same rule.

† In the literature the indices are often omitted on four-vectors, and their squares and scalar products are written as A^2, AB. We shall not use this notation in the present text.

‡ One should remember that the law for transformation of a four-vector expressed in covariant components differs (in signs) from the same law expressed for contravariant components. Thus, instead of (6.1), one will have:

$$A_0 = \frac{A'_0 - \frac{V}{c}A'_1}{\sqrt{1 - \frac{V^2}{c^2}}}, \quad A_1 = \frac{A'_1 - \frac{V}{c}A'_0}{\sqrt{1 - \frac{V^2}{c^2}}}, \quad A_2 = A'_2, \quad A_3 = A'_3.$$

The component A^0 is called the *time component,* and A^1, A^2, A^3 the *space components* of the four-vector (in analogy to the radius four-vector). The square of a four-vector can be positive, negative, or zero; such vectors are called, *timelike, spacelike,* and *null-vectors,* respectively (again in analogy to the terminology for intervals).†

Under purely spatial rotations (i.e. transformations not affecting the time axis) the three space components of the four-vector A^i form a three-dimensional vector **A**. The time component of the four-vector is a three-dimensional scalar (with respect to these transformations). In enumerating the components of a four-vector, we shall often write them as

$$A^i = (A^0, \mathbf{A}).$$

The covariant components of the same four-vector are $A_i = (A^0, -\mathbf{A})$, and the square of the four-vector is $A^i A_i = (A^0)^2 - \mathbf{A}^2$. Thus, for the radius four-vector:

$$x^i = (ct, \mathbf{r}), \qquad x_i = (ct, -\mathbf{r}), \qquad x^i x_i = c^2 t^2 - \mathbf{r}^2.$$

For three-dimensional vectors (with coordinates x, y, z) there is no need to distinguish between contra- and covariant components. Whenever this can be done without causing confusion, we shall write their components as $A_\alpha (\alpha = x, y, z)$ using Greek letters for subscripts. In particular we shall assume a summation over x, y, z for any repeated index (for example, $\mathbf{A} \cdot \mathbf{B} = A_\alpha B_\alpha$).

A *four-dimensional tensor* (*four-tensor*) of the second rank is a set of sixteen quantities A^{ik}, which under coordinate transformations transform like the products of components of two four-vectors. We similarly define four-tensors of higher rank.

The components of a second-rank tensor can be written in three forms: covariant, A_{ik}, contravariant, A^{ik}, and mixed, A_k^i (where, in the last case, one should distinguish between $A_k^{\ i}$ and $A_i^{\ k}$, i.e. one should be careful about which of the two is superscript and which a subscript). The connection between the different types of components is determined from the general rule: raising or lowering a space index $(1, 2, 3)$ changes the sign of the component, while raising or lowering the time index (0) does not. Thus:

$$A_{00} = A^{00}, \quad A_{01} = -A^{01}, \quad A_{11} = A^{11}, \dots,$$
$$A_0^{\ 0} = A^{00}, A_0^{\ 1} = A^{01}, A_1^{\ 0} = -A^{01}, A_1^{\ 1} = -A^{11}, \dots.$$

Under purely spatial transformations, the nine quantities A^{11}, A^{12}, ... form a three-tensor. The three components A^{01}, A^{02}, A^{03} and the three components A^{10}, A^{20}, A^{30} constitute three-dimensional vectors, while the component A^{00} is a three-dimensional scalar.

A tensor A^{ik} is said to be *symmetric* if $A^{ik} = A^{ki}$, and *antisymmetric* if $A^{ik} = -A^{ki}$. In an antisymmetric tensor, all the diagonal components (i.e. the components A^{00}, A^{11}, ...) are zero, since, for example, we must have $A^{00} = -A^{00}$. For a symmetric tensor A^{ik}, the mixed components A_k^i and $A_k^{\ i}$ obviously coincide; in such cases we shall simply write A_k^i, putting the indices one above the other.

In every tensor equation, the two sides must contain identical and identically placed (i.e. above or below) free indices (as distinguished from dummy indices). The free indices in tensor equations can be shifted up or down, but this must be done simultaneously in all terms in the equation. Equating covariant and contravariant components of different tensors is "illegal"; such an equation, even if it happened by chance to be valid in a particular reference system, would be violated on going to another frame.

---

† Null vectors are also said to be *isotropic.*

From the tensor components A^{ik} one can form a scalar by taking the sum

$$A^i{}_i = A^0{}_0 + A^1{}_1 + A^2{}_2 + A^3{}_3$$

(where, of course, $A_i{}^i = A^i{}_i$). This sum is called the *trace* of the tensor, and the operation for obtaining it is called *contraction*.

The formation of the scalar product of two vectors, considered earlier, is a contraction operation: it is the formation of the scalar $A^i B_i$ from the tensor $A^i B_k$. In general, contracting on any pair of indices reduces the rank of the tensor by 2. For example, $A^i{}_{kli}$ is a tensor of second rank $A^i{}_k B^k$ is a four-vector, $A^{ik}{}_{ik}$ is a scalar, etc.

The unit four-tensor δ^i_k satisfies the condition that for any four-vector A^i,

$$\delta^k_i A^i = A^k. \tag{6.3}$$

It is clear that the components of this tensor are

$$\delta^k_i = \begin{cases} 1, & \text{if} \quad i = k \\ 0, & \text{if} \quad i \neq k \end{cases} \tag{6.4}$$

Its trace is $\delta^i_i = 4$.

By raising the one index or lowering the other in δ^k_i, we can obtain the contra- or covariant tensor g^{ik} or g_{ik}, which is called the *metric tensor*. The tensors g^{ik} and g_{ik} have identical components, which can be written as a matrix:

$$(g^{ik}) = (g_{ik}) = \begin{pmatrix} 1 & 0 & 0 & 0 \\ 0 & -1 & 0 & 0 \\ 0 & 0 & -1 & 0 \\ 0 & 0 & 0 & -1 \end{pmatrix} \tag{6.5}$$

(the index i labels the rows, and k the columns, in the order 0, 1, 2, 3). It is clear that

$$g_{ik} A^k = A_i, \quad g^{ik} A_k = A^i. \tag{6.6}$$

The scalar product of two four-vectors can therefore be written in the form:

$$A^i A_i = g_{ik} A^i A^k = g^{ik} A_i A_k. \tag{6.7}$$

The tensors $\delta^i_k, g_{ik}, g^{ik}$ are special in the sense that their components are the same in all coordinate systems. The *completely antisymmetric unit tensor* of fourth rank, e^{iklm}, has the same property. This is the tensor whose components change sign under interchange of any pair of indices, and whose nonzero components are ± 1. From the antisymmetry it follows that all components in which two indices are the same are zero, so that the only non-vanishing components are those for which all four indices are different. We set

$$e^{0123} = +1 \tag{6.8}$$

(hence $e_{0123} = -1$). Then all the other nonvanishing components e^{iklm} are equal to +1 or −1, according as the numbers i, k, l, m can be brought to the arrangement 0, 1, 2, 3 by an even or an odd number of transpositions. The number of such components is 4! = 24. Thus,

$$e^{iklm} e_{iklm} = -24. \tag{6.9}$$

With respect to rotations of the coordinate system, the quantities e^{iklm} behave like the components of a tensor; but if we change the sign of one or three of the coordinates the components e^{iklm}, being defined as the same in all coordinate systems, do not change, whereas some of the components of a tensor should change sign. Thus e^{iklm} is, strictly speaking, not a tensor, but rather a *pseudotensor*. Pseudotensors of any rank, in particular *pseudoscalars*, behave like tensors under all coordinate transformations except those that cannot be reduced to rotations, i.e. reflections, which are changes in sign of the coordinates that are not reducible to a rotation.

The products $e^{iklm}e^{prst}$ form a four-tensor of rank 8, which is a true tensor; by contracting on one or more pairs of indices, one obtains tensors of rank 6, 4, and 2. All these tensors have the same form in all coordinate systems. Thus their components must be expressed as combinations of products of components of the unit tensor δ_k^i — the only true tensor whose components are the same in all coordinate systems. These combinations can easily be found by starting from the symmetries that they must possess under permutation of indices.†

If A^{ik} is an antisymmetric tensor, the tensor A^{ik} and the pseudotensor $A^{*ik} = \frac{1}{2}e^{iklm}A_{lm}$ are said to be *dual* to one another. Similarly, $e^{iklm}A_m$ is an antisymmetric pseudotensor of rank 3, dual to the vector A^i. The product $A^{ik}A_{ik}^*$ of dual tensors is obviously a pseudoscalar.

In this connection we note some analogous properties of three-dimensional vectors and tensors. The completely antisymmetric unit pseudotensor of rank 3 is the set of quantities $e_{\alpha\beta\gamma}$ which change sign under any transposition of a pair of indices. The only nonvanishing components of $e_{\alpha\beta\gamma}$ are those with three different indices. We set $e_{xyz} = 1$; the others are 1 or −1, depending on whether the sequence α, β, γ can be brought to the order x, y, z by an even or an odd number of transpositions.‡

† For reference we give the following formulas:

$$e^{iklm}e_{prst} = - \begin{vmatrix} \delta_p^i & \delta_r^i & \delta_s^i & \delta_t^i \\ \delta_p^k & \delta_r^k & \delta_s^k & \delta_t^k \\ \delta_p^l & \delta_r^l & \delta_s^l & \delta_t^l \\ \delta_p^m & \delta_r^m & \delta_s^m & \delta_t^m \end{vmatrix}, \quad e^{iklm}e_{prsm} = - \begin{vmatrix} \delta_p^i & \delta_r^i & \delta_s^i \\ \delta_p^k & \delta_r^k & \delta_s^k \\ \delta_p^l & \delta_r^l & \delta_s^l \end{vmatrix}$$

$$e^{iklm}e_{prlm} = - 2(\delta_p^i\delta_r^k - \delta_r^i\delta_p^k), \qquad e^{iklm}e_{prlm} = - 6\delta_p^i.$$

The overall coefficient in these formulas can be checked using the result of a complete contraction, which should give (6.9).

As a consequence of these formulas we have:

$$e^{prst}A_{lp}A_{kr}A_{is}A_{mt} = - Ae_{iklm}.$$

$$e^{iklm}e^{prst}A_{tp}A_{kr}A_{ls}A_{mt} = 24A.$$

where A is the determinant formed from the quantities A_{ik}.

‡ The fact that the components of the four-tensor e^{iklm} are unchanged under rotations of the four-dimensional coordinate system, and that the components of the three-tensor $e_{\alpha\beta\gamma}$ are unchanged by rotations of the space axes are special cases of a general rule: any completely antisymmetric tensor of rank equal to the number of dimensions of the space in which it is defined is invariant under rotations of the coordinate system in the space.

The products $e_{\alpha\beta\gamma}e_{\lambda\mu\nu}$ form a true three-dimensional tensor of rank 6, and are therefore expressible as combinations of products of components of the unit three-tensor $\delta_{\alpha\beta}$.†

Under a reflection of the coordinate system, i.e. under a change in sign of all the coordinates, the components of an ordinary vector also change sign. Such vectors are said to be *polar*. The components of a vector that can be written as the cross product of two polar vectors do not change sign under inversion. Such vectors are said to be *axial*. The scalar product of a polar and an axial vector is not a true scalar, but rather a pseudoscalar; it changes sign under a coordinate inversion. An axial vector is a pseudovector, dual to some antisymmetric tensor. Thus, if $\mathbf{C} = \mathbf{A} \times \mathbf{B}$, then

$$C_\alpha = \tfrac{1}{2}e_{\alpha\beta\gamma}C_{\beta\gamma}, \quad \text{where} \quad C_{\beta\gamma} = A_\beta B_\gamma - A_\gamma B_\beta.$$

Now consider four-tensors. The space components (i, k, = 1, 2, 3) of the antisymmetric tensor A^{ik} form a three-dimensional antisymmetric tensor with respect to purely spatial transformations; according to our statement its components can be expressed in terms of the components of a three-dimensional axial vector. With respect to these same transformations the components A^{01}, A^{02}, A^{03} form a three-dimensional polar vector. Thus the components of an antisymmetric four-tensor can be written as a matrix:

$$(A^{ik}) = \begin{vmatrix} 0 & p_x & p_y & p_z \\ -p_x & 0 & -a_z & a_y \\ -p_y & a_z & 0 & -a_x \\ -p_z & -a_y & a_x & 0 \end{vmatrix}, \tag{6.10}$$

where, with respect to spatial transformations, $\mathbf{p}$ and $\mathbf{a}$ are polar and axial vectors, respectively. In enumerating the components of an antisymmetric four-tensor, we shall write them in the form

$$A^{ik} = (\mathbf{p}, \mathbf{a});$$

then the covariant components of the same tensor are

$$A_{ik} = (-\mathbf{p}, \mathbf{a}).$$

Finally we consider certain differential and integral operations of four-dimensional tensor analysis.

The four-gradient of a scalar ϕ is the four-vector

† For reference, we give the appropriate formulas:

$$e_{\alpha\beta\gamma}e_{\lambda\mu\nu} = \begin{vmatrix} \delta_{\alpha\lambda} & \delta_{\alpha\mu} & \delta_{\alpha\nu} \\ \delta_{\beta\lambda} & \delta_{\beta\mu} & \delta_{\beta\nu} \\ \delta_{\gamma\lambda} & \delta_{\gamma\mu} & \delta_{\gamma\nu} \end{vmatrix}.$$

Contracting this tensor on one, two and three pairs of indices, we get:

$$e_{\alpha\beta\gamma}e_{\lambda\mu\gamma} = \delta_{\alpha\lambda}\delta_{\beta\mu} - \delta_{\alpha\mu}\delta_{\beta\lambda},$$

$$e_{\alpha\beta\gamma}e_{\lambda\beta\gamma} = 2\delta_{\alpha\lambda},$$

$$e_{\alpha\beta\gamma}e_{\alpha\beta\gamma} = 6.$$

$$\frac{\partial \phi}{\partial x^i} = \left(\frac{1}{c} \frac{\partial \phi}{\partial t}, \nabla \phi \right).$$

We must remember that these derivatives are to be regarded as the covariant components of the four-vector. In fact, the differential of the scalar

$$d\phi = \frac{\partial \phi}{\partial x^i} dx^i$$

is also a scalar; from its form (scalar product of two four-vectors) our assertion is obvious.

In general, the operators of differentiation with respect to the coordinates x^i, $\partial/\partial x^i$, should be regarded as the covariant components of the operator four-vector. Thus, for example, the divergence of a four-vector, the expression $\partial A^i/\partial x^i$, in which we differentiate the contravariant components A^i, is a scalar.†

In three-dimensional space one can extend integrals over a volume, a surface or a curve. In four-dimensional space there are four types of integrations:

(1) Integral over a curve in four-space. The element of integration is the line element, i.e. the four-vector dx^i.

(2) Integral over a (two-dimensional) surface in four-space. As we know, in three-space the projections of the area of the parallelogram formed from the vectors $d\mathbf{r}$ and $d\mathbf{r}'$ on the coordinate planes $x_\alpha x_\beta$ are $dx_\alpha dx_\beta' - dx_\beta dx_\alpha'$. Analogously, in four-space the infinitesimal element of surface is given by the antisymmetric tensor of second rank $df^{ik} = dx^i dx'^k - dx^k dx'^i$; its components are the projections of the element of area on the coordinate planes. In three-dimensional space, as we know, one uses as surface element in place of the tensor $df_{\alpha\beta}$ the vector df_α dual to the tensor $df_{\alpha\beta}$: $df_\alpha = \frac{1}{2} e_{\alpha\beta\gamma} df_{\beta\gamma}$. Geometrically this is a vector normal to the surface element and equal in absolute magnitude to the area of the element. In four-space we cannot construct such a vector, but we can construct the tensor df^{*ik} dual to the tensor df^{ik},

$$df^{*ik} = \frac{1}{2} e^{iklm} df_{lm}. \tag{6.11}$$

Geometrically it describes an element of surface equal to and "normal" to the element of

† If we differentiate with respect to the "covariant coordinates" x_i, then the derivatives

$$\frac{\partial \phi}{\partial x_i} = \left(\frac{1}{c} \frac{\partial \phi}{\partial t}, -\nabla \phi \right)$$

form the contravariant components of a four-vector. We shall use this form only in exceptional cases [for example, for writing the square of the four-gradient $(\partial \phi/\partial x^i)/(\partial \phi/\partial x_i)$].

We note that in the literature partial derivatives with respect to the coordinates are often abbreviated using the symbols.

$$\partial^i = \frac{\partial}{\partial x_i}, \quad \partial_i = \frac{\partial}{\partial x^i}.$$

In this form of writing of the differentiation operators, the co- or contravariant character of quantities formed with them is explicit. This same advantage exists for another abbreviated form for writing derivatives, using the index preceded by a comma:

$$\phi_{,i} - \frac{\partial \phi}{\partial x^i}, \quad \phi^{,i} = \frac{\partial \phi}{\partial x_i}.$$

surface df^{ik}; all segments lying in it are orthogonal to all segments in the element df^{ik}. It is obvious that $df^{ik}df_{ik}^* = 0$.

(3) Integral over a hypersurface, i.e. over a three-dimensional manifold. In three-dimensional space the volume of the parallelepiped spanned by three vectors is equal to the determinant of the third rank formed from the components of the vectors. One obtains analogously the projections of the volume of the parallelepiped (i.e. the "areas" of the hypersurface) spanned by three four-vectors dx^i, dx'^i, dx''^i; they are given by the determinants

$$dS^{ikl} = \begin{vmatrix} dx^i & dx'^i & dx''^i \\ dx^k & dx'^k & dx''^k \\ dx^l & dx'^l & dx''^l \end{vmatrix},$$

which form a tensor of rank 3, antisymmetric in all three indices. As element of integration over the hypersurface, it is more convenient to use the four-vector dS^i, dual to the tensor dS^{ikl}:

$$dS^i = -\frac{1}{6}e^{iklm}dS_{klm}, \quad dS_{klm} = e_{nklm}dS^n. \tag{6.12}$$

Here

$$dS^0 = dS^{123}, \quad dS^1 = dS^{023}, \ldots$$

Geometrically dS^i is a four-vector equal in magnitude to the "areas" of the hypersurface element, and normal to this element (i.e. perpendicular to all lines drawn in the hypersurface element). In particular, $dS^0 = dx\, dy\, dz$, i.e. it is the element of three-dimensional volume dV, the projection of the hypersurface element on the hyperplane $x^0 = $ const.

(4) Integral over a four-dimensional volume; the element of integration is the scalar

$$d\Omega = dx^0 dx^1 dx^2\, dx^3 = cdtdV. \tag{6.13}$$

The element is a scalar: it is obvious that the volume of a portion of four-space is unchanged by a rotation of the coordinate system.†

Analogous to the theorems of Gauss and Stokes in three-dimensional vector analysis, there are theorems that enable us to transform four-dimensional integrals.

The integral over a closed hypersurface can be transformed into an integral over the four-volume contained within it by replacing the element of integration dS_i by the operator

$$dS_i \to d\Omega\frac{\partial}{\partial x^i}. \tag{6.14}$$

For example, for the integral of a vector A^i we have:

† Under a transformation from the integration variables x^0, x^1, x^2, x^3 to new variables x'^0, x'^1, x'^2, x'^3, the element of integration changes to $J\, d\Omega'$, where $d\Omega' = dx'^0\, dx'^1\, dx'^2\, dx'^3$

$$J = \frac{\partial(x'^0, x'^1, x'^2, x'^3)}{\partial(x^0, x^1, x^2, x^3)}.$$

is the Jacobian of the transformation. For a linear transformation of the form $x'^i = a_k^i x^k$, the Jacobian J coincides with the determinant $|a_k^i|$ and is equal to unity for rotations of the coordinate system; this shows the invariance of $d\Omega$.

$$\oint A^i dS_i = \int \frac{\partial A^i}{\partial x^i} \, d\Omega. \tag{6.15}$$

This formula is the generalization of Gauss' theorem.

An integral over a two-dimensional surface is transformed into an integral over the hypersurface "spanning" it by replacing the element of integration df_{ik}^* by the operator

$$df_{ik}^* \to dS_i \frac{\partial}{\partial x^k} - dS_k \frac{\partial}{\partial x^i}. \tag{6.16}$$

For example, for the integral of an antisymmetric tensor A^{ik} we have:

$$\frac{1}{2} \int A^{ik} df_{ik}^* = \frac{1}{2} \int \left(dS_i \frac{\partial A^{ik}}{\partial x^k} - dS_k \frac{\partial A^{ik}}{\partial x^i} \right) = \int dS_i \frac{\partial A^{ik}}{\partial x^k}. \tag{6.17}$$

The integral over a four-dimensional closed curve is transformed into an integral over the surface spanning it by the substitution:

$$dx^i \to df^{ki} \frac{\partial}{\partial x^k}. \tag{6.18}$$

Thus for the integral of a vector, we have:

$$\oint A_i dx^i = \int df^{ki} \frac{\partial A_i}{\partial x^k} = \frac{1}{2} \int df^{ik} \left(\frac{\partial A_k}{\partial x^i} - \frac{\partial A_i}{\partial x^k} \right), \tag{6.19}$$

which is the generalization of Stokes' theorem.

PROBLEMS

1. Find the law of transformation of the components of a symmetric four-tensor A^{ik} under Lorentz transformations (6.1).

Solution: Considering the components of the tensor as products of components of two four-vectors, we get:

$$A^{00} = \frac{1}{1 - \frac{V^2}{c^2}} \left(A'^{00} + 2\frac{V}{c} A'^{01} + \frac{V^2}{c^2} A'^{11} \right), \quad A^{11} = \frac{1}{1 - \frac{V^2}{c^2}} \left(A'^{11} + 2\frac{V}{c} A'^{01} + \frac{V^2}{c^2} A'^{00} \right),$$

$$A^{22} = A'^{22}, \quad A^{23} = A'^{23}, \quad A^{12} = \frac{1}{\sqrt{1 - \frac{V^2}{c^2}}} \left(A'^{12} + \frac{V}{c} A'^{02} \right),$$

$$A^{01} = \frac{1}{1 - \frac{V^2}{c^2}} \left[A'^{01} \left(1 + \frac{V^2}{c^2} \right) + \frac{V}{c} A'^{00} + \frac{V}{c} + A'^{11} \right],$$

$$A^{02} = \frac{1}{\sqrt{1 - \frac{V^2}{c^2}}} \left(A'^{02} + \frac{V}{c} A'^{12} \right),$$

and analogous formulas for A^{33}, A^{13} and A^{03}.

2. The same for the antisymmetric tensor A^{ik}.

Solution: Since the coordinates x^2 and x^3 do not change, the tensor component A^{23} does not change, while the components A^{12}, A^{13} and A^{02}, A^{03} transform like x^1 and x^0:

$$A^{23} = A'^{23}, \quad A^{12} = \frac{A'^{12} + \dfrac{V}{c} A'^{02}}{\sqrt{1 - \dfrac{V^2}{c^2}}}, \quad A^{02} = \frac{A'^{02} + \dfrac{V}{c} A'^{12}}{\sqrt{1 - \dfrac{V^2}{c^2}}}$$

and similarly for A^{13}, A^{03}.

With respect to rotations of the two-dimensional coordinate system in the plane $x^0 x^1$ (which are the transformations we are considering) the components $A^{01} = -A^{10}$, $A^{00} = A^{11} = 0$, form an antisymmetric of tensor of rank two, equal to the number of dimensions of the space. Thus, (see the remark on p. 19) these components are not changed by the transformations:

$$A^{01} = A'^{01}.$$

§ 7. Four-dimensional velocity

From the ordinary three-dimensional velocity vector one can form a four-vector. This four-dimensional velocity (*four-velocity*) of a particle is the vector

$$u^i = \frac{dx^i}{ds}. \tag{7.1}$$

To find its components, we note that according to (3.1),

$$ds = cdt \sqrt{1 - \frac{v^2}{c^2}},$$

where v is the ordinary three-dimensional velocity of the particle. Thus

$$u^1 = \frac{dx^1}{ds} = \frac{dx}{cdt \sqrt{1 - \dfrac{v^2}{c^2}}} = \frac{v_x}{c \sqrt{1 - \dfrac{v^2}{c^2}}},$$

etc. Thus

$$u^i = \left(\frac{1}{\sqrt{1 - \dfrac{v^2}{c^2}}}, \frac{\mathbf{v}}{c \sqrt{1 - \dfrac{v^2}{c^2}}} \right). \tag{7.2}$$

Note that the four-velocity is a dimensionless quantity.

The components of the four-velocity are not independent. Noting that $dx_i dx^i = ds^2$, we have

$$u^i u_i = 1. \tag{7.3}$$

Geometrically, u^i is a unit four-vector tangent to the world line of the particle.

Similarly to the definition of the four-velocity, the second derivative

$$w^i = \frac{d^2 x^i}{ds^2} = \frac{du^i}{ds}$$

may be called the four-acceleration. Differentiating formula (7.3), we find:

$$u_i w^i = 0, \qquad (7.4)$$

i.e. the four-vectors of velocity and acceleration are "mutually perpendicular".

PROBLEM

Determine the relativistic uniformly accelerated motion, i.e. the rectilinear motion for which the acceleration w in the proper reference frame (at each instant of time) remains constant.

Solution: In the reference frame in which the particle velocity is $v = 0$, the components of the four-acceleration $w^i = (0, w/c^2, 0, 0)$ (where w is the ordinary three-dimensional acceleration, which is directed along the x axis). The relativistically invariant condition for uniform acceleration must be expressed by the constancy of the four-scalar which coincides with w^2 in the proper reference frame:

$$w^i w_i = \text{const} \equiv -\frac{w^2}{c^4}.$$

In the "fixed" frame, with respect to which the motion is observed, writing out the expression for $w^i w_i$ gives the equation

$$\frac{d}{dt} \frac{v}{\sqrt{1 - \dfrac{v^2}{c^2}}} = w, \quad \text{or} \quad \frac{v}{\sqrt{1 - \dfrac{v^2}{c^2}}} = wt + \text{const.}$$

Setting $v = 0$ for $t = 0$, we find that const = 0, so that

$$v = \frac{wt}{\sqrt{1 + \dfrac{w^2 t^2}{c^2}}}.$$

Integrating once more and setting $x = 0$ for $t = 0$, we find:

$$x = \frac{c^2}{w} \left(\sqrt{1 + \frac{w^2 t^2}{c^2}} - 1 \right).$$

For $wt \ll c$, these formulas go over the classical expressions $v = wt$, $x = wt^2/2$. For $wt \to \infty$, the velocity tends toward the constant value c.

The proper time of a uniformly accelerated particle is given by the integral

$$\int_0^t \sqrt{1 - \frac{v^2}{c^2}} \, dt = \frac{c}{w} \sinh^{-1} \left(\frac{wt}{c} \right).$$

As $t \to \infty$, it increases much more slowly than t, according to the law $c/w \ln (2wt/c)$.

CHAPTER 2

RELATIVISTIC MECHANICS

§ 8. The principle of least action

In studying the motion of material particles, we shall start from the Principle of Least Action. The *principle of least action* is defined, as we know, by the statement that for each mechanical system there exists a certain integral S, called the *action*, which has a minimum value for the actual motion, so that its variation δS is zero.†

To determine the action integral for a free material particle (a particle not under the influence of any external force), we note that this integral must not depend on our choice of reference system, that is, it must be invariant under Lorentz transformations. Then it follows that it must depend on a scalar. Furthermore, it is clear that the integrand must be a differential of the first order. But the only scalar of this kind that one can construct for a free particle is the interval ds, or $\alpha\,ds$, where α is some constant. So for a free particle the action must have the form

$$S = -\alpha \int_a^b ds,$$

where $\int_a^b$ is an integral along the world line of the particle between the two particular events of the arrival of the particle at the initial position and at the final position at definite times t_1 and t_2, i.e. between two given world points; and α is some constant characterizing the particle. It is easy to see that α must be a positive quantity for all particles. In fact, as we saw in § 3, $\int_a^b ds$ has its maximum value along a straight world line; by integrating along a curved world line we can make the integral arbitrarily small. Thus the integral $\int_a^b ds$ with the positive sign cannot have a minimum; with the opposite sign it clearly has a minimum, along the straight world line.

The action integral can be represented as an integral with respect to the time

$$S = \int_{t_1}^{t_2} L\,dt.$$

The coefficient L of dt represents the *Lagrange function* of the mechanical system. With the aid of (3.1), we find:

---

† Strictly speaking, the principle of least action asserts that the integral S must be a minimum only for infinitesimal lengths of the path of integration. For paths of arbitrary length we can say only that S must be an extremum, not necessarily a minimum. (See *Mechanics*, § 2.)

25

$$S = - \int_{t_1}^{t_2} \alpha c \sqrt{1 - \frac{v^2}{c^2}} \; dt,$$

where v is the velocity of the material particle. Consequently the Lagrangian for the particle is

$$L = - \alpha c \sqrt{1 - v^2/c^2} \; .$$

The quantity α, as already mentioned, characterizes the particle. In classical mechanics each particle is characterized by its mass m. Let us find the relation between α and m. It can be determined from the fact that in the limit as $c \to \infty$, our expression for L must go over into the classical expression $L = mv^2/2$. To carry out this transition we expand L in powers of v/c. Then, neglecting terms of higher order, we find

$$L = - \alpha c \sqrt{1 - \frac{v^2}{c^2}} \approx - \alpha c + \frac{\alpha v^2}{2c}.$$

Constant terms in the Lagrangian do not affect the equation of motion and can be omitted. Omitting the constant αc in L and comparing with the classical expression $L = mv^2/2$, we find that $\alpha = mc$.

Thus the action for a free material point is

$$S = - mc \int_a^b ds \tag{8.1}$$

and the Lagrangian is

$$L = -mc^2 \sqrt{1 - \frac{v^2}{c^2}}. \tag{8.2}$$

§ 9. Energy and momentum

By the *momentum* of a particle we can mean the vector $\mathbf{p} = \partial L/\partial \mathbf{v}$ ($\partial L/\partial \mathbf{v}$ is the symbolic representation of the vector whose components are the derivatives of L with respect to the corresponding components of $\mathbf{v}$). Using (8.2), we find;

$$\mathbf{p} = \frac{m\mathbf{v}}{\sqrt{1 - \frac{v^2}{c^2}}}. \tag{9.1}$$

For small velocities ($v \ll c$) or, in the limit as $c \to \infty$, this expression goes over into the classical $\mathbf{p} = m\mathbf{v}$. For $v = c$, the momentum becomes infinite.

The time derivative of the momentum is the force acting on the particle. Suppose the velocity of the particle changes only in direction, that is, suppose the force is directed perpendicular to the velocity. Then

$$\frac{d\mathbf{p}}{dt} = \frac{m}{\sqrt{1 - \frac{v^2}{c^2}}} \frac{d\mathbf{v}}{dt}. \tag{9.2}$$

If the velocity changes only in magnitude, that is, if the force is parallel to the velocity, then

$$\frac{d\mathbf{p}}{dt} = \frac{m}{\left(1 - \dfrac{v^2}{c^2}\right)^{\frac{1}{2}}} + \frac{d\mathbf{v}}{dt}. \tag{9.3}$$

We see that the ratio of force to acceleration is different in the two cases.

The *energy* $\mathscr{E}$ of the particle is defined as the quantity †

$$\mathscr{E} = \mathbf{p} \cdot \mathbf{v} - L.$$

Substituting the expressions (8.2) and (9.1) for L and $\mathbf{p}$, we find

$$\mathscr{E} = \frac{mc^2}{\sqrt{1 - \dfrac{v^2}{c^2}}}. \tag{9.4}$$

This very important formula shows, in particular, that in relativistic mechanics the energy of a free particle does not go to zero for $v = 0$, but rather takes on a finite value

$$\mathscr{E} = mc^2. \tag{9.5}$$

This quantity is called the *rest energy* of the particle.

For small velocities ($v/c \ll 1$), we have, expanding (9.4) in series in powers of v/c,

$$\mathscr{E} \approx mc^2 + \frac{mv^2}{2},$$

which, except for the rest energy, is the classical expression for the kinetic energy of a particle.

We emphasize that, although we speak of a "particle", we have nowhere made use of the fact that it is "elementary". Thus the formulas are equally applicable to any composite body consisting of many particles, where by m we mean the total mass of the body and by v the velocity of its motion as a whole. In particular, formula (9.5) is valid for any body which is at rest as a whole. We call attention to the fact that in relativistic mechanics the energy of a free body (i.e. the energy of any closed system) is a completely definite quantity which is always positive and is directly related to the mass of the body. In this connection we recall that in classical mechanics the energy of a body is defined only to within an arbitrary constant, and can be either positive or negative.

The energy of a body at rest contains, in addition to the rest energies of its constituent particles, the kinetic energy of the particles and the energy of their interactions with one another. In other words, mc^2 is not equal to $\sum m_a c^2$ (where m_a are the masses of the particles), and so m is not equal to $\sum m_a$. Thus in relativistic mechanics the law of conservation of mass does not hold: the mass of a composite body is not equal to the sum of the masses of its parts. Instead only the law of conservation of energy, in which the rest energies of the particles are included, is valid.

Squaring (9.1) and (9.4) and comparing the results, we get the following relation between the energy and momentum of particle:

---

† *See Mechanics*, § 6.

$$\frac{\mathscr{E}^2}{c^2} = p^2 + m^2 c^2.$$ (9.6)

The energy expressed in terms of the momentum is called the Hamiltonian function $\mathscr{H}$:

$$\mathscr{H} = c\sqrt{p^2 + m^2 c^2}.$$ (9.7)

For low velocities, $p \ll mc$, and we have approximately

$$\mathscr{H} \approx mc^2 + \frac{p^2}{2m},$$

i.e., except for the rest energy we get the familiar classical expression for the Hamiltonian.

From (9.1) and (9.4) we get the following relation between the energy, momentum, and velocity of a free particle:

$$\mathbf{p} = \mathscr{E}\,\frac{\mathbf{v}}{c^2}.$$ (9.8)

For $v = c$, the momentum and energy of the particle become infinite. This means that a particle with mass m different from zero cannot move with the velocity of light. Nevertheless, in relativistic mechanics, particles of zero mass moving with the velocity of light can exist.†
From (9.8) we have for such particles:

$$P = \frac{\mathscr{E}}{c}.$$ (9.9)

The same formula also holds approximately for particles with nonzero mass in the so-called *ultrarelativistic* case, when the particle energy $\mathscr{E}$ is large compared to its rest energy mc^2.

We now write all our formulas in four-dimensional form. According to the principle of least action,

$$\delta S = -mc\delta \int_a^b ds = 0.$$

To set up the expression for δS, we note that $ds = \sqrt{dx_i dx^i}$ and therefore

$$\delta S = -mc \int_a^b \frac{dx_i \delta dx^i}{ds} = -mc \int_a^b u_i d\delta x^i.$$

Integrating by parts, we obtain

$$\delta S = -mc u_i \delta x^i \Big|_a^b + mc \int_a^b \delta x^i \frac{du_i}{ds} ds.$$ (9.10)

As we know, to get the equations of motion we compare different trajectories between the same two points, i.e. at the limits $(\delta x^i)_a = (\delta x^i)_b = 0$. The actual trajectory is then determined

† For example, light quanta and neutrinos.

from the condition $\delta S = 0$. From (9.10) we thus obtain the equations $du_i/ds = 0$; that is, a constant velocity for the free particle in four-dimensional form.

To determine the variation of the action as a function of the coordinates, one must consider the point a as fixed, so that $(\delta x^i)_a = 0$. The second point is to be considered as variable, but only actual trajectories are admissible, i.e., those which satisfy the equations of motion. Therefore the integral in expression (9.10) for δS is zero. In place of $(\delta x^i)_b$ we may write simply δx^i, and thus obtain

$$\delta S = - mcu_i\delta x^i. \tag{9.11}$$

The four-vector

$$p_i = - \frac{\partial S}{\partial x^i} \tag{9.12}$$

is called the *momentum four-vector.* As we know from mechanics, the derivatives $\partial S/\partial x$, $\partial S/\partial y$, $\partial S/\partial z$ are the three components of the momentum vector $\mathbf{p}$ of the particle, while the derivative $-\partial S/\partial t$ is the particle energy $\mathscr{E}$. Thus the covariant components of the four-mementum- are $p_i = (\mathscr{E}/c, - \mathbf{p})$, while the contravariant components are†

$$p^i = (\mathscr{E}/c, \mathbf{p}). \tag{9.13}$$

From (9.11) we see that the components of the four-momentum of a free particle are:

$$p^i = mcu^i. \tag{9.14}$$

Substituting the components of the four-velocity from (7.2), we see that we actually get expressions (9.1) and (9.4) for $\mathbf{p}$ and $\mathscr{E}$.

Thus, in relativistic mechanics, momentum and energy are the components of a single four-vector. From this we immediately get the formulas for transformation of momentum and energy from one inertial system to another. Substituting (9.13) in the general formulas (6.1) for transformation of four-vectors, we find:

$$p_x = \frac{p'_x + \frac{V}{c^2}\mathscr{E}'}{\sqrt{1 - \frac{V^2}{c^2}}}, \quad p_y = p'_y, \quad p_z = p'_z, \quad \mathscr{E} = \frac{\mathscr{E}' + Vp'_x}{\sqrt{1 - \frac{V^2}{c^2}}}, \tag{9.15}$$

where p_x, p_y, p_z are the components of the three-dimensional vector $\mathbf{p}$.

From the definition (9.14) of the four-momentum, and the identity $u^iu_i = 1$, we have, for the square of the four-momentum of a free particle:

$$p_ip^i = m^2c^2. \tag{9.16}$$

Substituting the expressions (9.13), we get back (9.6).

By analogy with the usual definition of the force, the force four-vector is defined as the derivative:

$$g^i = \frac{dp^i}{ds} = mc\frac{du^i}{ds}. \tag{9.17}$$

---

† We call attention to a mnemonic for remembering the definition of the physical four-vectors: the *contravariant* components are related to the corresponding three-dimensional vectors ($\mathbf{r}$ for x^i, $\mathbf{p}$ for p^i) with the "right", positive sign.

Its components satisfy the identity $g_i u^i = 0$. The components of this four-vector are expressed in terms of the usual three-dimensional force vector $\mathbf{f} = d\mathbf{p}/dt$:

$$g^i = \left(\frac{\mathbf{f} \cdot \mathbf{v}}{c^2 \sqrt{1 - \frac{v^2}{c^2}}}, \frac{\mathbf{f}}{c \sqrt{1 - \frac{v^2}{c^2}}} \right). \tag{9.18}$$

The time component is related to the work done by the force.

The relativistic Hamilton–Jacobi equation is obtained by substituting the derivatives $-\partial S/\partial x^i$ for p_i in (9.16):

$$\frac{\partial S}{\partial x_i} \frac{\partial S}{\partial x^i} \equiv g^{ik} \frac{\partial S}{\partial x^i} \frac{\partial S}{\partial x^k} = m^2 c^2, \tag{9.19}$$

or, writing the sum explicitly:

$$\frac{1}{c^2} \left(\frac{\partial S}{\partial t} \right)^2 - \left(\frac{\partial S}{\partial x} \right)^2 - \left(\frac{\partial S}{\partial y} \right)^2 - \left(\frac{\partial S}{\partial z} \right)^2 = m^2 c^2. \tag{9.20}$$

The transition to the limiting case of classical mechanics in equation (9.19) is made as follows. First of all we must notice that just as in the corresponding transition with (9.7), the energy of a particle in relativistic mechanics contains the term mc^2, which it does not in classical mechanics. Inasmuch as the action S is related to the energy by $\mathscr{E} = -(\partial S/\partial t)$, in making the transition to classical mechanics we must in place of S substitute a new action S' according to the relation:

$$S = S' - mc^2 t.$$

Substituting this in (9.20), we find

$$\frac{1}{2m} \left[\left(\frac{\partial S'}{\partial x} \right)^2 + \left(\frac{\partial S'}{\partial y} \right)^2 + \left(\frac{\partial S'}{\partial z} \right)^2 \right] - \frac{1}{2mc^2} \left(\frac{\partial S'}{\partial t} \right)^2 + \frac{\partial S'}{\partial t} = 0.$$

In the limit as $c \to \infty$, this equation goes over into the classical Hamilton–Jacobi equation.

§ 10. Transformation of distribution functions

In various physical problems we have to deal with distribution functions for the momenta of particles: $f(\mathbf{p}) dp_x dp_y dp_z$ is the number of particles having momenta with components in given intervals dp_x, dp_y, dp_z (or, as we say for brevity, the number of particles in a given volume element $d^3 p \equiv dp_x dp_y dp_z$ in "momentum space"). We are then faced with the problem of finding the law of transformation of the distribution function $f(\mathbf{p})$ when we transform from one reference system to another.

To solve this problem, we first determine the properties of the "volume element" $dp_x dp_y dp_z$ with respect to Lorentz transformations. If we introduce a four-dimensional coordinate system, on whose axes are marked the components of the four-momentum of a particle, then $dp_x dp_y dp_z$ can be considered as the zeroth component of an element of the hypersurface defined by the equation $p^i p_i = m^2 c^2$. The element of hypersurface is a four-vector directed along the normal to the hypersurface; in our case the direction of the normal obviously coincides with the direction of the four-vector p_i. From this it follows that the ratio

$$\frac{dp_x dp_y dp_z}{\mathscr{E}} \tag{10.1}$$

is an invariant quantity, since it is the ratio of corresponding components of two parallel four-vectors †

The number of particles, $f dp_x dp_y dp_z$, is also obviously an invariant, since it does not depend on the choice of reference frame. Writing it in the form

$$f(\mathbf{p}) \mathscr{E} \frac{dp_x dp_y dp_z}{\mathscr{E}}$$

and using the invariance of the ratio (10.1), we conclude that the product $f(\mathbf{p})\mathscr{E}$ is invariant. Thus the distribution function in the K' system is related to the distribution function in the K system by the formula

$$f'(\mathbf{p}') = \frac{f(\mathbf{p})\mathscr{E}}{\mathscr{E}'}, \tag{10.2}$$

where $\mathbf{p}$ and $\mathscr{E}$ must be expressed in terms of $\mathbf{p}'$ and $\mathscr{E}'$ by using the transformation formulas (9.15).

Let us now return to the invariant expression (10.1). If we introduce "spherical coordinates" in momentum space, the volume element $dp_x dp_y dp_z$ becomes $p^2 dp\, do$, where do is the element of solid angle around the direction of the vector $\mathbf{p}$. Noting that $pdp = \mathscr{E}d\mathscr{E}/c^2$ [from (9.6)], we have:

$$\frac{p^2 dp\, do}{\mathscr{E}} = \frac{pd\mathscr{E}\, do}{c^2}.$$

Thus we find that the expression

$$pd\,\mathscr{E}\, do \tag{10.3}$$

is also invariant.

The notion of a distribution function appears in a different aspect in the kinetic theory of gases: the product $f(\mathbf{r}, \mathbf{p}) dp_x dp_y dp_z dV$ is the number of particles lying in a given volume element dV and having momenta in definite intervals dp_x, dp_y, dp_z. The function $f(\mathbf{r}, \mathbf{p})$ is

---

† The integration with respect to the element (10.1) can be expressed in four-dimensional form by means of the δ-function (cf. the footnote on p. 74) as an integration with respect to

$$\frac{2}{c} \delta(p_i p^i - m^2 c^2) d^4 p, \qquad d^4 p = dp^0 dp^1 dp^2 dp^3, \tag{10.1a}$$

The four components p^i are treated as independent variables (with p^0 taking on only positive values). Formula (10.1a) is obvious from the following representation of the delta function appearing in it:

$$\delta(p^i p_i - m^2 c^2) = \delta\left((p_0)^2 - \frac{\mathscr{E}^2}{c^2}\right) = \frac{c}{2\mathscr{E}}\left[\delta\left(p_0 + \frac{\mathscr{E}}{c}\right) + \delta\left(p_0 - \frac{\mathscr{E}}{c}\right)\right], \tag{10.1b}$$

where $\mathscr{E} = c\sqrt{p^2 + m^2 c^2}$. This formula in turn follows from formula (V) of the footnote on p. 74.

called the distribution function in *phase space* (the space of the coordinates and momenta of the particle), and the product of differentials $d\tau = d^3p\, dV$ is the element of volume of this space. We shall find the law of transformation of this function.

In addition to the two reference systems K and K', we also introduce the frame K_0 in which the particles with the given momentum are at rest; the proper volume dV_0 of the element occupied by the particles is defined relative to this system. The velocities of the systems K and K' relative to the system K_0 coincide, by definition, with the velocities v and v' which these particles have in the systems K and K'. Thus, according to (4.6), we have

$$dV = dV_0 \sqrt{1 - \frac{v^2}{c^2}}, \quad dV' = dV_0 \sqrt{1 - \frac{v'^2}{c^2}},$$

from which

$$\frac{dV}{dV'} = \frac{\mathscr{E}'}{\mathscr{E}}.$$

Multiplying this equation by the equation $d^3p/d^3p' = \mathscr{E}/\mathscr{E}'$, we find that

$$d\tau = d\tau', \tag{10.4}$$

i.e. the element of phase volume is invariant. Since the number of particles $f\, d\tau$ is also invariant, by definition, we conclude that the distribution function in phase space is an invariant:

$$f'(\mathbf{r}', \mathbf{p}') = f(\mathbf{r}, \mathbf{p}), \tag{10.5}$$

where $\mathbf{r}'$, $\mathbf{p}'$ are related to $\mathbf{r}$, $\mathbf{p}$ by the formulas for the Lorentz transformation.

§ 11. Decay of particles

Let us consider the spontaneous decay of a body of mass M into two parts with masses m_1 and m_2. The law of conservation of energy in the decay, applied in the system of reference in which the body is at rest, gives†

$$M = \mathscr{E}_{10} + \mathscr{E}_{20}. \tag{11.1}$$

where $\mathscr{E}_{10}$ and $\mathscr{E}_{20}$ are the energies of the emerging particles. Since $\mathscr{E}_{10} > m_1$ and $\mathscr{E}_{20} > m_2$, the equality (11.1) can be satisfied only if $M > m_1 + m_2$, i.e. a body can disintegrate spontaneously into parts the sum of whose masses is less than the mass of the body. On the other hand, if $M < m_1 + m_2$, the body is stable (with respect to the particular decay) and does not decay spontaneously. To cause the decay in this case, we would have to supply to the body from outside an amount of energy at least equal to its "binding energy" $(m_1 + m_2 - M)$.

Momentum as well as energy must be conserved in the decay process. Since the initial momentum of the body was zero, the sum of the momenta of the emerging particles must be zero: $\mathbf{p}_{10} + \mathbf{p}_{20} = 0$. Consequently $p_{10}^2 = p_{20}^2$, or

---

† In §§ 11–13 we set $c = 1$. In other words the velocity of light is taken as the unit of velocity (so that the dimensions of length and time become the same). This choice is a natural one in relativistic mechanics and greatly simplifies the writing of formulas. However, in this book (which also contains a considerable amount of nonrelativistic theory) we shall not usually use this system of units, and will explicitly indicate when we do.

If c has been set equal to unity in formulas, it is easy to convert back to ordinary units: the velocity is introduced to assure correct dimensions.

$$\mathscr{E}_{10}^2 - m_1^2 = \mathscr{E}_{20}^2 - m_2^2. \tag{11.2}$$

The two equations (11.1) and (11.2) uniquely determine the energies of the emerging particles:

$$\mathscr{E}_{10} = \frac{M^2 + m_1^2 - m_2^2}{2M}, \qquad \mathscr{E}_{20} = \frac{M^2 - m_1^2 + m_2^2}{2M}. \tag{11.3}$$

In a certain sense the inverse of this problem is the calculation of the total energy M of two colliding particles in the system of reference in which their total momentum is zero. (This is abbreviated as the "system of the centre of inertia" or the "C-system".) The computation of this quantity gives a criterion for the possible occurrence of various inelastic collision processes, accompanied by a change in state of the colliding particles, or the "creation" of new particles. A process of this type can occur only if the sum of the masses of the "reaction products" does not exceed M.

Suppose that in the initial reference system (the "laboratory" system) a particle with mass m_1 and energy $\mathscr{E}_1$ collides with a particle of mass m_2 which is at rest. The total energy of the two particles is

$$\mathscr{E} = \mathscr{E}_1 + \mathscr{E}_2 = \mathscr{E}_1 + m_2,$$

and their total momentum is $\mathbf{p} = \mathbf{p}_1 + \mathbf{p}_2 = \mathbf{p}_1$. Considering the two particles together as a single composite system, we find the velocity of its motion as a whole from (9.8):

$$\mathbf{V} = \frac{\mathbf{p}}{\mathscr{E}} = \frac{\mathbf{p}_1}{\mathscr{E}_1 + m_2}. \tag{11.4}$$

This quantity is the velocity of the C-system with respect to the laboratory system (the L-system).

However, in determining the mass M, there is no need to transform from one reference frame to the other. Instead we can make direct use of formula (9.6), which is applicable to the composite system just as it is to each particle individually. We thus have

$$M^2 = \mathscr{E}^2 - p^2 = (\mathscr{E}_1 + m_2)^2 - (\mathscr{E}_1^2 - m_1^2),$$

from which

$$M^2 = m_1^2 + m_2^2 + 2m_2\mathscr{E}_1. \tag{11.5}$$

PROBLEMS

1. A particle moving with velocity V dissociates "in flight" into two particles. Determine the relation between the angles of emergence of these particles and their energies.

Solution: Let $\mathscr{E}_0$ be the energy of one of the decay particles in the C-system [i.e. $\mathscr{E}_{10}$ or $\mathscr{E}_{20}$ in (11.3)], $\mathscr{E}$ the energy of this same particle in the L-system, and θ its angle of emergence in the L-system (with respect to the direction of $\mathbf{V}$). By using the transformation formulas we find:

$$\mathscr{E}_0 = \frac{\mathscr{E} - Vp \cos \theta}{\sqrt{1 - V^2}},$$

so that

$$\cos \theta = \frac{\mathscr{E} - \mathscr{E}_0 \sqrt{1 - V^2}}{V\sqrt{\mathscr{E}^2 - m^2}}. \tag{1}$$

For the determination of $\mathscr{E}$ from $\cos\theta$ we then get the quadratic equation

$$\mathscr{E}^2(1 - V^2\cos^2\theta) - 2\mathscr{E}\mathscr{E}_0\sqrt{1-V^2} + \mathscr{E}_0^2(1-V^2) + V^2m^2\cos^2\theta = 0, \tag{2}$$

which has one positive root (if the velocity v_0 of the decay particle in the C-system satisfies $v_0 > V$) or two positive roots (if $v_0 < V$).

The source of this ambiguity is clear from the following graphical construction. According to (9.15), the momentum components in the L-system are expressed in terms of quantities referring to the C-system by the formulas

$$p_z = \frac{p_0\cos\theta_0 + \mathscr{E}_0 V}{\sqrt{1-V^2}}, \quad p_y = p_0\sin\theta_0.$$

Eliminating θ_0, we get

$$p_y^2 + (p_x\sqrt{1-V^2} - \mathscr{E}_0 V)^2 = p_0^2.$$

With respect to the variables p_x, p_y, this is the equation of an ellipse with semiaxes $p_0/\sqrt{1-V^2}, p_0$, whose centre (the point O in Fig. 3) has been shifted a distance $\mathscr{E}_0 V/\sqrt{1-V^2}$ from the point $\mathbf{p} = 0$ (point A in Fig. 3).†

(a) $V < v_0$ (b) $V > v_0$

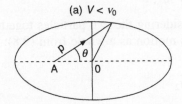   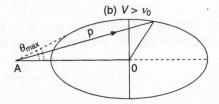

FIG. 3.

If $V > p_0/\mathscr{E}_0 = v_0$, the point A lies outside the ellipse (Fig. 3b), so that for a fixed angle θ the vector $\mathbf{p}$ (and consequently the energy $\mathscr{E}$) can have two different values. It is also clear from the construction that in this case the angle θ cannot exceed a definite value θ_{max} (corresponding to the position of the vector $\mathbf{p}$ in which it is tangent to the ellipse). The value of θ_{max} is most easily determined analytically from the condition that the discriminant of the quadratic equation (2) go to zero:

$$\sin\theta_{max} = \frac{p_0\sqrt{1-V^2}}{mV}.$$

2. Find the energy distribution of the decay particles in the L-system.

Solution: In the C-system the decay particles are distributed isotropically in direction, i.e. the number of particles within the element of solid angle $do_0 = 2\pi\sin\theta_0\,d\theta_0$ is

$$dN = \frac{1}{4\pi}do_0 = \frac{1}{2}|d\cos\theta_0|. \tag{1}$$

The energy in the L-system is given in terms of quantities referring to the C-system by

$$\mathscr{E} = \frac{\mathscr{E}_0 + p_0 V\cos\theta_0}{\sqrt{1-V^2}}$$

and runs through the range of values from

$$\frac{\mathscr{E}_0 - Vp_0}{\sqrt{1-V^2}} \quad \text{to} \quad \frac{\mathscr{E}_0 + Vp_0}{\sqrt{1-V^2}}.$$

† In the classical limit, the ellipse reduces to a circle. (See *Mechanics*, § 16.)

Expressing $d \,|\cos \theta_0|$ in terms of $d\mathscr{E}$, we obtain the normalized energy distribution (for each of the two types of decay particles):

$$dN = \frac{1}{2Vp_0} \sqrt{1 - V^2} \, d\mathscr{E}.$$

3. Determine the range of values in the L-system for the angle between the two decay particles (their separation angle) for the case of decay into two identical particles.

Solution: In the C-system, the particles fly off in opposite directions, so that $\theta_{10} = \pi - \theta_{20} \equiv \theta_0$. According to (5.4), the connection between angles in the C- and L-systems is given by the formulas:

$$\cot \theta_1 = \frac{v_0 \cos \theta_0 + V}{v_0 \sin \theta_0 \sqrt{1 - V^2}}, \qquad \cot \theta_2 = \frac{-v_0 \cos \theta_0 + V}{v_0 \sin \theta_0 \sqrt{1 - V^2}}$$

(since $v_{10} = v_{20} = v_0$ in the present case). The required separation angle is $\Theta = \theta_1 + \theta_2$, and a simple calculation gives:

$$\cot \Theta = \frac{V^2 - v_0^2 + V^2 v_0^2 \sin^2 \theta_0}{2Vv_0 \sqrt{1 - V^2} \sin \theta_0}.$$

An examination of the extreme for this expression gives the following ranges of possible values of Θ:

$$\text{for } V < v_0 : 2 \tan^{-1} \left(\frac{v_0}{V} \sqrt{1 - V^2} \right) < \Theta < \pi;$$

$$\text{for } v_0 < V < \frac{v_0}{\sqrt{1 - v_0^2}} : 0 < \Theta < \sin^{-1} \sqrt{\frac{1 - V^2}{1 - v_0^2}} < \frac{\pi}{2};$$

$$\text{for } V > \frac{v_0}{\sqrt{1 - v_0^2}} : 0 < \Theta < 2 \tan^{-1} \left(\frac{v_0}{V} \sqrt{1 - V^2} \right) < \frac{\pi}{2}.$$

4. Find the angular distribution in the L-system for decay particles of zero mass.

Solution: According to (5.6) the connection between the angles of emergence in the C- and L-systems for particles with $m = 0$ is

$$\cos \theta_0 = \frac{\cos \theta - V}{1 - V \cos \theta}.$$

Substituting this expression in formula (1) of Problem 2, we find:

$$dN = \frac{(1 - V^2) \, do}{4\pi (1 - V \cos \theta)^2}.$$

5. Find the distribution of separation angles in the L-system for a decay into two particles of zero mass.

Solution: The relation between the angles of emergence, θ_1, θ_2 in the L-system and the angles $\theta_{10} \equiv \theta_0$, $\theta_{20} = \pi - \theta_0$ in the C-system is given by (5.6), so that we have for the separation angle $\Theta = \theta_1 + \theta_2$:

$$\cos \Theta = \frac{2V^2 - 1 - V^2 \cos^2 \theta_0}{1 - V^2 \cos^2 \theta_0}$$

and conversely,

$$\cos \theta_0 = \sqrt{1 - \frac{1 - V^2}{V^2} \cot^2 \frac{\Theta}{2}}.$$

Substituting this expression in formula (1) of problem 2, we find:

$$dN = \frac{1 - V^2}{16 \pi V \sin^3 \frac{\Theta}{2} \sqrt{V^2 - \cos^2 \frac{\Theta}{2}}} do.$$

The angle Θ takes on values from π to $\Theta_{min} = 2 \cos^{-1} V$.

6. Determine the maximum energy which can be carried off by one of the decay particles, when a particle of mass M at rest decays into three particles with masses m_1, m_2, and m_3.

Solution: The particle m_1 has its maximum energy if the system of the other two particles m_2 and m_3 has the least possible mass; the latter is equal to the sum $m_2 + m_3$ (and corresponds to the case where the two particles move together with the same velocity). Having thus reduced the problem to the decay of a body into two parts, we obtain from (11.3):

$$\mathscr{E}_{1\,max} = \frac{M^2 + m_1^2 - (m_2 + m_3)^2}{2M}.$$

§ 12. Invariant cross-section

Collision processes are characterized by their *invariant cross-sections,* which determine the number of collisions (of the particular type) occurring between beams of colliding particles.

Suppose that we have two colliding beams; we denote by n_1 and n_2 the particle densities in them (i.e. the numbers of particles per unit volume) and by $\mathbf{v}_1$ and $\mathbf{v}_2$ the velocities of the particles. In the reference system in which particle 2 is at rest (or, as one says, in the *rest frame* of particle 2), we are dealing with the collision of the beam of particles 1 with a stationary target. Then according to the usual definition of the cross-section σ, the number of collisions occurring in volume dV in time dt is

$$d\nu = \sigma v_{rel} n_1 n_2 dV dt.$$

where v_{rel} is the velocity of particle 1 in the rest system of particle 2 (which is just the definition of the relative velocity of two particles in relativistic mechanics).

The number $d\nu$ is by its very nature an invariant quantity. Let us try to express it in a form which is applicable in any reference system:

$$d\nu = A n_1 n_2 dV dt, \tag{12.1}$$

where A is a number to be determined, for which we know that its value in the rest frame of one of the particles is $v_{rel} \sigma$. We shall always mean by σ precisely the cross-section in the rest frame of one of the particles, i.e. by definition, an invariant quantity. From its definition, the relative velocity v_{rel} is also invariant.

In the expression (12.1) the product $dV dt$ is an invariant. Therefore the product $A n_1 n_2$ must also be an invariant.

The law of transformation of the particle density n is easily found by noting that the number of particles in a given volume element dV, $n dV$, is invariant. Writing $n dV = n_0 dV_0$ (the index 0 refers to the rest frame of the particles) and using formula (4.6) for the transformation of the volume, we find:

$$n = \frac{n_0}{\sqrt{1 - v^2}} \tag{12.2}$$

or $n = n_0 \mathscr{E}/m$, where $\mathscr{E}$ is the energy and m the mass of the particles.

Thus the statement that $A n_1 n_2$ is invariant is equivalent to the invariance of the expression $A \mathscr{E}_1 \mathscr{E}_2$. This condition is more conveniently represented in the form

$$A \frac{\mathscr{E}_1 \mathscr{E}_2}{p_{1i} p_2^i} = A \frac{\mathscr{E}_1 \mathscr{E}_2}{\mathscr{E}_1 \mathscr{E}_2 - \mathbf{p}_1 \cdot \mathbf{p}_2} = \text{inv}, \tag{12.3}$$

where the denominator is an invariant—the product of the four-momenta of the two particles.

In the rest frame of particle 2, we have $\mathscr{E}_2 = m_2$, $\mathbf{p}_2 = 0$, so that the invariant quantity (12.3) reduces to A. On the other hand, in this frame $A = \sigma v_{\text{rel}}$. Thus in an arbitrary reference system,

$$A = \sigma v_{\text{rel}} \frac{p_{1i} p_2^i}{\mathscr{E}_1 \mathscr{E}_2}. \tag{12.4}$$

To give this expression its final form, we express v_{rel} in terms of the momenta or velocities of the particles in an arbitrary reference frame. To do this we note that in the rest frame of particle 2,

$$p_{1i} p_2^i = \frac{m_1}{\sqrt{1 - v_{\text{rel}}^2}} m_2.$$

Then

$$v_{\text{rel}} = \sqrt{1 - \frac{m_1^2 m_2^2}{(p_{1i} p_2^i)^2}}. \tag{12.5}$$

Expressing the quantity $p_{1i} p_2^i = \mathscr{E}_1 \mathscr{E}_2 - \mathbf{p}_1 \cdot \mathbf{p}_2$ in terms of the velocities $\mathbf{v}_1$ and $\mathbf{v}_2$ by using formulas (9.1) and (9.4):

$$p_{1i} p_2^i = m_1 m_2 \frac{1 - \mathbf{v}_1 \cdot \mathbf{v}_2}{\sqrt{(1 - v_1^2)(1 - v_2^2)}},$$

and substituting in (12.5), after some simple transformations we get the following expression for the relative velocity:

$$v_{\text{rel}} = \frac{\sqrt{(\mathbf{v}_1 - \mathbf{v}_2)^2 - (\mathbf{v}_1 \times \mathbf{v}_2)^2}}{1 - \mathbf{v}_1 \cdot \mathbf{v}_2} \tag{12.6}$$

(we note that this expression is symmetric in $\mathbf{v}_1$ and $\mathbf{v}_2$, i.e. the magnitude of the relative velocity is independent of the choice of particle used in defining it).

Substituting (12.5) or (12.6) in (12.4) and then in (12.1), we get the final formulas for solving our problem:

$$dv = \sigma \frac{\sqrt{(p_{1i} p_2^i)^2 - m_1^2 m_2^2}}{\mathscr{E}_1 \mathscr{E}_2} n_1 n_2 \, dV dt \tag{12.7}$$

or

$$dv = \sigma \sqrt{(\mathbf{v}_1 - \mathbf{v}_2)^2 - (\mathbf{v}_1 \times \mathbf{v}_2)^2} \, n_1 n_2 \, dV dt \tag{12.8}$$

(W. Pauli, 1933).

If the velocities $\mathbf{v}_1$ and $\mathbf{v}_2$ are collinear, then $\mathbf{v}_1 \times \mathbf{v}_2 = 0$, so that formula (12.8) takes the form:

$$dv = \sigma | \mathbf{v}_1 - \mathbf{v}_2 | \, n_1 n_2 dV dt. \tag{12.9}$$

PROBLEM

Find the "element of length" in relativistic "velocity space".

Solution: The required line element dl_v is the relative velocity of two points with velocities $\mathbf{v}$ and $\mathbf{v} + d\mathbf{v}$. We therefore find from (12.6)

$$dl_v^2 = \frac{(d\mathbf{v})^2 - (\mathbf{v} \times d\mathbf{v})^2}{(1 - v^2)^2} = \frac{dv^2}{(1 - v^2)^2} + \frac{v^2}{1 - v^2}(d\theta^2 + \sin^2 \theta \cdot d\phi^2),$$

where θ, ϕ are the polar angle and azimuth of the direction of $\mathbf{v}$. If in place of v we introduce the new variable χ through the equation $v = \tanh \chi$, the line element is expressed as:

$$dl_v^2 = d\chi^2 + \sinh^2 \chi (d\theta^2 + \sin^2 \theta \cdot d\phi^2).$$

From the geometrical point of view this is the line element in three-dimensional Lobachevskii space—the space of constant negative curvature (see (111.12)).

§ 13. Elastic collisions of particles

Let us consider, from the point of view of relativistic mechanics, the *elastic collision* of particles. We denote the momenta and energies of the two colliding particles (with masses m_1 and m_2) by $\mathbf{p}_1$, $\mathscr{E}_1$ and $\mathbf{p}_2$, $\mathscr{E}_2$; we use primes for the corresponding quantities after collision. The laws of conservation of momentum and energy in the collision can be written together as the equation for conservation of the four-momentum:

$$p_1^i + p_2^i = p_1'^i + p_2'^i. \tag{13.1}$$

From this four-vector equation we construct invariant relations which will be helpful in further computations. To do this we rewrite (13.1) in the form:

$$p_1^i + p_2^i - p_1'^i + p_2'^i,$$

and square both sides (i.e. we write the scalar product of each side with itself). Noting that the squares of the four-momenta p_1^i and $p_1'^i$ are equal to m_1^2, and the squares of p_2^i and $p_2'^i$ are equal to m_2^2, we get:

$$m_1^2 + p_{1i}p_2^i - p_{1i}p_1'^i - p_{2i}p_1'^i = 0. \tag{13.2}$$

Similarly, squaring the equation $p_1^i + p_2^i - p_2'^i = p_1'^i$, we get:

$$m_2^2 + p_{1i}p_2^i - p_2^i p_2'^i - p_{1i}p_2'^i = 0. \tag{13.3}$$

Let us consider the collision in a reference system (the L-system) in which one of the particles (m_2) was at rest before the collision. Then $\mathbf{p}_2 = 0$, $\mathscr{E}_2 = m_2$, and the scalar products appearing in (13.2) are:

$$p_{1i}p_2^i = \mathscr{E}_1 m_2,$$

$$p_{2i}p_1'^i = m_2 \mathscr{E}_1', \tag{13.4}$$

$$p_{1i}p_1'^i = \mathscr{E}_1 \mathscr{E}_1' - \mathbf{p}_1 \cdot \mathbf{p}_1' = \mathscr{E}_1 \mathscr{E}_1' - p_1 p_1' \cos \theta_1,$$

where θ_1 is the angle of scattering of the incident particle m_1. Substituting these expressions in (13.2) we get:

$$\cos \theta_1 = \frac{\mathscr{E}_1'(\mathscr{E}_1 + m_2) - \mathscr{E}_1 m_2 - m_1^2}{p_1 p_1'}. \tag{13.5}$$

Similarly, we find from (13.3):

$$\cos \theta_2 = \frac{(\mathscr{E}_1 + m_2)(\mathscr{E}_2' - m_2)}{p_1 p_2'}, \tag{13.6}$$

where θ_2 is the angle between the transferred momentum $\mathbf{p}_2'$ and the momentum of the incident particle $\mathbf{p}_1$.

The formulas (13.5)–(13.6) relate the angles of scattering of the two particles in the L-system to the changes in their energy in the collision. Inverting these formulas, we can express the energies $\mathscr{E}_1'$, $\mathscr{E}_2'$ in terms of the angles θ_1 or θ_2. Thus, substituting in (13.6) $p_1 = \sqrt{\mathscr{E}_1^2 - m_1^2}$, $p_2' = \sqrt{(\mathscr{E}_2')^2 - m_2^2}$ and squaring both sides, we find after a simple computation:

$$\mathscr{E}_2' = m_2 \frac{(\mathscr{E}_1 + m_2)^2 + (\mathscr{E}_1^2 - m_1^2) \cos^2 \theta_2}{(\mathscr{E}_1 + m_2)^2 - (\mathscr{E}_1^2 - m_1^2) \cos^2 \theta_2}. \tag{13.7}$$

Inversion of formula (13.5) leads in the general case to a very complicated formula for $\mathscr{E}_1'$ in terms of θ_1.

We note that if $m_1 > m_2$, i.e. if the incident particle is heavier than the target particle, the scattering angle θ_1 cannot exceed a certain maximum value. It is easy to find by elementary computations that this value is given by the equation

$$\sin \theta_{1\,\text{max}} = \frac{m_2}{m_1}, \tag{13.8}$$

which coincides with the familiar classical result.

Formulas (13.5)–(13.6) simplify in the case when the incident particle has zero mass: $m_1 = 0$, and correspondingly $p_1 = \mathscr{E}_1$, $p_1' = \mathscr{E}_1'$. For this case let us write the formula for the energy of the incident particle after the collision, expressed in terms of its angle of deflection:

$$\mathscr{E}_1' = \frac{m_2}{1 - \cos \theta_1 + \dfrac{m_2}{\mathscr{E}_1}}. \tag{13.9}$$

Let us now turn once again to the general case of collision of particles of arbitrary mass. The collision is most simply treated in the C-system. Designating quantities in this system by the additional subscript 0, we have $\mathbf{p}_{10} = -\mathbf{p}_{20} \equiv \mathbf{p}_0$. From the conservation of momentum, during the collision the momenta of the two particles merely rotate, remaining equal in magnitude and opposite in direction. From the conservation of energy, the value of each of the momenta remains unchanged.

Let χ be the angle of scattering in the C-system—the angle through which the momenta $\mathbf{p}_{10}$ and $\mathbf{p}_{20}$ are rotated by the collision. This quantity completely determines the scattering process in the C-system, and therefore also in any other reference system. It is also convenient in describing the collision in the L-system and serves as the single parameter which remains undetermined after the conservation of momentum and energy are applied.

We express the final energies of the two particles in the L-system in terms of this parameter. To do this we return to (13.2), but this time write out the product $p_{1i} p_1'^i$ in the C-system:

$$p_{1i} p_1'^i = \mathscr{E}_{10} \mathscr{E}_{10}' - \mathbf{p}_{10} \cdot \mathbf{p}_{10}' = \mathscr{E}_{10}^2 - p_0^2 \cos \chi = p_0^2 (1 - \cos \chi) + m_1^2$$

(in the C-system the energies of the particles do not change in the collision: $\mathscr{E}_{10}' = \mathscr{E}_{10}$). We write out the other two products in the L-system, i.e we use (13.4). As a result we get: $\mathscr{E}_1' - \mathscr{E}_1 = -(p_0^2/m_2)(1 - \cos \chi)$. We must still express p_0^2 in terms of quantities referring to the L-system. This is easily done by equating the values of the invariant $p_{1i} p_2^i$ in the L- and C-systems:

$$\mathscr{E}_{10} \mathscr{E}_{20} - \mathbf{p}_{10} \cdot \mathbf{p}_{20} = \mathscr{E}_1 m_2,$$

or

$$\sqrt{(p_0^2 + m_1^2)(p_0^2 + m_2^2)} = \mathscr{E}_1 m_2 - p_0^2.$$

Solving the equation for p_0^2, we get:

$$p_0^2 = \frac{m_2^2 (\mathscr{E}_1^2 - m_1^2)}{m_1^2 + m_2^2 + 2m_2 \mathscr{E}_1}. \tag{13.10}$$

Thus, we finally have:

$$\mathscr{E}_1' = \mathscr{E}_1 - \frac{m_2(\mathscr{E}_1^2 - m_1^2)}{m_1^2 + m_2^2 + 2m_2 \mathscr{E}_1} (1 - \cos \chi). \tag{13.11}$$

The energy of the second particle is obtained from the conservation law: $\mathscr{E}_1 + m_2 = \mathscr{E}_1' + \mathscr{E}_2'$. Therefore

$$\mathscr{E}_2' = m_2 + \frac{m_2(\mathscr{E}_1^2 - m_1^2)}{m_1^2 + m_2^2 + 2m_2 \mathscr{E}_1} (1 - \cos \chi). \tag{13.12}$$

The second terms in these formulas represent the energy lost by the first particle and transferred to the second particle. The maximum energy transfer occurs for $\chi = \pi$, and is equal to

$$\mathscr{E}_{2\,\text{max}}' - m_2 = \mathscr{E}_1 - \mathscr{E}_{1\,\text{min}}' = \frac{2m_2(\mathscr{E}_1^2 - m_1^2)}{m_1^2 + m_2^2 + 2m_2 \mathscr{E}_1}. \tag{13.13}$$

The ratio of the minimum kinetic energy of the incident particle after collision to its initial energy is:

$$\frac{\mathscr{E}_{1\,\text{min}}' - m_1}{\mathscr{E}_1 - m_1} = \frac{(m_1 - m_2)^2}{m_1^2 + m_2^2 + 2m_2 \mathscr{E}_1}. \tag{13.14}$$

In the limiting case of low velocities (when $\mathscr{E} \approx m + mv^2/2$), this relation tends to a constant limit, equal to

$$\left(\frac{m_1 - m_2}{m_1 + m_2} \right)^2.$$

In the opposite limit of large energies $\mathscr{E}_1$, relation (13.14) tends to zero; the quantity $\mathscr{E}_{1\,\text{min}}'$ tends to a constant limit. This limit is

$$\mathscr{E}_{1\,\text{min}}' = \frac{m_1^2 + m_2^2}{2m_2}.$$

Let us assume that $m_2 \gg m_1$, i.e. the mass of the incident particle is small compared to the mass of the particle at rest. According to classical mechanics the light particle could transfer only a negligible part of its energy (see *Mechanics*, § 17). This is not the case in relativistic mechanics. From formula (13.14) we see that for sufficiently large energies $\mathscr{E}_1$ the fraction of the energy transferred can reach the order of unity. For this it is not sufficient that the velocity of m_1 be of order 1, but one must have $\mathscr{E}_1 \sim m_2$, i.e. the light particle must have an energy of the order of the rest energy of the heavy particle.

A similar situation occurs for $m_2 \ll m_1$, i.e. when a heavy particle is incident on a light one. Here too, according to classical mechanics, the energy transfer would be insignificant. The fraction of the energy transferred begins to be significant only for energies $\mathscr{E}_1 \sim m_1^2/m_2$. We note that we are not taking simply of velocities of the order of the light velocity, but of energies large compared to m_1, i.e. we are dealing with the ultrarelativistic case.

PROBLEMS

1. The triangle ABC in Fig. 4 is formed by the momentum vector $\mathbf{p}$ of the impinging particle and the momenta $\mathbf{p}_1'$, $\mathbf{p}_2'$ of the two particles after the collision. Find the locus of the points C corresponding to all possible values of $\mathbf{p}_1'$, $\mathbf{p}_2'$.

Solution: The required curve is an ellipse whose semiaxes can be found by using the formulas obtained in problem 1 of § 11. In fact, the construction given there determined the locus of the vectors $\mathbf{p}$ in the L-system which are obtained from arbitrarily directed vectors $\mathbf{p}_0$ with given length p_0 in the C-system.

(a) $m_1 > m_2$ (b) $m_1 < m_2$

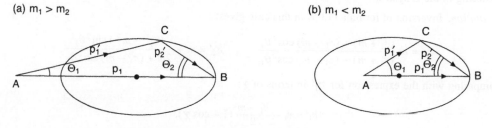

FIG. 4.

Since the absolute values of the momenta of the colliding particles are identical in the C-system, and do not change in the collision, we are dealing with a similar construction for the vector $\mathbf{p}_1'$, for which

$$p_0 \equiv p_{10} = p_{20} = \frac{m_2 V}{\sqrt{1 - V^2}}$$

in the C-system where V is the velocity of particle m_2 in the C-system, coincides in magnitude with the velocity of the centre of inertia, and is equal to $V = p_1/(\mathscr{E}_1 + m_2)$ (see (11.4)). As a result we find that the minor and major semiaxes of the ellipse are

$$p_0 = \frac{m_2 p_1}{\sqrt{m_1^2 + m_2^2 + 2m_2\mathscr{E}_1}}, \qquad \frac{p_0}{\sqrt{1 - V^2}} = \frac{m_2 p_1(\mathscr{E}_1 + m_2)}{m_1^2 + m_2^2 + 2m_2\mathscr{E}_1}$$

(the first of these is, of course, the same as (13.10)).

For $\theta_1 = 0$, the vector $\mathbf{p}_1'$ coincides with $\mathbf{p}_1$, so that the distance AB is equal to p_1. Comparing p_1 with the length of the major axis of the ellipse, it is easily shown that the point A lies outside the ellipse if $m_1 > m_2$ (Fig. 4a), and inside it if $m_1 < m_2$ (Fig. 4b).

2. Determine the minimum separation angle $\Theta_{\min}$ of two particles after collision of the masses of the two particles are the same ($m_1 = m_2 \equiv m$).

Solution: If $m_1 = m_2$, the point A of the diagram lies on the ellipse, while the minimum separation angle corresponds to the situation where point C is at the end of the minor axis (Fig. 5). From the construction it is clear that $\tan (\Theta_{min}/2)$ is the ratio of the lengths of the semiaxes, and we find:

$$\tan \frac{\Theta_{min}}{2} = \sqrt{\frac{2m}{\mathscr{E}_1 + m}},$$

or

$$\cos \Theta_{min} = \frac{\mathscr{E}_1 - m}{\mathscr{E}_1 + 3m}.$$

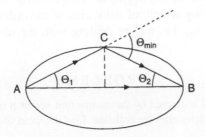

FIG. 5.

3. For the collision of two particles of equal mass m, express $\mathscr{E}_1'$, $\mathscr{E}_2'$, χ in terms of the angle θ_1 of scattering in the L-system.

Solution: Inversion of formula (13.5) in this case gives:

$$\mathscr{E}_1' = m \frac{(\mathscr{E}_1 + m) + (\mathscr{E}_1 - m) \cos^2\theta_1}{(\mathscr{E}_1 + m) - (\mathscr{E}_1 - m) \cos^2\theta_1}, \qquad \mathscr{E}_2' = m + \frac{(\mathscr{E}_1^2 - m^2) \sin^2\theta_1}{2m + (\mathscr{E}_1 - m) \sin^2\theta_1}.$$

Comparing with the expression for $\mathscr{E}_1'$ in terms of χ:

$$\mathscr{E}_1' = \mathscr{E}_1 - \frac{\mathscr{E}_1 - m}{2} (1 - \cos \chi),$$

we find the angle of scattering in the C-system:

$$\cos \chi = \frac{2m - (\mathscr{E}_1 + 3m) \sin^2\theta_1}{2m + (\mathscr{E}_1 - m) \sin^2\theta_1}.$$

§ 14. Angular momentum

As is well known from classical mechanics, for a closed system, in addition to conservation of energy and momentum, there is conservation of angular momentum, that is, of the vector

$$\mathbf{M} = \sum \mathbf{r} \times \mathbf{p}$$

where $\mathbf{r}$ and $\mathbf{p}$ are the radius vector and momentum of the particle; the summation runs over all the particles making up the system. The conservation of angular momentum is a consequence of the fact that because of the isotropy of space, the Lagrangian of a closed system does not change under a rotation of the system as a whole.

By carrying through a similar derivation in four-dimensional form, we obtain the relativistic expression for the angular momentum. Let x^i be the coordinates of one of the particles of the system. We make an infinitesimal rotation in the four-dimensional space. Under such a transformation, the coordinates x^i take on new values x'^i such that the differences $x'^i - x^i$ are linear functions

$$x'^i - x^i = x_k \delta\Omega^{ik} \tag{14.1}$$

with infinitesimal coefficients $\delta\Omega_{ik}$. The components of the four-tensor $\delta\Omega_{ik}$ are connected to one another by the relations resulting from the requirement that, under a rotation, the length of the radius vector must remain unchanged, that is, $x_i' x'^i = x_i x^i$. Substituting for x'^i from (14.1) and dropping terms quadratic in $\delta\Omega_{ik}$, as infinitesimals of higher order, we find

$$x^i x^k \delta\Omega_{ik} = 0.$$

This equation must be fulfilled for arbitrary x^i. Since $x^i x^k$ is a symmetric tensor, $\delta\Omega_{ik}$ must be an antisymmetric tensor (the product of a symmetrical and an antisymmetrical tensor is clearly identically zero). Thus we find that

$$\delta\Omega_{ki} = -\,\delta\Omega_{ik}. \tag{14.2}$$

The change in the action for an infinitesimal change of coordinates of the initial point a and the final point b of the trajectory has the form (see 9.11):

$$\delta S = -\,\Sigma p^i\, \delta x_i \Big|_a^b$$

(the summation extends over all the particles of the system). In the case of rotation which we are now considering, $\delta x_i = \delta\Omega_{ik} x^k$, and so

$$\delta S = -\,\delta\Omega_{ik}\, \Sigma\, p^i x^k \Big|_a^b.$$

If we resolve the tensor $\Sigma p^i x^k$ into symmetric and antisymmetric parts, then the first of these when multiplied by an antisymmetric tensor gives identically zero. Therefore, taking the antisymmetric part of $\Sigma p^i x^k$, we can write the preceding equality in the form

$$\delta S = -\,\delta\Omega_{ik}\, \tfrac{1}{2} \Sigma\, (p^i x^k - p^k x^i)\, \Big|_a^b. \tag{14.3}$$

For a closed system the action, being an invariant, is not changed by a rotation in 4-space. This means that the coefficients of $\delta\Omega_{ik}$ in (14.3) must vanish:

$$\Sigma(p^i x^k - p^k x^i)_b = \Sigma(p^i x^k - p^k x^i)_a.$$

Consequently we see that for a closed system the tensor

$$M^{ik} = \Sigma(x^i p^k - x^k p^i). \tag{14.4}$$

This antisymmetric tensor is called the *four-tensor of angular momentum.* The space components of this tensor are the components of the three-dimensional angular momentum vector $\mathbf{M} = \Sigma \mathbf{r} \times \mathbf{p}$:

$$M^{23} = M_x, \quad -M^{13} = M_y, \quad M^{12} = M_z.$$

The components M^{01}, M^{02}, M^{03} form a vector $\Sigma(t\mathbf{p} - \mathscr{E}\,\mathbf{r}/c^2)$. Thus, we can write the components of the tensor M^{ik} in the form:

$$M^{ik} = \left[c\,\Sigma \left(t\mathbf{p} - \frac{\mathscr{E}\,\mathbf{r}}{c^2} \right), -\mathbf{M} \right]. \tag{14.5}$$

(Compare (6.10).)

Because of the conservation of M^{ik} for a closed system, we have, in particular,

$$\Sigma \left(t\mathbf{p} - \frac{\mathscr{E}\mathbf{r}}{c^2} \right) = \text{const.}$$

Since, on the other hand, the total energy $\Sigma \mathscr{E}$ is also conserved, this equality can be written in the form

$$\frac{\Sigma \mathscr{E}\mathbf{r}}{\Sigma \mathscr{E}} - \frac{c^2 \Sigma \mathbf{p}}{\Sigma \mathscr{E}} t = \text{const.}$$

(Quantities referring to different particles are taken at the same time t).

From this we see that the point with the radius vector

$$\mathbf{R} = \frac{\Sigma \mathscr{E}\mathbf{r}}{\Sigma \mathscr{E}} \tag{14.6}$$

moves uniformly with the velocity

$$\mathbf{V} = \frac{c^2 \Sigma \mathbf{p}}{\Sigma \mathscr{E}}, \tag{14.7}$$

which is none other than the velocity of motion of the system as a whole. [It relates the total energy and momentum, according to formula (9.8).] Formula (14.6) gives the relativistic definition of the coordinates of the *centre of inertia* of the system. If the velocities of all the particles are small compared to c, we can approximately set $\mathscr{E} \approx mc^2$ so that (14.6) goes over into the usual classical expression

$$\mathbf{R} = \frac{\Sigma m\mathbf{r}}{\Sigma m}. \; \dagger$$

We note that the components of the vector (14.6) do not constitute the space components of any four-vector, and therefore under a transformation of reference frame they do not transform like the coordinates of a point. Thus we get different points for the centre of inertia of a given system with respect to different reference frames.

PROBLEM

Find the connection between the angular momentum $\mathbf{M}$ of a body (system of particles) in the reference frame K in which the body moves with velocity $\mathbf{V}$, and its angular momentum $\mathbf{M}^{(0)}$ in the frame K_0 in which the body is at rest as a whole; in both cases the angular momentum is defined with respect to the same point—the centre of inertia of the body in the system K_0.‡

---

† We note that whereas the classical formula for the centre of inertia applies equally well to interacting and non-interacting particles, formula (14.6) is valid only if we neglect interaction. In relativistic mechanics, the definition of the centre of inertia of a system of interacting particles requires us to include explicitly the momentum and energy of the field produced by the particles.

‡ We remind the reader that although in the system K_0 (in which $\Sigma \mathbf{p} = 0$) the angular momentum is independent of the choice of the point with respect to which it is defined, in the K system (in which $\Sigma \mathbf{p} \neq 0$) the angular momentum does depend on this choice (see *Mechanics*, § 9).

Solution: The K_0 system moves relative to the K system with velocity $\mathbf{V}$; we choose its direction for the x axis. The components of M^{ik} that we want transform according to the formulas (see problem 2 in § 6):

$$M^{12} = \frac{M^{(0)12} + \dfrac{V}{c} M^{(0)02}}{\sqrt{1 - \dfrac{V^2}{c^2}}}, \quad M^{13} = \frac{M^{(0)13} + \dfrac{V}{c} M^{(0)03}}{\sqrt{1 - \dfrac{V^2}{c^2}}}, \quad M^{23} = M^{(0)23}.$$

Since the origin of coordinates was chosen at the centre of inertia of the body (in the K_0 system), in that system $\sum \mathscr{E} \mathbf{r} = 0$, and since in that system $\sum \mathbf{p} = 0$, $M^{(0)02} = M^{(0)03} = 0$. Using the connection between the components of M^{ik} and the vector $\mathbf{M}$, we find for the latter:

$$M_z = M_x^{(0)}, \quad M_y = \frac{M_y^{(0)}}{\sqrt{1 - \dfrac{V^2}{c^2}}}, \quad M_z = \frac{M_z^{(0)}}{\sqrt{1 - \dfrac{V^2}{c^2}}}.$$

CHAPTER 3

CHARGES IN ELECTROMAGNETIC FIELDS

§ 15. Elementary particles in the theory of relativity

The interaction of particles can be described with the help of the concept of a *field* of force. Namely, instead of saying that one particle acts on another, we may say that the particle creates a field around itself; a certain force then acts on every other particle located in this field. In classical mechanics, the field is merely a mode of description of the physical phenomenon—the interaction of particles. In the theory of relativity, because of the finite velocity of propagation of interactions, the situation is changed fundamentally. The forces acting on a particle at a given moment are not determined by the positions at that same moment. A change in the position of one of the particles influences other particles only after the lapse of a certain time interval. This means that the field itself acquires physical reality. We cannot speak of a direct interaction of particles located at a distance from one another. Interactions can occur at any one moment only between neighbouring points in space (contact interaction). Therefore we must speak of the interaction of the one particle with the field, and of the subsequent interaction of the field with the second particle.

We shall consider two types of fields, gravitational and electromagnetic. The study of gravitational fields is left to Chapters 10 to 14 and in the other chapters we consider only electromagnetic fields.

Before considering the interactions of particles with the electromagnetic field, we shall make some remarks concerning the concept of a "particle" in relativistic mechanics.

In classical mechanics one can introduce the concept of a rigid body, i.e., a body which is not deformable under any conditions. In the theory of relativity it should follow similarly that we would consider as rigid those bodies whose dimensions all remain unchanged in the reference system in which they are at rest. However, it is easy to see that the theory of relativity makes the existence of rigid bodies impossible in general.

Consider, for example, a circular disk rotating around its axis, and let us assume that it is rigid. A reference frame fixed in the disk is clearly not inertial. It is possible, however, to introduce for each of the infinitesimal elements of the disk an inertial system in which this element would be at rest at the moment; for different elements of the disk, having different velocities, these systems will, of course, also be different. Let us consider a series of line elements, lying along a particular radius vector. Because of the rigidity of the disk, the length of each of these segments (in the corresponding inertial system of reference) will be the same as it was when the disk was at rest. This same length would be measured by an observer at rest, past whom this radius swings at the given moment, since each of its segments is perpendicular to its velocity and consequently a Lorentz contraction does not occur. Therefore the total length of the radius as measured by the observer at rest, being the sum of its segments, will be the same as when the disk was at rest. On the other hand, the length of each element of the circumference of the disk, passing by the observer at rest at a given moment, undergoes a Lorentz contraction, so that the length of the whole circumference (measured by the observer at rest as the sum of the lengths of its various segments) turns out to be smaller than the length of the circumference of the disk at rest. Thus we arrive at the result that due to the rotation of the disk, the ratio of circumference to radius (as measured by an observer at rest) must change, and not remain equal to 2π. The absurdity of this result shows that actually the disk cannot be rigid, and that in rotation it must necessarily undergo some complex deformation depending on the elastic properties of the material of the disk.

The impossibility of the existence of rigid bodies can be demonstrated in another way. Suppose some solid body is set in motion by an external force acting at one of its points. If the body were rigid, all of its points would have to be set in motion at the same time as the point to which the force is applied; if this were not so the body would be deformed. However, the theory of relativity makes this impossible, since the force at the particular point is transmitted to the others with a finite velocity, so that all the points cannot begin moving simultaneously.

From this discussion we can draw certain conclusions concerning the treatment of "*elementary*" particles, i.e. particles whose state we assume to be described completely by giving its three coordinates and the three components of its velocity as a whole. It is obvious that if an elementary particle had finite dimensions, i.e. if it were extended in space, it could not be deformable, since the concept of deformability is related to the possibility of independent motion of individual parts of the body. But, as we have seen, the theory of relativity shows that it is impossible for absolutely rigid bodies to exist.

Thus we come to the conclusion that in classical (non-quantum) relativistic mechanics, we cannot ascribe finite dimensions to particles which we regard as elementary. In other words, within the framework of classical theory elementary particles must be treated as points.†

§ 16. Four-potential of a field

For a particle moving in a given electromagnetic field, the action is made up of two parts: the action (8.1) for the free particle, and a term describing the interaction of the particle with the field. The latter term must contain quantities characterizing the particle and quantities characterizing the field.

It turns out‡ that the properties of a particle with respect to interaction with the electro-magnetic field are determined by a single parameter—the *charge e* of the particle, which can be either positive or negative (or equal to zero). The properties of the field are characterized by a four-vector A_i, the *four-potential*, whose components are functions of the coordinates and time. These quantities appear in the action function in the term

---

† Quantum mechanics makes a fundamental change in this situation, but here again relativity theory makes it extremely difficult to introduce anything other than point interactions.

‡ The assertions which follow should be regarded as being, to a certain extent, the consequence of experimental data. The form of the action for a particle in an electromagnetic field cannot be fixed on the basis of general considerations alone (such as, for example, the requirement of relativistic invariance). The latter would permit the occurrence in formula (16.1) of terms of the form $\int A \, ds$, where A is a scalar function.

To avoid any misunderstanding, we repeat that we are considering classical (and not quantum) theory, and therefore do not include effects which are related to the spins of particles.

$$-\frac{e}{c} \int_a^b A_i dx^i,$$

where the functions A_i are taken at points on the world line of the particle. The factor $1/c$ has been introduced for convenience. It should be pointed out that, so long as we have no formulas relating the charge or the potentials with already known quantities, the units for measuring these new quantities can be chosen arbitrarily.†

Thus the action function for a charge in an electromagnetic field has the form

$$S = \int_a^b \left(-mcds - \frac{e}{c} A_i dx^i \right).$$ (16.1)

The three space components of the four-vector A^i form a three-dimensional vector $\mathbf{A}$ called the *vector potential* of the field. The time component is called the *scalar potential;* we denote it by $A^0 = \phi$. Thus

$$A^i = (\phi, \mathbf{A}).$$ (16.2)

Therefore the action integral can be written in the form

$$S = \int_a^b \left(-mcds + \frac{e}{c} \mathbf{A} \cdot d\mathbf{r} - e\phi dt \right).$$

Introducing $d\mathbf{r}/dt = \mathbf{v}$, and changing to an integration over t,

$$S = \int_{t_1}^{t_2} \left(-mc^2 \sqrt{1 - \frac{v^2}{c^2}} + \frac{e}{c} \mathbf{A} \cdot \mathbf{v} - e\phi \right) dt.$$ (16.3)

The integrand is just the Lagrangian for a charge in an electromagnetic field:

$$L = -mc^2 \sqrt{1 - \frac{v^2}{c^2}} + \frac{e}{c} \mathbf{A} \cdot \mathbf{v} - e\phi.$$ (16.4)

This function differs from the Lagrangian for a free particle (8.2) by the terms $(e/c)\,\mathbf{A} \cdot \mathbf{v} - e\phi$, which describe the interaction of the charge with the field.

The derivative $\partial L/\partial \mathbf{v}$ is the generalized momentum of the particle; we denote it by $\mathbf{P}$. Carrying out the differentiation, we find

$$\mathbf{P} = \frac{m\mathbf{v}}{\sqrt{1 - \frac{v^2}{c^2}}} + \frac{e}{c}\mathbf{A} = \mathbf{p} + \frac{e}{c}\mathbf{A}.$$ (16.5)

Here we have denoted by $\mathbf{p}$ the ordinary momentum of the particle, which we shall refer to simply as its momentum.

From the Lagrangian we can find the Hamiltonian function for a particle in a field from the general formula

† Concerning the establishment of these units, see § 27.

$$\mathcal{H} = \mathbf{v} \cdot \frac{\partial L}{\partial \mathbf{v}} - L.$$

Substituting (16.4), we get

$$\mathcal{H} = \frac{mc^2}{\sqrt{1 - \dfrac{v^2}{c^2}}} + e\phi. \tag{16.6}$$

However, the Hamiltonian must be expressed not in terms of the velocity, but rather in terms of the generalized momentum of the particle.

From (16.5) and (16.6) it is clear that the relation between $\mathcal{H} - e\phi$ and $\mathbf{P} - (e/c)\mathbf{A}$ is the same as the relation between $\mathcal{H}$ and $\mathbf{p}$ in the absence of the field, i.e.

$$\left(\frac{\mathcal{H} - e\phi}{c}\right)^2 = m^2 c^2 + \left(\mathbf{P} - \frac{e}{c}\mathbf{A}\right)^2, \tag{16.7}$$

or else

$$\mathcal{H} = \sqrt{m^2 c^4 + c^2 \left(\mathbf{P} - \frac{e}{c}\mathbf{A}\right)^2} + e\phi. \tag{16.8}$$

For low velocities, i.e. for classical mechanics, the Lagrangian (16.4) goes over into

$$L = \frac{mv^2}{2} + \frac{e}{c}\mathbf{A} \cdot \mathbf{v} - e\phi. \tag{16.9}$$

In this approximation

$$\mathbf{p} = m\mathbf{v} = \mathbf{P} - \frac{e}{c}\mathbf{A},$$

and we find the following expression for the Hamiltonian:

$$\mathcal{H} = \frac{1}{2m}\left(\mathbf{P} - \frac{e}{c}\mathbf{A}\right)^2 + e\phi. \tag{16.10}$$

Finally we write the Hamilton–Jacobi equation for a particle in an electromagnetic field. It is obtained by replacing, in the equation for the Hamiltonian, $\mathbf{P}$ by $\partial S/\partial \mathbf{r}$, and $\mathcal{H}$ by $-(\partial S/\partial t)$. Thus we get from (16.7)

$$\left(\nabla S - \frac{e}{c}\mathbf{A}\right)^2 - \frac{1}{c^2}\left(\frac{\partial S}{\partial t} + e\phi\right)^2 + m^2 c^2 = 0. \tag{16.11}$$

§ 17. Equations of motion of a charge in a field

A charge located in a field not only is subjected to a force exerted by the field, but also in turn acts on the field, changing it. However, if the charge e is not large, the action of the charge on the field can be neglected. In this case, when considering the motion of the charge in a given field, we may assume that the field itself does not depend on the coordinates or the velocity of the charge. The precise conditions which the charge must fulfil in order to be considered as small in the present sense, will be clarified later on (see § 75). In what follows we shall assume that this condition is fulfilled.

So we must find the equations of motion of a charge in a given electromagnetic field. These equations are obtained by varying the action, i.e. they are given by the Lagrange equations

$$\frac{d}{dt}\left(\frac{\partial L}{\partial \mathbf{v}}\right) = \frac{\partial L}{\partial \mathbf{r}},\tag{17.1}$$

where L is given by formula (16.4).

The derivative $\partial L/\partial \mathbf{v}$ is the generalized momentum of the particle (16.5). Further, we write

$$\frac{\partial L}{\partial \mathbf{r}} = \nabla L = \frac{e}{c}\,\mathrm{grad}\,\mathbf{A}\cdot\mathbf{v} - e\,\mathrm{grad}\,\phi.$$

But from a formula of vector analysis.

$$\mathrm{grad}\,(\mathbf{a}\cdot\mathbf{b}) = (\mathbf{a}\cdot\nabla)\mathbf{b} + (\mathbf{b}\cdot\nabla)\mathbf{a} + \mathbf{b}\times\mathrm{curl}\,\mathbf{a} + \mathbf{a}\times\mathrm{curl}\,\mathbf{b},$$

where $\mathbf{a}$ and $\mathbf{b}$ are two arbitrary vectors. Applying this formula to $\mathbf{A}\cdot\mathbf{v}$, and remembering that differentiation with respect to $\mathbf{r}$ is carried out for constant $\mathbf{v}$, we find

$$\frac{\partial L}{\partial \mathbf{r}} = \frac{e}{c}(\mathbf{v}\cdot\nabla)\mathbf{A} + \frac{e}{c}\,\mathbf{v}\times\mathrm{curl}\,\mathbf{A} - e\,\mathrm{grad}\,\phi.$$

So the Lagrange equation has the form:

$$\frac{d}{dt}\left(\mathbf{p} + \frac{e}{c}\mathbf{A}\right) = \frac{e}{c}(\mathbf{v}\cdot\nabla)\mathbf{A} + \frac{e}{c}\,\mathbf{v}\times\mathrm{curl}\,\mathbf{A} - e\,\mathrm{grad}\,\phi.$$

But the total differential $(d\mathbf{A}/dt)\,dt$ consists of two parts: the change $(\partial\mathbf{A}/\partial t)\,dt$ of the vector potential with time at a fixed point in space, and the change due to motion from one point in space to another at distance $d\mathbf{r}$. This second part is equal to $(d\mathbf{r}\cdot\nabla)\mathbf{A}$. Thus

$$\frac{d\mathbf{A}}{dt} = \frac{\partial\mathbf{A}}{\partial t} + (\mathbf{v}\cdot\nabla)\mathbf{A}.$$

Substituting this in the previous equation, we find

$$\frac{d\mathbf{p}}{dt} = -\frac{e}{c}\frac{\partial\mathbf{A}}{\partial t} - e\,\mathrm{grad}\,\phi + \frac{e}{c}\,\mathbf{v}\times\mathrm{curl}\,\mathbf{A}.\tag{17.2}$$

This is the equation of motion of a particle in an electromagnetic field. On the left side stands the derivative of the particle's momentum with respect to the time. Therefore the expression on the right of (17.2) is the force exerted on the charge in an electromagnetic field. We see that this force consists of two parts. The first part (first and second terms on the right side of 17.2) does not depend on the velocity of the particle. The second part (third term) depends on the velocity, being proportional to the velocity and perpendicular to it.

The force of the first type, per unit charge, is called the *electric field intensity;* we denote it by $\mathbf{E}$. So by definition,

$$\mathbf{E} = -\frac{1}{c}\frac{\partial \mathbf{A}}{\partial t} - \text{grad } \phi. \tag{17.3}$$

The factor of $\mathbf{v}/c$ in the force of the second type, per unit charge, is called the *magnetic field intensity*. We designate it by $\mathbf{H}$. So by definition,

$$\mathbf{H} = \text{curl } \mathbf{A}. \tag{17.4}$$

If in an electromagnetic field, $\mathbf{E} \neq 0$ but $\mathbf{H} = 0$, then we speak of an electric field; if $\mathbf{E} = 0$ but $\mathbf{H} \neq 0$, then the field is said to be magnetic. In general, the electromagnetic field is a superposition of electric and magnetic fields.

We note that $\mathbf{E}$ is a polar vector while $\mathbf{H}$ is an axial vector.

The equation of motion of a charge in an electromagnetic field can now be written as

$$\frac{d\mathbf{p}}{dt} = e\mathbf{E} + \frac{e}{c}\mathbf{v} \times \mathbf{H}. \tag{17.5}$$

The expression on the right is called the *Lorentz force*. The first term (the force which the electric field exerts on the charge) does not depend on the velocity of the charge, and is along the direction of $\mathbf{E}$. The second part (the force exerted by the magnetic field on the charge) is proportional to the velocity of the charge and is directed perpendicular to the velocity and to the magnetic field $\mathbf{H}$.

For velocities small compared with the velocity of light, the momentum $\mathbf{p}$ is approximately equal to its classical expression $m\mathbf{v}$, and the equation of motion (17.5) becomes

$$m\frac{d\mathbf{v}}{dt} = e\mathbf{E} + \frac{e}{c}\mathbf{v} \times \mathbf{H}, \tag{17.6}$$

Next we derive the equation for the rate of change of the kinetic energy of the particle[†] with time, i.e. the derivative

$$\frac{d\mathscr{E}_{\text{kin}}}{dt} = \frac{d}{dt}\left(\frac{mc^2}{\sqrt{1 - \dfrac{v^2}{c^2}}}\right).$$

It is easy to check that

$$\frac{d\mathscr{E}_{\text{kin}}}{dt} = \mathbf{v} \cdot \frac{d\mathbf{p}}{dt}.$$

Substituting $d\mathbf{p}/dt$ from (17.5) and noting that $\mathbf{v} \times \mathbf{H} \cdot \mathbf{v} = 0$, we have

$$\frac{d\mathscr{E}_{\text{kin}}}{dt} = e\mathbf{E} \cdot \mathbf{v}. \tag{17.7}$$

The rate of change of the kinetic energy is the work done by the field on the particle per unit time. From (17.7) we see that this work is equal to the product of the velocity by the force which the electric field exerts on the charge. The work done by the field during a time dt, i.e. during a displacement of the charge by $d\mathbf{r}$, is clearly equal to $e\mathbf{E} \cdot d\mathbf{r}$.

---

† By "kinetic" we mean the energy (9.4), which includes the rest energy.

We emphasize the fact that work is done on the charge only by the electric field; the magnetic field does no work on a charge moving in it. This is connected with the fact that the force which the magnetic field exerts on a charge is always perpendicular to the velocity of the charge.

The equations of mechanics are invariant with respect to a change in sign of the time, that is, with respect to interchange of future and past. In other words, in mechanics the two time directions are equivalent. This means that if a certain motion is possible according to the equations of mechanics, then the reverse motion is also possible, in which the system passes through the same states in reverse order.

It is easy to see that this is also valid for the electromagnetic field in the theory of relativity. In this case, however, in addition to changing t into $-t$, we must reverse the sign of the magnetic field. In fact it is easy to see that the equations of motion (17.5) are not altered if we make the changes

$$t \to -t, \quad \mathbf{E} \to \mathbf{E}, \quad \mathbf{H} \to -\mathbf{H}. \tag{17.8}$$

According to (17.3) and (17.4), this does not change the scalar potential, while the vector potential changes sign:

$$\phi \to \phi, \quad \mathbf{A} \to -\mathbf{A}. \tag{17.9}$$

Thus, if a certain motion is possible in an electromagnetic field, then the reversed motion is possible in a field in which the direction of $\mathbf{H}$ is reversed.

PROBLEM

Express the acceleration of a particle in terms of its velocity and the electric and magnetic field intensities.

Solution: Substitute in the equation of motion (17.5) $\mathbf{p} = \mathbf{v}\,\mathscr{E}_{kin}/c^2$, and take the expression for $d\mathscr{E}_{kin}/dt$ from (17.7). As a result, we get

$$\dot{\mathbf{v}} = \frac{e}{m}\sqrt{1 - \frac{v^2}{c^2}}\left\{\mathbf{E} + \frac{1}{c}\mathbf{v} \times \mathbf{H} - \frac{1}{c^2}\mathbf{v}(\mathbf{v}\cdot\mathbf{E})\right\}.$$

§ 18. Gauge invariance

Let us consider to what extent the potentials are uniquely determined. First of all we call attention to the fact that the field is characterized by the effect which it produces on the motion of a charge located in it. But in the equation of motion (17.5) there appear not the potentials, but the field intensities $\mathbf{E}$ and $\mathbf{H}$. Therefore two fields are physically identical if they are characterized by the same vectors $\mathbf{E}$ and $\mathbf{H}$.

If we are given potentials $\mathbf{A}$ and ϕ, then these uniquely determine (according to (17.3) and (17.4)) the fields $\mathbf{E}$ and $\mathbf{H}$. However, to one and the same field there can correspond different potentials. To show this, let us add to each component of the potential the quantity $-\partial f/\partial x^k$, where f is an arbitrary function of the coordinates and the time. Then the potential A_k goes over into

$$A_k' = A_k - \frac{\partial f}{\partial x^k}. \tag{18.1}$$

As a result of this change there appears in the action integral (16.1) the additional term

$$\frac{e}{c}\frac{\partial f}{dx^k}\,dx^k = d\left(\frac{e}{c}f\right), \tag{18.2}$$

which is a total differential and has no effect on the equations of motion. (See *Mechanics*, § 2.)

If in place of the four-potential we introduce the scalar and vector potentials, and in place of x^i, the coordinates ct, x, y, z, then the four equations (18.1) can be written in the form

$$\mathbf{A}' = \mathbf{A} + \mathrm{grad}\,f, \quad \phi' = \phi - \frac{1}{c}\frac{\partial f}{\partial t}. \tag{18.3}$$

It is easy to check that electric and magnetic fields determined from equations (17.3) and (17.4) actually do not change upon replacement of $\mathbf{A}$ and ϕ by $\mathbf{A}'$ and ϕ', defined by (18.3). Thus the transformation of potentials (18.1) does not change the fields. The potentials are therefore not uniquely defined; the vector potential is determined to within the gradient of an arbitrary function, and the scalar potential to within the time derivative of the same function.

In particular, we see that we can add an arbitrary constant vector to the vector potential, and an arbitrary constant to the scalar potential. This is also clear directly from the fact that the definitions of $\mathbf{E}$ and $\mathbf{H}$ contain only derivatives of $\mathbf{A}$ and ϕ, and therefore the addition of constants to the latter does not affect the field intensities.

Only those quantities have physical meaning which are invariant with respect to the transformation (18.3) of the potentials; in particular all equations must be invariant under this transformation. This invariance is called *gauge invariance* (in German, *eichinvarianz*).†

This nonuniqueness of the potentials gives us the possibility of choosing them so that they fulfil one auxiliary condition chosen by us. We emphasize that we can set one condition, since we may choose the function f in (18.3) arbitrarily. In particular, it is always possible to choose the potentials so that the scalar potential ϕ is zero. If the vector potential is not zero, then it is not generally possible to make it zero, since the condition $\mathbf{A} = 0$ represents three auxiliary conditions (for the three components of $\mathbf{A}$).

§ 19. Constant electromagnetic field

By a constant electromagnetic field we mean a field which does not depend on the time. Clearly the potentials of a constant field can be chosen so that they are functions only of the coordinates and not of the time. A constant magnetic field is equal, as before, to $\mathbf{H} = \mathrm{curl}\,\mathbf{A}$. A constant electric field is equal to

$$\mathbf{E} = -\,\mathrm{grad}\,\phi. \tag{19.1}$$

Thus a constant electric field is determined only by the scalar potential and a constant magnetic field only by the vector potential.

We saw in the preceding section that the potentials are not uniquely determined. However, it is easy to convince oneself that if we describe the constant electromagnetic field in terms of potentials which do not depend on the time, then we can add to the scalar potential, without changing the fields, only an arbitrary constant (not depending on either the coordinates

---

† We emphasize that this is related to the assumed constancy of e in (18.2). Thus the gauge invariance of the equations of electrodynamics (see below) and the conservation of charge are closely related to one another.

or the time). Usually ϕ is subjected to the additional requirement that it has a definite value at some particular point in space; most frequently ϕ is chosen to be zero at infinity. Thus the arbitrary constant previously mentioned is determined, and the scalar potential of the constant field is thus determined uniquely.

On the other hand, just as before, the vector potential is not uniquely determined even for the constant electromagnetic field; namely, we can add to it the gradient of an arbitrary function of the coordinates.

We now determine the energy of a charge in a constant electromagnetic field. If the field is constant, then the Lagrangian for the charge also does not depend explicitly on the time. As we know, in this case the energy is conserved and coincides with the Hamiltonian.

According to (16.6), we have

$$\mathscr{E} = \frac{mc^2}{\sqrt{1 - \dfrac{v^2}{c^2}}} + e\phi. \tag{19.2}$$

Thus the presence of the field adds to the energy of the particle the term $e\phi$, the potential energy of the charge in the field. We note the important fact that the energy depends only on the scalar and not on the vector potential. This means that the magnetic field does not affect the energy of the charge. Only the electric field can change the energy of the particle. This is related to the fact that the magnetic field, unlike the electric field, does no work on the charge.

If the field intensities are the same at all points in space, then the field is said to be uniform. The scalar potential of a uniform electric field can be expressed in terms of the field intensity as

$$\phi = -\mathbf{E} \cdot \mathbf{r}. \tag{19.3}$$

In fact, since $\mathbf{E} = $ const, $\nabla(\mathbf{E} \cdot \mathbf{r}) = (\mathbf{E} \cdot \nabla)\mathbf{r} = \mathbf{E}$.

The vector potential of a uniform magnetic field can be expressed in terms of its field intensity as

$$\mathbf{A} = \tfrac{1}{2}\mathbf{H} \times \mathbf{r}. \tag{19.4}$$

In fact, recalling that $\mathbf{H} = $ const, we obtain with the aid of well-known formulas of vector analysis:

$$\text{curl } (\mathbf{H} \times \mathbf{r}) = \mathbf{H} \text{ div } \mathbf{r} - (\mathbf{H} \cdot \nabla)\mathbf{r} = 2\mathbf{H}$$

(noting that div $\mathbf{r} = 3$).

The vector potential of a uniform magnetic field can also be chosen in the form

$$A_x = -Hy, \quad A_y = A_z = 0 \tag{19.5}$$

(the z axis is along the direction of $\mathbf{H}$). It is easily verified that with this choice for $\mathbf{A}$ we have $\mathbf{H} = \text{curl } \mathbf{A}$. In accordance with the transformation formulas (18.3), the potentials (19.4) and (19.5) differ from one another by the gradient of some function: formula (19.5) is obtained from (19.4) by adding ∇f, where $f = -xyH/2$.

PROBLEM

Give the variational principle for the trajectory of a particle (Maupertuis' principle) in a constant electromagnetic field in relativistic mechanics.

Solution: Maupertuis' principle consists in the statement that if the energy of a particle is conserved (motion in a constant field), then its trajectory can be determined from the variational equation

$$\delta \int \mathbf{P} \cdot d\mathbf{r} = 0,$$

where $\mathbf{P}$ is the generalized momentum of the particle, expressed in terms of the energy and the coordinate differentials, and the integral is taken along the trajectory of the particle.† Substituting $\mathbf{P} = \mathbf{p} + (e/c)\mathbf{A}$ and noting that the directions of $\mathbf{p}$ and $d\mathbf{r}$ coincide, we have

$$\delta \int \left(p\,dl + \frac{e}{c}\mathbf{A} \cdot d\mathbf{r} \right) = 0,$$

where $dl = \sqrt{d\mathbf{r}^2}$ is the element of arc. Determining p from

$$p^2 + m^2c^2 = \left(\frac{\mathscr{E} - e\phi}{c} \right)^2,$$

we obtain finally

$$\delta \int \left\{ \sqrt{\left(\frac{\mathscr{E} - e\phi}{c} \right)^2 - m^2c^2}\, dl + \frac{e}{c}\mathbf{A} \cdot d\mathbf{r} \right\} = 0.$$

§ 20. Motion in a constant uniform electric field

Let us consider the motion of a charge e in a uniform constant electric field $\mathbf{E}$. We take the direction of the field as the X axis. The motion will obviously proceed in a plane, which we choose as the XY plane. Then the equations of motion (17.5) become

$$\dot{p}_x = eE, \quad \dot{p}_y = 0$$

(where the dot denotes differentiation with respect to t), so that

$$p_x = eEt, \quad p_y = p_0. \tag{20.1}$$

The time reference point has been chosen at the moment when $p_x = 0$; p_0 is the momentum of the particle at that moment.

The kinetic energy of the particle (the energy omitting the potential energy in the field) is $\mathscr{E}_{\text{kin}} = c\sqrt{m^2c^2 + p^2}$. Substituting (20.1), we find in our case

$$\mathscr{E}_{\text{kin}} = \sqrt{m^2c^4 + c^2p_0^2 + (ceEt)^2} = \sqrt{\mathscr{E}_0^2 + (ceEt)^2}, \tag{20.2}$$

where $\mathscr{E}_0$ is the energy at $t = 0$.

According to (9.8) the velocity of the particle is $\mathbf{v} = \mathbf{p}c^2/\mathscr{E}_{\text{kin}}$. For the velocity $v_x = \dot{x}$ we have therefore

$$\frac{dx}{dt} = \frac{p_x c^2}{\mathscr{E}_{\text{kin}}} = \frac{c^2 eEt}{\sqrt{\mathscr{E}_0^2 + (ceEt)^2}}.$$

Integrating, we find

† See *Mechanics,* § 44.

$$x = \frac{1}{eE} \sqrt{\mathscr{E}_0^2 + (ceEt)^2}. \tag{20.3}$$

The constant of integration we set equal to zero.†

For determining y, we have

$$\frac{dy}{dt} = \frac{p_y c^2}{\mathscr{E}_{\text{kin}}} = \frac{p_0 c^2}{\sqrt{\mathscr{E}_0^2 + (ceEt)^2}},$$

from which

$$y = \frac{p_0 c}{eE} \sinh^{-1}\left(\frac{ceEt}{\mathscr{E}_0}\right). \tag{20.4}$$

We obtain the equation of the trajectory by expressing t in terms of y from (20.4) and substituting in (20.3). This gives:

$$x = \frac{\mathscr{E}_0}{eE} \cosh \frac{eE_y}{p_0 c}. \tag{20.5}$$

Thus in a uniform electric field a charge moves along a catenary curve.

If the velocity of the particle is $v \ll c$, then we can set $p_0 = m v_0$, $\mathscr{E}_0 = mc^2$, and expand (20.5) in series in powers of $1/c$. Then we get, to within terms of higher order,

$$x = \frac{eE}{2m v_0^2} y^2 + \text{const},$$

that is, the charge moves along a parabola, a result well known from classical mechanics.

§ 21. Motion in a constant uniform magnetic field

We now consider the motion of a charge e in a uniform magnetic field **H**. We choose the direction of the field as the Z axis. We rewrite the equation of motion

$$\dot{\mathbf{p}} = \frac{e}{c} \mathbf{v} \times \mathbf{H}$$

in another form, by substituting for the momentum, from (9.8),

$$\mathbf{p} = \frac{\mathscr{E}\mathbf{v}}{c^2},$$

where $\mathscr{E}$ is the energy of the particle, which is constant in the magnetic field. The equation of motion then goes over into the form

$$\frac{\mathscr{E}}{c^2} \frac{d\mathbf{v}}{dt} = \frac{e}{c} \mathbf{v} \times \mathbf{H} \tag{21.1}$$

or, expressed in terms of components,

$$\dot{v}_x = \omega v_y, \quad \dot{v}_y = -\omega v_x, \quad \dot{v}_z = 0, \tag{21.2}$$

† This result (for $p_0 = 0$) coincides with the solution of the problem of relativistic motion with constant "proper acceleration" $w_0 = eE/m$ (see the problem in § 7). For the present case, the constancy of the acceleration is related to the fact that the electric field does not change for Lorentz transformations having velocities **V** along the direction of the field (see § 24).

where we have introduced the notation

$$\omega = \frac{ecH}{\mathscr{E}}. \tag{21.3}$$

We multiply the second equation of (21.2) by i, and add it to the first:

$$\frac{d}{dt}(v_x + iv_y) = -i\omega(v_x + iv_y),$$

so that

$$v_x + iv_y = ae^{-i\omega t},$$

where a is a complex constant. This can be written in the form $a = v_{0t}e^{-i\alpha}$ where v_{0t} and α are real. Then

$$v_x + iv_y = v_{0t}e^{-i(\omega t + \alpha)}$$

and, separating real and imaginary parts, we find

$$v_x = v_{0t}\cos(\omega t + \alpha), \quad v_y = -v_{0t}\sin(\omega t + \alpha). \tag{21.4}$$

The constants v_{0t}, and α are determined by the initial conditions; α is the initial phase, and as for v_{0t}, from (21.4) it is clear that

$$v_{0t} = \sqrt{v_x^2 + v_y^2},$$

that is, v_{0t}, is the velocity of the particle in the XY plane, and stays constant throughout the motion.

From (21.4) we find, integrating once more,

$$x = x_0 + r\sin(\omega t + \alpha), \quad y = y_0 + r\cos(\omega t + \alpha), \tag{21.5}$$

where

$$r = \frac{v_{0t}}{\omega} = \frac{v_{0t}\mathscr{E}}{ecH} = \frac{cp_t}{eH} \tag{21.6}$$

(p_t is the projection of the momentum on the XY plane). From the third equation of (21.2), we find $v_z = v_{0z}$ and

$$z = z_0 + v_{0z}t. \tag{21.7}$$

From (21.5) and (21.7), it is clear that the charge moves in a uniform magnetic field along a helix having its axis along the direction of the magnetic field and with a radius r given by (21.6). The velocity of the particle is constant. In the special case where $v_{0z} = 0$, that is, the charge has no velocity component along the field, it moves along a circle in the plane perpendicular to the field.

The quantity ω, as we see from the formulas, is the angular frequency of rotation of the particle in the plane perpendicular to the field.

If the velocity of the particle is low, then we can approximately set $\mathscr{E} = mc^2$. Then the frequency ω is changed to

$$\omega = \frac{eH}{mc}. \tag{21.8}$$

We shall now assume that the magnetic field remains uniform but varies slowly in magnitude and direction. Let us see how the motion of a charged particle changes in this case.

We know that when the conditions of the motion are changed slowly, certain quantities called adiabatic invariants remain constant. Since the motion in the plane perpendicular to the magnetic field is periodic, the adiabatic invariant is the integral

$$I = \frac{1}{2\pi} \oint \mathbf{P}_t \cdot d\mathbf{r},$$

taken over a complete period of the motion, i.e. over the circumference of a circle in the present case ($\mathbf{P}_t$ is the projection of the generalized momentum on the plane perpendicular to $\mathbf{H}$†). Substituting $\mathbf{P}_t = \mathbf{p}_t + (e/c)\,\mathbf{A}$, we have:

$$I = \frac{1}{2\pi} \oint \mathbf{P}_t \cdot d\mathbf{r} = \frac{1}{2\pi} \oint \mathbf{p}_t \cdot d\mathbf{r} + \frac{e}{2\pi c} \oint \mathbf{A} \cdot d\mathbf{r}.$$

In the first term we note that p_t is constant in magnitude and directed along $d\mathbf{r}$; we apply Stokes' theorem to the second term and write curl $\mathbf{A} = \mathbf{H}$:*

$$I = rp_t - \frac{e}{2c} Hr^2 = \frac{cp_t^2}{2eH}. \qquad (21.9)$$

From this we see that, for slow variation of H, the tangential momentum p_t varies proportionally to $\sqrt{H}$.

This result can also be applied to another case, when the particle moves along a helical path in a magnetic field that is not strictly homogeneous (so that the field varies little over distances comparable with the radius and step of the helix). Such a motion can be considered as a motion in a circular orbit that shifts in the course of time, while relative to the orbit the field appears to change in time but remain uniform. One can then state that the component of the momentum transverse to the direction of the field varies according to the law: $p_t = \sqrt{CH}$, where C is a constant and H is a given function of the coordinates. On the other hand, just as for the motion in any constant magnetic field, the energy of the particle (and consequently the square of its momentum p^2) remains constant. Therefore the longitudinal component of the momentum varies according to the formula:

$$p_l^2 = p^2 - p_t^2 = p^2 - CH(x, y, z). \qquad (21.10)$$

Since we should always have $p_l^2 \geq 0$, we see that penetration of the particle into regions of sufficiently high field ($CH > p^2$) is impossible. During motion in the direction of increasing field, the radius of the helical trajectory decreases proportionally to p_t/H (i.e. proportionally

---

† See *Mechanics*, § 49. In general the integrals $\oint p\,dq$, taken over a period of the particular coordinate q, are adiabatic invariants. In the present case the periods for the two coordinates in the plane perpendicular to $\mathbf{H}$ coincide, and the integral I which we have written is the sum of the two corresponding adiabatic invariants. However, each of these invariants individually has no special significance, since it depends on the (non-unique) choice of the vector potential of the field. The nonuniqueness of the adiabatic invariants which results from this is a reflection of the fact that, when we regard the magnetic field as uniform over all of space, we cannot in principle determine the electric field which results from changes in $\mathbf{H}$, since it will actually depend on the specific conditions at infinity.

*By inspecting the direction of motion of a charge along the orbit for a given direction of $\mathbf{H}$, we observe that it is counterclockwise if we look along $\mathbf{H}$. Hence the negative sign in the second term.

to $1/\sqrt{H}$), and the step proportionally to p_l. On reaching the boundary where p_l vanishes, the particle is reflected; while continuing to rotate in the same direction it begins to move opposite to the gradient of the field.

Inhomogeneity of the field also leads to another phenomenon—a slow transverse shift (*drift*) of the *guiding centre* of the helical trajectory of the particle (the name given to the centre of the circular orbit); problem 3 of the next section deals with this question.

PROBLEM

Determine the frequency of vibration of a charged spatial oscillator, placed in a constant, uniform magnetic field; the proper frequency of vibration of the oscillator (in the absence of the field) is ω_0.

Solution: The equations of forced vibration of the oscillator in a magnetic field (directed along the z axis) are:

$$\ddot{x} + \omega_0^2 x = \frac{eH}{me}\dot{y}, \quad \ddot{y} + \omega_0^2 y = \frac{eH}{mc}\dot{x}, \quad \ddot{z} + \omega_0^2 z = 0.$$

Multiplying the second equation by i and combining with the first, we find

$$\ddot{\zeta} + \omega_0^2 \zeta = -i\frac{eH}{mc}\dot{\zeta},$$

where $\zeta = x + iy$. From this we find that the frequency of vibration of the oscillator in a plane perpendicular to the field is

$$\omega = \sqrt{\omega_0^2 + \frac{1}{4}\left(\frac{cH}{mc}\right)^2} \pm \frac{eH}{2mc}.$$

If the field H is weak, this formula goes over into

$$\omega = \omega_0 \pm eH/2mc.$$

The vibration along the direction of the field remains unchanged.

§ 22. Motion of a charge in constant uniform electric and magnetic fields

Finally we consider the motion of a charge in the case where there are present both electric and magnetic fields, constant and uniform. We limit ourselves to the case where the velocity of the charge $v \ll c$, so that its momentum $\mathbf{p} = m\mathbf{v}$; as we shall see later, it is necessary for this that the electric field be small compared to the magnetic.

We choose the direction of $\mathbf{H}$ as the Z axis, and the plane passing through $\mathbf{H}$ and $\mathbf{E}$ as the YZ plane. Then the equation of motion

$$m\dot{\mathbf{v}} = e\mathbf{E} + \frac{e}{c}\mathbf{v} \times \mathbf{H}$$

can be written in the form

$$m\ddot{x} = \frac{e}{c}\dot{y}H, \quad m\ddot{y} = eE_y - \frac{e}{c}\dot{x}H, \quad m\ddot{z} = eE_z. \tag{22.1}$$

From the third equation we see that the charge moves with uniform acceleration in the Z direction, that is,

$$z = \frac{eE_z}{2m}t^2 + v_{0z}t. \tag{22.2}$$

Multiplying the second equation of (22.1) by i and combining with the first, we find

$$\frac{d}{dt}(\dot{x} + i\dot{y}) + i\omega(\dot{x} + i\dot{y}) = i\frac{e}{m}E_y$$

($\omega = eH/mc$). The integral of this equation, where $\dot{x} + i\dot{y}$ is considered as the unknown, is equal to the sum of the integral of the same equation without the right-hand term and a particular integral of the equation with the right-hand term. The first of these is $ae^{-i\omega t}$, the second is $eE_y/m\omega = cE_y/H$. Thus

$$\dot{x} + i\dot{y} = ae^{-i\omega t} + \frac{cE_y}{H}.$$

The constant a is in general complex. Writing it in the form $a = be^{i\alpha}$, with real b and α, we see that since a is multiplied by $e^{-i\omega t}$, we can, by a suitable choice of the time origin, give the phase α any arbitrary value. We choose this so that a is real. Then breaking up $\dot{x} + i\dot{y}$ into real and imaginary parts, we find

$$\dot{x} = a\cos\omega t + \frac{cE_y}{H}, \quad \dot{y} = -a\sin\omega t. \tag{22.3}$$

At $t = 0$ the velocity is along the X axis.

We see that the components of the velocity of the particle are periodic functions of the time. Their average values are:

$$\bar{\dot{x}} = \frac{cE_y}{H}, \quad \bar{\dot{y}} = 0.$$

This average velocity of motion of a charge in crossed electric and magnetic fields is often called the electrical *drift* velocity. Its direction is perpendicular to both fields and independent of the sign of the charge. It can be written in vector form as:

$$\bar{\mathbf{v}} = \frac{c\mathbf{E} \times \mathbf{H}}{H^2}. \tag{22.4}$$

All the formulas of this section assume that the velocity of the particle is small compared with the velocity of light; we see that for this to be so, it is necessary in particular that the electric and magnetic fields satisfy the condition

$$\frac{E_y}{H} \ll 1, \tag{22.5}$$

while the absolute magnitudes of E_y and H can be arbitrary.

Integrating equation (22.3) again, and choosing the constant of integration so that at $t = 0$, $x = y = 0$, we obtain

$$x = \frac{a}{\omega}\sin\omega t + \frac{cE_y}{H}t; \quad y = \frac{a}{\omega}(\cos\omega t - 1). \tag{22.6}$$

Considered as parametric equations of a curve, these equations define a trochoid. Depending on whether a is larger or smaller in absolute value than the quantity cE_y/H, the projection of the trajectory on the plane XY has the forms shown in Figs. 6a and 6b, respectively.

If $a = -cE_y/H$, then (22.6) becomes

$$x = \frac{cE_y}{\omega H}(\omega t - \sin\omega t),$$

$$y = \frac{cE_y}{\omega H}(1 - \cos \omega t) \tag{22.7}$$

that is, the projection of the trajectory on the XY plane is a cycloid (Fig. 6c).

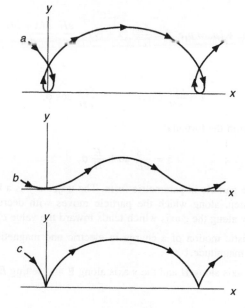

Fig. 6.

PROBLEMS

1. Determine the relativistic motion of a charge in parallel uniform electric and magnetic fields.

Solution: The magnetic field has no influence on the motion along the common direction of **E** and **H** (the z axis), which therefore occurs under the influence of the electric field alone; therefore according to § 20 we find:

$$z = \frac{\mathscr{E}_{kin}}{eE}, \quad \mathscr{E}_{kin} = \sqrt{\mathscr{E}_0^2 + (ceEt)^2} .$$

For the motion in the xy plane we have the equation

$$\dot{p}_x = \frac{e}{c} H v_y, \quad \dot{p}_y = -\frac{e}{c} H v_x,$$

or

$$\frac{d}{dt}(p_x + ip_y) = -i\frac{eH}{c}(v_x + iv_y) = -\frac{ieHc}{\mathscr{E}_{kin}}(p_x + ip_y).$$

Consequently

$$p_x + ip_y = p_t e^{-i\phi},$$

where p_t is the constant value of the projection of the momentum on the xy plane, and the auxiliary quantity ϕ is defined by the relation

$$d\phi = eHc\frac{dt}{\mathscr{E}_{kin}},$$

from which

$$ct = \frac{\mathscr{E}_0}{eE} \sinh \frac{E}{H} \phi. \tag{1}$$

Furthermore we have:

$$p_x + ip_y = p_t e^{-i\phi} = \frac{\mathscr{E}_{\text{kin}}}{c^2} (\dot{x} + i\dot{y}) = \frac{eH}{c} \frac{d(x + iy)}{d\phi},$$

so that

$$x = \frac{cp_t}{eH} \sin \phi, \qquad y = \frac{cp_t}{eH} \cos \phi. \tag{2}$$

Formulas (1), (2) together with the formula

$$z = \frac{\mathscr{E}_0}{eE} \cosh \frac{E}{H} \phi, \tag{3}$$

determine the motion of the particle in parametric form. The trajectory is a helix with radius cp_t/eH and monotonically increasing step, along which the particle moves with decreasing angular velocity $\phi = eHc/\mathscr{E}_{\text{kin}}$ and with a velocity along the z axis which tends toward the value c.

2. Determine the relativistic motion of a charge in electric and magnetic fields which are mutually perpendicular and equal in magnitude.†

Solution: Choosing the z axis along **H** and the y axis along **E** and setting $E = H$, we write the equations of motion:

$$\frac{dp_x}{dt} = \frac{e}{c} E v_y, \qquad \frac{dp_y}{dt} = eE\left(1 - \frac{v_x}{c}\right), \qquad \frac{dp_z}{dt} = 0$$

and, as a consequence of them, formula (17.7),

$$\frac{d\mathscr{E}_{\text{kin}}}{dt} = eEv_y.$$

From these equations we have:

$$p_z = \text{const}, \quad \mathscr{E}_{\text{kin}} - cp_x = \text{const} \equiv \alpha.$$

Also using the equation

$$\mathscr{E}_{\text{kin}}^2 - c^2 p_x^2 = (\mathscr{E}_{\text{kin}} + cp_x)(\mathscr{E}_{\text{kin}} - cp_x) = c^2 p_y^2 + \varepsilon^2$$

(where $\varepsilon^2 = m^2 c^4 + c^2 p_z^2 = \text{const}$), we find:

$$\mathscr{E}_{\text{kin}} + cp_x = \frac{1}{\alpha}(c^2 p_y^2 + \varepsilon^2),$$

and so

$$\mathscr{E}_{\text{kin}} = \frac{\alpha}{2} + \frac{c^2 p_y^2 + \varepsilon^2}{2\alpha},$$

$$p_x = -\frac{\alpha}{2c} + \frac{c^2 p_y^2 + \varepsilon^2}{2\alpha c}.$$

† The problem of motion in mutually perpendicular fields **E** and **H** which are not equal in magnitude can, by a suitable transformation of the reference system, be reduced to the problem of motion in a pure electric or a pure magnetic field (see § 25).

Furthermore, we write

$$\mathscr{E}_{\text{kin}} \frac{dp_y}{dt} = eE\left(\mathscr{E}_{\text{kin}} - \frac{\mathscr{E}_{\text{kin}} v_x}{c}\right) = eE(\mathscr{E}_{\text{kin}} - cp_x) = eE\alpha,$$

from which

$$2eEt = \left(1 + \frac{\varepsilon^2}{\alpha^2}\right)p_y + \frac{c^2}{3\alpha^2}p_y^3. \tag{1}$$

To determine the trajectory, we make a transformation of variables in the equations

$$\frac{dx}{dt} = \frac{c^2 p_x}{\mathscr{E}_{\text{kin}}}, \dots$$

to the variable p_y by using the relation $dt = \mathscr{E}_{\text{kin}} dp_y/eE\alpha$, after which integration gives the formulas:

$$x = \frac{c}{2eE}\left(-1 + \frac{\varepsilon^2}{\alpha^2}\right)p_y + \frac{c^3}{6\alpha^2 eE}p_y^3, \tag{2}$$

$$y = \frac{c^2}{2\alpha eE}p_y^2, \qquad z = \frac{p_z c^2}{eE\alpha}p_y.$$

Formulas (1) and (2) completely determine the motion of the particle in parametric form (parameter p_y). We call attention to the fact that the velocity increases most rapidly in the direction perpendicular to **E** and **H** (the x axis).

3. Determine the velocity of drift of the guiding centre of the orbit of a nonrelativistic charged particle in a quasihomogeneous magnetic field (H. Alfven, 1940).

Solution: We assume first that the particle is moving in a circular orbit, i.e. its velocity has no longitudinal component (along the field). We write the equation of the trajectory in the form $\mathbf{r} = \mathbf{R}(t) + \boldsymbol{\zeta}(t)$, where $\mathbf{R}(t)$ is the radius vector of the guiding centre (a slowly varying function of the time), while $\boldsymbol{\zeta}(t)$ is a rapidly oscillating quantity describing the rotational motion about the guiding centre. We average the force $(e/c)\,\dot{\mathbf{r}} \times \mathbf{H}(\mathbf{r})$ acting on the particle over a period of the oscillatory (circular) motion (compare *Mechanics*, § 30). We expand the function $\mathbf{H}(\mathbf{r})$ in this expression in powers of $\boldsymbol{\zeta}$:

$$\mathbf{H}(\mathbf{r}) = \mathbf{H}(\mathbf{R}) + (\boldsymbol{\zeta} \cdot \nabla)\mathbf{H}(\mathbf{R}).$$

On averaging, the terms of first order in $\boldsymbol{\zeta}(t)$ vanish, while the second-degree terms give rise to an additional force

$$\mathbf{f} = \frac{e}{c}\,\dot{\boldsymbol{\zeta}} \times (\boldsymbol{\zeta} \cdot \nabla)\mathbf{H}.$$

For a circular orbit

$$\dot{\boldsymbol{\zeta}} = \omega\boldsymbol{\zeta} \times \mathbf{n}, \qquad \zeta = \frac{v_\perp}{\omega},$$

where **n** is a unit vector along **H**; the frequency $\omega = eH/mc$; $v_\perp$ is the velocity of the particle in its circular motion. The average values of products of components of the vector $\boldsymbol{\zeta}$, rotating in a plane (the plane perpendicular to **n**), are:

$$\overline{\zeta_\alpha \zeta_\beta} = \tfrac{1}{2}\zeta^2 \delta_{\alpha\beta},$$

where $\delta_{\alpha\beta}$ is the unit tensor in this plane. As a result we find:

$$\mathbf{f} = -\frac{mv_\perp^2}{2H}(\mathbf{n} \times \nabla) \times \mathbf{H}.$$

Because of the equations div $\mathbf{H} = 0$ and curl $\mathbf{H} = 0$ which the constant field $\mathbf{H(R)}$ satisfies, we have:

$$(\mathbf{n} \times \nabla) \times \mathbf{H} = -\mathbf{n} \operatorname{div} \mathbf{H} + (\mathbf{n} \cdot \nabla)\mathbf{H} + \mathbf{n} \times (\nabla \times \mathbf{H}) = (\mathbf{n} \cdot \nabla)\mathbf{H} = H(\mathbf{n} \cdot \nabla)\mathbf{n} + \mathbf{n}(\mathbf{n} \cdot \nabla H).$$

We are interested in the force transverse to $\mathbf{n}$, giving rise to a shift of the orbit; it is equal to

$$\mathbf{f} = -\frac{m v_\perp^2}{2}(\mathbf{n} \cdot \nabla)\,\mathbf{n} = \frac{m v_\perp^2}{2\rho}\,\mathbf{v},$$

where ρ is the radius of curvature of the force line of the field at the given point, and v is a unit vector directed from the centre of curvature to this point.

The case where the particle also has a longitudinal velocity $v_\parallel$ (along $\mathbf{n}$) reduces to the previous case if we go over to a reference frame which is rotating about the instantaneous centre of curvature of the force line (which is the trajectory of the guiding centre) with angular velocity $v_\parallel/\rho$. In this reference system the particle has no longitudinal velocity, but there is an additional transverse force, the centrifugal force $m v_\parallel^2/\rho$. Thus the total transverse force is

$$\mathbf{f}_\perp = \mathbf{v}\frac{m}{\rho}\left(v_\parallel^2 + \frac{v_\perp^2}{2} \right).$$

This force is equivalent to a constant electric field of strength $\mathbf{f}_\perp/e$. According to (22.4) it causes a drift of the guiding center of the orbit with a velocity

$$\mathbf{v}_\mathrm{d} = \frac{1}{\omega\rho}\left(v_\parallel^2 + \frac{v_\perp^2}{2} \right)\mathbf{v} \times \mathbf{n}.$$

The sign of this velocity depends on the sign of the charge.

§ 23. The electromagnetic field tensor

In § 17, we derived the equation of motion of a charge in a field, starting from the Lagrangian (16.4) written in three-dimensional form. We now derive the same equation directly from the action (16.1) written in four-dimensional notation.

The principle of least action states

$$\delta S = \delta \int_a^b \left(-mc\, ds - \frac{e}{c}\, A_i dx^i \right) = 0. \tag{23.1}$$

Noting that $ds = \sqrt{dx_i dx^i}$, we find (the limits of integration a and b are omitted for brevity):

$$\delta S = -\int \left(mc\frac{dx_i d\delta x^i}{ds} + \frac{e}{c}A_i d\delta x^i + \frac{e}{c}\delta A_i dx^i \right) = 0.$$

We integrate the first two terms in the integrand by parts. Also, in the first term we set $dx_i/ds = u_i$, where u_i are the components of the four-velocity. Then

$$\int \left(mc\, du_i\, \delta x^i + \frac{e}{c}\, \delta x^i\, dA_i - \frac{e}{c}\, \delta A_i\, dx^i \right) - \left[\left(mcu_i + \frac{e}{c}A_i \right)\delta x^i \right] = 0. \tag{23.2}$$

The second term in this equation is zero, since the integral is varied with fixed coordinate values at the limits. Furthermore:

$$\delta A_i = \frac{\partial A_i}{\partial x^k}\, \delta x^k, \qquad dA_i = \frac{\partial A_i}{\partial x^k}\, dx^k,$$

and therefore

$$\int \left(mc\,du_i\delta x^i + \frac{e}{c}\frac{\partial A_i}{\partial x^k}\delta x^i dx^k - \frac{e}{c}\frac{\partial A_i}{\partial x^k}\,dx^i\delta x^k \right) = 0.$$

In the first term we write $du_i = (du_i/ds)\,ds$, in the second and third, $dx^i = u^i ds$. In addition, in the third term we interchange the indices i and k (this changes nothing since the indices i and k are summed over). Then

$$\int \left[mc\,\frac{du_i}{ds} - \frac{e}{c}\left(\frac{\partial A_k}{\partial x^i} - \frac{\partial A_i}{\partial x^k} \right)u^k \right] \delta x^i ds = 0.$$

In view of the arbitrariness of δx^i, it follows that the integrand is zero, that is,

$$mc\,\frac{du_i}{ds} = \frac{e}{c}\left(\frac{\partial A_k}{\partial x^i} - \frac{\partial A_i}{\partial x^k} \right)u^k.$$

We now introduce the notion

$$F_{ik} = \frac{\partial A_k}{\partial x^i} - \frac{\partial A_i}{\partial x^k}. \tag{23.3}$$

The antisymmetric tensor F_{ik} is called the *electromagnetic field tensor*. The equation of motion then takes the form:

$$mc\,\frac{du^i}{ds} = \frac{e}{c}\,F^{ik}u_k. \tag{23.4}$$

These are the equations of motion of a charge in four-dimensional form.

The meaning of the individual components of the tensor F_{ik} is easily seen by substituting the values $A_i = (\phi, -\mathbf{A})$ in the definition (23.3). The result can be written as a matrix in which the index $i = 0, 1, 2, 3$ labels the rows, and the index k the columns:

$$F_{ik} = \begin{pmatrix} 0 & E_x & E_y & E_z \\ -E_x & 0 & -H_z & H_y \\ -E_y & H_z & 0 & -H_x \\ -E_z & -H_y & H_x & 0 \end{pmatrix}, \quad F^{ik} = \begin{pmatrix} 0 & -E_x & -E_y & -E_z \\ E_x & 0 & -H_z & H_y \\ E_y & H_z & 0 & -H_x \\ E_z & -H_y & H_x & 0 \end{pmatrix}. \tag{23.5}$$

More briefly, we can write (see § 6):

$$F_{ik} = (\mathbf{E}, \mathbf{H}), \quad F^{ik} = (-\mathbf{E}, \mathbf{H}).$$

Thus the components of the electric and magnetic field strengths are components of the same electromagnetic field four-tensor.

Changing to three dimensional notation, it is easy to verify that the three space components ($i = 1, 2, 3$) of (23.4) are identical with the vector equation of motion (17.5), while the time component ($i = 0$) gives the work equation (17.7). The latter is a consequence of the equations of motion; the fact that only three of the four equations are independent can also easily be found directly by multiplying both sides of (23.4) by u^i. Then the left side of the equation vanishes because of the orthogonality of the four-vectors u^i and du_i/ds, while the right side vanishes because of the antisymmetry of F_{ik}.

If we admit only possible trajectories when we vary S, the first term in (23.2) vanishes identically. Then the second term, in which the upper limit is considered as variable, gives the differential of the action as a function of the coordinates. Thus

$$\delta S = - \left(mcu_i + \frac{e}{c} A_i \right) \delta x^i .$$ (23.6)

Then

$$-\frac{\partial S}{\partial x^i} = mcu_i + \frac{e}{c} A_i = p_i + \frac{e}{c} A_i .$$ (23.7)

The four-vector $-\delta S/\partial x^i$ is the four-vector P_i of the generalized momentum of the particle. Substituting the values of the components p_i and A_i, we find that

$$P^i = \left(\frac{\mathscr{E}_{kin} + e\phi}{c}, \ \mathbf{p} + \frac{e}{c}\mathbf{A} \right).$$ (23.8)

As expected, the space components of the four-vector form the three-dimensional generalized momentum vector (16.5), while the time component is $\mathscr{E}/c$, where $\mathscr{E}$ is the total energy of the charge in the field.

§ 24. Lorentz transformation of the field

In this section we find the transformation formulas for fields, that is, formulas by means of which we can determine the field in one inertial system of reference, knowing the same field in another system.

The formulas for transformation of the potentials are obtained directly from the general formulas for transformation of four-vectors (6.1). Remembering that $A^i = (\phi, \mathbf{A})$, we get easily

$$\phi = \frac{\phi' + \frac{V}{c}A'_x}{\sqrt{1 - \frac{V^2}{c^2}}}, \quad A_x = \frac{A'_x + \frac{V}{c}\phi'}{\sqrt{1 - \frac{V^2}{c^2}}}, \quad A_y = A'_y, \quad A_z = A'_z.$$ (24.1)

The transformation formulas for an antisymmetric second-rank tensor (like F^{ik}) were found in problem 2 of § 6: the components F^{23} and F^{01} do not change, while the components F^{02}, F^{03}, and F^{12}, F^{13} transform like x^0 and x^1, respectively. Expressing the components of F^{ik} in terms of the components of the fields $\mathbf{E}$ and $\mathbf{H}$, according to (23.5), we then find the following formulas of transformation for the electric field:

$$E_x = E'_x, \ E_y = \frac{E'_y + \frac{V}{c}H'_z}{\sqrt{1 - \frac{V^2}{c^2}}}, \quad E_z = \frac{E'_z - \frac{V}{c}H'_y}{\sqrt{1 - \frac{V^2}{c}}},$$ (24.2)

and for the magnetic field:

$$H_x = H'_x, \ H_y = \frac{H'_y - \frac{V}{c}E'_z}{\sqrt{1 - \frac{V^2}{c^2}}}, \quad H_z = \frac{H'_z - \frac{V}{c}E'_y}{\sqrt{1 - \frac{V^2}{c^2}}},$$ (24.3)

Thus the electric and magnetic fields, like the majority of physical quantities, are relative; that is, their properties are different in different reference systems. In particular, the electric or the magnetic field can be equal to zero in one reference system and at the same time be present in another system.

The formulas (24.2), (24.3) simplify considerably for the case $V \ll c$. To terms of order V/c, we have:

$$E'_x = E'_x, E_y = E'_y + \frac{V}{c} H'_z, E_z = E'_z - \frac{V}{c} H'_y;$$

$$H_x = H'_x, H_y = H'_y - \frac{V}{c} E'_z, H_z = H'_z + \frac{V}{c} E'_y.$$

These formulas can be written in vector form

$$\mathbf{E} = \mathbf{E}' + \frac{1}{c} \mathbf{H}' \times \mathbf{V}, \mathbf{H} = \mathbf{H}' - \frac{1}{c} \mathbf{E}' \times \mathbf{V}. \tag{24.4}$$

The formulas for the inverse transformation from K' to K are obtained directly from (24.2)–(24.4) by changing the sign of V and shifting the prime.

If the magnetic field $\mathbf{H}' = 0$ in the K' system, then, as we easily verify on the basis of (24.2) and (24.3), the following relation exists between the electric and magnetic fields in the K system:

$$\mathbf{H} = \frac{1}{c} \mathbf{V} \times \mathbf{E}. \tag{24.5}$$

If in the K' system, $\mathbf{E}' = 0$, then in the K system

$$\mathbf{E} = -\frac{1}{c} \mathbf{V} \times \mathbf{H}. \tag{24.6}$$

Consequently, in both cases, in the K system the magnetic and electric fields are mutually perpendicular.

These formulas also have a significance when used in the reverse direction: if the fields $\mathbf{E}$ and $\mathbf{H}$ are mutually perpendicular (but not equal in magnitude) in some reference system K, then there exists a reference system K' in which the field is pure electric or pure magnetic. The velocity $\mathbf{V}$ of this system (relative to K) is perpendicular to $\mathbf{E}$ and $\mathbf{H}$ and equal in magnitude to cH/E in the first case (where we must have $H < E$) and to cE/H in the second case (where $E < H$).

§ 25. Invariants of the field

From the electric and magnetic field intensities we can form invariant quantities, which remain unchanged in the transition from one inertial reference system to another.

The form of these invariants is easily found starting from the four dimensional representation of the field using the antisymmetric four-tensor F^{ik}. It is obvious that we can form the following invariant quantities from the components of this tensor:

$$F_{ik}F^{ik} = \text{inv}, \tag{25.1}$$

$$e^{iklm}F_{ik}F_{lm} = \text{inv}, \tag{25.2}$$

where e^{iklm} is the completely antisymmetric unit tensor of the fourth rank (cf. § 6). The first quantity is a scalar, while the second is a pseudoscalar (the product of the tensor F^{ik} with its dual tensor.†

Expressing F^{ik} in terms of the components of **E** and **H** using (23.5), it is easily shown that, in three-dimensional form, these invariants have the form:

$$H^2 - E^2 = \text{inv}, \qquad (25.3)$$

$$\mathbf{E} \cdot \mathbf{H} = \text{inv}. \qquad (25.4)$$

The pseudoscalar character of the second of these is here apparent from the fact that it is the product of the polar vector **E** with the axial vector **H** (whereas its square $(\mathbf{E} \cdot \mathbf{H})^2$ is a true scalar).

From the invariance of the two expressions presented, we get the following theorems. If the electric and magnetic fields are mutually perpendicular in any reference system, that is, $\mathbf{E} \cdot \mathbf{H} = 0$, then they are also perpendicular in every other inertial reference system. If the absolute values of **E** and **H** are equal to each other in any reference system, then they are the same in any other system.

The following inequalities are also clearly valid. If in any reference system $E > H$ (or $H > E$), then in every other system we will have $E > H$ (or $H > E$). If in any system of reference the vectors **E** and **H** make an acute (or obtuse) angle, then they will make an acute (or obtuse) angle in every other reference system.

By means of a Lorentz transformation we can always give **E** and **H** any arbitrary values, subject only to the condition that $E^2 - H^2$ and $\mathbf{E} \cdot \mathbf{H}$ have fixed values. In particular, we can always find an inertial system in which the electric and magnetic fields are parallel to each other at a given point. In this system $\mathbf{E} \cdot \mathbf{H} = EH$, and from the two equations

$$E^2 - H^2 = E_0^2 - H_0^2, \qquad EH = \mathbf{E}_0 \cdot \mathbf{H}_0.$$

we can find the values of **E** and **H** in this system of reference ($\mathbf{E}_0$ and $\mathbf{H}_0$ are the electric and magnetic fields in the original system of reference).

The case where both invariants are zero is excluded. In this case, **E** and **H** are equal and mutually perpendicular in all reference systems.

If $\mathbf{E} \cdot \mathbf{H} = 0$, then we can always find a reference system in which $\mathbf{E} = 0$ or $\mathbf{H} = 0$ (according as $E^2 - H^2 < \text{or} > 0$), that is, the field is purely magnetic or purely electric. Conversely, if in any reference system $\mathbf{E} = 0$ or $\mathbf{H} = 0$, then they are mutually perpendicular in every other system, in accordance with the statement at the end of the preceding section.

We shall give still another approach to the problem of finding the invariants of an antisymmetric four-tensor. From this method we shall, in particular, see that (25.3)–(25.4) are actually the only two independent invariants and at the same time we will explain some instructive mathematical properties of the Lorentz transformations when applied to such a four-tensor.

Let us consider the complex vector

$$\mathbf{F} = \mathbf{E} + i\mathbf{H}. \qquad (25.5)$$

---

† We also note that the pseudoscalar (25.2) can also be expressed as a four-divergence:

$$e^{iklm} F_{ik} F_{im} = 4 \frac{\partial}{\partial x^i} \left(e^{iklm} A_k \frac{\partial}{\partial x^l} A_m \right),$$

as can be easily verified by using the antisymmetry of e^{iklm}.

Using formulas (24.2)–(24.3), it is easy to see that a Lorentz transformation (along the x axis) for this vector has the form

$$F_x = F_x', F_y = F_y' \cosh \phi - iF_z' \sinh \phi = F_y' \cos i\phi - F_z' \sin i\phi.$$

$$F_z = F_z' \cos i\phi + F_y' \sin i\phi, \tanh \phi = \frac{V}{c}. \quad (25.6)$$

We see that a rotation in the x, t plane in four-space (which is what this Lorentz transformation is) for the vector $\mathbf{F}$ is equivalent to a rotation in the y, z plane through an imaginary angle in three-dimensional space. The set of all possible rotations in four-space (including also the simple rotations around the x, y, and z axes) is equivalent to the set of all possible rotations, through complex angles in three-dimensional space (where the six angles of rotation in four-space correspond to the three complex angles of rotation of the three-dimensional system).

The only invariant of a vector with respect to rotation is its square: $\mathbf{F}^2 = E^2 - H^2 + 2i \, \mathbf{E} \cdot \mathbf{H}$; thus the real quantities $E^2 - H^2$ and $\mathbf{E} \cdot \mathbf{H}$ are the only two independent invariants of the tensor F_{ik}.

If $\mathbf{F}^2 \neq 0$, the vector $\mathbf{F}$ can be written as $\mathbf{F} = a\mathbf{n}$, where $\mathbf{n}$ is a complex unit vector ($\mathbf{n}^2 = 1$). By a suitable complex rotation we can point $\mathbf{n}$ along one of the coordinate axes; it is clear that then $\mathbf{n}$ becomes real and determines the directions of the two vectors $\mathbf{E}$ and $\mathbf{H}$: $\mathbf{F} = (E + iH)\mathbf{n}$; in other words we get the result that $\mathbf{E}$ and $\mathbf{H}$ become parallel to one another.

PROBLEM

Determine the velocity of the system of reference in which the electric and magnetic fields are parallel.

Solution: Systems of reference K', satisfying the required condition, exist in infinite numbers. If we have found one such, then the same property will be had by any other system moving relative to the first with its velocity directed along the common direction of $\mathbf{E}$ and $\mathbf{H}$. Therefore it is sufficient to find one of these systems which has a velocity perpendicular to both fields. Choosing the direction of the velocity as the x axis, and making use of the fact that in K': $E_z' = H_z' = 0$, $E_y'H_z' - F_z'H_y' = 0$, we obtain with the aid of formulas (24.2) and (24.3) for the velocity $\mathbf{V}$ of the K' system relative to the original system the following equation:

$$\frac{\dfrac{\mathbf{V}}{c}}{1 + \dfrac{\mathbf{V}^2}{c^2}} = \frac{\mathbf{E} \times \mathbf{H}}{E^2 + H^2}$$

(we must choose that root of the quadratic equation for which $V < c$).

CHAPTER 4

THE ELECTROMAGNETIC FIELD EQUATIONS

§ 26. The first pair of Maxwell's equations

From the expressions

$$\mathbf{H} = \text{curl } \mathbf{A}, \quad \mathbf{E} = -\frac{1}{c}\frac{\partial \mathbf{A}}{\partial t} - \text{grad } \phi$$

it is easy to obtain equations containing only $\mathbf{E}$ and $\mathbf{H}$. To do this we find curl $\mathbf{E}$:

$$\text{curl } \mathbf{E} = -\frac{1}{c}\frac{\partial}{\partial t}\text{ curl } \mathbf{A} - \text{curl grad } \phi.$$

But the curl of any gradient is zero. Consequently,

$$\text{curl } \mathbf{E} = -\frac{1}{c}\frac{\partial \mathbf{H}}{\partial t}. \tag{26.1}$$

Taking the divergence of both sides of the equation curl $\mathbf{A} = \mathbf{H}$, and recalling that div curl $= 0$, we find

$$\text{div } \mathbf{H} = 0. \tag{26.2}$$

The equations (26.1) and (26.2) are called the first pair of Maxwell's equations.† We note that these two equations still do not completely determine the properties of the fields. This is clear from the fact that they determine the change of the magnetic field with time (the derivative $\partial \mathbf{H}/\partial t$), but do not determine the derivative $\partial \mathbf{E}/\partial t$.

Equations (26.1) and (26.2) can be written in integral form. According to Gauss'theorem

$$\int \text{div } \mathbf{H}dV = \oint \mathbf{H} \cdot d\mathbf{f},$$

where the integral on the right goes over the entire closed surface surrounding the volume over which the integral on the left is extended. On the basis of (26.2), we have

$$\oint \mathbf{H} \cdot d\mathbf{f} = 0. \tag{26.3}$$

---

† Maxwell's equations (the fundamental equations of electrodynamics) were first formulated by him in the 1860's.

The integral of a vector over a surface is called the *flux of the vector* through the surface. Thus the flux of the magnetic field through every closed surface is zero.

According to Stokes' theorem,

$$\int \operatorname{curl} \mathbf{E} \cdot d\mathbf{f} = \oint \mathbf{E} \cdot d\mathbf{l},$$

where the integral on the right is taken over the closed contour bounding the surface over which the left side is integrated. From (26.1) we find, integrating both sides for any surface,

$$\oint \mathbf{E} \cdot d\mathbf{l} = \frac{1}{c} \frac{\partial}{\partial t} \int \mathbf{H} \cdot d\mathbf{f}. \tag{26.4}$$

The integral of a vector over a closed contour is called the *circulation* of the vector around the contour. The circulation of the electric field is also called the *electromotive force* in the given contour. Thus the electromotive force in any contour is equal to minus the time derivative of the magnetic flux through a surface bounded by this contour.

The Maxwell equations (26.1) and (26.2) can be expressed in four-dimensional notation. Using the definition of the electromagnetic field tensor

$$F_{ik} = \partial A_k/\partial x^i - \partial A_i/\partial x^k,$$

it is easy to verify that

$$\frac{\partial F_{ik}}{\partial x^l} + \frac{\partial F_{kl}}{\partial x^i} + \frac{\partial F_{li}}{\partial x^k} = 0. \tag{26.5}$$

The expression on the left is a tensor of third rank, which is antisymmetric in all three indices. The only components which are not identically zero are those with $i \neq k \neq l$. Thus there are altogether four different equations which we can easily show [by substituting from (23.5)] coincide with equations (26.1) and (26.2).

We can construct the four-vector which is dual to this antisymmetric four-tensor of rank three by multiplying the tensor by e^{iklm} and contracting on three pairs of indices (see § 6). Thus (26.5) can be written in the form

$$e^{iklm} \frac{\partial F_{lm}}{\partial x^k} = 0, \tag{26.6}$$

which shows explicitly that there are only four independent equations.

§ 27. The action function of the electromagnetic field

The action function S for the whole system, consisting of an electromagnetic field as well as the particles located in it, must consist of three parts:

$$S - S_f + S_m + S_{mf}, \tag{27.1}$$

where S_m is that part of the action which depends only on the properties of the particles, that is, just the action for free particles. For a single free particle, it is given by (8.1). If there are several particles, then their total action is the sum of the actions for each of the individual particles. Thus,

$$S_m = - \Sigma \, mc \int ds. \tag{27.2}$$

The quantity S_{mf} is that part of the action which depends on the interaction between the particles and the field. According to § 16, we have for a system of particles:

$$S_{mf} = -\sum \frac{e}{c} \int A_k \, dx^k. \tag{27.3}$$

In each term of this sum, A_k is the potential of the field at that point of specetime at which the corresponding particle is located. The sum $S_m + S_{mf}$ is already familiar to us as the action (16.1) for charges in a field.

Finally S_f is that part of the action which depends only on the properties of the field itself, that is, S_f is the action for a field in the absence of charges. Up to now, because we were interested only in the motion of charges in a *given* electromagnetic field, the quantity S_f, which does not depend on the particles, did not concern us, since this term cannot affect the motion of the particles. Nevertheless this term is necessary when we want to find equations determining the field itself. This corresponds to the fact that from the parts $S_m + S_{mf}$ of the action we found only two equations for the field, (26.1) and (26.2), which are not yet sufficient for complete determination of the field.

To establish the form of the action S_f for the field, we start from the following very important property of electromagnetic fields. As experiment shows, the electromagnetic field satisfies the so-called *principle of superposition*. This principle consists in the statement that the field produced by a system of charges is the result of a simple composition of the fields produced by each of the particles individually. This means that the resultant field intensity at each point is equal to the vector sum of the individual field intensities at that point.

Every solution of the field equations gives a field that can exist in nature. According to the principle of superposition, the sum of any such fields must be a field that can exist in nature, that is, must satisfy the field equations.

As is well known, linear differential equations have just this property, that the sum of any solutions is also a solution. Consequently the field equations must be linear differential equations.

From the discussion, it follows that under the integral sign for the action S_f there must stand an expression quadratic in the field. Only in this case will the field equations be linear; the field equations are obtained by varying the action, and in the variation the degree of the expression under the integral sign decreases by unity.

The potentials cannot enter into the expression for the action S_f, since they are not uniquely determined (in S_{mf} this lack of uniqueness was not important). Therefore S_f must be the integral of some function of the electromagnetic field tensor F_{ik}. But the action must be a scalar and must therefore be the integral of some scalar. The only such quantity is the product $F_{ik}F^{ik}$.†

---

† The function in the integrand of S_f must not include derivatives of F_{ik}, since the Lagrangian can contain, aside from the coordinates, only their first time derivatives. The role of "coordinates" (i.e., parameters to be varied in the principle of least action) is in this case played by the field potential A_k; this is analogous to the situation in mechanics where the Lagrangian of a mechanical system contains only the coordinates of the particles and their first time derivatives.

As for the quantity $e^{iklm}F_{ik}F_{lm}$ (§ 25), as pointed out in the footnote on p. 68, it is a complete four-divergence, so that adding it to the integrand in S_f would have no effect on the "equations of motion". It is interesting that this quantity is already excluded from the action for a reason independent of the fact that it is a pseudoscalar and not a true scalar.

Thus S_f must have the form:

$$S_f = a \iint F_{ik} F^{ik} \, dV dt, \quad dV = dx \, dy \, dz,$$

where the integral extends over all of space and the time between two given moments; a is some constant. Under the integral stands $F_{ik} F^{ik} = 2(H^2 - E^2)$. The field $\mathbf{E}$ contains the derivative $\partial \mathbf{A}/\partial t$; but it is easy to see that $(\partial \mathbf{A}/\partial t)^2$ must appear in the action with the positive sign (and therefore E^2 must have a positive sign). For if $(\partial \mathbf{A}/\partial t)^2$ appeared in S_f with a minus sign, then sufficiently rapid change of the potential with time (in the time interval under consideration) could always make S_f a negative quantity with arbitrarily large absolute value. Consequently S_f could not have a minimum, as is required by the principle of least action. Thus, a must be negative.

The numerical value of a depends on the choice of units for measurement of the field. We note that after the choice of a definite value for a and for the units of measurement of field, the units for measurement of all other electromagnetic quantities are determined.

From now on we shall use the *Gaussian system of units;* in this system a is a dimensionless quantity, equal to $-(1/16\pi)$.†

Thus the action for the field has the form

$$S_f = -\frac{1}{16\pi c} \int F_{ik} F^{ik} \, d\Omega, \quad d\Omega = c \, dt \, dx \, dy \, dz. \tag{27.4}$$

In three-dimensional form:

$$S_f = \frac{1}{8\pi} \int (E^2 - H^2) dV dt. \tag{27.5}$$

In other words, the Lagrangian for the field is

$$L_f = \frac{1}{8\pi} \int (E^2 - H^2) dV. \tag{27.6}$$

The action for field plus particles has the form

$$S = -\Sigma \int mc \, ds - \Sigma \int \frac{e}{c} A_k \, dx^k - \frac{1}{16\pi c} \int f_{ik} F^{ik} \, d\Omega. \tag{27.7}$$

We emphasize that now the charges are not assumed to be small, as in the derivation of the equation of motion of a charge in a given field. Therefore A_k and F_{ik} refer to the actual field, that is, the external field plus the field produced by the particles themselves; A_k and F_{ik} now depend on the positions and velocities of the charges.

§ 28. The four-dimensional current vector

Instead of treating charges as points, for mathematical convenience we frequently consider them to be distributed continuously in space. Then we can introduce the "charge density" ϱ

---

† In addition to the Gaussian system, one also uses the Heaviside system, in which $a = -\frac{1}{4}$. In this system of units the field equations have a more convenient form (4π does not appear) but on the other hand, π appears in the Coulomb law. Conversely, in the Gaussian system the field equations contain 4π, but the Coulomb law has a simple form.

such that ϱdV is the charge contained in the volume dV. The density ϱ is in general a function of the coordinates and the time. The integral of ϱ over a certain volume is the charge contained in that volume.

Here we must remember that charges are actually pointlike, so that the density ϱ is zero everywhere except at points where the point charges are located, and the integral $\int \varrho dV$ must be equal to the sum of the charges contained in the given volume. Therefore ϱ can be expressed with the help of the δ-function in the following form†:

$$\varrho = \sum_a e_a \delta(\mathbf{r} - \mathbf{r}_a) \tag{28.1}$$

where the sum goes over all the charges and $\mathbf{r}_a$ is the radius vector of the charge e_a.

The charge on a particle is, from its very definition, an invariant quantity, that is, it does not depend on the choice of reference system. On the other hand, the density ϱ is not generally an invariant—only the product ϱdV is invariant.

Multiplying the equality $de = \varrho dV$ on both sides with dx^i:

$$de\, dx^i = \varrho dV dx^i = \varrho dV dt\, \frac{dx^i}{dt}.$$

† The δ-function $\delta(x)$ is defined as follows: $\delta(x) = 0$, for all nonzero values of x; for $x = 0$, $\delta(0) = \infty$, in such a way that the integral

$$\int_{-\infty}^{+\infty} \delta(x)dx = 1. \tag{I}$$

From this definition there result the following properties: if $f(x)$ is any continuous function, then

$$\int_{-\infty}^{+\infty} f(x)\delta(x - a)\, dx = f(a), \tag{II}$$

and in particular,

$$\int_{-\infty}^{+\infty} f(x)\delta(x)\, dx = \mathbf{f}(0). \tag{III}$$

(The limits of integration, it is understood, need not be $\pm \infty$; the range of integration can be arbitrary, provided it includes the point at which the δ-function does not vanish.)

The meaning of the following equalities is that the left and right sides give the same result when introduced as factors under an integral sign:

$$\delta(-x) = \delta(x), \qquad \delta(ax) = \frac{1}{|a|}\delta(x). \tag{IV}$$

The last equality is a special case of the more general relation

$$\delta[\phi(x)] = \sum_i \frac{1}{|\phi'(a_i)|}\delta(x - a_i), \tag{V}$$

where $\phi(x)$ is a single-valued function (whose inverse need not be single-valued) and the a_i are the roots of the equation $\phi(x) = 0$.

Just as $\delta(x)$ was defined for one variable x, we can introduce a three-dimensional δ-function, $\delta(\mathbf{r})$, equal to zero everywhere except at the origin of the three-dimensional coordinate system, and whose integral overall space is unity. As such a function we can clearly use the product $\delta(x)\, \delta(y)\, \delta(z)$.

On the left stands a four-vector (since de is a scalar and dx^i is a four-vector). This means that the right side must be a four-vector. But $dV\, dt$ is a scalar, and so $\varrho(dx^i/dt)$ is a four-vector. This vector (we denote it by j^i) is called the *current four-vector:*

$$j^i = \varrho \frac{dx^i}{dt}. \tag{28.2}$$

The space components of this vector form the *current density vector,*

$$\mathbf{j} = \varrho\mathbf{v}, \tag{28.3}$$

where $\mathbf{v}$ is the velocity of the charge at the given point. The time component of the four-vector (28.2) is $c\varrho$. Thus

$$j^i = (c\varrho, \mathbf{j}). \tag{28.4}$$

The total charge present in all of space is equal to the integral $\int \varrho dV$ over all space. We can write this integral in four-dimensional form:

$$\int \varrho dV = \frac{1}{c} \int j^0 dV = \frac{1}{c} \int j^i dS_i, \tag{28.5}$$

where the integral is taken over the entire four-dimensional hyperplane perpendicular to the x^0 axis (clearly this integration means integration over the whole three-dimensional space). Generally, the integral

$$\frac{1}{c} \int j^i dS_i$$

over an arbitrary hypersurface is the sum of the charges whose world lines pass through this surface.

Let us introduce the current four-vector into the expression (27.7) for the action and transform the second term in that expression. Introducing in place of the point charges e a continuous distribution of charge with density ϱ, we must write this term as

$$-\frac{1}{c} \int \varrho A_i dx^i dV,$$

replacing the sum over the charges by an integral over the whole volume. Rewriting in the form

$$-\frac{1}{c} \int \varrho \frac{dx^i}{dt} A_i dV dt,$$

we see that this term is equal to

$$-\frac{1}{c^2} \int A_i j^i d\Omega.$$

Thus the action S takes the form

$$S = -\Sigma \int mc\, ds - \frac{1}{c^2} \int A_i j^i d\Omega - \frac{1}{16\pi c} \int F_{ik} F^{ik} d\Omega. \tag{28.6}$$

§ 29. The equation of continuity

The change with time of the charge contained in a certain volume is determined by the derivative

$$\frac{\partial}{\partial t} \int \varrho \, dV.$$

On the other hand, the change in unit time, say, is determined by the quantity of charge which in unit time leaves the volume and goes to the outside or, conversely, passes to its interior. The quantity of charge which passes in unit time through the element $d\mathbf{f}$ of the surface bounding our volume is equal to $\varrho \mathbf{v} \cdot d\mathbf{f}$, where $\mathbf{v}$ is the velocity of the charge at the point in space where the element $d\mathbf{f}$ is located. The vector $d\mathbf{f}$ is directed, as always, along the external normal to the surface, that is, along the normal toward the outside of the volume under consideration. Therefore $\varrho \mathbf{v} \cdot d\mathbf{f}$ is positive if charge leaves the volume, and negative if charge enters the volume. The total amount of charge leaving the given volume per unit time is consequently $\oint \varrho \mathbf{v} \cdot d\mathbf{f}$, where the integral extends over the whole of the closed surface bounding the volume.

From the equality of these two expressions, we get

$$\frac{\partial}{\partial t} \int \varrho \, dV = - \oint \varrho \mathbf{v} \cdot d\mathbf{f}. \tag{29.1}$$

The minus sign appears on the right, since the left side is positive if the total charge in the given volume increases. The equation (29.1) is the so-called *equation of continuity*, expressing the conservation of charge in integral form. Noting that $\varrho \mathbf{v}$ is the current density, we can rewrite (29.1) in the form

$$\frac{\partial}{\partial t} \int \varrho \, dV = - \oint \mathbf{j} \cdot d\mathbf{f}. \tag{29.2}$$

We also write this equation in differential form. To do this we apply Gauss' theorem to (29.2):

$$\oint \mathbf{j} \cdot d\mathbf{f} = \int \operatorname{div} \mathbf{j} \, dV.$$

and we find

$$\int \left(\operatorname{div} \mathbf{j} + \frac{\partial \varrho}{\partial t} \right) dV = 0.$$

Since this must hold for integration over an arbitrary volume, the integrand must be zero:

$$\operatorname{div} \mathbf{j} + \frac{\partial \varrho}{\partial t} = 0. \tag{29.3}$$

This is the equation of continuity in differential form.

It is easy to check that the expression (28.1) for ϱ in δ-function form automatically satisfies the equation (29.3). For simplicity we assume that we have altogether only one charge, so that

$$\varrho = e\delta(\mathbf{r} - \mathbf{r}_0).$$

The current $\mathbf{j}$ is then

$$\mathbf{j} = e\mathbf{v}\,\delta(\mathbf{r} - \mathbf{r}_0),$$

where $\mathbf{v}$ is the velocity of the charge. We determine the derivative $\partial\varrho/\partial t$. During the motion of the charge its coordinates change, that is, the vector $\mathbf{r}_0$ changes. Therefore

$$\frac{\partial\varrho}{\partial t} = \frac{\partial\varrho}{\partial\mathbf{r}_0} \cdot \frac{\partial\mathbf{r}_0}{\partial t}.$$

But $\partial\mathbf{r}_0/\partial t$ is just the velocity $\mathbf{v}$ of the charge. Furthermore, since ϱ is a function of $\mathbf{r} - \mathbf{r}_0$,

$$\frac{\partial\varrho}{\partial\mathbf{r}_0} = -\frac{\partial\varrho}{\partial\mathbf{r}}.$$

Consequently

$$\frac{\partial\varrho}{\partial t} = -\mathbf{v}\cdot\operatorname{grad}\varrho = -\operatorname{div}(\varrho\mathbf{v})$$

(the velocity $\mathbf{v}$ of the charge of course does not depend on $\mathbf{r}$). Thus we arrive at the equation (29.3).

It is easily verified that, in four-dimensional form, the continuity equation (29.3) is expressed by the statement that the four-divergence of the current four-vector is zero:

$$\frac{\partial j^i}{\partial x^i} = 0. \tag{29.4}$$

In the preceding section we saw that the total charge present in all of space can be written as

$$\frac{1}{c}\int j^i\,dS_i,$$

where the integration is extended over the hyperplane $x^0 = $ const. At each moment of time, the total charge is given by such an integral taken over a different hyperplane perpendicular to the x^0 axis. It is easy to verify that the equation (29.4) actually leads to conservation of charge, that is, to the result that the integral $\int j^i dS_i$ is the same no matter what hyperplane $x^0 = $ const we integrate over. The difference between the integrals $\int j^i dS_i$ taken over two such hyperplanes can be written in the form $\oint j^i dS_i$, where the integral is taken over the whole closed hypersurface surrounding the four-volume between the two hyperplanes under consideration (this integral differs from the required integral because of the presence of the integral over the infinitely distant "sides" of the hypersurface which, however, drop out, since there are no charges at infinity). Using Gauss' theorem (6.15) we can transform this to an integral over the four-volume between the two hyperplanes and verify that

$$\oint j^i\,dS_i = \int \frac{\partial j^i}{\partial x^i}\,d\Omega = 0. \tag{29.5}$$

The proof presented clearly remains valid also for any two integrals $\int j^i dS_i$, in which the integration is extended over any two infinite hypersurfaces (and not just the hyperplanes $x^0 = $ const) which each contain all of three-dimensional space. From this it follows that the integral

$$\frac{1}{c} \int j^i \, dS_i$$

is actually identical in value (and equal to the total charge in space) no matter over what such hypersurface the integration is taken.

We have already mentioned (see the footnote on p. 53) the close connection between the gauge invariance of the equations of electrodynamics and the law of conservation of charge. Let us show this once again using the expression for the action in the form (28.6). On replacing A_i by $A_i - (\partial f/\partial x^i)$, the integral

$$\frac{1}{c^2} \int j^i \frac{\partial f}{\partial x^i} \, d\Omega$$

is added to the second term in this expression. It is precisely the conservation of charge, as expressed in the continuity equation (29.4), that enables us to write the integrand as a four-divergence $\partial (f j^i)/\partial x^i$, after which, using Gauss' theorem, the integral over the four-volume is transformed into an integral over the bounding hypersurface; on varying the action, these integrals drop out and thus have no effect on the equations on motion.

§ 30. The second pair of Maxwell equations

In finding the field equations with the aid of the principle of least action we must assume the motion of the charges to be given and vary only the potentials (which serve as the "coordinates" of the system); on the other hand, to find the equations of motion we assumed the field to be given and varied the trajectory of the particle.

Therefore the variation of the first term in (28.6) is zero, and in the second we must not vary the current j^i. Thus,

$$\delta S = - \int \frac{1}{c} \left\{ \frac{1}{c} j^i \delta A_i + \frac{1}{8\pi} F^{ik} \delta F_{ik} \right\} d\Omega = 0.$$

(where we have used the fact that $F^{ik} \delta F_{ik} \equiv F_{ik} \delta F^{ik}$). Substituting $F_{ik} = \partial A_k/\partial x^i - \partial A_i/\partial x^k$ we have

$$\delta S = - \int \frac{1}{c} \left\{ \frac{1}{c} j^i \delta A_i + \frac{1}{8\pi} F^{ik} \frac{\partial}{\partial x^i} \delta A_k - \frac{1}{8\pi} F^{ik} \frac{\partial}{\partial x^k} \delta A_i \right\} d\Omega.$$

In the second term we interchange the indices i and k, over which the expressions are summed, and in addition replace F^{ik} by $-F^{ik}$. Then we obtain

$$\delta S = - \int \frac{1}{c} \left\{ \frac{1}{c} j^i \delta A_i - \frac{1}{4\pi} F^{ik} \frac{\partial}{\partial x^k} \delta A_i \right\} d\Omega.$$

The second of these integrals we integrate by parts, that is, we apply Gauss' theorem:

$$\delta S = - \frac{1}{c} \int \left\{ \frac{1}{c} j^i + \frac{1}{4\pi} \frac{\partial F^{ik}}{\partial x^k} \right\} \delta A_i \, d\Omega - \frac{1}{4\pi c} \int F^{ik} \delta A_i \, dS_k \bigg|. \tag{30.1}$$

In the second term we must insert the values at the limits of integration. The limits for the coordinates are at infinity, where the field is zero. At the limits of the time integration, that is, at the given initial and final time values, the variation of the potentials is zero, since in accord with the principle of least action the potentials are given at these times. Thus the second term in (30.1) is zero, and we find

$$\int \left(\frac{1}{c} j^i + \frac{1}{4\pi} \frac{\partial F^{ik}}{\partial x^k} \right) \delta A_i \, d\Omega = 0.$$

Since according to the principle of least action, the variations δA_i are arbitrary, the coefficients of the δA_i must be set equal to zero.

$$\frac{\partial F^{ik}}{\partial x^k} = -\frac{4\pi}{c} j^i. \tag{30.2}$$

Let us express these four ($i = 0, 1, 2, 3$) equations in three-dimensional form. For $i = 1$:

$$\frac{\partial F^{11}}{\partial x} + \frac{\partial F^{12}}{\partial y} + \frac{\partial F^{13}}{\partial z} + \frac{1}{c} \frac{\partial F^{10}}{\partial t} = -\frac{4\pi}{c} j^1.$$

Substituting the values for the components of F^{ik}, we find

$$\frac{\partial H_z}{\partial y} - \frac{\partial H_y}{\partial z} - \frac{1}{c} \frac{\partial E_x}{\partial t} = \frac{4\pi}{c} j_x.$$

This together with the two succeeding equations ($i = 2, 3$) can be written as one vector equation:

$$\operatorname{curl} \mathbf{H} = \frac{1}{c} \frac{\partial \mathbf{E}}{\partial t} + \frac{4\pi}{c} \mathbf{j}. \tag{30.3}$$

Finally, the fourth equation ($i = 0$) gives

$$\operatorname{div} \mathbf{E} = 4\pi \varrho. \tag{30.4}$$

Equations (30.3) and (30.4) are the second pair of Maxwell equations.† Together with the first pair of Maxwell equations they completely determine the electromagnetic field, and are the fundamental equations of the theory of such fields, i.e. of *electrodynamics*.

Let us write these equations in integral form. Integrating (30.4) over a volume and applying Gauss' theorem

$$\int \operatorname{div} \mathbf{E} \, dV = \oint \mathbf{E} \cdot d\mathbf{f},$$

we get

$$\oint \mathbf{E} \cdot d\mathbf{f} = 4\pi \int \varrho \, dV. \tag{30.5}$$

Thus the flux of the electric field through a closed surface is equal to 4π times the total charge contained in the volume bounded by the surface.

Integrating (30.3) over an open surface and applying Stokes' theorem

† The Maxwell equations in a form applicable to point charges in the electromagnetic field in vacuum were formulated by H. A. Lorentz.

$$\int \text{curl } \mathbf{H} \cdot d\mathbf{f} = \oint \mathbf{H} \cdot d\mathbf{l},$$

we find

$$\oint \mathbf{H} \cdot d\mathbf{l} = \frac{1}{c} \frac{\partial}{\partial t} \int \mathbf{E} \cdot d\mathbf{f} + \frac{4\pi}{c} \int \mathbf{j} \cdot d\mathbf{f}. \tag{30.6}$$

The quantity

$$\frac{1}{4\pi} \frac{\partial \mathbf{E}}{\partial t} \tag{30.7}$$

is called the *"displacement current"*. From (30.6) written in the form

$$\oint \mathbf{H} \cdot d\mathbf{l} = \frac{4\pi}{c} \int \left(\mathbf{j} + \frac{1}{4\pi} \frac{\partial \mathbf{E}}{\partial t} \right) \cdot d\mathbf{f}, \tag{30.8}$$

we see that the circulation of the magnetic field around any contour is equal to $4\pi/c$ times the sum of the true current and displacement current passing through a surface bounded by this contour.

From the Maxwell equations we can obtain the already familiar continuity equation (29.3). Taking the divergence of both sides of (30.3), we find

$$\text{div curl } \mathbf{H} = \frac{1}{c} \frac{\partial}{\partial t} \text{ div } \mathbf{E} + \frac{4\pi}{c} \text{ div } \mathbf{j}.$$

But div curl $\mathbf{H} = 0$ and div $\mathbf{E} = 4\pi\varrho$, according to (30.4). Thus we arrive once more at equation (29.3). In four-dimensional form, from (30.2), we have:

$$\frac{\partial^2 F^{ik}}{\partial x^i \partial x^k} = - \frac{4\pi}{c} \frac{\partial j^i}{\partial x^i}.$$

But when the operator $\partial^2/\partial x^i \partial x^k$, which is symmetric in the indices i and k, is applied to the antisymmetric tensor F^{ik}, it gives zero identically and we arrive at the continuity equation (29.4) expressed in four-dimensional form.

§ 31. Energy density and energy flux

Let us multiply both sides of (30.3) by $\mathbf{E}$ and both sides of (26.1) by $\mathbf{H}$ and combine the resultant equations. Then we get

$$\frac{1}{c} \mathbf{E} \cdot \frac{\partial \mathbf{E}}{\partial t} + \frac{1}{c} \mathbf{H} \cdot \frac{\partial \mathbf{H}}{\partial t} = - \frac{4\pi}{c} \mathbf{j} \cdot \mathbf{E} - (\mathbf{H} \cdot \text{curl } \mathbf{E} - \mathbf{E} \cdot \text{curl } \mathbf{H}).$$

Using the well-known formula of vector analysis,

$$\text{div } (\mathbf{a} \times \mathbf{b}) = \mathbf{b} \cdot \text{curl } \mathbf{a} - \mathbf{a} \cdot \text{curl } \mathbf{b},$$

we rewrite this relation in the form

$$\frac{1}{2c} \frac{\partial}{\partial t} (E^2 + H^2) = - \frac{4\pi}{c} \mathbf{j} \cdot \mathbf{E} - \text{div}(\mathbf{E} \times \mathbf{H})$$

or

$$\frac{\partial}{\partial t}\left(\frac{E^2 + H^2}{8\pi}\right) = -\mathbf{j} \cdot \mathbf{E} - \text{div } \mathbf{S}. \tag{31.1}$$

The vector

$$\mathbf{S} = \frac{c}{4\pi} \mathbf{E} \times \mathbf{H} \tag{31.2}$$

is called the *Poynting vector*.

We integrate (31.1) over a volume and apply Gauss' theorem to the second term on the right. Then we obtain

$$\frac{\partial}{\partial t} \int \frac{E^2 + H^2}{8\pi} \, dV = -\int \mathbf{j} \cdot \mathbf{E} \, dV - \oint \mathbf{S} \cdot d\mathbf{f}. \tag{31.3}$$

If the integral extends over all space, then the surface integral vanishes (the field is zero at infinity). Furthermore, we can express the integral $\int \mathbf{j} \cdot \mathbf{E} dV$ as a sum $\sum e\mathbf{v} \cdot \mathbf{E}$ over all the charges, and substitute from (17.7):

$$e\mathbf{v} \cdot \mathbf{E} = \frac{d}{dt} \mathscr{E}_{\text{kin}}.$$

Then (31.3) becomes

$$\frac{d}{dt}\left\{\int \frac{E^2 + H^2}{8\pi} \, dV + \sum \mathscr{E}_{\text{kin}}\right\} = 0. \tag{31.4}$$

Thus for the closed system consisting of the electromagnetic field and particles present in it, the quantity in brackets in this equation is conserved. The second term in this expression is the kinetic energy (including the rest energy of all the particles; see the footnote on p. 51), the first term is consequently the energy of the field itself. We can therefore call the quantity

$$W = \frac{E^2 + H^2}{8\pi} \tag{31.5}$$

the *energy density* of the electromagnetic field; it is the energy per unit volume of the field.

If we integrate over any finite volume, then the surface integral in (31.3) generally does not vanish, so that we can write the equation in the form

$$\frac{\partial}{\partial t}\left\{\int \frac{E^2 + H^2}{8\pi} \, dV + \sum \mathscr{E}_{\text{kin}}\right\} = -\oint \mathbf{S} \cdot d\mathbf{f}, \tag{31.6}$$

where now the second term in the brackets is summed only over the particles present in the volume under consideration. On the left stands the change in the total energy of field and particles per unit time. Therefore the integral $\oint \mathbf{S} \cdot d\mathbf{f}$ must be interpreted as the flux of field energy across the surface bounding the given volume, so that the Poynting vector $\mathbf{S}$ is this flux density—the amount of field energy passing through unit area of the surface in unit time.†

† We assume that at the given moment there are no charges on the surface itself. If this were not the case, then on the right we would have to include the energy flux transported by particles passing through the surface.

§ 32. The energy-momentum tensor

In the preceding section we derived an expression for the energy of the electromagnetic field. Now we derive this expression, together with one for the field momentum, in four-dimensional form. In doing this we shall for simplicity consider for the present an electromagnetic field without charges. Having in mind later applications (to the gravitational field), and also to simplify the calculation, we present the derivation in a general form, not specializing the nature of the system. So we consider any system whose action integral has the form

$$S = \int \Lambda \left(q, \frac{\partial q}{\partial x^i} \right) dV \, dt = \frac{1}{c} \int \Lambda \, d\Omega, \tag{32.1}$$

where Λ is some function of the quantities q, describing the state of the system, and of their first derivatives with respect to coordinates and time (for the electromagnetic field the components of the four-potential are the quantities q); for brevity we write here only one of the q's. We note that the space integral $\int \Lambda \, dV$ is the Lagrangian of the system, so that Λ can be considered as the *Lagrangian "density"*. The mathematical expression of the fact that the system is closed is the absence of any explicit dependence of Λ on the x^i, similarly to the situation for a closed system in mechanics, where the Lagrangian does not depend explicitly on the time.

The "equations of motion" (i.e. the field equations, if we are dealing with some field) are obtained in accordance with the principle of least action by varying S. We have (for brevity we write $q_{,i} \equiv \partial q / \partial x^i$),

$$\delta S = \frac{1}{c} \int \left(\frac{\partial \Lambda}{\partial q} \delta q + \frac{\partial \Lambda}{\partial q_{,i}} \delta q_{,i} \right) d\Omega$$

$$= \frac{1}{c} \int \left[\frac{\partial \Lambda}{\partial q} \delta q + \frac{\partial}{\partial x^i} \left(\frac{\partial \Lambda}{\partial q_{,i}} \delta q \right) - \delta q \frac{\partial}{\partial x^i} \frac{\partial \Lambda}{\partial q_{,i}} \right] d\Omega = 0.$$

The second term in the integrand, after transformation by Gauss' theorem, vanishes upon integration over all space, and we then find the following "equations of motion":

$$\frac{\partial}{\partial x^i} \frac{\partial \Lambda}{\partial q_{,i}} - \frac{\partial \Lambda}{\partial q} = 0 \tag{32.2}$$

(it is, of course, understood that we sum over any repeated index).

The remainder of the derivation is similar to the procedure in mechanics for deriving the conservation of energy. Namely, we write:

$$\frac{\partial \Lambda}{\partial x^i} = \frac{\partial \Lambda}{\partial q} \frac{\partial q}{\partial x^i} + \frac{\partial \Lambda}{\partial q_{,k}} \frac{\partial q_{,k}}{\partial x^i}.$$

Substituting (32.2) and noting that $q_{,k,i} = q_{,i,k}$, we find

$$\frac{\partial \Lambda}{\partial x^i} = \frac{\partial}{\partial x^k} \left(\frac{\partial \Lambda}{\partial q_{,k}} \right) q_{,i} + \frac{\partial \Lambda}{\partial q_{,k}} \frac{\partial q_{,i}}{\partial x^k} = \frac{\partial}{\partial x^k} \left(q_{,i} \frac{\partial \Lambda}{\partial q_{,k}} \right).$$

On the other hand, we can write

$$\frac{\partial \Lambda}{\partial x^i} = \delta_i^k \frac{\partial \Lambda}{\partial x^k},$$

so that, introducing the notation

$$T_i^k = q_{,i} \frac{\partial \Lambda}{\partial q_{,k}} - \delta_i^k \Lambda, \tag{32.3}$$

we can express the relation in the form

$$\frac{\partial T_i^k}{\partial x^k} = 0. \tag{32.4}$$

We note that if there is not one but several quantities $q^{(l)}$, then in place of (32.3) we must write

$$T_i^k = \sum_l q_{,i}^{(l)} \frac{\partial \Lambda}{\partial q_{,k}^{(l)}} - \delta_i^k \Lambda. \tag{32.5}$$

But in § 29 we saw that an equation of the form $\partial A^k / \partial x^k = 0$, i.e. the vanishing of the four-divergence of a vector, is equivalent to the statement that the integral $\int A^k dS_k$ of the vector over a hypersurface which contains all of three-dimensional space is conserved. It is clear that an analogous result holds for the divergence of a tensor; the equation (32.4) asserts that the vector $P^i = \text{const} \int T^{ik} dS_k$ is conserved.

This vector must be identified with the four-vector of momentum of the system. We choose the constant factor in front of the integral so that, in accord with our previous definition, the time component P^0 is equal to the energy of the system multiplied by $1/c$. To do this we note that

$$P^0 = \text{const} \int T^{0k} dS_k = \text{const} \int T^{00} dV$$

if the integration is extended over the hyperplane $x^0 = \text{const}$. On the other hand, according to (32.3),

$$T^{00} = \dot{q} \frac{\partial \Lambda}{\partial \dot{q}} - \Lambda. \quad \left(\dot{q} \equiv \frac{\partial q}{\partial t} \right)$$

Comparing with the usual formulas relating the energy and the Lagrangian, we see that this quantity must be considered as the energy density of the system, and therefore $\int T^{00} dV$ is the total energy of the system. Thus we must set $\text{const} = 1/c$, and we get finally for the four-momentum of the system the expression

$$P^i = \frac{1}{c} \int T^{ik} dS_k. \tag{32.6}$$

The tensor T^{ik} is called the *energy-momentum tensor* of the system.

It is necessary to point out that the definition of the tensor T^{ik} is not unique. In fact, if T^{ik} is defined by (32.3), then any other tensor of the form

$$T^{ik} + \frac{\partial}{\partial x^l} \psi^{ikl}, \quad \psi^{ikl} = -\psi^{ilk} \tag{32.7}$$

will also satisfy equation (32.4), since we have identically $\partial^2 \psi^{ikl}/\partial x^k \partial x^l = 0$. The total four-momentum of the system does not change, since according to (6.17) we can write

$$\int \frac{\partial \psi^{ikl}}{\partial x^l} dS_k = \frac{1}{2} \int \left(dS_k \frac{\partial \psi^{ikl}}{\partial x^l} - dS_l \frac{\partial \psi^{ikl}}{\partial x^k} \right) = \frac{1}{2} \int \psi^{ikl} df_{kl}^*,$$

where the integration on the right side of the equation is extended over the (ordinary) surface which "bounds" the hypersurface over which the integration on the left is taken. This surface is clearly located at infinity in the three-dimensional space, and since neither field nor particles are present at infinity this integral is zero. Thus the four-momentum of the system is, as it must be, a uniquely determined quantity. To define the tensor T^{ik} uniquely we can use the requirement that the four-tensor of angular momentum (see § 14) of the system be expressed in terms of the four-momentum by

$$M^{ik} = \int (x^i dP^k - x^k dP^i) = \frac{1}{c} \int (x^i T^{kl} - x^k T^{il}) dS_l, \qquad (32.8)$$

that is its "density" is expressed in terms of the "density" of momentum by the usual formula.

It is easy to determine what conditions the energy-momentum tensor must satisfy in order that this be valid. We note that the law of conservation of angular momentum can be expressed, as we already know, by setting equal to zero the divergence of the expression under the integral sign in M^{ik}. Thus

$$\frac{\partial}{\partial x^l} (x^i T^{kl} - x^k T^{il}) = 0. \qquad (32.9)$$

Noting that $\partial x^i/\partial x^l = \delta_l^i$ and that $\partial T^{kl}/\partial x^l = 0$, we find from this

$$\delta_l^i T^{kl} - \delta_l^k T^{il} = T^{ki} - T^{ik} = 0$$

or

$$T^{ik} = T^{ki}, \qquad (32.10)$$

that is, the energy-momentum tensor must be symmetric.

We note that T^{ik}, defined by formula (32.5), is generally speaking not symmetric, but can be made so by transformation (32.7) with suitable ψ^{ikl}. Later on (§ 94) we shall see that there is a direct method for obtaining a symmetric tensor T^{ik}.

As we mentioned above, if we carry out the integration in (32.6) over the hyperplane x^0 = const., then $\mathbf{P}^i$ takes on the form

$$P^i = \frac{1}{c} \int T^{i0} dV, \qquad (32.11)$$

where the integration extends over the whole (three-dimensional) space. The space components of P^i form the three-dimensional momentum vector of the system and the time component is its energy multiplied by $1/c$. Thus the vector with components

$$\frac{1}{c} T^{10}, \quad \frac{1}{c} T^{20}, \quad \frac{1}{c} T^{30}$$

may be called the "*momentum density*", and the quantity

$$W = T^{00}$$

the *"energy density"*.

To clarify the meaning of the remaining components of T^{ik}, we separate the conservation equation (32.4) into space and time parts:

$$\frac{1}{c}\frac{\partial T^{00}}{\partial t} + \frac{\partial T^{0\alpha}}{\partial x^{\alpha}} = 0, \qquad \frac{1}{c}\frac{\partial T^{\alpha 0}}{\partial t} + \frac{\partial T^{\alpha\beta}}{\partial x^{\beta}} = 0. \tag{32.12}$$

We integrate these equations over a volume V in space. From the first equation

$$\frac{1}{c}\frac{\partial}{\partial t}\int T^{00} dV + \int \frac{\partial T^{0\alpha}}{\partial x^{\alpha}} dV = 0$$

or, transforming the second integral by Gauss' theorem,

$$\frac{\partial}{\partial t}\int T^{00} dV = -c\oint T^{0\alpha} df_{\alpha}, \tag{32.13}$$

where the integral on the right is taken over the surface surrounding the volume V (df_x, df_y, df_z are the components of the three-vector of the surface element $d\mathbf{f}$). The expression on the left is the rate of change of the energy contained in the volume V; from this it is clear that the expression on the right is the amount of energy transferred across the boundary of the volume V, and the vector $\mathbf{S}$ with components

$$cT^{01}, \ cT^{02}, \ cT^{03}$$

is its flux density—the amount of energy passing through unit surface in unit time. Thus we arrive at the important conclusion that the requirements of relativistic invariance, as expressed by the tensor character of the quantities T^{ik}, automatically lead to a definite connection between the energy flux and the momentum density: the energy flux density is equal to the momentum density multiplied by c^2.

From the second equation in (32.12) we find similarly:

$$\frac{\partial}{\partial t}\int \frac{1}{c}T^{\alpha 0} dV = -\oint T^{\alpha\beta} df_{\beta}. \tag{32.14}$$

On the left is the change of the momentum of the system in volume V per unit time, therefore $\oint T^{\alpha\beta} df_{\beta}$ is the momentum emerging from the volume V per unit time. Thus the components $T^{\alpha\beta}$ of the energy-momentum tensor constitute the three-dimensional tensor of momentum flux density; we denote it by $-\sigma_{\alpha\beta}$, where $\sigma_{\alpha\beta}$ is the stress tensor. The energy flux density is a vector; the density of flux of momentum, which is itself a vector, must obviously be a tensor (the component $T_{\alpha\beta}$ of this tensor is the amount of the α-component of the momentum passing per unit time through unit surface perpendicular to the x^{β} axis).

We give a table indicating the meanings of the individual components of the energy-momentum tensor:

$$T_d^{ik} = \begin{bmatrix} W & S_x/c & S_y/c & S_z/c \\ S_x/c & -\sigma_{xx} & -\sigma_{xy} & -\sigma_{xz} \\ S_y/c & -\sigma_{yx} & -\sigma_{yy} & -\sigma_{yz} \\ S_z/c & -\sigma_{zx} & -\sigma_{zy} & -\sigma_{zz} \end{bmatrix} \tag{32.15}$$

§ 33. Energy-momentum tensor of the electromagnetic field

We now apply the general relations obtained in the previous section to the electromagnetic field. For the electromagnetic field, the quantity standing under the integral sign in (32.1) is equal, according to (27.4), to

$$\Lambda = -\frac{1}{16\pi} F_{kl} F^{kl}.$$

The quantities q are the components of the four-potential of the field, A_k, so that the definition (32.5) of the tensor T_i^k becomes

$$T_i^{\,k} = \frac{\partial A_l}{\partial x^i} \frac{\partial \Lambda}{\partial \left(\dfrac{\partial A_l}{\partial x^k} \right)} - \delta_i^k \Lambda.$$

To calculate the derivatives of Λ which appear here, we find the variation $\delta\Lambda$. We have

$$\delta\Lambda = -\frac{1}{8\pi} F^{kl} \delta F_{kl} = -\frac{1}{8\pi} F^{kl} \left(\delta \frac{\partial A_l}{\partial x^k} - \delta \frac{\partial A_k}{\partial x^l} \right)$$

or, interchanging indices and making use of the fact that $F_{kl} = - F_{lk}$,

$$\delta\Lambda = -\frac{1}{4\pi} F^{kl} \delta \frac{\partial A_l}{\partial x^k}.$$

From this we see that

$$\frac{\partial \Lambda}{\partial \left(\dfrac{\partial A_l}{\partial x^k} \right)} = -\frac{1}{4\pi} F^{kl},$$

and therefore

$$T_i^{\,k} = -\frac{1}{4\pi} \frac{\partial A_l}{\partial x^i} F^{kl} + \frac{1}{16\pi} \delta_i^k F_{lm} F^{lm},$$

or, for the contravariant components:

$$T^{ik} = -\frac{1}{4\pi} \frac{\partial A^l}{\partial x_i} F_l^{\,k} + \frac{1}{16\pi} g^{ik} F_{lm} F^{lm}.$$

But this tensor is not symmetric. To symmetrize it we add the quantity

$$\frac{1}{4\pi} \frac{\partial A^i}{\partial x_l} F_l^{\,k}.$$

According to the field equation (30.2) in the absence of charges, $\partial F_l^{\,k} / \partial x_l = 0$, and therefore

$$\frac{1}{4\pi} \frac{\partial A^i}{\partial x_l} F_l^{\,k} = \frac{1}{4\pi} \frac{\partial}{\partial x^l} (A^i F^{kl}),$$

so that the change made in T^{ik} is of the form (32.7) and is admissible. Since $\partial A^l / \partial x_i$ –

$\partial A^i / \partial x_l = F^{il}$, we get finally the following expression for the energy-momentum tensor of the electromagnetic field:

$$T^{ik} = \frac{1}{4\pi}\left(-F^{il}F^k_l + \frac{1}{4}g^{ik}F_{lm}F^{lm}\right). \tag{33.1}$$

This tensor is obviously symmetric. In addition it has the property that

$$T^i_i = 0, \tag{33.2}$$

i.e. the sum of its diagonal terms is zero.

Let us express the components of the tensor T^{ik} in terms of the electric and magnetic field intensities. By using the values (23.5) for the components F^{ik}, we easily verify that the quantity T^{00} coincides with the energy density (31.5), while the components $cT^{0\alpha}$ are the same as the components of the Poynting vector (31.2). The space components $T^{\alpha\beta}$ form a three-dimensional tensor with components

$$-\sigma_{xx} = \frac{1}{8\pi}(E_y^2 + E_z^2 - E_x^2 + H_y^2 + H_z^2 - H_x^2),$$

$$-\sigma_{xy} = -\frac{1}{4\pi}(E_x E_y + H_x H_y),$$

etc., or

$$\sigma_{\alpha\beta} = \frac{1}{4\pi}\left\{+ E_\alpha E_\beta + H_\alpha H_\beta - \frac{1}{2}\delta_{\alpha\beta}(E^2 + H^2)\right\}. \tag{33.3}$$

This tensor is called the *Maxwell strees tensor.*

To bring the tensor T_{ik} to diagonal form, we must transform to a reference system in which the vectors **E** and **H** (at the given point in space and moment in time) are parallel to one another or where one of them is equal to zero; as we know (§ 25), such a transformation is always possible except when **E** and **H** are mutually perpendicular and equal in magnitude. It is easy to see that after the transformation the only non-zero components of T^{ik} will be

$$T^{00} = -T^{11} = T^{22} = T^{33} = W$$

(the x axis has been taken along the direction of the field).

But if the vectors **E** and **H** are mutually perpendicular and equal in magnitude, the tensor T^{ik} cannot be brought to diagonal form.† The non-zero components in this case are

$$T^{00} = T^{33} = T^{30} = W$$

(where the x axis is taken along the direction of **E** and the y axis along **H**).

Up to now we have considered fields in the absence of charges. When charged particles are present, the energy-momentum tensor of the whole system is the sum of the energy-momentum tensors for the electromagnetic field and for the particles, where in the latter the particles are assumed not to interact with one another.

To determine the form of the energy-momentum tensor of the particles we must describe their mass distribution in space by using a "mass density" in the same way as we describe

---

† The fact that the reduction of the symmetric tensor T^{ik} to principal axes may be impossible is related to the fact that the four-space is pseudo-euclidean. (See also the problem in § 94.)

a distribution of point charges in terms of their density. Analogously to formula (28.1) for the charge density, we can write the mass density in the form

$$\mu = \sum_a m_a \, \delta(\mathbf{r} - \mathbf{r}_a), \tag{33.4}$$

where $\mathbf{r}_a$ are the radius-vectors of the particles, and the summation extends over all the particles of the system.

The "four-momentum density" of the particles is given by $\mu c u_i$. We know that this density is the component $T^{0\alpha}/c$ of the energy-momentum tensor, i.e. $T^{0x} = \mu c^2 u^\alpha (\alpha = 1, 2, 3)$. But the mass density is the time component of the four-vector $\mu/c(dx^k/dt)$ (in analogy to the charge density; see § 28). Therefore the energy-momentum tensor of the system of non-interacting particles is

$$T^{ik} = \mu c \frac{dx^i}{ds} \frac{dx^k}{dt} = \mu c \, u^i \, u^k \frac{ds}{dt}. \tag{33.5}$$

As expected, this tensor is symmetric.

We verify by a direct computation that the energy and momentum of the system, defined as the sum of the energies and momenta of field and particles, are actually conserved. In other words we shall verify the equation

$$\frac{\partial}{\partial x_k}(T^{(f)k}_i + T^{(p)k}_i) = 0, \tag{33.6}$$

which expresses these conservation laws.

Differentiating (33.1), we write

$$\frac{\partial T^{(f)k}_i}{\partial x^k} = \frac{1}{4\pi}\left(\frac{1}{2} F^{lm} \frac{\partial F_{lm}}{\partial x^i} - \frac{\partial F_{il}}{\partial x^k} F^{kl} - \frac{\partial F^{kl}}{\partial x^k} F_{il} \right).$$

Substituting from the Maxwell equations (26.5) and (30.2),

$$\frac{\partial F^{kl}}{\partial x^k} = \frac{4\pi}{c} j^l, \qquad \frac{\partial F_{lm}}{\partial x^i} = -\frac{\partial F_{mi}}{\partial x^l} - \frac{\partial F_{il}}{\partial x^m},$$

we have:

$$\frac{\partial T^{(f)k}_i}{\partial x^k} = \frac{1}{4\pi}\left(-\frac{1}{2}\frac{\partial F_{mi}}{\partial x^l} F^{lm} - \frac{1}{2}\frac{\partial F_{il}}{\partial x^m} F^{lm} - \frac{\partial F_{il}}{\partial x^k} F^{kl} - \frac{4\pi}{c} F_{il} j^l \right).$$

By permuting the indices, we easily show that the first three terms on the right cancel one another, and we arrive at the result:

$$\frac{\partial T^{(f)k}_i}{\partial x_k} = -\frac{1}{c} F_{ik} j^k. \tag{33.7}$$

Differentiating the expression (33.5) for the energy-momentum tensor of the particles gives

$$\frac{\partial T^{(p)k}_i}{\partial x^k} = cu_i \frac{\partial}{\partial x^k}\left(\mu \frac{dx^k}{dt} \right) + \mu c \frac{dx^k}{dt} \frac{\partial u_i}{\partial x^k}.$$

The first term in this expression is zero because of the conservation of mass for non-interacting particles. In fact, the quantities $\mu(dx^k/dt)$ constitute the "mass current" four-vector, analogous to the charge current four-vector (28.2); the conservation of mass is expressed by equating to zero the divergence of this four-vector:

$$\frac{\partial}{\partial x^k}\left(\mu\,\frac{dx^k}{dt}\right) = 0, \tag{33.8}$$

just as the conservation of charge is expressed by equation (29.4).

Thus we have:

$$\frac{\partial T^{(p)k}_i}{\partial x^k} = \mu c\,\frac{dx^k}{dt}\,\frac{\partial u_i}{\partial x^k} = \mu c\,\frac{du_i}{dt}.$$

Next we use the equation of motion of the charges in the field, expressed in the four-dimensional form (23.4).

$$mc\,\frac{du_i}{ds} = \frac{e}{c}\,F_{ik}u^k.$$

Changing to continuous distributions of charge and mass, we have, from the definitions of the densities μ and ϱ: $\mu/m = \varrho/e$. We can therefore write the equation of motion in the form

$$\mu c\,\frac{du_i}{ds} = \frac{\varrho}{c}\,F_{ik}u^k$$

or

$$\mu c\,\frac{du_i}{dt} = \frac{1}{c}\,F_{ik}\,\varrho\,u^k\,\frac{ds}{dt} = \frac{1}{c}\,F_{ik}\,j^k.$$

Thus,

$$\frac{\partial T^{(p)k}_i}{\partial x^k} = \frac{1}{c}\,F_{ik}\,j^k. \tag{33.9}$$

Combining this with (33.7), we find that we actually get zero, i.e. we arrive at equation (33.6).

PROBLEM

Find the law of transformation of the energy density, the energy flux density, and the components of the stress tensor under a Lorentz transformation.

Solution: Suppose that the K′ coordinate system moves relative to the K system along the x axis with velocity V. Applying the formulas of problem 1, § 6 to the symmetric tensor T^{ik}, we find:

$$W = \frac{1}{1 - \dfrac{V^2}{c^2}}\left(W' + \frac{V}{c^2}\,S'_x - \frac{V^2}{c^2}\,\sigma'_{xx}\right),$$

$$S_x = \frac{1}{1 - \dfrac{V^2}{c^2}}\left[\left(1 + \frac{V^2}{c^2}\right)S'_x - VW' - V\sigma'_{xx}\right],$$

$$S_y = \frac{1}{\sqrt{1 - \dfrac{V^2}{c^2}}} (S_y' - V\sigma_{xy}'),$$

$$\sigma_{xx} = \frac{1}{1 - \dfrac{V^2}{c^2}} \left(\sigma_{xx}' - 2\frac{V}{c^2} S_x' - \frac{V^2}{c^2} W' \right),$$

$$\sigma_{yy} = \sigma_{yy}', \quad \sigma_{zz} = \sigma_{zz}', \quad \sigma_{yz} = \sigma_{yz}',$$

$$\sigma_{xy} = \frac{1}{\sqrt{1 - \dfrac{V^2}{c^2}}} \left(\sigma_{xy}' - \frac{V}{c^2} S_y' \right)$$

and similar formulas for S_z and σ_{xz}.

§ 34. The virial theorem

Since the sum of the diagonal terms of the energy-momentum tensor of the electromagnetic field is equal to zero, the sum T_i^i for any system of interacting particles reduces to the trace of the energy-momentum tensor for the particles alone. Using (33.5), we therefore have:

$$T_i^i = T^{(p)i}_i = \mu c u_i u^i \frac{ds}{dt} = \mu c \frac{ds}{dt} = \mu c^2 \sqrt{1 - \frac{v^2}{c^2}}.$$

Let us rewrite this result, shifting to a summation over the particles, i.e. writing μ as the sum (33.4). We then get finally:

$$T_i^i = \sum_a m_a c^2 \sqrt{1 - \frac{v_a^2}{c^2}} \, \delta(\mathbf{r} - \mathbf{r}_a). \tag{34.1}$$

We note that, according to this formula, we have for every system:

$$T_i^i \geq 0, \tag{34.2}$$

where the equality sign holds only for the electromagnetic field without charges.

Let us consider a closed system of charged particles carrying out a finite motion, in which all the quantities (coordinates, momenta) characterizing the system vary over finite ranges.†

We average the equation

$$\frac{1}{c} \frac{\partial T^{\alpha 0}}{\partial t} + \frac{\partial T^{\alpha \beta}}{\partial x^\beta} = 0$$

[see (32.11)] with respect to the time. The average of the derivative $\partial T^{\alpha 0}/\partial t$, like the average of the derivative of any bounded quantity, is zero.‡ Therefore we get

---

† Here we also assume that the electromagnetic field of the system goes to zero sufficiently rapidly at infinity. In specific cases this condition may require the neglect of radiation of electromagnetic waves by the system.

‡ Let $f(t)$ be such a quantity. Then the average value of the derivative df/dt over a certain time interval T is

$$\frac{\overline{df}}{dt} = \frac{1}{T} \int_0^T \frac{df}{dt} dt = \frac{f(T) - f(0)}{T}.$$

Since $f(t)$ varies only within finite limits, then as T increases without limit, the average value of df/dt clearly goes to zero.

$$\frac{\partial}{\partial x^\beta} \overline{T_\alpha^\beta} = 0.$$

We multiply this equation by x^α and integrate over all space. We transform the integral by Gauss' theorem, keeping in mind that at infinity $T_\alpha^\beta = 0$, and so the surface integral vanishes:

$$\int x^\alpha \frac{\overline{\partial T_\alpha^\beta}}{\partial x^\beta} dV = -\int \frac{\partial x^\alpha}{\partial x^\beta} \overline{T_\alpha^\beta} \, dV = -\int \delta_\beta^\alpha \overline{T_\alpha^\beta} \, dV = 0,$$

or finally,

$$\int \overline{T_\alpha^\alpha} \, dV = 0. \tag{34.3}$$

On the basis of this equality we can write for the integral of $\overline{T_i^i} = \overline{T_\alpha^\alpha} + \overline{T_0^0}$:

$$\int \overline{T_i^i} \, dV = \int \overline{T_0^0} \, dV = \mathcal{E},$$

where $\mathcal{E}$ is the total energy of the system.

Finally, substituting (34.1) we get:

$$\mathcal{E} = \sum_a m_a c^2 \sqrt{1 - \frac{v_a^2}{c^2}}. \tag{34.4}$$

This relation is the relativistic generalization of the *virial theorem* of classical mechanics. (See *Mechanics*, § 10.) For low velocities, it becomes

$$\mathcal{E} - \sum_a m_a c^2 = -\sum_a \frac{\overline{m_a v_a^2}}{2},$$

that is, the total energy (minus the rest energy) is equal to the negative of the average value of the kinetic energy—in agreement with the result given by the classical virial theorem for a system of charged particles (interacting according to the Coulomb law).

We must point out that our formulas have a quite formal character and need to be made more precise. The point is that the electromagnetic field energy contains terms that give an infinite contribution to the electromagnetic self-energy of point charges (see § 37). To give meaning to the corresponding expressions we should omit these terms, considering that the intrinsic electromagnetic energy is already included in the kinetic energy of the particle (9.4). This means that we should "renormalize" the energy making the replacement

$$\mathcal{E} \to \mathcal{E} - \sum_a \int \frac{\mathbf{E}_a^2 + \mathbf{H}_a^2}{8\pi} dV$$

in (34.4), where $\mathbf{E}_a$ and $\mathbf{H}_a$ are the fields produced by the a'th particle. Similarly in (34.3) we should make the replacement†

$$\int T_a^a \, dV \to \int T_a^a \, dV + \sum_a \frac{\mathbf{E}_a^2 + \mathbf{H}_a^2}{8\pi} dV.$$

---

† Note that without this change the expression $-\int T_a^a \, dV = \int \frac{\mathbf{E}_a^2 + \mathbf{H}_a^2}{8\pi} dV + \sum_a \frac{m_a v_a^2}{\sqrt{1 - v_a^2/c^2}}$ is essentially positive and cannot vanish.

§ 35. The energy-momentum tensor for macroscopic bodies

In addition to the energy-momentum tensor for a system of point particles (33.5), we shall also need the expression for this tensor for macroscopic bodies which are treated as being continuous.

The flux of momentum through the element $d\mathbf{f}$ of the surface of the body is just the force acting on this surface element. Therefore $-\sigma_{\alpha\beta}\, df_\beta$ is the α-component of the force acting on the element. Now we introduce a reference system in which a given element of volume of the body is at rest. In such a reference system, Pascal's law is valid, that is, the pressure p applied to a given portion of the body is transmitted equally in all directions and is everywhere perpendicular to the surface on which it acts.† Therefore we can write $\sigma_{\alpha\beta}\, df_\beta = -p df_\alpha$, so that the stress tensor is $\sigma_{\alpha\beta} = -p\delta_{\alpha\beta}$. As for the components $T^{\alpha 0}$, which represent the momentum density, they are equal to zero for the given volume element in the reference system we are using. The component T^{00} is as always the energy density of the body, which we denote by ε; ε/c^2 is then the mass density of the body, i.e. the mass per unit volume. We emphasize that we are talking here about the unit "proper" volume, that is, the volume in the reference system in which the given portion of the body is at rest.

Thus, in the reference system under consideration, the energy-momentum tensor (for the given portion of the body) has the form:

$$T^{ik} = \begin{pmatrix} \varepsilon & 0 & 0 & 0 \\ 0 & p & 0 & 0 \\ 0 & 0 & p & 0 \\ 0 & 0 & 0 & p \end{pmatrix} \tag{35.1}$$

Now it is easy to find the expression for the energy-momentum tensor in an arbitrary reference system. To do this we introduce the four-velocity u^i for the macroscopic motion of an element of volume of the body. In the rest frame of the particular element, $u^i = (1, 0)$. The expression for T^{ik} must be chosen so that in this reference system it takes on the form (35.1). It is easy to verify that this is

$$T^{ik} = (p + \varepsilon)u^i u^k - pg^{ik}, \tag{35.2}$$

or, for the mixed components,

$$T_i^k = (p + \varepsilon)u_i u^k - p\delta_i^k.$$

This expression gives the energy-momentum tensor for a macroscopic body. The expressions for the energy density W, energy flow vector $\mathbf{S}$ and stress tensor $\sigma_{\alpha\beta}$ are:

$$W = \frac{\varepsilon + p\dfrac{v^2}{c^2}}{1 - \dfrac{v^2}{c^2}}, \qquad \mathbf{S} = \frac{(p + \varepsilon)\mathbf{v}}{1 - \dfrac{v^2}{c^2}}, \tag{35.3}$$

---

† Strictly speaking, Pascal's law is valid for liquids and gases. However, for solid bodies the maximum possible difference in the stress in different directions is negligible in comparison with the stresses which can play a role in the theory of relativity, so that its consideration is of no interest.

$$\sigma_{\alpha\beta} = -\frac{(p + \varepsilon)v_\alpha v_\beta}{c^2\left(1 - \dfrac{v^2}{c^2}\right)} - p\delta_{\alpha\beta}.$$

If the velocity v of the macroscopic motion is small compared with the velocity of light, then we have approximately:

$$\mathbf{S} = (p + \varepsilon)\mathbf{v}.$$

Since S/c^2 is the momentum density, we see that in this case the sum $(p + \varepsilon)/c^2$ plays the role of the mass density of the body.

The expression for T^{ik} simplifies in the case where the velocities of all the particles making up the body are small compared with the velocity of light (the velocity of the macroscopic motion itself can be arbitrary). In this case we can neglect, in the energy density ε, all terms small compared with the rest energy, that is, we can write $\mu_0 c^2$ in place of ε, where μ_0 is the sum of the masses of the particles present in unit (proper) volume of the body (we emphasize that in the general case, μ_0 must differ from the actual mass density ε/c^2 of the body, which includes also the mass corresponding to the energy of microscopic motion of the particles in the body and the energy of their interactions). As for the pressure determined by the energy of microscopic motion of the molecules, in the case under consideration it is also clearly small compared with the rest energy $\mu_0 c^2$. Thus we find

$$T^{ik} = \mu_0 c^2 u^i u^k. \tag{35.4}$$

From the expression (35.2), we get

$$T_i^i = \varepsilon - 3p. \tag{35.5}$$

The general property (34.2) of the energy-momentum tensor of an arbitrary system now shows that the following inequality is always valid for the pressure and density of a macroscopic body:

$$p < \frac{\varepsilon}{3}. \tag{35.6}$$

Let us compare the relation (35.5) with the general formula (34.1) which we saw was valid for an arbitrary system. Since we are at present considering a macroscopic body, the expression (34.1) must be averaged over all the values of $\mathbf{r}$ in unit volume. We obtain the result

$$\varepsilon - 3p = \sum_a m_a c^2 \sqrt{1 - \frac{v_a^2}{c^2}} \tag{35.7}$$

(the summation extends over all particles in unit volume).

The right side of this equation tends to zero in the ultrarelativistic limit, so in this limit the equation of state of matter is: †

$$p = \frac{\varepsilon}{3}. \tag{35.8}$$

---

† This limiting equation of state is obtained here assuming an electromagnetic interaction between the particles. We shall assume (when this is needed in Chapter 14) that it remains valid for the other possible interactions between particles, though there is at present no proof of this assumption.

We apply our formula to an ideal gas, which we assume to consist of identical particles. Since the particles of an ideal gas do not interact with one another, we can use formula (33.5) after averaging it. Thus for an ideal gas,

$$T^{ik} = nmc \, \overline{\frac{dx^i}{dt} \cdot \frac{dx^k}{ds}},$$

where n is the number of particles in unit volume and the dash means an average over all the particles. If there is no macroscopic motion in the gas then we can use for T^{ik} the expression (35.1). Comparing the two formulas, we arrive at the equations:

$$\varepsilon = nm \, \overline{\left(\frac{c^2}{\sqrt{1 - \frac{v^2}{c^2}}} \right)}, \qquad p = \frac{nm}{3} \, \overline{\left(\frac{v^2}{\sqrt{1 - \frac{v^2}{c^2}}} \right)}. \qquad (35.9)$$

These equations determine the density and pressure of a relativistic ideal gas in terms of the velocity of its particles; the second of these replaces the well-known formula $p = nm\overline{v^2}/3$ of the nonrelativistic kinetic theory of gases.

CHAPTER 5

CONSTANT ELECTROMAGNETIC FIELDS

§ 36. Coulomb's law

For a constant electric, or as it is usually called, *electrostatic* field, the Maxwell equations have the form:

$$\text{div } \mathbf{E} = 4\pi\rho, \tag{36.1}$$

$$\text{curl } \mathbf{E} = 0. \tag{36.2}$$

The electric field $\mathbf{E}$ is expressed in terms of the scalar potential alone by the relation

$$\mathbf{E} = -\text{ grad } \phi. \tag{36.3}$$

Substituting (36.3) in (36.1), we get the equation which is satisfied by the potential of a constant electric field:

$$\Delta\phi = -4\pi\rho. \tag{36.4}$$

This equation is called the *Poisson equation*. In particular, in vacuum, i.e., for $\varrho = 0$, the potential satisfies the *Laplace equation*

$$\Delta\phi = 0. \tag{36.5}$$

From the last equation it follows, in particular, that the potential of the electric field can nowhere have a maximum or a minimum. For in order that ϕ have an extreme value, it would be necessary that the first derivatives of ϕ with respect to the coordinates be zero, and that the second derivatives $\partial^2\phi/\partial x^2$, $\partial^2\phi/\partial y^2$, $\partial^2\phi/\partial z^2$ all have the same sign. The last is impossible, since in that case (36.5) could not be satisfied.

We now determine the field produced by a point charge. From symmetry considerations, it is clear that it is directed along the radius-vector from the point at which the charge e is located. From the same consideration it is clear that the value E of the field depends only on the distance R from the charge. To find this absolute value, we apply equation (36.1) in the integral form (30.5). The flux of the electric field through a spherical surface of radius R circumscribed around the charge e is equal to $4\pi R^2 E$; this flux must equal $4\pi e$. From this we get

$$E = \frac{e}{R^2}.$$

In vector notation:

$$E = \frac{e\mathbf{R}}{R^3}. \tag{36.6}$$

Thus the field produced by a point charge is inversely proportional to the square of the distance from the charge. This is the *Coulomb law*. The potential of this field is, clearly,

$$\phi = \frac{e}{R}. \tag{36.7}$$

If we have a system of charges, then the field produced by this system is equal, according to the principle of superposition, to the sum of the fields produced by each of the particles individually. In particular, the potential of such a field is

$$\phi = \sum_a \frac{e_a}{R_a},$$

where R_a is the distance from the charge e_a to the point at which we are determining the potential. If we introduce the charge density ϱ, this formula takes on the form

$$\phi = \int \frac{\varrho}{R} dV, \tag{36.8}$$

where R is the distance from the volume element dV to the given point of the field.

We note a mathematical relation which is obtained from (36.4) by substituting the values of ϱ and ϕ for a point charge, i.e. $\varrho = e\delta(\mathbf{R})$ and $\phi = e/R$. We then find

$$\Delta\left(\frac{1}{R}\right) = -4\pi\delta(\mathbf{R}). \tag{36.9}$$

§ 37. Electrostatic energy of charges

We determine the energy of a system of charges. We start from the enegy of the field, that is, from the expression (31.5) for the energy density. Namely, the energy of the system of charges must be equal to

$$U = \frac{1}{8\pi} \int E^2 \, dV,$$

where $\mathbf{E}$ is the field produced by these charges, and the integral goes over all space. Substituting $\mathbf{E} = -$ grad ϕ, U can be changed to the following form:

$$U = -\frac{1}{8\pi} \int \mathbf{E} \cdot \text{grad } \phi \, dV = -\frac{1}{8\pi} \int \text{div} (\mathbf{E}\phi) \, dV + \frac{1}{8\pi} \int \phi \, \text{div } \mathbf{E} \, dV.$$

According to Gauss' theorem, the first integral is equal to the integral of $\mathbf{E}\phi$ over the surface bounding the volume of integration, but since the integral is taken over all space and since the field is zero at infinity, this integral vanishes. Substituting in the second integral, div $\mathbf{E}$ $= 4\pi\varrho$, we find the following expression for the energy of a system of charges:

$$U = \frac{1}{2} \int \varrho\phi \, dV. \tag{37.1}$$

For a system of point charges, e_a, we can write in place of the integral a sum over the charges

$$U = \frac{1}{2}\sum_a e_a\phi_a, \tag{37.2}$$

where ϕ_a is the potential of the field produced by all the charges, at the point where the charge e_a is located.

If we apply our formula to a single elementary charged particle (say, an electron), and the field which the charge itself produces, we arrive at the result that the charge must have a certain "self"-potential energy equal to $e\phi/2$, where ϕ is the potential of the field produced by the charge at the point where it is located. But we know that in the theory of relativity every elementary particle must be considered as pointlike. The potential $\phi = e/R$ of its field becomes infinite at the point $R = 0$. Thus according to electrodynamics, the electron would have to have an infinite "self-energy", and consequently also an infinite mass. The physical absurdity of this result shows that the basic principles of electrodynamics itself lead to the result that its application must be restricted to definite limits.

We note that in view of the infinity obtained from electrodynamics for the self-energy and mass, it is impossible within the framework of classical electrodynamics itself to pose the question whether the total mass of the electron is electrodynamic (that is, associated with the electromagnetic self-energy of the particle).†

Since the occurrence of the physically meaningless infinite self-energy of the elementary particle is related to the fact that such a particle must be considered as pointlike, we can conclude that electrodynamics as a logically closed physical theory presents internal contradictions when we go to sufficiently small distances. We can pose the question as to the order of magnitude of such distances. We can answer this question by noting that for the electromagnetic self-energy of the electron we should obtain a value of the order of the rest energy mc^2. If, on the other hand, we consider an electron as possessing a certain radius R_0, then its self-potential energy would be of order e^2/R_0. From the requirement that these two quantities be of the same order, $e^2/R_0 \sim mc^2$, we find

$$R_0 \sim \frac{e^2}{mc^2}. \tag{37.3}$$

This dimension (the "radius" of the electron) determines the limit of applicability of electrodynamics to the electron, and follows already from its fundamental principles. We must, however, keep in mind that actually the limits of applicability of the classical electrodynamics which is presented here lie must higher, because of the occurrence of quantum phenomena.‡

We now turn again to formula (37.2). The potentials ϕ_a which appear there are equal, from Coulomb's law, to

$$\phi_a = \Sigma \, \frac{e_b}{R_{ab}}, \tag{37.4}$$

where R_{ab} is the distance between the charges e_a, e_b. The expression for the energy (37.2) consists of two parts. First, it contains an infinite constant, the self-energy of the charges, not depending on their mutual separations. The second part is the energy of interaction of the charges, depending on their separations. Only this part has physical interest. It is equal to

$$U' = \tfrac{1}{2} \Sigma \, e_a \phi_a', \tag{37.5}$$

---

† From the purely formal point of view, the finiteness of the electron mass can be handled by introducing an infinite negative mass of nonelectromagnetic origin which compensates the infinity of the electromagnetic mass (mass "renormalization"). However, we shall see later (§ 75) that this does not eliminate all the internal contradictions of classical electrodynamics.

‡ Quantum effects become important for distances of the order of $\hbar/mc$, where $\hbar$ is Planck's constant. The ratio of these distances to R_0 is of order $\hbar c/e^2 \sim 137$.

where

$$\phi'_a = \sum_{b \neq a} \frac{e_b}{R_{ab}} \qquad (37.6)$$

is the potential at the point of location of e_a, produced by all the charges other than e_a. In other words, we can write

$$U' = \frac{1}{2} \sum_{a \neq b} \frac{e_a e_b}{R_{ab}}. \qquad (37.7)$$

In particular, the energy of interaction of two charges is

$$U' = \frac{e_1 e_2}{R_{12}}. \qquad (37.8)$$

§ 38. The field of a uniformly moving charge

We determine the field produced by a charge e, moving uniformly with velocity V. We call the laboratory frame the system K; the system of reference moving with the charge is the K' system. Let the charge be located at the origin of coordinates of the K' system. The system K' moves relative to K along the X axis; the axes Y and Z are parallel to Y' and Z'. At the time $t = 0$ the origins of the two systems coincide. The coordinates of the charge in the K system are consequently $x = Vt$, $y = z = 0$. In the K' system, we have a constant electric field with vector potential $\mathbf{A}' = 0$, and scalar potential equal to $\phi' = e/R'$, where $R'^2 = x'^2 + y'^2 + z'^2$. In the K system, according to (24.1) for $\mathbf{A}' = 0$,

$$\phi = \frac{\phi'}{\sqrt{1 - \dfrac{V^2}{c^2}}} = \frac{e}{R'\sqrt{1 - \dfrac{V^2}{c^2}}}. \qquad (38.1)$$

We must now express R' in terms of the coordinates x, y, z, in the K system. According to the formulas for the Lorentz transformation

$$x' = \frac{x - Vt}{\sqrt{1 - \dfrac{V^2}{c^2}}}, \qquad y' = y, \qquad z' = z,$$

from which

$$R'^2 = \frac{(x - Vt)^2 + \left(1 - \dfrac{V^2}{c^2}\right)(y^2 + z^2)}{1 - \dfrac{V^2}{c^2}}. \qquad (38.2)$$

Substituting this in (38.1) we find

$$\phi = \frac{e}{R^*} \qquad (38.3)$$

where we have introduced the notation

$$R^{*2} = (x - Vt)^2 + \left(1 - \frac{V^2}{c^2}\right)(y^2 + z^2).\tag{38.4}$$

The vector potential in the K system is equal to

$$\mathbf{A} = \phi\frac{\mathbf{V}}{c} = \frac{e\mathbf{V}}{cR^*}.\tag{38.5}$$

In the K' system the magnetic field $\mathbf{H}'$ is absent and the electric field is

$$\mathbf{E}' = \frac{e\mathbf{R}'}{R'^3}.$$

From formula (24.2), we find

$$E_x = E'_x = \frac{ex'}{R'^3}, \quad E_y = \frac{E'_y}{\sqrt{1 - \frac{V^2}{c^2}}} = \frac{ey'}{R'^3\sqrt{1 - \frac{V^2}{c^2}}},$$

$$E_z = \frac{ez'}{R'^3\sqrt{1 - \frac{V^2}{c^2}}}.$$

Substituting for R', x', y', z', their expressions in terms of x, y, z, we obtain

$$\mathbf{E} = \left(1 - \frac{V^2}{c^2}\right)\frac{e\mathbf{R}}{R^{*3}},\tag{38.6}$$

where $\mathbf{R}$ is the radius vector from the charge e to the field point with coordinates x, y, z (its components are $x - Vt$, y, z).

This expression for $\mathbf{E}$ can be written in another form by introducing the angle θ between the direction of motion and the radius vector $\mathbf{R}$. It is clear that $y^2 + z^2 = R^2 \sin^2\theta$, and therefore R^{*2} can be written in the form:

$$R^{*2} = R^2\left(1 - \frac{V^2}{c^2}\sin^2\theta\right).\tag{38.7}$$

Then we have for $\mathbf{E}$,

$$\mathbf{E} = \frac{e\mathbf{R}}{R^3}\frac{1 - \frac{V^2}{c^2}}{\left(1 - \frac{V^2}{c^2}\sin^2\theta\right)^{3/2}}.\tag{38.8}$$

For a fixed distance R from the charge, the value of the field E increases as θ increases from 0 to $\pi/2$ (or as θ decreases from π to $\pi/2$). The field along the direction of motion ($\theta = 0, \pi$) has the smallest value; it is equal to

$$E_{\parallel} = \frac{e}{R^2}\left(1 - \frac{V^2}{c^2}\right).$$

The largest field is that perpendicular to the velocity ($\theta = \pi/2$), equal to

$$E_\perp = \frac{e}{R^2} \frac{1}{\sqrt{1 - \dfrac{V^2}{c^2}}}.$$

We note that as the velocity increases, the field $E_\parallel$ decreases, while $E_\perp$ increases. We can describe this pictorially by saying that the electric field of a moving charge is "contracted" in the direction of motion. For velocities V close to the velocity of light, the denominator in formula (38.8) is close to zero in a narrow interval of values θ around the value $\theta = \pi/2$. The "width" of this interval is, in order of magnitude,

$$\Delta\theta \sim \sqrt{1 - \frac{V^2}{c^2}}.$$

Thus the electric field of a rapidly moving charge at a given distance from it is large only in a narrow range of angles in the neighbourhood of the equatorial plane, and the width of this interval decreases with increasing V like $\sqrt{1 - (V^2/c^2)}$.

The magnetic field in the K system is

$$\mathbf{H} = \frac{1}{c} \mathbf{V} \times \mathbf{E} \qquad\qquad (38.9)$$

[see (24.5)]. In particular, for $V \ll c$ the electric field is given approximately by the usual formula for the Coulomb law, $\mathbf{E} = e\mathbf{R}/R^3$, and the magnetic field is

$$\mathbf{H} = \frac{e}{c} \frac{\mathbf{V} \times \mathbf{R}}{R^3}, \qquad\qquad (38.10)$$

PROBLEM

Determine the force (in the K system) between two charges moving with the same velocity $\mathbf{V}$.

Solution: We shall determine the force $\mathbf{F}$ by computing the force acting on one of the charges (e_1) in the field produced by the other (e_2). Using (38.9), we have

$$\mathbf{F} = e_1 \mathbf{E}_2 + \frac{e_1}{c} \mathbf{V} \times \mathbf{H}_2 = e_1 \left(1 - \frac{V^2}{c^2}\right) \mathbf{E}_2 + \frac{e_1}{c^2} \mathbf{V}(\mathbf{V} \cdot \mathbf{E}_2).$$

Substituting for $\mathbf{E}_2$ from (38.8), we get for the components of the force in the direction of motion (F_x) and perpendicular to it (F_y):

$$F_x = \frac{e_1 e_2}{R^2} \frac{\left(1 - \dfrac{V^2}{c^2}\right)\cos\theta}{\left(1 - \dfrac{V^2}{c^2}\sin^2\theta\right)^{3/2}}, \qquad F_y = \frac{e_1 e_2}{R^2} \frac{\left(1 - \dfrac{V^2}{c^2}\right)^2 \sin\theta}{\left(1 - \dfrac{V^2}{c^2}\sin^2\theta\right)^{3/2}},$$

where $\mathbf{R}$ is the radius vector from e_2 to e_1, and θ is the angle between $\mathbf{R}$ and $\mathbf{V}$.

§ 39. Motion in the Coulomb field

We consider the motion of a particle with mass m and charge e in the field produced by a second charge e'; we assume that the mass of this second charge is so large that it can be considered as fixed. Then our problem becomes the study of the motion of a charge e in a centrally symmetric electric field with potential $\phi = e'/r$.

The total energy $\mathscr{E}$ of the particle is equal to

$$\mathscr{E} = c\sqrt{p^2 + m^2 c^2} + \frac{\alpha}{r},$$

where $\alpha = ee'$. If we use polar coordinates in the plane of motion of the particle, then as we know from mechanics,

$$p^2 = (M^2/r^2) + p_r^2,$$

where p_r is the radial component of the momentum, and M is the constant angular momentum of the particle. Then

$$\mathscr{E} = c\sqrt{p_r^2 + \frac{M^2}{r^2} + m^2 c^2} + \frac{\alpha}{r}. \tag{39.1}$$

We discuss the question whether the particle during its motion can approach arbitrarily close to the centre. First of all, it is clear that this is never possible if the charges e and e' repel each other, that is, if e and e' have the same sign. Furthermore, in the case of attraction (e and e' of opposite sign), arbitrarily close approach to the centre is not possible if $Mc > |\alpha|$, for in this case the first term in (39.1) is always large than the second, and for $r \to 0$ the right side of the equation would approach infinity. On the other hand, if $Mc < |\alpha|$, then as $r \to 0$, this expression can remain finite (here it is understood that p_r approaches infinity). Thus, if

$$cM < |\alpha|, \tag{39.2}$$

the particle during its motion "falls in" toward the charge attracting it, in contrast to non-relativistic mechanics, where for the Coulomb field such a collapse is generally impossible (with the exception of the one case $M = 0$, where the particle e moves on a line toward the particle e').

A complete determination of the motion of a charge in a Coulomb field starts most conveniently from the Hamilton–Jacobi equation. We choose polar coordinates r, ϕ, in the plane of the motion. The Hamilton–Jacobi equation (16.11) has the form

$$-\frac{1}{c^2}\left(\frac{\partial S}{\partial t} + \frac{\alpha}{r}\right)^2 + \left(\frac{\partial S}{\partial r}\right)^2 + \frac{1}{r^2}\left(\frac{\partial S}{\partial \phi}\right)^2 + m^2 c^2 = 0.$$

We seek an S of the form

$$S = -\mathscr{E}t + M\phi + f(r),$$

where $\mathscr{E}$ and M are the constant energy and angular momentum of the moving particle. The result is

$$S = -\mathscr{E}t + M\phi + \int \sqrt{\frac{1}{c^2}\left(\mathscr{E} - \frac{\alpha}{r}\right)^2 - \frac{M^2}{r^2} - m^2 c^2}\; dr. \tag{39.3}$$

The trajectory is determined by the equation $\partial S/\partial M = \text{const}$. Integration of (39.3) leads to the following results for the trajectory:

(a) If $Mc > |\alpha|$,

$$(c^2 M^2 - \alpha^2)\frac{1}{r} = c\sqrt{(M\mathscr{E})^2 - m^2 c^2 (M^2 c^2 - \alpha^2)} \, \cos\left(\phi\sqrt{1 - \frac{\alpha^2}{c^2 M^2}}\right) - \mathscr{E}\alpha. \quad (39.4)$$

(b) If $Mc < |\alpha|$,

$$(\alpha^2 - M^2 c^2)\frac{1}{r} = \pm c\sqrt{(M\mathscr{E})^2 + m^2 c^2 (\alpha^2 - M^2 c^2)} \, \cosh\left(\phi\sqrt{\frac{\alpha^2}{c^2 M^2} - 1}\right) + \mathscr{E}\alpha.$$

$$(39.5)$$

(c) If $Mc = |\alpha|$,

$$\frac{2\mathscr{E}\alpha}{r} = \mathscr{E}^2 - m^2 c^4 - \phi^2 \left(\frac{\mathscr{E}\alpha}{cM}\right)^2. \quad (39.6)$$

The integration constant is contained in the arbitrary choice of the reference line for measurement of the angle ϕ.

In (39.4) the ambiguity of sign in front of the square root is unimportant, since it already contains the arbitrary reference origin of the angle ϕ under the cos. In the case of attraction ($\alpha < 0$) the trajectory corresponding to this equation lies entirely at finite values of r (finite motion), if $\mathscr{E} < mc^2$. If $\mathscr{E} > mc^2$, then r can go to infinity (infinite motion). The finite motion corresponds to motion in a closed orbit (ellipse) in nonrelativistic mechanics. From (39.4) it is clear that in relativistic mechanics the trajectory can never be closed; when the angle ϕ changes by 2π, the distance r from the centre does not return to its initial value. In place of ellipses we here get orbits in the form of open "rosettes". Thus, whereas in nonrelativistic mechanics the finite motion in a Coulomb field leads to a closed orbit, in relativistic mechanics the Coulomb field loses this property.

In (39.5) we must choose the positive sign for the root in case $\alpha < 0$, and the negative sign if $\alpha > 0$ [the opposite choice of sign would correspond to a reversal of the sign of the root in (39.1)].

For $\alpha < 0$ the trajectories (39.5) and (39.6) are spirals in which the distance r approaches 0 as $\phi \to \infty$. The time required for the "falling in" of the charge to the coordinate origin is finite. This can be verified by noting that the dependence of the coordinate r on the time is determined by the equation $\partial S/\partial\mathscr{E} = \text{const}$; substituting (39.3), we see that the time is determined by an integral which converges for $r \to 0$.

PROBLEMS

1. Determine the angle of deflection of a charge passing through a repulsive Coulomb field ($a > 0$).

Solution: The angle of deflection χ equals $\chi = \pi - 2\phi_0$, where $2\phi_0$ is the angle between the two asymptotes of the trajectory (39.4). We find

$$\chi = \pi - \frac{2cM}{\sqrt{c^2 M^2 - a^2}} \tan^{-1}\left(\frac{v\sqrt{c^2 M^2 - a^2}}{ca}\right),$$

where v is the velocity of the charge at infinity.

2. Determine the effective scattering cross section at small angles for the scattering of particles in a Coulomb field.

Solution: The effective cross section $d\sigma$ is the ratio of the number of particles scattered per second into a given element do of solid angle to the flux density of impinging particles (i.e., to the number of particles crossing one square centimetre, per second, of a surface perpendicular to the beam of particles).

Since the angle of deflection χ of the particle during its passage through the field is determined by the *impact parameter* ϱ (i.e. the distance from the centre to the line along which the particle would move in the absence of the field),

$$d\sigma = 2\pi\varrho d\varrho = 2\pi\varrho \frac{d\varrho}{d\chi} d\chi = \varrho \frac{d\varrho}{d\chi} \frac{do}{\sin\chi},$$

where $do = 2\pi \sin \chi d\chi$.† The angle of deflection (for small angles) can be taken equal to the ratio of the change in momentum to its initial value. The change in momentum is equal to the time integral of the force acting on the charge, in the direction perpendicular to the direction of motion; it is approximately $(a/r^2) \cdot (\varrho/r)$. Thus we have

$$\chi = \frac{1}{p} \int_{-\infty}^{+\infty} \frac{a\varrho \, dt}{(\varrho^2 + v^2 t^2)^{3/2}} = \frac{2a}{p\varrho v}$$

(v is the velocity of the particles). From this we find the effective cross section for small χ:

$$d\sigma = 4 \left(\frac{a}{pv} \right)^2 \frac{do}{\chi^4}.$$

In the nonrelativistic case, $p \cong mv$, and the expression coincides with the one obtained from the Rutherford formula‡ for small χ.

§ 40. The dipole moment

We consider the field produced by a system of charges at large distances, that is, at distances large compared with the dimensions of the system.

We introduce a coordinate system with origin anywhere within the system of charges. Let the radius vectors of the various charges be $\mathbf{r}_a$. The potential of the field produced by all the charges at the point having the radius vector $\mathbf{R}_0$ is

$$\phi = \sum_a \frac{e_a}{|\mathbf{R}_0 - \mathbf{r}_a|} \tag{40.1}$$

(the summation goes over all charges); here $\mathbf{R}_0 - \mathbf{r}_a$ are the radius vectors from the charges e_a to the point where we are finding the potential.

We must investigate this expansion for large $\mathbf{R}_0$ ($\mathbf{R}_0 \gg \mathbf{r}_a$). To do this, we expand it in powers or $\mathbf{r}_a/\mathbf{R}_0$, using the formula

$$f(\mathbf{R}_0 - \mathbf{r}) = f(\mathbf{R}_0) - \mathbf{r} \cdot \operatorname{grad} f(\mathbf{R}_0)$$

(in the grad, the differentiation applies to the coordinates of the vector $\mathbf{R}_0$). To terms of first order,

$$\phi = \frac{\sum e_a}{R_0} - \sum e_a \mathbf{r}_a \cdot \operatorname{grad} \frac{1}{R_0}. \tag{40.2}$$

---

† See *Mechanics*, § 18.
‡ See *Mechanics*, § 19.

The sum

$$d = \sum e_a r_a \qquad (40.3)$$

is called the *dipole moment* of the system of charges. It is important to note that if the sum of all the charges, $\sum e_a$, is zero, then the dipole moment does not depend on the choice of the origin of coordinates, for the radius vectors r_a and r_a' of one and the same charge in two different coordinate systems are related by

$$r_a' = r_a + a,$$

where a is some constant vector. Therefore if $\sum e_a = 0$, the dipole moment is the same in both systems:

$$d' = \sum e_a r_a' = \sum e_a r_a + a \sum e_a = d.$$

If we denote by e_a^+, r_a^+ and e_a^-, r_a^- the positive and negative charges of the system and their radius vectors, then we can write the dipole moment in the form

$$d = \sum e_a^+ r_a^+ - \sum e_a^- r_a^- = R_a^+ \sum e_a^+ - R_a^- \sum e_a^- \qquad (40.4)$$

where

$$R^+ \frac{\sum e_a^+ r_a^+}{\sum e_a^+}, \qquad R^- = \frac{\sum e_a^- r_a^-}{\sum e_a^-} \qquad (40.5)$$

are the radius vectors of the "charge centres" for the positive and negative charges. If $\sum e_a^+ = \sum e_a^- = e$, then

$$d = e R_{+-}, \qquad (40.6)$$

where $R_{+-} = R^+ - R^-$ is the radius vector from the centre of negative to the centre of positive charge. In particular, if we have altogether two charges, then R_{+-} is the radius vector between them.

If the total charge of the system is zero, then the potential of the field of this system at large distances is

$$\phi = -d \cdot \nabla \frac{1}{R_0} = \frac{d \cdot R_0}{R_0^3}. \qquad (40.7)$$

The field intensity is:

$$E = -\operatorname{grad} \frac{d \cdot R_0}{R_0^3} = -\frac{1}{R_0^3} \operatorname{grad}(d \cdot R_0) - (d \cdot R_0) \operatorname{grad} \frac{1}{R_0^3},$$

or finally,

$$E = \frac{3(n \cdot d)n - d}{R_0^3}, \qquad (40.8)$$

where n is a unit vector along R_0. Another useful expression for the field is

$$E = (d \cdot \nabla)\nabla \frac{1}{R_0}, \qquad (40.9)$$

Thus the potential of the field at large distances produced by a system of charges with total charge equal to zero is inversely proportional to the square of the distance, and the field intensity is inversely proportional to the cube of the distance. This field has axial symmetry around the direction of $\mathbf{d}$. In a plane passing through this direction (which we choose as the z axis), the components of the vector $\mathbf{E}$ are:

$$E_z = d\frac{3\cos^2\theta - 1}{R_0^3}, \quad E_x = d\frac{3\sin\theta\cos\theta}{R_0^3}. \tag{40.10}$$

The radial and tangential components in this plane are

$$E_R = d\frac{2\cos\theta}{R_0^3}, \quad E_\theta = -d\frac{\sin\theta}{R_0^3}. \tag{40.11}$$

§ 41. Multipole moments

In the expansion of the potential in powers of $1/R_0$,

$$\phi = \phi^{(0)} + \phi^{(1)} + \phi^{(2)} + \ldots, \tag{41.1}$$

the term $\phi^{(n)}$ is proportional to $1/R_0^{n+1}$. We saw that the first term, $\phi^{(0)}$, is determined by the sum of all the charges; the second term, $\phi^{(1)}$, sometimes called the dipole potential of the system, is determined by the dipole moment of the system.

The third term in the expansion is

$$\phi^{(2)} = \frac{1}{2}\sum ex_\alpha x_\beta \frac{\partial^2}{\partial X_\alpha \partial X_\beta}\left(\frac{1}{R_0}\right), \tag{41.2}$$

where the sum goes over all charges; we here drop the index numbering the charges; x_α are the components of the vector $\mathbf{r}$, and X_α those of the vector $\mathbf{R}_0$. This part of the potential is usually called the *quadrupole potential*. If the sum of the charges and the dipole moment of the system are both equal to zero, the expansion begins with $\phi^{(2)}$.

In the expression (41.2) there enter the six quantities $\sum ex_\alpha x_\beta$. However, it is easy to see that the field depends not on six independent quantities, but only on five. This follows from the fact that the function $1/R_0$ satisfies the Laplace equation, that is,

$$\Delta\left(\frac{1}{R_0}\right) = \delta_{\alpha\beta}\frac{\partial^2}{\partial X_\alpha \partial X_\beta}\left(\frac{1}{R_0}\right) = 0.$$

We can therefore write $\phi^{(2)}$ in the form

$$\phi^{(2)} = \frac{1}{2}\sum e\left(x_\alpha x_\beta - \frac{1}{3}r^2\delta_{\alpha\beta}\right)\frac{\partial^2}{\partial X_\alpha \partial X_\beta}\left(\frac{1}{R_0}\right).$$

The tensor

$$D_{\alpha\beta} = \sum e(3x_\alpha x_\beta - r^2\delta_{\alpha\beta}) \tag{41.3}$$

is called the *quadrupole moment* of the system. From the definition of $D_{\alpha\beta}$ it is clear that the sum of its diagonal elements is zero:

$$D_{\alpha\alpha} = 0. \tag{41.4}$$

Therefore the symmetric tensor $D_{\alpha\beta}$ has altogether five independent components. With the aid of $D_{\alpha\beta}$, we can write

$$\phi^{(2)} = \frac{D_{\alpha\beta}}{6} \frac{\partial^2}{\partial X_\alpha \partial X_\beta} \left(\frac{1}{R_0} \right), \tag{41.5}$$

or, performing the differentiation,

$$\frac{\partial^2}{\partial X_\alpha \partial X_\beta} \frac{1}{R_0} = \frac{3 X_\alpha X_\beta}{R_0^5} - \frac{\delta_{\alpha\beta}}{R_0^3},$$

and using the fact that $\delta_{\alpha\beta} D_{\alpha\beta} = D_{\alpha\alpha} = 0$,

$$\phi^{(2)} = \frac{D_{\alpha\beta} n_\alpha n_\beta}{2 R_0^3}. \tag{41.6}$$

Like every symmetric three-dimensional tensor, the tensor $D_{\alpha\beta}$ can be brought to principal axes. Because of (41.4), in general only two of the three principal values will be independent. If it happens that the system of charges is symmetric around some axis (the z axis)† then this axis must be one of the principal axes of the tensor $D_{\alpha\beta}$, the location of the other two axes in the x, y plane is arbitrary, and the three principal values are related to one another:

$$D_{xx} = D_{yy} = -\frac{1}{2} D_{zz}. \tag{41.7}$$

Denoting the component D_{zz} by D (in this case it is simply called the quadrupole moment), we get for the potential

$$\phi^{(2)} = \frac{D}{4 R_0^3} (3 \cos^2 \theta - 1) = \frac{D}{2 R_0^3} P_2 (\cos \theta), \tag{41.8}$$

where θ is the angle between $\mathbf{R}_0$ and the z axis, and P_2 is a Legendre polynomial.

Just as we did for the dipole moment in the preceding section, we can easily show that the quadrupole moment of a system does not depend on the choice of the coordinate origin, if both the total charge and the dipole moment of the system are equal to zero.

In similar fashion we could also write the succeeding terms of the expansion (41.1). The l'th term of the expansion defines a tensor (which is called the tensor of the 2^l-pole moment) of rank l, symmetric in all its indices and vanishing when contracted on any pair of indices; it can be shown that such a tensor has $2l + 1$ independent components.‡

We shall express the general term in the expansion of the potential in another form, by using the well-known formula of the theory of spherical harmonics

$$\frac{1}{|\mathbf{R}_0 - \mathbf{r}|} = \frac{1}{\sqrt{R_0^2 + r^2 - 2 r R_0 \cos \chi}} = \sum_{l=0}^{\infty} \frac{r^l}{R_0^{l+1}} P_l (\cos \chi), \tag{41.9}$$

where χ is the angle between $\mathbf{R}_0$ and $\mathbf{r}$. We introduce the spherical angles Θ, Φ and θ, ϕ, formed by the vectors $\mathbf{R}_0$ and $\mathbf{r}$, respectively, with the fixed coordinate axes, and use the addition theorem for the spherical harmonics:

---

† We are assuming a symmetry axis of any order higher than the second.

‡ Such a tensor is said to be *irreducible*. The vanishing on contraction means that no tensor of lower rank can be formed from the components.

$$P_l(\cos \chi) = \sum_{m=-l}^{l} \frac{(l - |m|)!}{(l + |m|)!} P_l^{|m|}(\cos \Theta) P_l^{|m|}(\cos \theta) e^{-im(\Phi - \phi)}, \tag{41.10}$$

where the P_l^m are the associated Legendre polynomials.

We also introduce the spherical functions†

$$Y_{lm}(\theta, \phi) = (-1)^{m} i^{l} \sqrt{\frac{2l + 1(l - m)!}{4\pi(l + m)!}} P_l^m(\cos \theta) e^{im\phi}, \quad m \geq 0,$$

$$Y_{l,-|m|}(\theta, \phi) = (-1)^{l-m} Y_{l,|m|}^{*}. \tag{41.11}$$

Then the expansion (41.9) takes the form:

$$\frac{1}{|\mathbf{R}_0 - \mathbf{r}|} = \sum_{l=0}^{\infty} \sum_{m=-l}^{l} \frac{r^l}{R_0^{l+1}} \frac{4\pi}{2l + 1} Y_{lm}^{*}(\Theta, \Phi) Y_{lm}(\theta, \Phi).$$

Carrying out this expansion in each term of (40.1), we finally get the following expression for the l'th term of the expansion of the potential:

$$\phi^{(l)} = \frac{1}{R_0^{l+1}} \sum_{m=-l}^{l} \sqrt{\frac{4\pi}{2l + 1}} Q_m^{(l)} Y_{lm}^{*}(\Theta, \Phi), \tag{41.12}$$

where

$$Q_m^{(l)} = \sum_a e_a r_a^l \sqrt{\frac{4\pi}{2l + 1}} Y_{lm}(\theta_a, \phi_a). \tag{41.13}$$

The set of $2l + 1$ quantities $Q_m^{(l)}$ form the 2^l-pole moment of the system of charges.

The quantities $Q_m^{(1)}$ defined in this way are related to the components of the dipole moment vector **d** by the formulas

$$Q_0^{(1)} = id_z, \quad Q_{\pm 1}^{(1)} = \mp \frac{i}{\sqrt{2}}(d_x \pm id_y). \tag{41.14}$$

The quantities $Q_m^{(2)}$ are related to the tensor components $D_{\alpha\beta}$ by the relations

$$Q_0^{(2)} = -\frac{1}{2}D_{zz}, \quad Q_{\pm 1}^{(2)} = \pm \frac{1}{\sqrt{6}}(D_{xz} \pm iD_{yz}),$$

$$Q_{\pm 2}^{(2)} = -\frac{1}{2\sqrt{6}}(D_{xx} - D_{yy} \pm 2iD_{xy}). \tag{41.15}$$

PROBLEM

Determine the quadrupole moment of a uniformly charged ellipsoid with respect to its centre.

Solution: Replacing the summation in (41.3) by an integration over the volume of the ellipsoid, we have:

$$D_{xx} = \rho \iiint (2x^2 - y^2 - z^2) dx \, dy \, dz, \quad \text{etc.}$$

† In accordance with the definition used to quantum mechanics.

Let us choose the coordinate axes along the axes of the ellipsoid with the origin at its centre; from symmetry considerations it is obvious that these axes are the principal axes of the tensor $D_{\alpha\beta}$. By means of the transformation

$$x = x'a, \quad y = y'b, \quad z = z'c$$

the integration over the volume of the ellipsoid

$$\frac{x^2}{a^2} + \frac{y^2}{b^2} + \frac{z^2}{c^2} = 1$$

is reduced to integration over the volume of the unit sphere

$$x'^2 + y'^2 + z'^2 = 1.$$

As a result we obtain:

$$D_{xx} = \frac{e}{5}(2a^2 - b^2 - c^2), \quad D_{yy} = \frac{e}{5}(2b^2 - a^2 - c^2),$$

$$D_{zz} = \frac{e}{5}(2c^2 - a^2 - b^2),$$

where $e = (4\pi/3)abc\varrho$ is the total charge of the ellipsoid.

§ 42. System of charges in an external field

We now consider a system of charges located in an external electric field. We designate the potential of this external field by $\phi(\mathbf{r})$. The potential energy of each of the charges is $e_a\phi(\mathbf{r}_a)$, and the total potential energy of the system is

$$U = \sum_a e_a \phi(\mathbf{r}_a). \tag{42.1}$$

We introduce another coordinate system with its origin anywhere within the system of charges; $\mathbf{r}_a$ is the radius vector of the charge e_a in these coordinates.

Let us assume that the external field changes slowly over the region of the system of charges, i.e. is quasiuniform with respect to the system. Then we can expand the energy U in powers of $\mathbf{r}_a$:

$$U = U^{(0)} + U^{(1)} + U^{(2)} + \ldots; \tag{42.2}$$

in this expansion the first term is

$$U^{(0)} = \phi_0 \sum e_a, \tag{42.3}$$

where ϕ_0 is the value of the potential at the origin. In this approximation, the energy of the system is the same as it would be if all the charges were located at one point (the origin).

The second term in the expansion is

$$U^{(1)} = (\operatorname{grad} \phi)_0 \cdot \sum e_a \mathbf{r}_a$$

Introducing the field intensity $\mathbf{E}_0$ at the origin and the dipole moment $\mathbf{d}$ of the system, we have

$$U^{(1)} = -\mathbf{d} \cdot \mathbf{E}_0. \tag{42.4}$$

The total force acting on the system in the external quasiuniform field is, to the order we are considering,

$$\mathbf{F} = \mathbf{E}_0 \sum e_a + [\nabla(\mathbf{d} \cdot \mathbf{E})]_0.$$

If the total charge is zero, the first term vanishes, and

$$\mathbf{F} = (\mathbf{d} \cdot \nabla)\mathbf{E}, \tag{42.5}$$

i.e. the force is determined by the derivatives of the field intensity (taken at the origin). The total moment of the forces acting on the system is

$$\mathbf{K} = \sum (\mathbf{r}_a \times e_a \mathbf{E}_0) = \mathbf{d} \times \mathbf{E}_0, \tag{42.6}$$

i.e. to lowest order it is determined by the field intensity itself.

Let us assume that there are two systems, each having total charge zero, and with dipole moments $\mathbf{d}_1$ and $\mathbf{d}_2$, respectively. Their mutual distance is assumed to be large in comparison with their internal dimensions. Let us determine their potential energy of interaction, U. To do this we regard one of the systems as being in the field of the other. Then

$$U = - \mathbf{d}_2 \cdot \mathbf{E}_1.$$

where $\mathbf{E}_1$ is the field of the first system. Substituting (40.8) for $\mathbf{E}_1$, we find:

$$U = \frac{(\mathbf{d}_1 \cdot \mathbf{d}_2)R^2 - 3(\mathbf{d}_1 \cdot \mathbf{R})(\mathbf{d}_2 \cdot \mathbf{R})}{R^5}, \tag{42.7}$$

where $\mathbf{R}$ is the vector separation between the two systems.

For the case where one of the systems has a total charge different from zero (and equal to e), we obtain similarly

$$U = e \frac{\mathbf{d} \cdot \mathbf{R}}{R^3}, \tag{42.8}$$

where $\mathbf{R}$ is the vector directed from the dipole to the charge.

The next term in the expansion (42.1) is

$$U^{(2)} = \frac{1}{2} \sum e x_\alpha x_\beta \frac{\partial^2 \phi_0}{\partial x_\alpha \partial x_\beta}.$$

Here, as in § 41, we omit the index numbering the charge; the value of the second derivative of the potential is taken at the origin; but the potential ϕ satisfies Laplace's equation,

$$\frac{\partial^2 \phi}{\partial x_\alpha^2} = \delta_{\alpha\beta} \frac{\partial^2 \phi}{\partial x_\alpha \partial x_\beta} = 0.$$

Therefore we can write

$$U^{(2)} = \frac{1}{2} \frac{\partial^2 \phi_0}{\partial x_\alpha \partial x_\beta} \sum e \left(x_\alpha x_\beta - \frac{1}{3} \delta_{\alpha\beta} r^2 \right)$$

or, finally,

$$U^{(2)} = \frac{D_{\alpha\beta}}{6} \frac{\partial^2 \phi_0}{\partial x_\alpha \partial x_\beta}. \tag{42.9}$$

The general term in the series (42.2) can be expressed in terms of the 2^l-pole moments $D_m^{(l)}$ defined in the preceding section. To do this, we first expand the potential $\phi(\mathbf{r})$ in spherical harmonics; the general form of this expansion is

$$\phi(\mathbf{r}) = \sum_{l=0}^{\infty} r^l \sum_{m=-l}^{l} \sqrt{\frac{4\pi}{2l+1}} \, a_{lm} Y_{lm}(\theta, \phi), \tag{42.10}$$

where r, θ, ϕ are the spherical coordinates of a point and the a_{lm} are constants. Forming the sum (42.1) and using the definition (41.13), we obtain:

$$U^{(l)} = \sum_{m=-l}^{l} a_{lm} Q_m^{(l)}. \tag{42.11}$$

§ 43. Constant magnetic field

Let us consider the magnetic field produced by charges which perform a finite motion, in which the particles are always within a finite region of space and the momenta also always remain finite. Such a motion has a "stationary" character, and it is of interest to consider the time average magnetic field $\overline{\mathbf{H}}$, produced by the charges; this field will now be a function only of the coordinates and not of the time, that is, it will be constant.

In order to find equations for the average magnetic field $\overline{\mathbf{H}}$, we take the time average of the Maxwell equations

$$\operatorname{div} \mathbf{H} = 0, \quad \operatorname{curl} \mathbf{H} = \frac{1}{c} \frac{\partial \mathbf{E}}{\partial t} + \frac{4\pi}{c} \mathbf{j}.$$

The first of these gives simply

$$\operatorname{div} \overline{\mathbf{H}} = 0. \tag{43.1}$$

In the second equation the average value of the derivative $\partial \mathbf{E}/\partial t$, like the derivative of any quantity which varies over a finite range, is zero (cf. the footnote on p. 90). Therefore the second Maxwell equation becomes

$$\operatorname{curl} \overline{\mathbf{H}} = \frac{4\pi}{c} \overline{\mathbf{j}}. \tag{43.2}$$

These two equations determine the constant field $\overline{\mathbf{H}}$.

We introduce the average vector potential $\overline{\mathbf{A}}$ in accordance with

$$\operatorname{curl} \overline{\mathbf{A}} = \overline{\mathbf{H}}.$$

We substitute this in equation (43.2). We find

$$\operatorname{grad} \operatorname{div} \overline{\mathbf{A}} - \Delta \overline{\mathbf{A}} = \frac{4\pi}{c} \overline{\mathbf{j}}.$$

But we know that the vector potential of a field is not uniquely defined, and we can impose an arbitrary auxiliary condition on it. On this basis, we choose the potential $\overline{\mathbf{A}}$ so that

$$\operatorname{div} \overline{\mathbf{A}} = 0. \tag{43.3}$$

Then the equation defining the vector potential of the constant magnetic field becomes

$$\Delta \overline{\mathbf{A}} = -\frac{4\pi}{c} \overline{\mathbf{j}}. \tag{43.4}$$

It is easy to find the solution of this equation by noting that (43.4) is completely analogous to the Poisson equation (36.4) for the scalar potential of a constant electric field, where in place of the charge density ϱ we here have the current density $\bar{\mathbf{j}}/c$. By analogy with the solution (36.8) of the Poisson equation, we can write

$$\overline{\mathbf{A}} = \frac{1}{c} \int \frac{\bar{\mathbf{j}}}{R} \, dV, \tag{43.5}$$

where R is the distance from the field point to the volume element dV.

In formula (43.5) we can go over from the integral to a sum over the charges, by substituting in place of $\mathbf{j}$ the product $\varrho \mathbf{v}$, and recalling that all the charges are pointlike. In this we must keep in mind that in the integral (43.5), R is simply an integration variable, and is therefore not subject to the averaging process. If we write in place of the integral

$$\int \frac{\mathbf{j}}{R} \, dV, \quad \text{the sum} \quad \sum \frac{e_a \mathbf{v}_a}{R_a},$$

then R_a here are the radius vectors of the various particles, which change during the motion of the charges. Therefore we must write

$$\overline{\mathbf{A}} = \frac{1}{c} \sum \overline{\frac{e_a \mathbf{v}_a}{R_a}}, \tag{43.6}$$

where we average the whole expression under the summation sign.

Knowing $\overline{\mathbf{A}}$, we can also find the magnetic field,

$$\overline{\mathbf{H}} = \operatorname{curl} \overline{\mathbf{A}} = \operatorname{curl} \frac{1}{c} \int \frac{\bar{\mathbf{j}}}{R} \, dV.$$

The curl operator refers to the coordinates of the field point. Therefore the curl can be brought under the integral sign and $\bar{\mathbf{j}}$ can be treated as constant in the differentiation. Applying the well-known formula

$$\operatorname{curl} f\mathbf{a} = f \operatorname{curl} \mathbf{a} + \operatorname{grad} f \times \mathbf{a},$$

where f and $\mathbf{a}$ are an arbitrary scalar and vector, to the product $\bar{\mathbf{j}} \cdot 1/R$, we get

$$\operatorname{curl} \frac{\bar{\mathbf{j}}}{R} = \operatorname{grad} \frac{1}{R} \times \bar{\mathbf{j}} = \frac{\bar{\mathbf{j}} \times \mathbf{R}}{R^3},$$

and consequently,

$$\overline{\mathbf{H}} = \frac{1}{c} \int \frac{\bar{\mathbf{j}} \times \mathbf{R}}{R^3} \, dV \tag{43.7}$$

(the radius vector $\mathbf{R}$ is directed from dV to the field point). This is the *law of Biot and Savart*.

§ 44. Magnetic moments

Let us consider the average magnetic field produced by a system of charges in stationary motion, at large distances from the system.

We introduce a coordinate system with its origin anywhere within the system of charges, just as we did in § 40. Again we denote the radius vectors of the various charges by $\mathbf{r}_a$, and the radius vector of the point at which we calculate the field by $\mathbf{R}_0$. Then $\mathbf{R}_0 - \mathbf{r}_a$ is the radius vector from the charge e_a to the field point. According to (43.6), we have for the vector potential:

$$\overline{\mathbf{A}} = \frac{1}{c} \sum \frac{e_a \mathbf{v}_a}{|\mathbf{R}_0 - \mathbf{r}_a|}. \tag{44.1}$$

As in § 40, we expand this expression in powers of $\mathbf{r}_a$. To terms of first order (we omit the index a), we have

$$\overline{\mathbf{A}} = \frac{1}{cR_0} \sum e\overline{\mathbf{v}} - \frac{1}{c} \sum \overline{e\mathbf{v}\left(\mathbf{r} \cdot \nabla \frac{1}{R_0}\right)}.$$

In the first term we can write

$$\sum e\overline{\mathbf{v}} = \overline{\frac{d}{dt} \sum e\mathbf{r}}.$$

But the average value of the derivative of a quantity changing within a finite interval (like $\sum e\mathbf{r}$) is zero. Thus there remains for $\overline{\mathbf{A}}$ the expression

$$\overline{\mathbf{A}} = -\frac{1}{c} \sum \overline{e\mathbf{v}\left(\mathbf{r} \cdot \nabla \frac{1}{R_0}\right)} = \frac{1}{cR_0^3} \sum \overline{e\mathbf{v}(\mathbf{r} \cdot \mathbf{R}_0)}.$$

We transform this expression as follows. Noting that $\mathbf{v} = \dot{\mathbf{r}}$, we can write (remembering that $\mathbf{R}_0$ is a constant vector)

$$\sum e(\mathbf{R}_0 \cdot \mathbf{r})\mathbf{v} = \frac{1}{2}\frac{d}{dt} \sum e\mathbf{r}(\mathbf{r} \cdot \mathbf{R}_0) + \frac{1}{2} \sum e[\mathbf{v}(\mathbf{r} \cdot \mathbf{R}_0) - \mathbf{r}(\mathbf{v} \cdot \mathbf{R}_0)].$$

Upon substitution of this expression in $\overline{\mathbf{A}}$, the average of the first term (containing the time derivative) again goes to zero, and we get

$$\overline{\mathbf{A}} = \frac{1}{2cR_0^3} \sum \overline{e[\mathbf{v}(\mathbf{r} \cdot \mathbf{R}_0) - \mathbf{r}(\mathbf{v} \cdot \mathbf{R}_0)]}.$$

We introduce the vector

$$m = \frac{1}{2c} \sum e\mathbf{r} \times \mathbf{v}, \tag{44.2}$$

which is called the *magnetic moment* of the system. Then we get for $\overline{\mathbf{A}}$:

$$\overline{\mathbf{A}} = \frac{\overline{m} \times \mathbf{R}_0}{R_0^3} = \nabla \frac{1}{R_0} \times \overline{m} \tag{44.3}$$

Knowing the vector potential, it is easy to find the magnetic field. With the aid of the formula

$$\text{curl } (\mathbf{a} \times \mathbf{b}) = (\mathbf{b} \cdot \nabla)\mathbf{a} - (\mathbf{a} \cdot \nabla)\mathbf{b} + \mathbf{a} \text{ div } \mathbf{b} - \mathbf{b} \text{ div } \mathbf{a},$$

we find

$$\overline{\mathbf{H}} = \text{curl } \overline{\mathbf{A}} = \text{curl}\left(\frac{\overline{m} \times \mathbf{R}_0}{R_0^3}\right) = \overline{m} \text{ div } \frac{\mathbf{R}_0}{R_0^3} - (\overline{m} \cdot \nabla)\frac{\mathbf{R}_0}{R_0^3}.$$

Furthermore,

$$\text{div} \, \frac{\mathbf{R}_0}{R_0^3} = \mathbf{R}_0 \cdot \text{grad} \, \frac{1}{R_0^3} + \frac{1}{R_0^3} \, \text{div} \, \mathbf{R}_0 = 0$$

and

$$(\overline{m} \cdot \nabla) \frac{\mathbf{R}_0}{R_0^3} = \frac{1}{R_0^3} \, (\overline{m} \cdot \nabla) \mathbf{R}_0 + \mathbf{R}_0 (\overline{m} \quad \nabla) \frac{1}{R_0^3} = \frac{\overline{m}}{R_0^3} - \frac{3 \mathbf{R}_0 (\overline{m} \cdot \mathbf{R}_0)}{R_0^5}.$$

Thus,

$$\overline{\mathbf{H}} = \frac{3 \mathbf{n} (\overline{m} \cdot \mathbf{n}) - \overline{m}}{R_0^3}, \tag{44.4}$$

where $\mathbf{n}$ is again the unit vector along $\mathbf{R}_0$. We see that the magnetic field is expressed in terms of the magnetic moment by the same formula by which the electric field was expressed in terms of the dipole moment [see (40.8)].

If all the charges of the system have the same ratio of charge to mass, then we can write

$$\overline{m} = \frac{1}{2c} \cdot \Sigma \, e\mathbf{r} \times \mathbf{v} = \frac{e}{2mc} \Sigma \, \overline{m} \mathbf{r} \times \mathbf{v}.$$

If the velocities of all the charges $v \ll c$ then $m\mathbf{v}$ is the momentum $\mathbf{p}$ of the charge and we get

$$\overline{m} = \frac{e}{2mc} \Sigma \, \mathbf{r} \times \mathbf{p} = \frac{e}{2mc} \mathbf{M}, \tag{44.5}$$

where $\mathbf{M} = \Sigma \, \mathbf{r} \times \mathbf{p}$ is the mechanical angular momentum of the system. Thus in this case, the ratio of magnetic moment to the angular momentum is constant and equal to $e/2mc$.

PROBLEM

Find the ratio of the magnetic moment to the angular momentum for a system of two charges (velocities $v \ll c$).

Solution: Choosing the origin of coordinates as the centre of mass of the two particles we have $m_1 \mathbf{r}_1 + m_2 \mathbf{r}_2 = 0$ and $\mathbf{p}_1 = -\mathbf{p}_2 = \mathbf{p}$, where $\mathbf{p}$ is the momentum of the relative motion. With the aid of these relations, we find

$$\overline{m} = \frac{1}{2c} \left(\frac{e_1}{m_1^2} + \frac{e_2}{m_2^2} \right) \frac{m_1 m_2}{m_1 + m_2} \mathbf{M}.$$

§ 45. Larmor's theorem

Let us consider a system of charges in an external constant uniform magnetic field. The time average of the force acting on the system,

$$\overline{\mathbf{F}} = \Sigma \frac{e}{c} \overline{\mathbf{v} \times \mathbf{H}} = \frac{d}{dt} \Sigma \frac{e}{c} \mathbf{r} \times \mathbf{H},$$

is zero, as is the time average of the time derivative of any quantity which varies over a finite range. The average value of the moment of the forces is

$$\overline{\mathbf{K}} = \Sigma \frac{e}{c} \overline{(\mathbf{r} \times (\mathbf{v} \times \mathbf{H}))}$$

and is different from zero. It can be expressed in terms of the magnetic moment of the system, by expanding the vector triple product:

$$\mathbf{K} = \Sigma \frac{e}{c} \{\mathbf{v}(\mathbf{r} \cdot \mathbf{H}) - \mathbf{H}(\mathbf{v} \cdot \mathbf{r})\} = \Sigma \frac{e}{c} \left\{\mathbf{v}(\mathbf{r} \cdot \mathbf{H}) - \frac{1}{2}\mathbf{H}\frac{d}{dt}\mathbf{r}^2\right\}.$$

The second term gives zero after averaging, so that

$$\overline{\mathbf{K}} = \Sigma \frac{e}{c} \overline{\mathbf{v}(\mathbf{r} \cdot \mathbf{H})} = \frac{1}{2c} \Sigma e\{\overline{\mathbf{v}(\mathbf{r} \cdot \mathbf{H})} - \overline{\mathbf{r}(\mathbf{v} \cdot \mathbf{H})}\}$$

[the last transformation is analogous to the one used in deriving (44.3)], or finally

$$\overline{\mathbf{K}} = \overline{m} \times \mathbf{H}. \tag{45.1}$$

We call attention to the analogy with formula (42.6) for the electrical case.

The Lagrangian for a system of charges in an external constant uniform magnetic field contains (compared with the Lagrangian for a closed system) the additional term

$$L_H = \Sigma \frac{e}{c} \mathbf{A} \cdot \mathbf{v} = \Sigma \frac{e}{2c}(\mathbf{H} \times \mathbf{r}) \cdot \mathbf{v} = \Sigma \frac{e}{2c}(\mathbf{r} \times \mathbf{v}) \cdot \mathbf{H} \tag{45.2}$$

[where we have used the expression (19.4) for the vector potential of a uniform field]. Introducing the magnetic moment of the system, we have:

$$L_H = m \cdot \mathbf{H}. \tag{45.3}$$

We call attention to the analogy with the electric field; in a uniform electric field, the Lagrangian of a system of charges with total charge zero contains the term

$$L_E = \mathbf{d} \cdot \mathbf{E},$$

which in that case is the negative of the potential energy of the charge system (see § 42).

We now consider a system of charges performing a finite motion (with velocities $v \ll c$) in the centrally symmetric electric field produced by a certain fixed charge. We transform from the laboratory coordinate system to a system rotating uniformly around an axis passing through the fixed particle. From the well-known formula, the velocity $\mathbf{v}$ of the particle in the new coordinate system is related to its velocity $\mathbf{v}'$ in the old system by the relation

$$\mathbf{v}' = \mathbf{v} + \mathbf{\Omega} \times \mathbf{r},$$

where $\mathbf{r}$ is the radius vector of the particle and $\mathbf{\Omega}$ is the angular velocity of the rotating co-ordinate system. In the fixed system the Lagrangian of the system of charges is

$$L = \Sigma \frac{m\mathbf{v}'^2}{2} - U,$$

where U is the potential energy of the charges in the external field plus the energy of their mutual interactions. The quantity U is a function of the distances of the charges from the fixed particle and of their mutual separations; when transformed to the rotating system it obviously remains unchanged. Therefore in the new system the Lagrangian is

$$L = \Sigma \frac{m}{2}(\mathbf{v} + \mathbf{\Omega} \times \mathbf{r})^2 - U.$$

Let us assume that all the charges have the same charge-to-mass ratio e/m, and set

$$\boldsymbol{\Omega} = \frac{e}{2mc}\,\mathbf{H}. \tag{45.4}$$

Then for sufficiently small H (when we can neglect terms in H^2) the Lagrangian becomes:

$$L = \sum \frac{mv^2}{2} + \frac{1}{2c}\sum e\mathbf{H} \times \mathbf{r} \cdot \mathbf{v} - U.$$

We see that it coincides with the Lagrangian which would have described the motion of the charges in the laboratory system of coordinates in the presence of a constant magnetic field (see (45.2)).

Thus we arrive at the result that, in the nonrelativistic case, the behaviour of a system of charges all having the same e/m, performing a finite motion in a centrally symmetric electric field and in a weak uniform magnetic field $\mathbf{H}$, is equivalent to the behaviour of the same system of charges in the same electric field in a coordinate system rotating uniformly with the angular velocity (45.4). This assertion is the content of the *Larmor theorem,* and the angular velocity $\Omega = eH/2mc$ is called the *Larmor frequency.*

We can approach this same problem from a different point of view. If the magnetic field $\mathbf{H}$ is sufficiently weak, the Larmor frequency will be small compared to the frequencies of the finite motion of the system of charges. Then we may consider the averages, over times small compared to the period $2\pi/\Omega$, of quantities describing the system. These new quantities will vary slowly in time (with frequency Ω).

Let us consider the change in the average angular momentum $\mathbf{M}$ of the system. According to a well-known equation of mechanics, the derivative of $\mathbf{M}$ is equal to the moment $\mathbf{K}$ of the forces acting on the system. We therefore have, using (45.1):

$$\frac{d\overline{\mathbf{M}}}{dt} = \overline{\mathbf{K}} = \overline{\boldsymbol{m}} \times \mathbf{H}.$$

If the e/m ratio is the same for all particles of the system, the angular momentum and magnetic moment are proportional to one another, and we find by using formulas (44.5) and (45.4):

$$\frac{d\overline{\mathbf{M}}}{dt} = -\boldsymbol{\Omega} \times \overline{\mathbf{M}}. \tag{45.5}$$

This equation states that the vector $\overline{\mathbf{M}}$ (and with it the megnetic moment $\overline{\boldsymbol{m}}$) rotates with angular velocity $-\Omega$ around the direction of the field, while its absolute magnitude and the angle which it makes with this direction remain fixed. (This motion is called the *Larmor precession.*)

CHAPTER 6

ELECTROMAGNETIC WAVES

§ 46. The wave equation

The electromagnetic field in vacuum is determined by the Maxwell equations in which we must set $\rho = 0$, $\mathbf{j} = 0$. We write them once more:

$$\operatorname{curl} \mathbf{E} = -\frac{1}{c} \frac{\partial \mathbf{H}}{\partial t}, \qquad \operatorname{div} \mathbf{H} = 0, \tag{46.1}$$

$$\operatorname{curl} \mathbf{H} = \frac{1}{c} \frac{\partial \mathbf{E}}{\partial t}, \qquad \operatorname{div} \mathbf{E} = 0. \tag{46.2}$$

These equations possess nonzero solutions. This means that an electromagnetic field can exist even in the absence of any charges.

Electromagnetic fields occurring in vacuum in the absence of charges are called *electromagnetic waves*. We now take up the study of the properties of such waves.

First of all we note that such fields must necessarily be time-varying. In fact, in the contrary case, $\partial \mathbf{H}/\partial t = \partial \mathbf{E}/\partial t = 0$ and the equations (46.1) and (46.2) go over into the equations (36.1), (36.2) and (43.1), (43.2) of a constant field in which, however, we now have $\rho = 0$, $\mathbf{j} = 0$. But the solution of these equations which is given by formulas (36.8) and (43.5) becomes zero for $\rho = 0$, $\mathbf{j} = 0$.

We derive the equations determining the potentials of electromagnetic waves.

As we already know, because of the ambiguity in the potentials we can always subject them to an auxiliary condition. For this reason, we choose the potentials of the electromagnetic wave so that the scalar potential is zero:

$$\phi = 0. \tag{46.3}$$

Then

$$\mathbf{E} = -\frac{1}{c} \frac{\partial \mathbf{A}}{\partial t}, \qquad \mathbf{H} = \operatorname{curl} \mathbf{A}. \tag{46.4}$$

Substituting these two expressions in the first of equations (46.2), we get

$$\operatorname{curl} \operatorname{curl} \mathbf{A} = -\Delta \mathbf{A} + \operatorname{grad} \operatorname{div} \mathbf{A} = -\frac{1}{c^2} \frac{\partial^2 \mathbf{A}}{\partial t^2}. \tag{46.5}$$

Despite the fact that we have already imposed one auxiliary condition on the potentials, the potential $\mathbf{A}$ is still not completely unique. Namely, we can add to it the gradient of an arbitrary function which does not depend on the time (meantime leaving ϕ unchanged). In particular, we can choose the potentials of the electromagnetic wave so that

$$\text{div } \mathbf{A} = 0. \tag{46.6}$$

In fact, substituting for $\mathbf{E}$ from (46.4) in div $\mathbf{E} \doteq 0$, we have

$$\text{div } \frac{\partial \mathbf{A}}{\partial t} = \frac{\partial}{\partial t} \text{ div } \mathbf{A} = 0,$$

that is, div $\mathbf{A}$ is a function only of the coordinates. This function can always be made zero by adding to $\mathbf{A}$ the gradient of a suitable time-independent function.

The equation (46.5) now becomes

$$\Delta \mathbf{A} - \frac{1}{c^2} \frac{\partial^2 \mathbf{A}}{\partial t^2} = 0. \tag{46.7}$$

This is the equation which determines the potentials of electromagnetic waves. It is called the *d'Alembert equation*, or the *wave equation*.†

Applying to (46.7) the operators curl and $\partial/\partial t$, we can verify that the electric and magnetic fields $\mathbf{E}$ and $\mathbf{H}$ satisfy the same wave equation.

We repeat the derivation of the wave equation in four-dimensional form. We write the second pair of Maxwell equations for the field in the absence of charges in the form

$$\frac{\partial F^{ik}}{\partial x^k} = 0$$

(This is equation (30.2) with $j^i = 0$.) Substituting F^{ik}, expressed in terms of the potentials,

$$F^{ik} = \frac{\partial A^k}{\partial x_i} - \frac{\partial A^i}{\partial x_k},$$

we get

$$\frac{\partial^2 A^k}{\partial x_i \partial x^k} - \frac{\partial^2 A^i}{\partial x_k \partial x^k} = 0. \tag{46.8}$$

We impose on the potentials the auxiliary condition:

$$\frac{\partial A^k}{\partial x^k} = 0. \tag{46.9}$$

(This condition is called the *Lorentz condition*, and potentials that satisfy it are said to be in the *Lorentz gauge*.) Then the first term in (46.8) drops out and there remains

$$\frac{\partial^2 A^i}{\partial x_k \partial x^k} \equiv g^{kl} \frac{\partial^2 A^i}{\partial x^k \partial x^l} = 0. \tag{46.10}$$

---

† The wave equation is sometimes written in the form $\square \mathbf{A} = 0$, where

$$\square = -\frac{\partial^2}{\partial x_i \partial x^i} = \triangle - \frac{1}{c^2} \frac{\partial^2}{\partial t^2}$$

is called the *d'Alembertian operator*.

This is the wave equation expressed in four-dimensional form.†

In three-dimensional form, the condition (46.9) is:

$$\frac{1}{c}\frac{\partial \phi}{\partial t} + \text{div } \mathbf{A} = 0.$$ (46.11)

It is more general than the conditions $\phi = 0$ and div $\mathbf{A} = 0$ that were used earlier; potentials that satisfy those conditions also satisfy (46.11). But unlike them the Lorentz condition has a relativistically invariant character: potentials satisfying it in one frame satisfy it in any other frame (whereas condition (46.6) is generally violated if the frame is changed).

§ 47. Plane waves

We consider the special case of electromagnetic waves in which the field depends only on one coordinate, say x (and on the time). Such waves are said to be *plane*. In this case the equation for the field becomes

$$\frac{\partial^2 f}{\partial t^2} - c^2 \frac{\partial^2 f}{\partial x^2} = 0,$$ (47.1)

where by f is understood any component of the vectors $\mathbf{E}$ or $\mathbf{H}$.

To solve this equation, we rewrite it in the form

$$\left(\frac{\partial}{\partial t} - c \frac{\partial}{\partial x}\right)\left(\frac{\partial}{\partial t} + c \frac{\partial}{\partial x}\right) f = 0,$$

and introduce new variables

$$\xi = t - \frac{x}{c}, \qquad \eta = t + \frac{x}{c}$$

so that $t = \frac{1}{2}(\eta + \xi), x = \frac{c}{2}(\eta - \xi)$. Then

$$\frac{\partial}{\partial \xi} = \frac{1}{2}\left(\frac{\partial}{\partial t} - c \frac{\partial}{\partial x}\right), \qquad \frac{\partial}{\partial \eta} = \frac{1}{2}\left(\frac{\partial}{\partial t} + c \frac{\partial}{\partial x}\right),$$

so that the equation for f becomes

$$\frac{\partial^2 f}{\partial \xi \partial \eta} = 0.$$

The solution obviously has the form $f = f_1(\xi) + f_2(\eta)$, where f_1 and f_2 are arbitrary functions. Thus

$$f = f_1\left(t - \frac{x}{c}\right) + f_2\left(t + \frac{x}{c}\right).$$ (47.2)

---

† It should be mentioned that the condition (46.9) still does not determine the choice of the potentials uniquely. We can add to $\mathbf{A}$ a term grad f, and subtract a term $1/c \, (\partial f/\partial t)$ from ϕ, where the function f is not arbitrary but must satisfy the wave equation $\Box f = 0$.

Suppose, for example, $f_2 = 0$, so that

$$f = f_1 \left(t - \frac{x}{c} \right).$$

Let us clarify the meaning of this solution. In each plane $x = $ const, the field changes with the time; at each given moment the field is different for different x. It is clear that the field has the same values for coordinates x and times t which satisfy the relation $t - (x/c) = $ const, that is,

$$x = \text{const} + ct.$$

This means that if, at some time $t = 0$, the field at a certain point x in space had some definite value, then after an interval of time t the field has that same value at a distance ct along the X axis from the original place. We can say that all the values of the electromagnetic field are propagated in space along the X axis with a velocity equal to the velocity of light, c.

Thus,

$$f_1 \left(t - \frac{x}{c} \right)$$

represents a plane wave moving in the positive direction along the X axis. It is easy to show that

$$f_2 \left(t + \frac{x}{c} \right)$$

represents a wave moving in the opposite, negative, direction along the X axis.

In § 46 we showed that the potentials of the electromagnetic wave can be chosen so that $\phi = 0$, and div $\mathbf{A} = 0$. We choose the potentials of the plane wave which we are now considering in this same way. The condition div $\mathbf{A} = 0$ gives in this case

$$\frac{\partial A_x}{\partial x} = 0,$$

since all quantities are independent of y and z. According to (47.1) we then have also $\partial^2 A_x / \partial t^2 = 0$, that is, $\partial A_x / \partial t = $ const. But the derivative $\partial \mathbf{A}/\partial t$ determines the electric field, and we see that the nonzero component A_x represents in this case the presence of a constant longitudinal electric field. Since such a field has no relation to the electromagnetic wave, we can set $A_x = 0$.

Thus the vector potential of the plane wave can always be chosen perpendicular to the X axis, i.e. to the direction of propagation of that wave.

We consider a plane wave moving in the positive direction of the X axis; in this wave, all quantities, in particular also $\mathbf{A}$, are functions only of $t - (x/c)$. From the formulas

$$\mathbf{E} = -\frac{1}{c} \frac{\partial \mathbf{A}}{\partial t}, \quad \mathbf{H} = \text{curl } \mathbf{A},$$

we therefore obtain

$$\mathbf{E} = -\frac{1}{c} \mathbf{A}', \quad \mathbf{H} = \nabla \times \mathbf{A} = \nabla \left(t - \frac{x}{c} \right) \times \mathbf{A}' = -\frac{1}{c} \mathbf{n} \times \mathbf{A}', \qquad (47.3)$$

where the prime denotes differentiation with respect to $t - (x/c)$ and $\mathbf{n}$ is a unit vector along the direction of propagation of the wave. Substituting the first equation in the second, we obtain

$$\mathbf{H} = \mathbf{n} \times \mathbf{E}. \tag{47.4}$$

We see that the electric and magnetic fields $\mathbf{E}$ and $\mathbf{H}$ of a plane wave are directed perpendicular to the direction of propagation of the wave. For this reason, electromagnetic waves are said to be *transverse*. From (47.4) it is clear also that the electric and magnetic fields of the plane wave are perpendicular to each other and equal to each other in absolute value.

The energy flux in the plane wave, i.e. its Poynting vector is

$$\mathbf{S} = \frac{c}{4\pi} \mathbf{E} \times \mathbf{H} = \frac{c}{4\pi} \mathbf{E} \times (\mathbf{n} \times \mathbf{E}),$$

and since $\mathbf{E} \cdot \mathbf{n} = 0$,

$$\mathbf{S} = \frac{c}{4\pi} E^2 \mathbf{n} = \frac{c}{4\pi} H^2 \mathbf{n}.$$

Thus the energy flux is directed along the direction of propagation of the wave. Since

$$W = \frac{1}{8\pi} (E^2 + H^2) = \frac{E^2}{4\pi}$$

is the energy density of the wave, we can write

$$\mathbf{S} = cW\mathbf{n}, \tag{47.5}$$

in accordance with the fact that the field propagates with the velocity of light.

The momentum per unit volume of the electromagnetic field is $\mathbf{S}/c^2$. For a plane wave this gives $(W/c)\mathbf{n}$. We call attention to the fact that the relation between energy W and momentum W/c for the electromagnetic wave is the same as for a particle moving with the velocity of light [see (9.9)].

The flux of momentum of the field is determined by the components $\sigma_{\alpha\beta}$ of the Maxwell stress tensor (33.3). Choosing the direction of propagation of the wave as the X axis, we find that the only nonzero component of $T^{\alpha\beta}$ is

$$T^{xx} = -\sigma_{xx} = W. \tag{47.6}$$

As it must be, the flux of momentum is along the direction of propagation of the wave, and is equal in magnitude to the energy density.

Let us find the law of transformation of the energy density of a plane electromagnetic wave when we change from one inertial reference system to another. To do this we start from the formula

$$W = \frac{1}{1 - \dfrac{V^2}{c^2}} \left(W' + 2\frac{V}{c^2} S_x' + \frac{V^2}{c^2} \sigma_{xx}' \right)$$

(see the problem in § 33) and must substitute

$$S_x' = cW' \cos \alpha', \quad \sigma_{xx}' = -W' \cos^2 \alpha',$$

where α' is the angle (in the K' system) between the X' axis (along which the velocity $\mathbf{V}$ is directed) and the direction of propagation of the wave. We find:

$$W = W' \frac{\left(1 + \dfrac{V}{c} \cos \alpha'\right)^2}{1 - \dfrac{V^2}{c^2}}.$$ (47.7)

Since $W = E^2/4\pi = H^2/4\pi$, the absolute values of the field intensities in the wave transform like $\sqrt{W}$.

PROBLEMS

1. Determine the force exerted on a wall from which an incident plane electromagnetic wave is reflected (with reflection coefficient R).

Solution: The force $\mathbf{f}$ acting on unit area of the wall is given by the flux of momentum through this area, i.e., it is the vector with components

$$f_\alpha = -\sigma_{\alpha\beta}N_\beta - \sigma'_{\alpha\beta}N_\beta,$$

where $\mathbf{N}$ is the vector normal to the surface of the wall, and $\sigma_{\alpha\beta}$ and $\sigma'_{\alpha\beta}$ are the components of the energy-momentum tensors for the incident and reflected waves. Using (47.6), we obtain:

$$\mathbf{f} = W\mathbf{n}(\mathbf{N}\cdot\mathbf{n}) + W'\mathbf{n}'(\mathbf{N}\cdot\mathbf{n}').$$

From the definition of the reflection coefficient, we have: $W' = RW$. Also introducing the angle of incidence θ (which is equal to the reflection angle) and writing out components, we find the normal force ("light pressure")

$$f_N = W(1 + R)\cos^2\theta$$

and the tangential force

$$f_t = W(1 - R)\sin\theta\cos\theta.$$

2. Use the Hamilton–Jacobi method to find the motion of a charge in the field of a plane electromagnetic wave with vector potential $\mathbf{A}[t - (x/c)]$.

Solution: We write the Hamilton–Jacobi equation in four-dimensional form:

$$g^{ik}\left(\frac{\partial S}{\partial x^i} + \frac{e}{c}A_i\right)\left(\frac{\partial S}{\partial x^k} + \frac{e}{c}A_k\right) = m^2c^2.$$ (1)

The fact that the field is a plane wave means that the A^i are functions of one independent variable, which can be written in the form $\xi = k_i x^i$, where k^i is a constant four-vector with its square equal to zero, $k_i k^i = 0$ (see the following section). We subject the potentials to the Lorentz condition

$$\frac{\partial A^i}{\partial x^i} = \frac{dA^i}{d\xi}k_i = 0;$$

for the variables field this is equivalent to the condition $A^i k_i = 0$.

We seek a solution of equation (1) in the form

$$S = -f_i x^i + F(\xi),$$

where $f^i = (f^0, \mathbf{f})$ is a constant vector satisfying the condition $f_i f^i = m^2 c^2$ ($S = -f_i x^i$ is the solution of the Hamilton–Jacobi equation for a free particle with four-momentum $p^i = f^i$). Substitution in (1) gives the equation

$$\frac{e^2}{c^2} A_i A^i - 2\gamma \frac{dF}{d\xi} - \frac{2e}{c} f_i A^i = 0,$$

where the constant $\gamma = k_i f^i$. Having determined F from this equation, we get

$$S = -f_i x^i - \frac{e}{c\gamma} \int f_i A^i d\xi + \frac{e^2}{2\gamma c^2} \int A_i A^i d\xi. \qquad (2)$$

Changing to three-dimensional notation with a fixed reference frame, we choose the direction of propagation of the wave as the x axis. Then $\xi = ct - x$, while the constant $\gamma = f^0 - f^1$. Denoting the two-dimensional vector f_y, f_z by $\mathbf{\kappa}$, we find from the condition $f_i f^i = (f^0)^2 - (f^1)^2 - \mathbf{\kappa}^2 = m^2 c^2$,

$$f^0 + f^1 = \frac{m^2 c^2 + \mathbf{\kappa}^2}{\gamma}.$$

We choose the potentials in the gauge in which $\phi = 0$, while $\mathbf{A}(\xi)$ lies in the yz plane. Then equation (2) takes the form:

$$S = \mathbf{\kappa} \cdot \mathbf{r} - \frac{\gamma}{2}(ct + x) - \frac{m^2 c^2 + \mathbf{\kappa}^2}{2\gamma}\xi + \frac{e}{c\gamma}\int \mathbf{\kappa} \cdot \mathbf{A} d\xi - \frac{e^2}{2\gamma c^2}\int \mathbf{A}^2 d\xi.$$

According to the general rules (*Mechanics,* § 47), to determine the motion we must equate the derivatives $\partial S/\partial \mathbf{\kappa}, \partial S/\partial \gamma$ to certain new constants, which can be made to vanish by a suitable choice of the coordinate and time origins. We thus obtain the parametric equations in ξ:

$$y = \frac{1}{\gamma} \kappa_y \xi - \frac{e}{c\gamma} \int A_y d\xi, \qquad z = \frac{1}{\gamma} \kappa_z \xi - \frac{e}{c\gamma} \int A_z d\xi,$$

$$x = \frac{1}{2}\left(\frac{m^2 c^2 + \mathbf{\kappa}^2}{\gamma^2} - 1\right)\xi - \frac{e}{c\gamma^2}\int \mathbf{\kappa} \cdot \mathbf{A} d\xi + \frac{e^2}{2\gamma^2 c^2}\int \mathbf{A}^2 d\xi, \qquad ct = \xi + x.$$

The generalized momentum $\mathbf{P} = \mathbf{p} + (e/c)\mathbf{A}$ and the energy $\mathscr{E}$ are found by differentiating the action with respect to the coordinates and the time; this gives:

$$p_y = \kappa_y - \frac{e}{c}A_y, \qquad p_z = \kappa_z - \frac{e}{c}A_z,$$

$$p_x = -\frac{\gamma}{2} + \frac{m^2 c^2 + \mathbf{\kappa}^2}{2\gamma} - \frac{e}{c\gamma}\mathbf{\kappa} \cdot \mathbf{A} + \frac{e^2}{2\gamma c^2}\mathbf{A}^2,$$

$$\mathscr{E} = (\gamma + p_x)c.$$

If we average these over the time, the terms of first degree in the periodic function $\mathbf{A}(\xi)$ vanish. We assume that the reference system has been chosen so that the particle is at rest in it on the average, i.e. so that its averaged momentum is zero. Then

$$\mathbf{\kappa} = 0, \qquad \gamma^2 = m^2 c^2 + \frac{e^2}{c^2}\overline{\mathbf{A}^2}.$$

The final formulas for determining the motion have the form:

$$x = \frac{e^2}{2\gamma^2 c^2}\int (\mathbf{A}^2 - \overline{\mathbf{A}^2})d\xi, \qquad y = -\frac{e}{c\gamma}\int A_y d\xi, \qquad z = -\frac{e}{c\gamma}\int A_z\, d\xi,$$

$$ct = \xi + \frac{e^2}{2\gamma^2 c^2} \int (\mathbf{A}^2 - \overline{\mathbf{A}^2})\, d\xi; \tag{3}$$

$$p_x = \frac{e^2}{2\gamma c^2} (\mathbf{A}^2 - \overline{\mathbf{A}^2}), \quad p_y = -\frac{e}{c} A_y, \quad p_z = -\frac{e}{c} A_z,$$

$$\mathcal{E} = c\gamma + \frac{e^2}{2\gamma c} (\mathbf{A}^2 - \overline{\mathbf{A}^2}). \tag{4}$$

§ 48. Monochromatic plane waves

A very important special case of electromagnetic waves is a wave in which the field is a simply periodic function of the time. Such a wave is said to be *monochromatic*. All quantities (potentials, field components) in a monochromatic wave depend on the time through a factor of the form cos ($\omega t + \alpha$). The quantity ω is called the *cyclic frequency* of the wave (we shall simply call it the *frequency*).

In the wave equation, the second derivative of the field with respect to the time is now $\partial^2 f / \partial t^2 = -\omega^2 f$, so that the distribution of the field in space is determined for a monochromatic wave by the equation

$$\Delta f + \frac{\omega^2}{c^2} f = 0. \tag{48.1}$$

In a plane wave (propagating along the x axis), the field is a function only of $t - (x/c)$. Therefore, if the plane wave is monochromatic, its field is a simply periodic function of $t - (x/c)$. The vector potential of such a wave is most conveniently written as the real part of a complex expression:

$$\mathbf{A} = \text{Re}\, \{\mathbf{A}_0 e^{-i\omega(t - \frac{x}{c})}\} \tag{48.2}$$

Here $\mathbf{A}_0$ is a certain constant complex vector. Obviously, the fields $\mathbf{E}$ and $\mathbf{H}$ of such a wave have analogous forms with the same frequency ω. The quantity

$$\lambda = \frac{2\pi c}{\omega} \tag{48.3}$$

is called the *wavelength;* it is the period of variation of the field with the coordinate x at a fixed time t.

The vector

$$\mathbf{k} = \frac{\omega}{c}\, \mathbf{n} \tag{48.4}$$

(where $\mathbf{n}$ is a unit vector along the direction of propagation of the wave) is called the *wave vector.* In terms of it we can write (48.2) in the form

$$\mathbf{A} = \text{Re}\, \{\mathbf{A}_0 e^{i(\mathbf{k}\cdot\mathbf{r} - \omega t)}\}, \tag{48.5}$$

which is independent of the choice of coordinate axes. The quantity which appears multiplied by i in the exponent is called the *phase* of the wave.

So long as we perform only linear operations, we can omit the sign Re for taking the real part, and operate with complex quantities as such.† Thus, substituting

$$\mathbf{A} = \mathbf{A}_0 e^{i(\mathbf{k} \cdot \mathbf{r} - \omega t)}$$

in (47.3), we find the relation between the intensities and the vector potential of a plane monochromatic wave in the form

$$\mathbf{E} = i k \mathbf{A}, \qquad \mathbf{H} = i \mathbf{k} \times \mathbf{A}. \tag{48.6}$$

We now treat in more detail the direction of the field of a monochromatic wave. To be specific, we shall talk of the electric field

$$\mathbf{E} = \mathrm{Re} \, \{\mathbf{E}_0 e^{i(\mathbf{k} \cdot \mathbf{r} - \omega t)}\}$$

(everything stated below applies equally well, of course, to the magnetic field). The quantity $\mathbf{E}_0$ is a certain complex vector. Its square $\mathbf{E}_0^2$ is (in general) a complex number. If the argument of this number is -2α (i.e. $\mathbf{E}_0^2 = |\, \mathbf{E}_0^2 \,| \, e^{-2i\alpha}$), the vector $\mathbf{b}$ defined by

$$\mathbf{E}_0 = \mathbf{b} e^{-i\alpha} \tag{48.7}$$

will have its square real, $\mathbf{b}^2 = |\, \mathbf{E}_0 \,|^2$. With this definition, we write:

$$\mathbf{E} = \mathrm{Re} \, \{\mathbf{b} e^{i(\mathbf{k} \cdot \mathbf{r} - \omega t - \alpha)}\}. \tag{48.8}$$

We write $\mathbf{b}$ in the form

$$\mathbf{b} = \mathbf{b}_1 + i\mathbf{b}_2,$$

where $\mathbf{b}_1$ and $\mathbf{b}_2$ are real vectors. Since $\mathbf{b}^2 = \mathbf{b}_1^2 - \mathbf{b}_2^2 + 2i\mathbf{b}_1 \cdot \mathbf{b}_2$ must be a real quantity, $\mathbf{b}_1 \cdot \mathbf{b}_2 = 0$, i.e. the vectors $\mathbf{b}_1$ and $\mathbf{b}_2$ are mutually perpendicular. We choose the direction of $\mathbf{b}_1$ as the y axis (and the x axis along the direction of propagation of the wave). We then have from (48.8):

$$E_y = b_1 \cos \, (\omega t - \mathbf{k} \cdot \mathbf{r} + \alpha),$$
$$E_z = \pm \, b_2 \sin \, (\omega t - \mathbf{k} \cdot \mathbf{r} + \alpha), \tag{48.9}$$

where we use the plus (minus) sign if $\mathbf{b}_2$ is along the positive (negative) z axis. From (48.9) it follows that

---

† If two quantities $\mathbf{A}(t)$ and $\mathbf{B}(t)$ are written in complex form

$$\mathbf{A}(t) = \mathbf{A}_0 e^{-i\omega t}, \qquad \mathbf{B}(t) = \mathbf{B}_0 e^{-i\omega t},$$

then in forming their product we must first, of course, separate out the real part. But if, as it frequently happens, we are interested only in the time average of this product, it can be computed as

$$\tfrac{1}{2} \, \mathrm{Re} \, \{\mathbf{A} \cdot \mathbf{B}^*\}.$$

In fact, we have:

$$\mathrm{Re} \, \mathbf{A} \cdot \mathrm{Re} \, \mathbf{B} = \tfrac{1}{4} (\mathbf{A}_0 e^{-i\omega t} + \mathbf{A}_0^* e^{i\omega t}) \cdot (\mathbf{B}_0 e^{-i\omega t} + \mathbf{B}_0^* e^{i\omega t}).$$

When we average, the terms containing factors $e^{\pm 2i\omega t}$ vanish, so that we are left with

$$\overline{\mathrm{Re} \, \mathbf{A} \cdot \mathrm{Re} \, \mathbf{B}} = \tfrac{1}{4} (\mathbf{A}_0 \cdot \mathbf{B}_0^* + \mathbf{A}_0^* \cdot \mathbf{B}_0) = \tfrac{1}{2} \, \mathrm{Re} \, (\mathbf{A} \cdot \mathbf{B}^*).$$

$$\frac{E_y^2}{b_1^2} + \frac{E_z^2}{b_2^2} = 1. \tag{48.10}$$

Thus we see that, at each point in space, the electric field vector rotates in a plane perpendicular to the direction of propagation of the wave, while its endpoint describes the ellipse (48.10). Such a wave is said to be *elliptically polarized.* The rotation occurs in the direction of (opposite to) a right-hand screw rotating along the x axis, if we have the plus (minus) sign in (48.9).

If $b_1 = b_2$, the ellipse (48.10) reduces to a circle, i.e. the vector **E** rotates while remaining constant in magnitude. In this case we say that the wave is *circularly polarized.* The choice of the directions of the y and z axes is now obviously arbitrary. We note that in such a wave the ratio of the y and z components of the complex amplitude $\mathbf{E}_0$ is

$$\frac{E_{0z}}{E_{0y}} = \pm i \tag{48.11}$$

for rotation in the same (opposite) direction as that of a right-hand screw *right* and *left* polarizations).†

Finally, if b_1 or b_2 equals zero, the field of the wave is everywhere and always parallel (or antiparallel) to one and the same direction. In this case the wave is said to be *linearly polarized,* or plane polarized. An elliptically polarized wave can clearly be treated as the superposition of two plane polarized waves.

Now let us turn to the definition of the wave vector and introduce the four-dimensional wave vector with components

$$k^i = \left(\frac{\omega}{c}, \mathbf{k}\right). \tag{48.12}$$

That these quantities actually form a four-vector is obvious from the fact that we get a scalar the phase of the wave) when we nultiply by x^i:

$$k_i x^i = \omega t - \mathbf{k} \cdot \mathbf{r}. \tag{48.13}$$

From the definitions (48.4) and (48.12) we see that the square of the wave four-vector is zero:

$$k^i k_i = 0. \tag{48.14}$$

This relation also follows directly from the fact that the expression

$$\mathbf{A} = \mathbf{A}_0 e^{-ik_i x^i}$$

must be a solution of the wave equation (46.10).

As is the case for every plane wave, in a monochromatic wave propagating along the x axis only the following components of the energy-momentum tensor are different from zero (see § 47):

$$T^{00} = T^{01} = T^{11} = W.$$

By means of the wave four-vector, these equations can be written in tensor form as

$$T^{ik} = \frac{Wc^2}{\omega^2} k^i k^k. \tag{48.15}$$

† We assume that the coordinate axes form a right-handed system.

Finally, by using the law of transformation of the wave four-vector we can easily treat the so-called *Doppler effect*—the change in frequency ω of the wave emitted by a source moving with respect to the observer, as compared to the "true" frequency ω_0 of the same source in the reference system (K_0) in which it is at rest.

Let V be the velocity of the source, i.e. the velocity of the K_0 system relative to K. According to the general formula for transformation of four-vectors, we have:

$$k^{(0)0} = \frac{k^0 - \frac{V}{c} k^1}{\sqrt{1 - \frac{V^2}{c^2}}}$$

(the velocity of the K system relative to K_0 is $-V$). Substituting $k^0 = \omega/c$, $k^1 = k \cos \alpha = \omega/c \cos \alpha$, where α is the angle (in the K system) between the direction of emission of the wave and the direction of motion of the source, and expressing ω in terms of ω_0, we obtain:

$$\omega = \omega_0 \frac{\sqrt{1 - \frac{V^2}{c^2}}}{1 - \frac{V}{c} \cos \alpha} \tag{48.16}$$

This is the required formula. For $V \ll c$, and if the angle α is not too close to $\pi/2$, it gives:

$$\omega \cong \omega_0 \left(1 + \frac{V}{c} \cos \alpha\right). \tag{48.17}$$

For $\alpha = \pi/2$, we have:

$$\omega = \omega_0 \sqrt{1 - \frac{V^2}{c^2}} \cong \omega_0 \left(1 - \frac{V^2}{2c^2}\right); \tag{48.18}$$

in this case the relative change in frequency is proportional to the square of the ratio V/c.

PROBLEMS

1. Determine the direction and magnitude of the axes of the polarization ellipse in terms of the complex amplitude $\mathbf{E}_0$.

Solution: The problem consists in determining the vector $\mathbf{b} = \mathbf{b}_1 + i\mathbf{b}_2$, whose square is real. We have from (48.7):

$$\mathbf{E}_0 \cdot \mathbf{E}_0^* = b_1^2 + b_2^2, \qquad \mathbf{E}_0 \times \mathbf{E}_0^* = -2i\mathbf{b}_1 \times \mathbf{b}_2, \tag{1}$$

or

$$b_1^2 \cdot b_2^2 = A^2 + B^2, \qquad b_1 b_2 = AB \sin \delta,$$

where we have introduced the notation

$$|E_{0y}| = A, \qquad |E_{0z}| = B, \qquad \frac{E_{0z}}{B} = \frac{E_{0y}}{A} e^{i\delta}$$

for the absolute values of E_{0y} and E_{0z} and for the phase difference δ between them. Then

$$2b_{1,2} = \sqrt{A^2 + B^2 + 2AB \sin \delta} \pm \sqrt{A^2 + B^2 - 2AB \sin \delta}, \tag{2}$$

from which we get the magnitudes of the semiaxes of the polarization ellipse.

To determine their directions (relative to the arbitrary initial axes y and z) we start from the equality

$$\text{Re} \{(\mathbf{E}_0 \cdot \mathbf{b}_1)(\mathbf{E}_0^* \cdot \mathbf{b}_2)\} = 0,$$

which is easily verified by substituting $\mathbf{E}_0 = (\mathbf{b}_1 + i\mathbf{b}_2) e^{-1a}$. Writing out this equality in the y, z coordinates, we get for the angle θ between the direction of $\mathbf{b}_1$ and the y axis:

$$\tan 2\theta = \frac{2AB \cos \delta}{A^2 - B^2}. \tag{3}$$

The direction of rotation of the field is determined by the sign of the x component of the vector $\mathbf{b}_1 \times \mathbf{b}_2$. Taking its expression from (1)

$$2i(\mathbf{b}_1 \times \mathbf{b}_2)_x = E_{0z}E_{0y}^* - E_{0z}^*E_{0y} = |E_{0y}|^2 \left\{ \left(\frac{E_{0z}}{E_{0y}}\right) - \left(\frac{E_{0z}}{E_{0y}}\right)^* \right\},$$

we see that the direction of $\mathbf{b}_1 \times \mathbf{b}_2$ (whether it is along or opposite to the positive direction of the x axis), and the sign of the rotation (whether in the same direction, or opposite to the direction of a right-hand screw along the x axis) are given by the sign of the imaginary part of the ratio E_{0z}/E_{0y} (plus for the first case and minus for the second). This is a generalization of the rule (48.11) for the case of circular polarization.

2. Determine the motion of a charge in the field of a plane monochromatic linearly polarized wave.

Solution: Choosing the direction of the field $\mathbf{E}$ of the wave as the y axis, we write:

$$E_y = E = E_0 \cos \omega\xi, \quad A_y = A = -\frac{cE_0}{\omega} \sin \omega\xi$$

($\xi = t - x/c$). From formulas (3) and (4) of problem 2, § 47, we find (in the reference system in which the particle is at rest on the average) the following representation of the motion in terms of the parameter $\eta = \omega\xi$):

$$x = -\frac{e^2 E_0^2 c}{8\gamma^2 \omega^3} \sin 2\eta, \quad y = -\frac{eE_0 c}{\gamma\omega^2} \cos \eta, \quad z = 0,$$

$$t = \frac{\eta}{\omega} - \frac{e^2 E_0^2}{8\gamma^2 \omega^3} \sin 2\eta, \quad \gamma^2 = m^2 c^2 + \frac{e^2 E_0^2}{2\omega^2};$$

$$p_x = -\frac{e^2 E_0^2}{4\gamma\omega^2} \cos 2\eta, \quad p_y = \frac{eE_0}{\omega} \sin \eta, \quad p_z = 0.$$

The charge moves in the x, y plane in a symmetric figure-8 curve with its longitudinal axis along the y axis. During a period of the motion, η varies from 0 to 2π.

3. Determine the motion of a charge in the field of a circularly polarized wave.

Solution: For the field of the wave we have:

$$E_y = E_0 \cos \omega\xi, \quad E_z = E_0 \sin \omega\xi,$$

$$A_y = -\frac{cE_0}{\omega} \sin \omega\xi, \quad A_z = \frac{cE_0}{\omega} \cos \omega\xi.$$

The motion is given by the formulas:

$$x = 0, \quad y = -\frac{ecE_0}{\gamma\omega^2} \cos \omega t, \quad z = -\frac{ecE_0}{\gamma\omega^2} \sin \omega t,$$

$$p_z = 0, \quad p_y = \frac{eE_0}{\omega} \sin \omega t, \quad p_z = -\frac{eE_0}{\omega} \cos \omega t,$$

$$\gamma^2 = m^2 c^2 + \frac{c^2 E_0^2}{\omega^2}.$$

Thus the charge moves in the y, z plane along a circle of radius $ecE_0/\gamma\omega^2$ with a momentum having the constant magnitude $p = eE_0/\omega$; at each instant the direction of the momentum $\mathbf{p}$ is opposite to the direction of the magnetic field $\mathbf{H}$ of the wave.

§ 49. Spectral resolution

Every wave can be subjected to the process of spectral resolution, i.e. can be represented as a superposition of monochromatic waves with various frequencies. The character of this expansion varies according to the character of the time dependence of the field.

One category consists of those cases where the expansion contains frequencies forming a discrete sequence of values. The simplest case of this type arises in the resolution of a purely periodic (though not monochromatic) field. This is the usual expansion in Fourier series; it contains the frequencies which are integral multiples of the "fundamental" frequency $\omega_0 = 2\pi/T$, where T is the period of the field. We write it in the form

$$f = \sum_{n=-\infty}^{\infty} f_n e^{-i\omega_0 n t} \tag{49.1}$$

(where f is any of the quantities describing the field). The quantities f_n are defined in terms of the function f by the integrals

$$f_n = \frac{1}{T} \int_{-T/2}^{T/2} f(t) e^{in\omega_0 t} dt. \tag{49.2}$$

Because $f(t)$ must be real,

$$f_{-n} = f_n^*. \tag{49.3}$$

In more complicated cases, the expansion may contain integral multiples (and sums of integral multiples) of several different incommensurable fundamental frequencies.

When the sum (49.1) is squared and averaged over the time, the products of terms with different frequencies give zero because they contain oscillating factors. Only terms of the form $f_n f_{-n} = |f_n|^2$ remain. Thus the average of the square of the field, i.e. the average intensity of the wave, is the sum of the intensities of its monochromatic components:

$$\overline{f^2} = \sum_{n=-\infty}^{\infty} |f_n|^2 = 2 \sum_{n=1}^{\infty} |f_n|^2. \tag{49.4}$$

(where it is assumed that the average of the function f over a period is zero, i.e. $f_0 = \bar{f} = 0$).

Another category consists of fields which are expandable in a Fourier integral containing a continuous distribution of different frequencies. For this to be possible, the function $f(t)$ must satisfy certain definite conditions; usually we consider functions which vanish for $t \rightarrow \pm \infty$. Such an expansion has the form

$$f(t) = \int_{-\infty}^{\infty} f_\omega e^{-i\omega t} \frac{d\omega}{2\pi}, \tag{49.5}$$

where the Fourier components are given in terms of the function $f(t)$ by the integrals

$$f_\omega = \int_{-\infty}^{\infty} f(t) e^{i\omega t} dt. \tag{49.6}$$

Analogously to (49.3),

$$f_{-\omega} = f_\omega^*. \tag{49.7}$$

Let us express the total intensity of the wave, i.e. the integral of f^2 over all time, in terms of the intensity of the Fourier components. Using (49.5) and (49.6), we have:

$$\int_{-\infty}^{\infty} f^2 dt = \int_{-\infty}^{\infty} \left\{ f \int_{-\infty}^{\infty} f_\omega e^{-i\omega t} \frac{d\omega}{2\pi} \right\} dt = \int_{-\infty}^{\infty} \left\{ f_\omega \int_{-\infty}^{\infty} f e^{-i\omega t} dt \right\} \frac{d\omega}{2\pi} = \int_{-\infty}^{\infty} f_\omega f_{-\omega} \frac{d\omega}{2\pi},$$

or, using (49.7),

$$\int_{-\infty}^{\infty} f^2 dt = \int_{-\infty}^{\infty} |f_\omega|^2 \frac{d\omega}{2\pi} = 2 \int_{0}^{\infty} |f_\omega|^2 \frac{d\omega}{2\pi}. \tag{49.8}$$

§ 50. Partially polarized light

Every monochromatic wave is, by definition, necessarily polarized. However we usually have to deal with waves which are only approximately monochromatic, and which contain frequencies in a small interval $\Delta\omega$. We consider such a wave, and let ω be some average frequency for it. Then its field (to be specific we shall consider the electric field **E**) at a fixed point in space can be writen in the form

$$\mathbf{E}_0(t) e^{-i\omega t},$$

where the complex amplitude $\mathbf{E}_0(t)$ is some slowly varying function of the time (for a strictly monochromatic wave $\mathbf{E}_0$ would be constant). Since $\mathbf{E}_0$ determines the polarization of the wave, this means that at each point of the wave, its polarization changes with time, such a wave is said to be *partially polarized*.

The polarization properties of electromagnetic waves, and of light in particular, are observed experimentally by passing the light to be investigated through various bodies† and then observing the intensity of the transmitted light. From the mathematical point of view this means that we draw conclusions concerning the polarization properties of the light from the values of certain quadratic functions of its field. Here of course we are considering the time averages of such functions.

Quadratic functions of the field are made up of terms proportional to the products $E_\alpha E_\beta$, $E_\alpha^* E_\beta^*$ or $E_\alpha E_\beta^*$. Products of the form

$$E_\alpha E_\beta = E_{0\alpha} E_{0\beta} e^{-2i\omega t}, \quad E_\alpha^* E_\beta^* = E_{0\alpha}^* E_{0\beta}^* e^{2i\omega t},$$

which contain the rapidly oscillating factors $e^{\pm 2i\omega t}$ give zero when the time average is taken. The products $E_\alpha E_\beta^* = E_{0\alpha} E_{0\beta}^*$ do not contain such factors, and so their averages are not

† For example, through a Nicol prism.

zero. Thus we see that the polarization properties of the light are completely characterized by the tensor

$$J_{\alpha\beta} = \overline{E_{0\alpha} E_{0\beta}^*} . \tag{50.1}$$

Since the vector $\mathbf{E}_0$ always lies in a plane perpendicular to the direction of the wave, the tensor $J_{\alpha\beta}$ has altogether four components (in this section the indices α, β are understood to take on only two values: α, $\beta = 1, 2$, corresponding to the y and z axes; the x axis is along the direction of propagation of the wave).

The sum of the diagonal elements of the tensor $J_{\alpha\beta}$ (we denote it by J) is a real quantity— the average value of the square modulus of the vector $\mathbf{E}_0$ (or $\mathbf{E}$):

$$J \equiv J_{\alpha\alpha} = \overline{\mathbf{E}_0 \cdot \mathbf{E}_0^*} . \tag{50.2}$$

This quantity determines the intensity of the wave, as measured by the energy flux density. To eliminate this quantity which is not directly related to the polarization properties, we introduce in place of $J_{\alpha\beta}$ the tensor

$$\rho_{\alpha\beta} = \frac{J_{\alpha\beta}}{J}, \tag{50.3}$$

for which $\rho_{\alpha\alpha} = 1$; we call it the *polarization tensor.*

From the definition (50.1) we see that the components of the tensor $J_{\alpha\beta}$, and consequently also $\rho_{\alpha\beta}$, are related by

$$\rho_{\alpha\beta} = \rho_{\beta\alpha}^* \tag{50.4}$$

(i.e. the tensor is hermitian). Consequently the diagonal components ρ_{11} and ρ_{22} are real (with $\rho_{11} + \rho_{22} = 1$) while $\rho_{21} = \rho_{12}^*$. Thus the polarization is characterized by three real parameters.

Let us study the conditions that the tensor $\rho_{\alpha\beta}$ must satisfy for completely polarized light. In this case $\mathbf{E}_0 = $ const, and so we have simply

$$J_{\alpha\beta} = J\rho_{\alpha\beta} = E_{0\alpha} E_{0\beta}^* \tag{50.5}$$

(without averaging), i.e. the components of the tensor can be written as products of components of some constant vector. The necessary and sufficient condition for this is that the determinant vanish:

$$|\rho_{\alpha\beta}| = \rho_{11}\rho_{22} - \rho_{12}\rho_{21} = 0. \tag{50.6}$$

The opposite case is that of unpolarized or *natural* light. Complete absence of polarization means that all directions (in the y_z plane) are equivalent. In other words the polarization tensor must have the form:

$$\rho_{\alpha\beta} = \tfrac{1}{2}\delta_{\alpha\beta} . \tag{50.7}$$

The determinant is $|\rho_{\alpha\beta}| = \tfrac{1}{4}$.

In the general case of arbitrary polarization the determinant has values between 0 and $\tfrac{1}{4}$.†

---

† The fact that the determinant is positive for any tensor of the form (50.1) is easily seen by considering the averaging, for simplicity, as a summation over discrete values, and using the well-known algebraic inequality

$$|\sum_{a,b} x_a y_b|^2 \leq \sum_a |x_a|^2 \sum_b |y_b|^2 .$$

By the *degree of polarization* we mean the positive quantity P, defined from

$$|\rho_{\alpha\beta}| = \tfrac{1}{4}(1 - P^2).$$ (50.8)

It runs from the value 0 for unpolarized to 1 for polarized light.

An arbitrary tensor $\rho_{\alpha\beta}$ can be split into two parts—a symmetric and an antisymmetric part. Of these, the first

$$S_{\alpha\beta} = \tfrac{1}{2}(\rho_{\alpha\beta} + \rho_{\beta\alpha})$$

is real because of the hermiticity of $\rho_{\alpha\beta}$. The antisymmetric part is pure imaginary. Like any antisymmetric tensor of rank equal to the number of dimensions, it reduces to a pseudo-scalar (see the footnote on p. 18):

$$\tfrac{1}{2}(\rho_{\alpha\beta} - \rho_{\beta\alpha}) = -\frac{i}{2}e_{\alpha\beta}A,$$

where A is a real pseudoscalar, $e_{\alpha\beta}$ is the unit antisymmetric tensor (with components $e_{12} = -e_{21} = 1$). Thus the polarization tensor has the form:

$$\rho_{\alpha\beta} = S_{\alpha\beta} - \frac{i}{2}\,e_{\alpha\beta}A, \quad S_{\alpha\beta} = S_{\beta\alpha},$$ (50.9)

i.e. it reduces to one real symmetric tensor and one pseudoscalar.

For a circularly polarized wave, the vector $\mathbf{E}_0 = \text{const}$, where

$$E_{02} = \pm iE_{01}.$$

It is easy to see that then $S_{\alpha\beta} = \tfrac{1}{2}\delta_{\alpha\beta}$, while $A = \pm 1$. On the other hand, for a linearly polarized wave the constant vector $\mathbf{E}_0$ can be chosen to be real, so that $A = 0$. In the general case the quantity A may be called the degree of circular polarization; it runs through values from $+1$ to -1, where the limiting values correspond to right- and left-circularly polarized waves, respectively.

The real symmetric tensor $S_{\alpha\beta}$, like any symmetric tensor, can be brought to principal axes, with different principal values which we denote by λ_1 and λ_2. The directions of the principal axes are mutually perpendicular. Denoting the unit vectors along these directions by $\mathbf{n}^{(1)}$ and $\mathbf{n}^{(2)}$, we can write $S_{\alpha\beta}$ in the form:

$$S_{\alpha\beta} = \lambda_1 n_\alpha^{(1)} n_\beta^{(1)} + \lambda_2 n_\alpha^{(2)} n_\beta^{(2)}, \quad \lambda_1 + \lambda_2 = 1.$$ (50.10)

The quantities λ_1 and λ_2 are positive and take on values from 0 to 1.

Suppose that $A = 0$, so that $\rho_{\alpha\beta} = S_{\alpha\beta}$. Each of the two terms in (50.10) has the form of a product of two components of a constant vector ($\sqrt{\lambda_1}\,\mathbf{n}^{(1)}$ or $\sqrt{\lambda_2}\,\mathbf{n}^{(2)}$). In other words, each of the terms corresponds to linearly polarized light. Furthermore, we see that there is no term in (50.10) containing products of components of the two waves. This means that the two parts can be regarded as physically independent of one another, or, as one says, they are *incoherent*. In fact, if two waves are independent, the average value of the product $E_\alpha^{(1)}E_\beta^{(2)}$ is equal to the product of the averages of each of the factors, and since each of them is zero,

$$\overline{E_\alpha^{(1)}E_\beta^{(2)}} = 0.$$

Thus we arrive at the result that in this case ($A = 0$) the partially polarized light can be represented as a superposition of two incoherent waves (with intensities proportional to λ_1 and λ_2), linearly polarized along mutually perpendicular directions.† (In the general case of a complex tensor $\rho_{\alpha\beta}$ one can show that the light can be represented as a superposition of two incoherent elliptically polarized waves, whose polarization ellipses are similar and mutually perpendicular (see problem 2).)

Let ϕ be the angle between the axis 1 (the y axis) and the unit vector $\mathbf{n}^{(1)}$; then

$$\mathbf{n}^{(1)} = (\cos\phi, \sin\phi), \quad \mathbf{n}^{(2)} = (-\sin\phi, \cos\phi).$$

Introducing the quantity $l = \lambda_1 - \lambda_2$ (assume $\lambda_1 > \lambda_2$), we write the components of the tensor (50.10) in the following form:

$$S_{\alpha\beta} = \frac{1}{2}\begin{pmatrix} 1 + l\cos 2\phi & l\sin 2\phi \\ l\sin 2\phi & 1 - l\cos 2\phi \end{pmatrix}. \tag{50.11}$$

Thus, for an arbitrary choice of the axes y and z, the polarization properties of the wave can be characterized by the following three real parameters: A—the degree of circular polarization, l—the degree of maximum linear polarization, and ϕ—the angle between the direction $\mathbf{n}^{(1)}$ of maximum polarization and the y axis.

In place of these parameters one can choose another set of three parameters:

$$\xi_1 = l\sin 2\phi, \quad \xi_2 = A, \quad \xi_3 = l\cos 2\phi \tag{50.12}$$

(the *Stokes parameters*). The polarization tensor is expressed in terms of them as

$$\rho_{\alpha\beta} = \frac{1}{2}\begin{pmatrix} 1 + \xi_3 & \xi_1 - i\xi_2 \\ \xi_1 + i\xi_2 & 1 - \xi_3 \end{pmatrix}. \tag{50.13}$$

All three parameters run through values from -1 to $+1$. The parameter ξ_3 characterizes the linear polarization along the y and z axes: the value $\xi_3 = 1$ corresponds to complete linear polarization along the y axis, and $\xi_3 = -1$ to complete polarization along the z axis. The parameter ξ_1 characterizes the linear polarization along directions making an angle of $45°$ with the y axis: the value $\xi_1 = 1$ means complete polarization at an angle $\phi = \pi/4$, while $\xi_1 = -1$ means complete polarization at $\phi = -\pi/4$.‡

The determinant of (50.13) is equal to

$$|\rho_{\alpha\beta}| = \tfrac{1}{4}(1 - \xi_1^2 - \xi_2^2 - \xi_3^2). \tag{50.14}$$

Comparing with (50.8), we see that

$$P = \sqrt{\xi_1^2 + \xi_2^2 + \xi_3^2}. \tag{50.15}$$

---

† The determinant $|S_{\alpha\beta}| = \lambda_1\lambda_2$; suppose that $\lambda_1 > \lambda_2$; then the degree of polarization, as defined in (50.8), is $P = 1 - 2\lambda_2$. In the present case ($A = 0$) one frequently characterizes the degree of polarization by using the *depolarization coefficient*, defined as the ratio λ_2/λ_1.

‡ For a completely elliptically polarized wave with axes of the ellipse $\mathbf{b}_1$ and $\mathbf{b}_2$ (see § 48), the Stokes parameters are:

$$\xi_1 = 0, \quad \xi_2 = \pm 2b_1 b_2, \quad \xi_3 = b_1^2 - b_2^2.$$

Here the y axis is along $\mathbf{b}_1$, while the two signs in ξ_2 correspond to directions of $\mathbf{b}_2$ along and opposite to the direction on the z axis.

Thus, for a given overall degree of polarization P, different types of polarization are possible, characterized by the values of the three quantities ξ_2, ξ_2, ξ_3, the sum of whose squares is fixed; they form a sort of vector of fixed length.

We note that the quantities $\xi_2 = A$ and $\sqrt{\xi_1^2 + \xi_3^2} = l$ are invariant under Lorentz transformations. This remark is already almost obvious from the very meaning of these quantities as degrees of circular and linear polarization.†

PROBLEMS

1. Resolve an arbitrary partially polarized light wave into its "natural" and "polarized" parts.

Solution: This resolution means the representation of the tensor $J_{\alpha\beta}$ in the form

$$J_{\alpha\beta} = \tfrac{1}{2} J^{(n)} \delta_{\alpha\beta} + E_{0\alpha}^{(p)} E_{0\beta}^{(p)*}.$$

The first term corresponds to the natural, and the second to the polarized parts of the light. To determine the intensities of the parts we note that the determinant

$$|J_{\alpha\beta} - \tfrac{1}{2} J^{(n)} \delta_{\alpha\beta}| = |E_{0\alpha}^{(p)} E_{0\beta}^{(p)*}| = 0.$$

Writing $J_{\alpha\beta} = J_{\rho\alpha\beta}$ in the form (50.13) and solving the equation, we get

$$J^{(n)} = J(1 - P).$$

The intensity of the polarized part is $J^{(p)} = |\mathbf{E}_0^{(p)}|^2 = J - J^{(n)} = JP$.

The polarized part of the light is in general an elliptically polarized wave, where the directions of the axes of the ellipse coincide with the principal axes of the tensor $S_{\alpha\beta}$. The lengths b_1 and b_2 of the axes of the ellipse and the angle ϕ formed by the axis $\mathbf{b}_1$ and the y axis are given by the equations:

$$b_1^2 + b_2^2 = JP, \qquad 2b_1 b_2 = JP\xi_2, \qquad \tan 2\phi = \frac{\xi_1}{\xi_3}.$$

2. Represent an arbitrary partially polarized wave as a superposition of two incoherent elliptically polarized waves.

Solution: For the hermitian tensor $\rho_{\alpha\beta}$ the "principal axes" are determined by two unit complex vectors $\mathbf{n}(\mathbf{n} \cdot \mathbf{n}^* = 1)$, satisfying the equations

$$\rho_{\alpha\beta} n_\beta = \lambda n_\alpha. \tag{1}$$

The principal values λ_1 and λ_2 are the roots of the equation

$$|\rho_{\alpha\beta} - \lambda \delta_{\alpha\beta}| = 0.$$

Multiplying (1) on both sides by n_α^*, we have:

$$\lambda = \rho_{\alpha\beta} n_\alpha^* n_\beta = \frac{1}{l} |E_{0\alpha} n_\alpha^*|^2,$$

---

† For a direct proof, we note that since the field of the wave is transverse in any reference frame, it is clear from the start that the tensor $\rho_{\alpha\beta}$ remains two-dimensional in any new frame. The transformation of $\rho_{\alpha\beta}$ into $\rho'_{\alpha\beta}$ leaves unchanged the sum of absolute squares $\rho_{\alpha\beta} \rho_{\alpha\beta}^*$ (in fact, the form of the transformation does not depend on the specific polarization properties of the light, while for a completely polarized wave this sum is 1 in any reference system). Because this transformation is real, the real and imaginary parts of the tensor $\rho_{\alpha\beta}$ (50.9) transform independently, so that the sums of the squares of the components of each separately remain constant, and are expressed in terms of l and A.

from which we see that λ_1, λ_2 are real and positive. Multiplying the equations

$$\rho_{\alpha\beta} n_\beta^{(1)} = \lambda_1 n_\alpha^{(1)}, \quad \rho_{\alpha\beta}^* n_\beta^{(2)*} = \lambda_2 n_\alpha^{(2)*}$$

for the first by $n_\alpha^{(2)*}$ and for the second by $n_\alpha^{(1)}$, taking the difference of the results and using the hermiticity of $\rho_{\alpha\beta}$, we get:

$$(\lambda_1 - \lambda_2) n_\alpha^{(1)} n_\alpha^{(2)*} = 0.$$

It then follows that $\mathbf{n}^{(1)} \cdot \mathbf{n}^{(2)*} = 0$, i.e. the unit vectors $\mathbf{n}^{(1)}$ and $\mathbf{n}^{(2)}$ are mutually orthogonal.

The expansion of the wave is provided by the formula

$$\rho_{\alpha\beta} = \lambda_1 n_\alpha^{(1)} n_\beta^{(1)*} + \lambda_2 n_\alpha^{(2)} n_\beta^{(2)*}.$$

One can always choose the complex amplitude so that, of the two mutually perpendicular components, one is real and the other imaginary (compare § 48). Setting

$$n_1^{(1)} = b_1, \quad n_2^{(1)} = ib_2$$

(where now b_1 and b_2 are understood to be normalized by the condition $b_1^2 + b_2^2 = 1$), we get from the equation $\mathbf{n}^{(1)} \cdot \mathbf{n}^{(2)*} = 0$:

$$n_1^{(2)} = ib_2, \quad n_2^{(2)} = b_1.$$

We then see that the ellipses of the two elliptically polarized vibrations are similar (have equal axis ratio), and one of them is turned through 90° relative to the other.

3. Find the law of transformation of the Stokes parameters for a rotation of the y, z axes through and angle ϕ.

Solution: The law is determined by the connection of the Stokes parameters to the components of the two-dimensional tensor in the yz plane, and is given by the formulas

$$\xi_1' = \xi_1 \cos 2\phi - \xi_3 \sin 2\phi, \qquad \xi_3' = \xi_1 \sin 2\phi + \xi_3 \cos 2\phi, \qquad \xi_2' = \xi_2.$$

§ 51. The Fourier resolution of the electrostatic field

The field produced by charges can also be formally expanded in plane waves (in a Fourier integral). This expansion, however, is essentially different from the expansion of electromagnetic waves in vacuum, for the field produced by charges does not satisfy the homogeneous wave equation, and therefore each term of this expansion does not satisfy the equation. From this it follows that for the plane waves into which the field of charges can be expanded, the relation $k^2 = \omega^2/c^2$, which holds for plane monochromatic electromagnetic waves, is not fulfilled.

In particular, if we formally represent the electrostatic field as a superposition of plane waves, then the "frequency" of these waves is clearly zero, since the field under consideration does not depend on the time. The wave vectors themselves are, of course, different from zero.

We consider the field produced by a point charge e, located at the origin of coordinates. The potential ϕ of this field is determined by the equation (see § 36)

$$\Delta\phi = -4\pi e \delta(\mathbf{r}). \tag{51.1}$$

We expand ϕ in a Fourier integral, i.e. we represent it in the form

$$\phi = \int\limits_{-\infty}^{+\infty} e^{i\mathbf{k}\cdot\mathbf{r}} \phi_{\mathbf{k}} \frac{d^3 k}{(2\pi)^3} \tag{51.2}$$

where $d^3 k$ denotes $dk_x\, dk_y\, dk_z$. In this formula $\phi_{\mathbf{k}} = \int \phi(\mathbf{r}) e^{-i\mathbf{k}\cdot\mathbf{r}} dV$. Applying the Laplace operator to both sides of (51.2), we obtain

$$\Delta\phi = - \int\limits_{-\infty}^{+\infty} k^2 e^{i\mathbf{k}\cdot\mathbf{r}} \phi_{\mathbf{k}} \frac{d^3 k}{(2\pi)^3},$$

so that the Fourier component of the expression $\Delta\phi$ is

$$(\Delta\phi)_{\mathbf{k}} = -k^2 \phi_{\mathbf{k}}.$$

On the other hand, we can find $(\Delta\phi)_{\mathbf{k}}$ by taking Fourier components of both sides of equation (51.1),

$$(\Delta\phi)_{\mathbf{k}} = - \int 4\pi e \delta(\mathbf{r}) e^{-i\mathbf{k}\cdot\mathbf{r}} dV = -4\pi e.$$

Equating the two expressions obtained for $(\Delta\phi)_{\mathbf{k}}$, we find

$$\phi_{\mathbf{k}} = \frac{4\pi e}{k^2}. \tag{51.3}$$

This formula solves our problem.

Just as for the potential ϕ, we can expand the field

$$\mathbf{E} = \int\limits_{-\infty}^{+\infty} \mathbf{E}_{\mathbf{k}} e^{i\mathbf{k}\cdot\mathbf{r}} \frac{d^3 k}{(2\pi)^3}. \tag{51.4}$$

With the aid of (51.2), we have

$$\mathbf{E} = - \operatorname{grad} \int\limits_{-\infty}^{+\infty} \phi_{\mathbf{k}} e^{i\mathbf{k}\cdot\mathbf{r}} \frac{d^3 k}{(2\pi)^3} = - \int i\mathbf{k}\phi_{\mathbf{k}} e^{i\mathbf{k}\cdot\mathbf{r}} \frac{d^3 k}{(2\pi)^3}.$$

Comparing with (51.4), we obtain

$$\mathbf{E}_{\mathbf{k}} = -i\mathbf{k}\phi_{\mathbf{k}} = -i \frac{4\pi e\mathbf{k}}{k^2}. \tag{51.5}$$

From this we see that the field of the waves, into which we have resolved the Coulomb field, is directed along the wave vector. Therefore these waves can be said to be *longitudinal*.

§ 52. Characteristic vibrations of the field

We consider an electromagnetic field (in the absence of charges) in some finite volume of space. To simplify further calculations we assume that this volume has the form of a rectangular parallelepiped with sides A, B, C, respectively. Then we can expand all quantities characterizing the field in this parallelepiped in a triple Fourier series[1] (for the three coordinates). This expansion can be written (e.g. for the vector potential) in the form:

$$\mathbf{A} = \sum_{\mathbf{k}} \mathbf{A}_{\mathbf{k}} e^{i\mathbf{k} \cdot \mathbf{r}} \tag{52.1}$$

explicitly indicating that $\mathbf{A}$ is real. The summation extends here over all possible values of the vector $\mathbf{k}$ whose components run through the values

$$k_x = \frac{2\pi n_x}{A}, \quad k_y = \frac{2\pi n_y}{B}, \quad k_z = \frac{2\pi n_z}{C}, \tag{52.2}$$

where n_x, n_y, n_z are positive or negative integers. Since $\mathbf{A}$ is real, the coefficients in the expansion (52.1) are related by the equations $\mathbf{A}_{-\mathbf{k}} = \mathbf{A}_{\mathbf{k}}^*$. From the equation div $\mathbf{A} = 0$ it follows that for each $\mathbf{k}$,

$$\mathbf{k} \cdot \mathbf{A}_{\mathbf{k}} = 0, \tag{52.3}$$

i.e., the complex vectors $\mathbf{A}_{\mathbf{k}}$ are "perpendicular" to the corresponding wave vectors $\mathbf{k}$. The vectors $\mathbf{A}_{\mathbf{k}}$ are, of course, functions of the time; from the wave equation (46.7), they satisfy the equation

$$\ddot{\mathbf{A}}_{\mathbf{k}} + c^2 k^2 \mathbf{A}_{\mathbf{k}} = 0. \tag{52.4}$$

If the dimensions A, B, C of the volume are sufficiently large, then neighbouring values of k_x, k_y, k_z (for which n_x, n_y, n_z differ by unity) are very close to one another. In this case we may speak of the number of possible values of k_x, k_y, k_z in the small intervals Δk_x, Δk_y, Δk_z.

Since to neighbouring values of, say, k_x, there correspond values of n_x differing by unity, the number Δn_x of possible values of k_x in the interval Δk_x is equal simply to the number of values of n_x in the corresponding interval. Thus, we obtain

$$\Delta n_x = \frac{A}{2\pi} \Delta k_x, \quad \Delta n_y = \frac{B}{2\pi} \Delta k_y, \quad \Delta n_z = \frac{C}{2\pi} \Delta k_z.$$

The total number Δn of possible values of the vector $\mathbf{k}$ with components in the intervals Δk_x, Δk_y, Δk_z is equal to the product $\Delta n_x \, \Delta n_y \, \Delta n_z$, that is,

$$\Delta n = \frac{V}{(2\pi)^3} \Delta k_x \Delta k_y \Delta k_z, \tag{52.5}$$

where $V = ABC$ is the volume of the field. It is easy to determine from this the number of possible values of the wave vector having absolute values in the interval Δk, and directed into the element of solid angle Δo. To get this we need only transform to polar coordinates in the "k space" and write in place of $\Delta k_x \Delta k_y \Delta k_z$ the element of volume in these coordinates. Thus

$$\Delta n = \frac{V}{(2\pi)^3} k^2 \Delta k \Delta o. \tag{52.6}$$

Replacing Δo by 4π, we find the number of possible values of $\mathbf{k}$ with absolute value in the interval Δk and pointing in all directions: $\Delta n = (V/2\pi^2)k^2\Delta k$.

We calculate the total energy

$$\mathscr{E} = \frac{1}{8\pi} \int (\mathbf{E}^2 + \mathbf{H}^2) dV$$

of the field, expressing it in terms of the quantities $\mathbf{A}_{\mathbf{k}}$. For the electric and magnetic fields we have

$$\mathbf{E} = -\frac{1}{c}\dot{\mathbf{A}} = -\frac{1}{c}\sum_{\mathbf{k}} \dot{\mathbf{A}}_{\mathbf{k}} e^{i\mathbf{k} \cdot \mathbf{r}},$$

$$\mathbf{H} = \text{curl } \mathbf{A} = i \sum_{\mathbf{k}} (\mathbf{k} \times \mathbf{A}_{\mathbf{k}}) e^{i\mathbf{k}\cdot\mathbf{r}}. \tag{52.7}$$

When calculating the squares of these sums, we must keep in mind that all products of terms with wave vectors $\mathbf{k}$ and $\mathbf{k}'$ such that $\mathbf{k} \neq \mathbf{k}'$ give zero on integration over the whole volume. In fact, such terms contain factors of the form $e^{i(\mathbf{k}+\mathbf{k}')\cdot\mathbf{r}}$, and the integral, e.g. of

$$\int_0^A e^{i\frac{2\pi}{A}n_x x} dx,$$

with integer n_x different from zero, gives zero. In those terms with $\mathbf{k}' = -\mathbf{k}$, the exponentials drop out and integration over dV gives just the volume V.

As a result, we obtain

$$\mathscr{E} = \frac{V}{8\pi} \sum_{\mathbf{k}} \left\{ \frac{1}{c^2} \dot{\mathbf{A}}_{\mathbf{k}} \cdot \dot{\mathbf{A}}_{\mathbf{k}}^* + (\mathbf{k} \times \mathbf{A}_{\mathbf{k}}) \cdot (\mathbf{k} \times \mathbf{A}_{\mathbf{k}}^*) \right\}.$$

From (52.3), we have

$$(\mathbf{k} \times \mathbf{A}_{\mathbf{k}}) \cdot (\mathbf{k} \times \mathbf{A}_{\mathbf{k}}^*) = k^2 \mathbf{A}_{\mathbf{k}} \cdot \mathbf{A}_{\mathbf{k}}^*,$$

so that

$$\mathscr{E} = \frac{V}{8\pi c^2} \sum_{\mathbf{k}} \{ \dot{\mathbf{A}}_{\mathbf{k}} \cdot \dot{\mathbf{A}}_{\mathbf{k}}^* + k^2 c^2 \mathbf{A}_{\mathbf{k}} \cdot \mathbf{A}_{\mathbf{k}}^* \}. \tag{52.8}$$

Each term of this sum corresponds to one of the terms of the expansion (52.1).

Because of (52.4), the vectors $\mathbf{A}_{\mathbf{k}}$ are harmonic functions of the time with frequencies $\omega_{\mathbf{k}} = ck$, depending only on the absolute value of the wave vector. Depending on the choice of these functions, the terms in the expansion (52.1) can represent standing or running plane waves. We shall write the expansion so that its terms describe running waves. To do this we write it in the form

$$\mathbf{A} = \sum_{\mathbf{k}} (\mathbf{a}_{\mathbf{k}} e^{i\mathbf{k}\cdot\mathbf{r}} + \mathbf{a}_{\mathbf{k}}^* e^{-i\mathbf{k}\cdot\mathbf{r}}) \tag{52.9}$$

which explicitly exhibits that $\mathbf{A}$ is real, and each of the vectors $\mathbf{a}_{\mathbf{k}}$ depends on the time according to the law

$$\mathbf{a}_{\mathbf{k}} \sim e^{-i\omega_{\mathbf{k}} t}, \quad \omega_{\mathbf{k}} = ck. \tag{52.10}$$

Then each individual term in the sum (52.9) will be a function only of the difference $\mathbf{k}\cdot\mathbf{r} - \omega_{\mathbf{k}}t$, which corresponds to a wave propagating in the $\mathbf{k}$ direction.

Comparing the expansions (52.9) and (52.1), we find that their coefficients are related by the formulas

$$\mathbf{A}_{\mathbf{k}} = \mathbf{a}_{\mathbf{k}} + \mathbf{a}_{-\mathbf{k}}^*,$$

and from (52.10) the time derivatives are related by

$$\dot{\mathbf{A}}_{\mathbf{k}} = -ick(\mathbf{a}_{\mathbf{k}} - \mathbf{a}_{-\mathbf{k}}^*).$$

Substituting in (52.8), we express the field energy in terms of the coefficients of the expansion (52.9). Terms with products of the form $\mathbf{a}_{\mathbf{k}}\cdot \mathbf{a}_{-\mathbf{k}}$ or $\mathbf{a}_{\mathbf{k}}^* \cdot \mathbf{a}_{-\mathbf{k}}^*$ cancel one another; also noting that the sums $\Sigma \mathbf{a_k} \cdot \mathbf{a_k^*}$ and $\Sigma \mathbf{a_{-k}} \mathbf{a_{-k}^*}$ differ only in the labelling of the summation index, and therefore coincide, we finally obtain:

$$\mathscr{E} = \sum_k \mathscr{E}_k, \quad \mathscr{E}_k = \frac{k^2 V}{2\pi} \mathbf{a_k} \cdot \mathbf{a_k^*}. \tag{52.11}$$

Thus the total energy of the field is expressed as a sum of the energies $\mathscr{E}_k$, associated with each of the plane waves individually.

In a completely analogous fashion, we can calculate the total momentum of the field,

$$\frac{1}{c^2} \int \mathbf{S} dV = \frac{1}{4\pi c} \int \mathbf{E} \times \mathbf{H} dV,$$

for which we obtain

$$\sum_k \frac{\mathbf{k}}{k} \frac{\mathscr{E}_k}{c}. \tag{52.12}$$

This result could have been anticipated in view of the relation between the energy and momentum of a plane wave (see § 47).

The expansion (52.9) succeeds in expressing the field in terms of a series of discrete parameters (the vectors $\mathbf{a_k}$), in place of the description in terms of a continuous series of parameters, which is essentially what is done when we give the potential $\mathbf{A}(x, y, z, t)$ at all points of space. We now make a transformation of the variables $\mathbf{a_k}$, which has the result that the equations of the field take on a form similar to the canonical equations (Hamilton equations) of mechanics.

We introduce the real "canonical variables" $\mathbf{Q_k}$ and $\mathbf{P_k}$ according to the relations

$$\mathbf{Q_k} = \sqrt{\frac{V}{4\pi c^2}} (\mathbf{a_k} + \mathbf{a_k^*}), \tag{52.13}$$

$$\mathbf{P_k} = -i\omega_k \sqrt{\frac{V}{4\pi c^2}} (\mathbf{a_k} - \mathbf{a_k^*}) = \dot{\mathbf{Q}}_k.$$

The Hamiltonian of the field is obtained by substituting these expressions in the energy (52.11):

$$\mathscr{H} = \sum_k \mathscr{H}_k = \sum_k \frac{1}{2} (\mathbf{P_k^2} + \omega_k^2 \mathbf{Q_k^2}). \tag{52.14}$$

Then the Hamilton equation $\partial \mathscr{H} / \partial \mathbf{P_k} = \dot{\mathbf{Q}}_k$ coincide with $\mathbf{P_k} = \dot{\mathbf{Q}}_k$, which is thus a consequence of the equations of motion. (This was achieved by an appropriate choice of the coefficient in (52.13).) The equations of motion, $\partial \mathscr{H} / \partial \mathbf{Q_k} = - \dot{\mathbf{P}}_k$, become the equations

$$\ddot{\mathbf{Q}}_k + \omega_k^2 \mathbf{Q_k} = 0, \tag{52.15}$$

that is, they are identical with the equations of the field.

Each of the vectors $\mathbf{Q_k}$ and $\mathbf{P_k}$ is perpendicular to the wave vector $\mathbf{k}$, i.e. has two independent components. The direction of these vectors determines the direction of polarization of the corresponding travelling wave. Denoting the two components of the vector $\mathbf{Q_k}$ (in the plane perpendicular to $\mathbf{k}$) by Q_{kj}, $j = 1, 2$, we have

$$\mathbf{Q_k^2} = \sum_j Q_{kj}^2,$$

and similarly for $\mathbf{P_k}$. Then

$$\mathcal{H} = \sum_{kj} \mathcal{H}_{kj}, \quad \mathcal{H}_{kj} = \tfrac{1}{2}(P_{kj}^2 + \omega_k^2 Q_{kj}^2). \tag{52.16}$$

We see that the Hamiltonian splits into a sum of independent terms $\mathcal{H}_{kj}$, each of which contains only one pair of the quantities Q_{kj}, P_{kj}. Each such term corresponds to a travelling wave with a definite wave vector and polarization. The quantity $\mathcal{H}_{kj}$ has the form of the Hamiltonian of a one-dimensional "oscillator", performing a simple harmonic vibration. For this reason, one sometimes refers to this result as the expansion of the field in terms of oscillators.

We give the formulas which express the field explicitly in terms of the variables $\mathbf{P_k}$, $\mathbf{Q_k}$. From (52.13), we have

$$\mathbf{a_k} = \frac{i}{k}\sqrt{\frac{\pi}{V}}\,(\mathbf{P_k} - i\omega_k \mathbf{Q_k}), \quad \mathbf{a_k^*} = -\frac{i}{k}\sqrt{\frac{\pi}{V}}\,(\mathbf{P_k} + i\omega_k \mathbf{Q_k}). \tag{52.17}$$

Substituting these expressions in (52.1), we obtain for the vector potential of the field:

$$\mathbf{A} = 2\sqrt{\frac{\pi}{V}}\sum_{k}\frac{1}{k}(ck\mathbf{Q_k}\cos\mathbf{k}\cdot\mathbf{r} - \mathbf{P_k}\sin\mathbf{k}\cdot\mathbf{r}). \tag{52.18}$$

For the electric and magnetic fields, we find

$$\mathbf{E} = -2\sqrt{\frac{\pi}{V}}\sum_{k}(ck\mathbf{Q_k}\sin\mathbf{k}\cdot\mathbf{r} + \mathbf{P_k}\cos\mathbf{k}\cdot\mathbf{r}),$$

$$\mathbf{H} = -2\sqrt{\frac{\pi}{V}}\sum_{k}\frac{1}{k}\{ck(\mathbf{k}\times\mathbf{Q_k})\sin\mathbf{k}\cdot\mathbf{r} + (\mathbf{k}\times\mathbf{P_k})\cos\mathbf{k}\cdot\mathbf{r}\}. \tag{52.19}$$

CHAPTER 7

THE PROPAGATION OF LIGHT

§ 53. Geometrical optics

A plane wave is characterized by the property that its direction of propagation and amplitude are the same everywhere. Arbitrary electromagnetic waves, of course, do not have this property. Nevertheless, a great many electromagnetic waves, which are not plane, have the property that within each small region of space they can be considered to be plane. For this, it is clearly necessary that the amplitude and direction of the wave remain practically constant over distances of the order of the wavelength. If this condition is satisfied, we can introduce the so-called *wave surface,* i.e. a surface at all of whose points the phase of the wave is the same (at a given time). (The wave surfaces of a plane wave are obviously planes perpendicular to the direction of propagation of the wave.) In each small region of space we can speak of a direction of propagation of the wave, normal to the wave surface. In this way we can introduce the concept of *rays*—curves whose tangents at each point coincide with the direction of propagation of the wave.

The study of the laws of propagation of waves in this case constitutes the domain of *geometrical optics.* Consequently, geometrical optics considers the propagation of waves, in particular of light, as the propagation of rays, completely divorced from their wave properties. In other words, geometrical optics corresponds to the limiting case of small wavelength, $\lambda \to 0$.

We now take up the derivation of the fundamental equation of geometrical optics—the equation determining the direction of the rays. Let f be any quantity describing the field of the wave (any component of **E** or **H**). For a plane monochromatic wave, f has the form

$$f = ae^{i(\mathbf{k} \cdot \mathbf{r} - \omega t + \alpha)} = ae^{i(-k_i x^i + \alpha)} \tag{53.1}$$

(we omit the Re; it is understood that we take the real part of all expressions).

We write the expression for the field in the form

$$f = ae^{i\psi}. \tag{53.2}$$

In case the wave is not plane, but geometrical optics is applicable, the amplitude a is, generally speaking, a function of the coordinates and time, and the phase ψ, which is called the *eikonal,* does not have a simple form, as in (53.1). It is essential, however, that ψ be a large quantity. This is clear immediately from the fact that it changes by 2π when we move through one wavelength, and geometrical optics corresponds to the limit $\lambda \to 0$.

Over small space regions and time intervals the eikonal ψ can be expanded in series; to terms of first order, we have

140

$$\psi = \psi_0 + \mathbf{r} \cdot \frac{\partial \psi}{\partial \mathbf{r}} + t \frac{\partial \psi}{\partial t}$$

(the origin for coordinates and time has been chosen within the space region and time interval under consideration; the derivatives are evaluated at the origin). Comparing this expression with (53.1), we can write

$$\mathbf{k} = \frac{\partial \psi}{\partial \mathbf{r}} \equiv \text{grad } \psi, \quad \omega = -\frac{\partial \psi}{\partial t}, \tag{53.3}$$

which corresponds to the fact that in each small region of space (and each small interval of time) the wave can be considered as plane. In four-dimensional form, the relation (53.3) is expressed as

$$k_i = -\frac{\partial \psi}{\partial x^i}, \tag{53.4}$$

where k_i is the wave four-vector.

We saw in § 48 that the components of the four-vector k^i are related by $k_i k^i = 0$. Substituting (53.4), we obtain the equation

$$\frac{\partial \psi}{\partial x_i} \frac{\partial \psi}{\partial x^i} = 0. \tag{53.5}$$

This equation, the *eikonal equation,* is the fundamental equation of geometrical optics.

The eikonal equation can also be derived by direct transition to the limit $\lambda \to 0$ in the wave equation. The field f satisfies the wave equation

$$\frac{\partial^2 f}{\partial x_i \partial x^i} = 0.$$

Substituting $f = ae^{i\psi}$, we obtain

$$\frac{\partial^2 a}{\partial x_i \partial x^i} e^{i\psi} + 2i \frac{\partial a}{\partial x_i} \frac{\partial \psi}{\partial x^i} e^{i\psi} + if \frac{\partial^2 \psi}{\partial x_i \partial x^i} - \frac{\partial \psi}{\partial x_i} \cdot \frac{\partial \psi}{\partial x^i} f = 0. \tag{53.6}$$

But the eikonal ψ, as we pointed out above, is a large quantity; therefore we can neglect the first three terms compared with the fourth, and we arrive once more at equation (53.5).

We shall give certain relations which, in their application to the propagation of light in vacuum, lead only to completely obvious results. Nevertheless, they are important because, in their general form, these derivations apply also to the propagation of light in material media.

From the form of the eikonal equation there results a remarkable analogy between geometrical optics and the mechanics of material particles. The motion of a material particle is determined by the Hamilton–Jacobi equation (16.11). This equation, like the eikonal equation, is an equation in the first partial derivatives and is of second degree. As we know, the action S is related to the momentum $\mathbf{p}$ and the Hamiltonian $\mathscr{H}$ of the particle by the relations

$$\mathbf{p} = \frac{\partial S}{\partial \mathbf{r}}, \quad \mathscr{H} = -\frac{\partial S}{\partial t}.$$

Comparing these formulas with the formulas (53.3), we see that the wave vector plays the same role in geometrical optics as the momentum of the particle in mechanics, while the frequency plays the role of the Hamiltonian, i.e., the energy of the particle. The absolute magnitude k of the wave vector is related to the frequency by the formula $k = \omega/c$. This relation is analogous to the relation $p = \mathscr{E}/c$ between the momentum and energy of a particle with zero mass and velocity equal to the velocity of light.

For a particle, we have the Hamilton equations

$$\dot{\mathbf{p}} = -\frac{\partial \mathscr{H}}{\partial \mathbf{r}}, \quad \mathbf{v} = \dot{\mathbf{r}} = \frac{\partial \mathscr{H}}{\partial \mathbf{p}}.$$

In view of the analogy we have pointed out, we can immediately write the corresponding equations for rays:

$$\dot{\mathbf{k}} = -\frac{\partial \omega}{\partial \mathbf{r}}, \quad \dot{\mathbf{r}} = \frac{\partial \omega}{\partial \mathbf{k}}. \tag{53.7}$$

In vacuum, $\omega = ck$, so that $\dot{\mathbf{k}} = 0$, $\mathbf{v} = c\mathbf{n}$ ($\mathbf{n}$ is a unit vector along the direction of propagation); in other words, as it must be, in vacuum the rays are straight lines, along which the light travels with velocity c.

The analogy between the wave vector of a wave and the momentum of a particle is made especially clear by the following consideration. Let us consider a wave which is a superposition of monochromatic waves with frequencies in a certain small interval and occupying some finite region in space (this is called a *wave packet*). We calculate the four-momentum of the field of this wave, using formula (32.6) with the energy-momentum tensor (48.15) (for each monochromatic component). Replacing k^i in this formula by some average value, we obtain an expression of the form

$$P^i = Ak^i, \tag{53.8}$$

where the coefficient of proportionality A between the two four-vectors P^i and k^i is some scalar. In three-dimensional form this relation gives:

$$\mathbf{P} = A\mathbf{k}, \quad \mathscr{E} = A\omega. \tag{53.9}$$

Thus we see that the momentum and energy of a wave packet transform, when we go from one reference system to another, like the wave vector and the frequency.

Pursuing the analogy, we can establish for geometrical optics a principle analogous to the principle of least action in mechanics. However, it cannot be written in Hamiltonian form as $\delta \int L \, dt = 0$, since it turns out to be impossible to introduce, for rays, a function analogous to the Lagrangian of a particle. Since the Lagrangian of a particle is related to the Hamiltonian $\mathscr{H}$ by the equation $L = \mathbf{p} \cdot \partial \mathscr{H}/\partial \mathbf{p} - \mathscr{H}$, replacing the Hamiltonian $\mathscr{H}$ by the frequency ω and the momentum by the wave vector $\mathbf{k}$, we should have to write for the Lagrangian in optics $\mathbf{k} \cdot \partial \omega/\partial \mathbf{k} - \omega$. But this expression is equal to zero, since $\omega = ck$. The impossibility of introducing a Lagrangian for rays is also clear directly from the consideration mentioned earlier that the propagation of rays is analogous to the motion of particles with zero mass.

If the wave has a definite constant frequency ω, then the time dependence of its field is given by a factor of the form $e^{-i\omega t}$. Therefore for the eikonal of such a wave we can write

$$\psi = -\omega t + \psi_0(x, y, z), \tag{53.10}$$

where ψ_0 is a function only of the coordinates. The eikonal equation (53.5) now takes the form

$$(\text{grad }\psi_0)^2 = \frac{\omega^2}{c^2}. \tag{53.11}$$

The wave surfaces are the surfaces of constant eikonal, i.e. the family of surfaces of the form $\psi_0(x, y, z) = \text{const}$. The rays themselves are at each point normal to the corresponding wave surface; their direction is determined by the gradient $\nabla\psi_0$.

As is well known, in the case where the energy is constant, the principle of least action for particles can also be written in the form of the so-called *principle of Maupertuis:*

$$\delta S = \delta \int \mathbf{p} \cdot d\mathbf{l} = 0,$$

where the integration extends over the trajectory of the particle between two of its points. In this expression the momentum is assumed to be a function of the energy and the coordinates. The analogous principle for rays is called *Fermat's principle*. In this case, we can write by analogy:

$$\delta\psi = \delta \int \mathbf{k} \cdot d\mathbf{l} = 0. \tag{53.12}$$

In vacuum, $\mathbf{k} = (\omega/c)\mathbf{n}$, and we obtain ($d\mathbf{l} \cdot \mathbf{n} = dl$):

$$\delta \int dl = 0, \tag{53.13}$$

which corresponds to rectilinear propagation of the rays.

§ 54. Intensity

In geometrical optics, the light wave can be considered as a bundle of rays. The rays themselves, however, determine only the direction of propagation of the light at each point; there remains the question of the distribution of the light intensity in space.

On some wave surface of the bundle of rays under consideration, we isolate an infinitesimal surface element. From differential geometry it is known that every surface has, at each of its points, two (generally different) principal radii of curvature. Let ac and bd (Fig. 7) be elements of the principal circles of curvature, constructed at a given element of the wave surface. Then the rays passing through a and c meet at the corresponding centre of curvature O_1, while the rays passing through b and d meet at the other centre of curvature O_2.

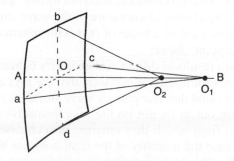

FIG. 7.

For fixed angular openings of the beams starting from O_1 and O_2, the lengths of the arcs ac and bd are, clearly, proportional to the corresponding radii of curvature R_1 and R_2 (i.e. to the lengths O_1O and O_2O). The area of the surface element is proportional to the product of the lengths ac and bd, i.e., proportional to R_1R_2. In other words, if we consider the element of the wave surface bounded by a definite set of rays, then as we move along them the area of the element will change proportionally to R_1R_2.

On the other hand, the intensity, i.e. the energy flux density, is inversely proportional to the surface area through which a given amount of light energy passes. Thus we arrive at the result that the intensity is

$$I = \frac{\text{const}}{R_1 R_2}. \tag{54.1}$$

This formula must be understood as follows. On each ray (AB in Fig. 7) there are definite points O_1 and O_2, which are the centres of curvature of all the wave surfaces intersecting the given ray. The distances OO_1 and OO_2 from the point O where the wave surface intersects the ray, to the points O_1 and O_2, are the radii of curvature R_1 and R_2 of the wave surface at the point O. Thus formula (54.1) determines the change in intensity of the light along a given ray as a function of the distances from definite points on this ray. We emphasize that this formula cannot be used to compare intensities at different points on a single wave surface.

Since the intensity is determined by the square modulus of the field, we can write for the change of the field itself along the ray

$$f = \frac{\text{const}}{\sqrt{R_1 R_2}} e^{ikR}, \tag{54.2}$$

where in the phase factor e^{ikR} we can write either e^{ikR_1} or e^{ikR_2}. The quantities e^{ikR_1} and e^{ikR_2}. (for a given ray) differ from each other only by a constant factor, since the difference $R_1 - R_2$, the distance between the two centres of curvature, is a constant.

If the two radii of curvature of the wave surface coincide, then (54.1) and (54.2) have the form:

$$I = \frac{\text{const}}{R^2}, \quad f = \frac{\text{const}}{R} e^{ikR}. \tag{54.3}$$

This happens always when the light is emitted from a point source (the wave surfaces are then concentric spheres and R is the distance from the light source).

From (54.1) we see that the intensity becomes infinite at the points $R_1 = 0$, $R_2 = 0$, i.e. at the centres of curvature of the wave surface. Applying this to all the rays in a bundle, we find that the intensity of the light in the given bundle becomes infinite, generally, on two surfaces—the geometrical loci of all the centres of curvature of the wave surfaces. These surfaces are called *caustics*. In the special case of a beam of rays with spherical wave surfaces, the two caustics fuse into a single point (*focus*).

We note from well-known results of differential geometry concerning the properties of the loci of centres of curvature of a family of surfaces, that the rays are tangent to the caustic.

It is necessary to keep in mind that (for convex wave surfaces) the centres of curvature of the wave surfaces can turn out to lie not on the rays themselves, but on their extensions beyond the optical system from which they emerge. In such cases we speak of *imaginary caustics* (or foci). In this case the intensity of the light does not become infinite anywhere.

As for the increase of intensity to infinity, in actuality we must understand that the intensity does become large at points on the caustic, but it remains finite (see the problem in § 59). The formal increase to infinity means that the approximation of geometrical optics is never applicable in the neighbourhood of the caustic. To this is related the fact that the change in phase along the ray can be determined from formula (54.2) only over sections of the ray which do not include its point of tangency to the caustic. Later (in § 59), we shall show that actually in passing through the caustic the phase of the field decreases by $\pi/2$. This means that if, on the section of the ray before its first intersection with the caustic, the field is proportional to the factor e^{ikx} (x is the coordinate along the ray), then after passage through the caustic the field will be proportional to $e^{i(kx-(\pi/2))}$. The same thing occurs in the neighbourhood of the point of tangency to the second caustic, and beyond that point the field is proportional to $e^{i(kx-\pi)}$.†

§ 55. The angular eikonal

A light ray travelling in vacuum and impinging on a transparent body will, on its emergence from this body, generally have a direction different from its initial direction. This change in direction will, of course, depend on the specific properties of the body and on its form. However, it turns out that one can derive general laws relating to the change in direction of a light ray on passage through an arbitrary material body. In this it is assumed only that geometrical optics is applicable to rays propagating in the interior of the body under consideration. As is customary, we shall call such transparent bodies, through which rays of light propagate, *optical systems*.

Because of the analogy mentioned in § 53, between the propagation of rays and the motion of particles, the same general laws are valid for the change in direction of motion of a particle, initially moving in a straight line in vacuum, then passing through some electromagnetic field, and once more emerging into vacuum. For definiteness, we shall, however, always speak later of the propagation of light rays.

We saw in a previous section that the eikonal equation, describing the propagation of the rays, can be written in the form (53.11) (for light of a definite frequency). From now on we shall, for convenience, designate by ψ the eikonal ψ_0 divided by the constant ω/c. Then the basic equation of geometrical optics has the form:

$$(\nabla\psi)^2 = 1. \tag{55.1}$$

Each solution of this equation describes a definite beam of rays, in which the direction of the rays passing through a given point in space is determined by the gradient of ψ at that point. However, for our purposes this description is insufficient, since we are seeking general relations determining the passage through an optical system not of a single definite bundle of rays, but of arbitrary rays. Therefore we must use an eikonal expressed in such a form that it describes all the generally possible rays of light, i.e. rays passing through any pair of points in space. In its usual form the eikonal $\psi(\mathbf{r})$ is the phase of the rays in a certain bundle passing through the point $\mathbf{r}$. Now we must introduce the eikonal as a function $\psi(\mathbf{r}, \mathbf{r}')$ of the coordinates of two points ($\mathbf{r}, \mathbf{r}'$ are the radius vectors of the initial and end points of the ray). A ray can pass through each pair of points $\mathbf{r}, \mathbf{r}'$, and $\psi(\mathbf{r}, \mathbf{r}')$ is the phase difference (or, as it is called, the *optical path length*) of this ray between the points $\mathbf{r}$ and $\mathbf{r}'$. From now on we shall always understand by $\mathbf{r}$ and $\mathbf{r}'$ the radius vectors to points on the ray before and after its passage through the optical system.

---

† Although formula (54.2) itself is not valid near the caustic, the change in phase of the field corresponds formally to a change in sign (i.e. multiplication by $e^{i\pi}$) of R_1 or R_2 in this formula.

If in $\psi(\mathbf{r}, \mathbf{r}')$ one of the radius vectors, say $\mathbf{r}'$, is fixed, then ψ as a function of $\mathbf{r}$ describes a definite bundle of rays, namely, the bundle of rays passing through the point $\mathbf{r}'$. Then ψ must satisfy equation (55.1), where the differentiations are applied to the components of $\mathbf{r}$. Similarly, if $\mathbf{r}$ is assumed fixed, we again obtain an equation for $\psi(\mathbf{r}, \mathbf{r}')$, so that

$$(\nabla_{\mathbf{r}}\psi)^2 = 1, \quad (\nabla_{\mathbf{r}'}\psi)^2 = 1. \tag{55.2}$$

The direction of the ray is determined by the gradient of its phase. Since $\psi(\mathbf{r}, \mathbf{r}')$ is the difference in phase at the points $\mathbf{r}$ and $\mathbf{r}'$, the direction of the ray at the point $\mathbf{r}'$ is given by the vector $\mathbf{n}' = \partial\psi/\partial\mathbf{r}'$, and at the point $\mathbf{r}$ by the vector $\mathbf{n} = -\partial\psi/\partial\mathbf{r}$. From (55.2) it is clear that $\mathbf{n}$ and $\mathbf{n}'$ are unit vectors:

$$\mathbf{n}^2 = \mathbf{n}'^2 = 1. \tag{55.3}$$

The four vectors $\mathbf{r}, \mathbf{r}', \mathbf{n}, \mathbf{n}'$ are interrelated, since two of them $(\mathbf{n}, \mathbf{n}')$ are derivatives of a certain function ψ with respect to the other two $(\mathbf{r}, \mathbf{r}')$. The function ψ itself satisfies the auxiliary conditions (55.2).

To obtain the relation between $\mathbf{n}, \mathbf{n}', \mathbf{r}, \mathbf{r}'$, it is convenient to introduce, in place of ψ, another quantity, on which no auxiliary condition is imposed (i.e., is not required to satisfy any differential equations). This can be done as follows. In the function ψ the independent variables are $\mathbf{r}$ and $\mathbf{r}'$, so that for the differential $d\psi$ we have

$$d\psi = \frac{\partial\psi}{\partial\mathbf{r}} \cdot d\mathbf{r} + \frac{\partial\psi}{\partial\mathbf{r}'} \cdot d\mathbf{r}' = -\mathbf{n} \cdot d\mathbf{r} + \mathbf{n}' \cdot d\mathbf{r}'.$$

We now make a Legendre transformation from $\mathbf{r}, \mathbf{r}'$ to the new independent variables $\mathbf{n}$, $\mathbf{n}'$, that is, we write

$$d\psi = -d(\mathbf{n} \cdot \mathbf{r}) + \mathbf{r} \cdot d\mathbf{n} + d(\mathbf{n}' \cdot \mathbf{r}'(-\mathbf{r}' \cdot d\mathbf{n}'.$$

from which, introducing the function

$$\chi = \mathbf{n}' \cdot \mathbf{r}' - \mathbf{n} \cdot \mathbf{r} - \psi, \tag{55.4}$$

we have

$$d\chi = -\mathbf{r} \cdot d\mathbf{n} + \mathbf{r}' \cdot d\mathbf{n}'. \tag{55.5}$$

The function χ is called the *angular eikonal*; as we see from (55.5), the independent variables in it are $\mathbf{n}$ and $\mathbf{n}'$. No auxiliary conditions are imposed on χ. In fact, equation (55.3) now states only a condition referring to the independent variables: of the three components n_x, n_y, n_z, of the vector $\mathbf{n}$ (and similarly for $\mathbf{n}'$), only two are independent. As independent variables we shall use n_y, n_z, n_y', n_z'; then

$$n_x = \sqrt{1 - n_y^2 - n_z^2}, \quad n_x' = \sqrt{1 - n_y'^2 - n_z'^2}.$$

Substituting these expressions in

$$d\chi = -x\,dn_x - y\,dn_y - z\,dn_z + x'dn_x' + y'\,dn_y' + z'\,dn_z',$$

we obtain for the differential $d\chi$:

$$d\chi = -\left(y - \frac{n_y}{n_x}x\right)dn_y - \left(z - \frac{n_z}{n_x}x\right)dn_z + \left(y' - \frac{n_y'}{n_x'}x'\right)dn_y' + \left(z' - \frac{n_z'}{n_x'}x'\right)dn_z'.$$

From this we obtain, finally, the following equations:

$$y - \frac{n_y}{n_x}x = -\frac{\partial \chi}{\partial n_y}, \quad z - \frac{n_z}{n_x}x = -\frac{\partial \chi}{\partial n_z},$$

$$y' - \frac{n_y'}{n_x'}x' = \frac{\partial \chi}{\partial n_y'}, \quad z' - \frac{n_z'}{n_x'}x' = \frac{\partial \chi}{\partial n_z'},$$

(55.6)

which is the relation sought between **n**, **n'**, **r**, **r'**. The function χ characterizes the special properties of the body through which the rays pass (or the properties of the field, in the case of the motion of a charged particle).

For fixed values of **n**, **n'**, each of the two pairs of equations (55.6) represent a straight line. These lines are precisely the rays before and after passage through the optical system. Thus the equation (55.6) directly determines the path of the ray on the two sides of the optical system.

§ 56. Narrow bundles of rays

In studying the passage of beams of rays through optical systems, special interest attaches to bundles whose rays all pass through one point (such bundles are said to be *homocentric*).

After passage through an optical system, homocentric bundles in general cease to be homocentric, i.e. after passing through a body the rays no longer come together in any one point. Only in exceptional cases will the rays starting from a luminous point come together after passage through an optical system and all meet at one point (the image of the luminous point).†

One can show (see § 57) that the only case for which all homocentric bundles remain strictly homocentric after passage through the optical system is the case of identical imaging, i.e. the case where the image differs from the object only in its position or orientation, or is mirror inverted.

Thus no optical system can give a completely sharp image of an object (having finite dimensions) except in the trivial case of identical imaging.‡ Only approximate, but not completely sharp images can be produced of an extended body, in any case other than for identical imaging.

The most important case where there is approximate transition of homocentric bundles into homocentric bundles is that of sufficiently narrow beams (i.e. beams with a small opening angle) passing close to a particular line (for a given optical system). This line is called the *optic axis* of the system.

Nevertheless, we must note that even infinitely narrow bundles of rays (in the three-dimensional case) are in general not homocentric; we have seen (Fig. 7) that even in such a bundle different rays intersect at different points (this phenomenon is called *astigmatism*). Exceptions are those points of the wave surface at which the two principal radii of curvature are equal—a small region of the surface in the neighbourhood of such points can be considered as spherical, and the corresponding narrow bundle of rays is homocentric.

---

† The point of intersection can lie either on the rays themselves or on their continuations; depending on this, the image is said to be *real* or *virtual*.

‡ Such imaging can be produced with a plane mirror.

We consider an optical system having axial symmetry.† The axis of symmetry of the system is also its optical axis. The wave surface of a bundle of rays travelling along this axis also has axial symmetry; as we know, surfaces of rotation have equal radii of curvature at their points of intersection with the symmetry axis. Therefore a narrow bundle moving in this direction remains homocentric.

To obtain general quantitative relations, determining image formation with the aid of narrow bundles, passing through an axially-symmetric optical system, we use the general equations (55.6) after determining first of all the form of the function χ in the case under consideration.

Since the bundles of rays are narrow and move in the neighbourhood of the optical axis, the vectors $\mathbf{n}$, $\mathbf{n}'$ for each bundle are directed almost along this axis. If we choose the optical axis as the X axis, then the components, n_y, n_z, n_y', n_z' will be small compared with unity. As for the components n_x, n_x'; $n_x \approx 1$ and n_x' can be approximately equal to either $+1$ or -1. In the first case the rays continue to travel almost in their original direction, emerging into the space on the other side of the optical system, which in this case is called a *lens*. In the second the rays change their direction to almost the reverse; such an optical system is called a *mirror*.

Making use of the smallness of n_y, n_z, n_y', n_z', we expand the angular eikonal $\chi(n_y, n_z, n_y', n_z')$ in series and stop at the first terms. Because of the axial symmetry of the whole system, χ must be invariant with respect to rotations of the coordinate system around the optical axis. From this it is clear that in the expansion of χ there can be no terms of first order, proportional to the first powers of the y- and z-components of the vectors $\mathbf{n}$ and $\mathbf{n}'$; such terms would not have the required invariance. The terms of second order which have the required property are the squares $\mathbf{n}^2$ and $\mathbf{n}'^2$ and the scalar product $\mathbf{n} \cdot \mathbf{n}'$. Thus, to terms of second order, the angular eikonal of an axially-symmetric optical system has the form

$$\chi = \text{const} + \frac{g}{2}(n_y^2 + n_z^2) + f(n_y n_y' + n_z n_z') + \frac{h}{2}(n_y'^2 + n_z'^2), \qquad (56.1)$$

where f, g, h are constants.

For definiteness, we now consider a lens, so that we set $n_x' \approx 1$; for a mirror, as we shall show later, all the formulas have a similar appearance. Now substituting the expression (56.1) in the general equations (55.6), we obtain:

$$n_y(x - g) - fn_y' = y, \qquad fn_y + n_y'(x' + h) = y',$$
$$n_z(x - g) - fn_z' = z, \qquad fn_z + n_z'(x' + h) = z'. \qquad (56.2)$$

We consider a homocentric bundle emanating from the point x, y, z; let the point x', y', z' be the point in which all the rays of the bundle intersect after passing through the lens. If the first and second pairs of equations (56.2) were independent, then these four equations, for given x, y, z, x', y', z', would determine one definite set of values n_y, n_z, n_y', n_z', that is, there would be just *one* ray starting from the point x, y, z, which would pass through the point x', y', z'. In order that all rays starting from x, y, z shall pass through x', y', z', it is consequently necessary that the equations (56.2) not be independent, that is, one pair of these equations must be a consequence of the other. The necessary condition for this dependence is that the

---

† It can be shown that the problem of image formation with the aid of narrow bundles, moving in the neighbourhood of the optical axis in a nonaxially-symmetric system, can be reduced to image formation in an axially-symmetric system plus a subsequent rotation of the image thus obtained, relative to the object.

coefficients in the one pair of equations be proportional to the coefficients of the other pair. Thus we must have

$$\frac{x-g}{f} = -\frac{f}{x'+h} = \frac{y}{y'} = \frac{z}{z'}. \tag{56.3}$$

In particular,

$$(x-g)(x'+h) = -f^2. \tag{56.4}$$

The equations we have obtained give the required connection between the coordinates of the image and object for image formation using narrow bundles.

The points $x = g$ and $x' = -h$ on the optical axis are called the *principal foci* of the optical system. Let us consider bundles of rays parallel to the optical axis. The source point of such rays is, clearly, located at infinity on the optical axis, that is, $x = \infty$. From (56.3) we see that in this case, $x' = -h$. Thus a parallel bundle of rays, after passage through the optical system, intersects at the principal focus. Conversely, a bundle of rays emerging from the principal focus becomes parallel after passage through the system.

In the equation (56.3) the coordinates x and x' are measured from the same origin of coordinates, lying on the optical axis. It is, however, more convenient to measure the coordinates of object and image from different origins, choosing them at the corresponding principal foci. As positive direction of the coordinates we choose the direction from the corresponding focus toward the side to which the light travels. Designating the new co-ordinates of object and image by capital letters, we have

$$X = x - g, \quad X' = x' + h, \quad Y = y, \quad Y' = y', \quad Z = z, \quad Z' = z'.$$

The equations of image formation (56.3) and (56.4) in the new coordinates take the form

$$XX' = -f^2, \tag{56.5}$$

$$\frac{Y'}{Y} = \frac{Z'}{Z} = \frac{f}{X} = -\frac{X'}{f}. \tag{56.6}$$

The quantity f is called the *principal focal length* of the system.

The ratio Y'/Y is called the *lateral magnification*. As for the *longitudinal magnification*, since the coordinates are not simply proportional to each other, it must be written in differential form, comparing the length of an element of the object (along the direction of the axis) with the length of the corresponding element in the image. From (56.5) we get for the "longitudinal magnification"

$$\left| \frac{dX'}{dX} \right| = \frac{f^2}{X^2} = \left(\frac{Y'}{Y} \right)^2. \tag{56.7}$$

We see from this that even for an infinitely small object, it is impossible to obtain a geometrically similar image. The longitudinal magnification is never equal to the transverse (except in the trivial case of identical imaging).

A bundle passing through the point $X = f$ on the optical axis intersects once more at the point $X' = -f$ on the axis; these two points are called *principal points*. From equation (56.2) $(n_y X - f n'_y = Y, n_z X - f n'_z = Z)$ it is clear that in this case $(X = f, Y = Z = 0)$, we have the equations $n_y = n'_y, n_z = n'_z$. Thus every ray starting from a principal point crosses the optical axis again at the other principal point in a direction parallel to its original direction.

If the coordinates of object and image are measured from the principal points (and not from the principal foci), then for these coordinates ξ and ξ', we have

$$\xi' = X' + f, \quad \xi = X - f.$$

Substituting in (56.5) it is easy to obtain the equations of image formation in the form

$$\frac{1}{\xi} - \frac{1}{\xi'} = -\frac{1}{f}. \tag{56.8}$$

One can show that for an optical system with small thickness (for example, a mirror or a thin lens), the two principal points almost coincide. In this case the equation (56.8) is particularly convenient, since in it ξ and ξ' are then measured practically from one and the same point.

If the focal distance is positive, then objects located in front of the focus ($X > 0$) are imaged erect ($Y'/Y > 0$); such optical systems are said to be *converging*. If $f < 0$, then for $X > 0$ we have $Y'/Y < 0$, that is, the object is imaged in inverted form; such systems are said to be *diverging*.

There is one limiting case of image formation which is not contained in the formulas (56.8); this is the case where all three coefficients f, g, h are infinite (i.e. the optical system has an infinite focal distance and its principal foci are located at infinity). Going to the limit of infinite f, g, h in (56.4) we obtain

$$x' = \frac{h}{g}x + \frac{f^2 - gh}{g}.$$

Since we are interested only in the case where the object and its image are located at finite distances from the optical system, f, g, h must approach infinity in such fashion that the ratios h/g, $(f^2 - gh)/g$ are finite. Denoting them, respectively, by α^2 and β, we have

$$x' = \alpha^2 x + \beta.$$

For the other two coordinates we now have from the general equation (56.7):

$$\frac{y'}{y} = \frac{z'}{z} = \pm \alpha.$$

Finally, again measuring the coordinates x and x' from different origins, namely from some arbitrary point on the axis and from the image of this point, respectively, we finally obtain the equations of image formation in the simple form

$$X' = \alpha^2 X, \quad Y' = \pm \alpha Y, \quad Z' = \pm \alpha Z. \tag{56.9}$$

Thus the longitudinal and transverse magnifications are constants (but not equal to each other). This case of image formation is called *telescopic*.

All the equations (56.5) through (56.9), derived by us for lenses, apply equally to mirrors, and even to an optical system without axial symmetry, if only the image formation occurs by means of narrow bundles of rays travelling near the optical axis. In this, the reference points for the x coordinates of object and image must always be chosen along the optical axis from corresponding points (principal foci or principal points) in the direction of propagation of the ray. In doing this, we must keep in mind that for an optical system not possessing axial symmetry, the directions of the optical axis in front of and beyond the system do not lie in the same plane.

PROBLEMS

1. Find the focal distance for image formation with the aid of two axially-symmetric optical systems whose optical axes, coincide.

Solution: Let f_1 and f_2 be the focal lengths of the two systems. For each system separately, we have

$$X_1 X_1' = -f_1^2, \qquad X_2 X_2' = -f_2^2.$$

Since the image produced by the first system acts as the object for the second, then denoting by l the distance between the rear principal focus of the first system and the front focus of the second, we have $X_2 = X_1' - l$; expressing X_2' in terms of X_1, we obtain

$$X_2' = \frac{X_1 f_2^2}{f_1^2 + l X_1}$$

or

$$\left(X_1 + \frac{f_1^2}{l} \right)\left(X_2' - \frac{f_2^2}{l} \right) = -\left(\frac{f_1 f_2}{l} \right)^2,$$

from which it is clear that the principal foci of the composite system are located at the points $X_1 = -f_1^2/l, X_2' = f_2^2/l$ and the focal length is

$$f = -\frac{f_1 f_2}{l}$$

(to choose the sign of this expression, we must write the corresponding equation for the transverse magnification).

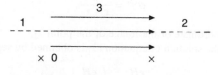

FIG. 8.

In case $l = 0$, the focal length $f = \infty$, that is, the composite system gives telescopic image formation. In this case we have $X_2' = X_1(f_2/f_1)^2$, that is, the parameter α in the general formula (56.9) is $\alpha = f_2/f_1$.

2. Find the focal length for charged particles of a "magnetic lens" in the form of a longitudinal homogeneous field in the section of length l (Fig. 8).†

Solution: The kinetic energy of the particle is conserved during its motion in a magnetic field; therefore the Hamilton–Jacobi equation for the reduced action $S_0(\mathbf{r})$ (where the total action is $S = -\mathscr{E}t + S_0$) is

$$\left(\nabla S_0 - \frac{e}{c}\mathbf{A} \right)^2 = p^2,$$

where

$$p^2 - \frac{\mathscr{E}^2}{c^2} - m^2 c^3 = \text{const.}$$

Using formula (19.4) for the vector potential of the homogeneous magnetic field, choosing the x axis along the field direction and considering this axis as the optical axis of an axially-symmetric optical system, we get the Hamilton–Jacobi equation in the form:

† This might be the field inside a long solenoid, when we neglect the disturbance of the homogeneity of the field near the ends of the solenoid.

$$\left(\frac{\partial S_0}{\partial x}\right)^2 + \left(\frac{\partial S_0}{\partial r}\right)^2 + \frac{e^2}{4c^2}H^2 r^2 = p^2, \tag{1}$$

where r is the distance from the x axis, and S_0 is a function of x and r.

For narrow beams of particles propagating close to the optical axis, the coordinate r is small, so that accordingly we try to find S_0 as a power series in r. The first two terms of this series are

$$S_0 = px + \tfrac{1}{2}\sigma(x)r^2, \tag{2}$$

where $\sigma(x)$ satisfies the equation

$$p\sigma'(x) + \sigma^2 - \frac{e^2}{4c^2}H^2 = 0. \tag{3}$$

In region 1 in front of the lens, we have:

$$\sigma^{(1)} = \frac{p}{x - x_1}$$

where $x_1 < 0$ is a constant. This solution corresponds to a free beam of particles, emerging along straight line rays from the point $x = x_1$ on the optical axis in region 1. In fact, the action function for the free motion of a particle with a momentum p in a direction out from the point $x = x_1$ is

$$S_0 = p\sqrt{r^2 + (x - x_1)^2} \cong p(x - x_1) + \frac{pr^2}{2(x - x_1)}.$$

Similarly, in region 2 behind the lens we write:

$$\sigma^{(2)} = \frac{p}{x - x_2},$$

where the constant x_2 is the coordinate of the image of the point x_1.

In region 3 inside the lens, the solution of equation (3) is obtained by separation of variables, and gives:

$$\sigma^{(3)} = \frac{eH}{2c} \cot\left(\frac{eH}{2cp}x + C\right),$$

where C is an arbitrary constant.

The constant C and x_2 (for given x_1) are determined by the requirements of continuity of $\sigma(x)$ for $x = 0$ and $x = l$:

$$-\frac{p}{x_1} = \frac{eH}{2c}\cot C, \quad \frac{p}{l - x_2} = \frac{eH}{2c}\cot\left(\frac{eH}{2cp}l + C\right).$$

Eliminating the constant C from these equations, we find:

$$(x_1 - g)(x_2 + h) = -f^2,$$

where†

$$g = -\frac{2cp}{eH}\cot\frac{eHl}{2cp}, \quad h = g + l,$$

$$f = \frac{2cp}{eH \sin\dfrac{eHl}{2cp}}.$$

† The value of f is given with the correct sign. However, to show this requires additional investigation.

§ 57. Image formation with broad bundles of rays

The formation of images with the aid of narrow bundles of rays, which was considered in the previous section, is approximate; it is the more exact (i.e. the sharper) the narrower the bundles. We now go over to the question of image formation with bundles of rays of arbitrary breadth.

In contrast to the formation of an image of an object by narrow beams, which can be achieved for any optical system having axial symmetry, image formation with broad beams is possible only for specially constituted optical systems. Even with this limitation, as already pointed out in § 56, image formation is not possible for all points in space.

The later derivations are based on the following essential remark. Suppose that all rays, starting from a certain point O and travelling through the optical system, intersect again at some other point O'. It is easy to see that the optical path length ψ is the same for all these rays. In the neighbourhood of each of the points O, O', the wave surfaces for the rays intersecting in them are spheres with centres at O and O', respectively, and, in the limit as we approach O and O', degenerate to these points. But the wave surfaces are the surfaces of constant phase, and therefore the change in phase along different rays, between their points of intersection with two given wave surfaces, is the same. From what has been said, it follows that the total change in phase between the points O and O' is the same (for the different rays).

Let us consider the conditions which must be fulfilled in order to have formation of an image of a small line segment using broad beams; the image is then also a small line segment. We choose the directions of these segments as the directions of the ξ and ξ' axes, with origins at any two corresponding points O and O' of the object and image. Let ψ be the optical path length for the rays starting from O and reaching O'. For the rays starting from a point infinitely near to O with coordinate $d\xi$, and arriving at a point of the image with coordinate $d\xi'$, the optical path length is $\psi + d\psi$, where

$$d\psi = \frac{\partial \psi}{\partial \xi} d\xi + \frac{\partial \psi}{\partial \xi'} d\xi'.$$

We introduce the "magnification"

$$\alpha_\xi = \frac{d\xi'}{d\xi}$$

as the ratio of the length $d\xi'$ of the element of the image to the length $d\xi$ of the imaged element. Because of the smallness of the line segment which is being imaged, the quantity α can be considered constant along the line segment. Writing, as usual, $\partial \psi / \partial \xi = -n_\xi$, $\partial \psi / \partial \xi' = n'_\xi$ (n_ξ, n'_ξ are the cosines of the angles between the directions of the ray and the corresponding axes ξ and ξ'), we obtain

$$d\psi = (\alpha_\xi n'_\xi - n_\xi)d\xi.$$

As for every pair of corresponding points of object and image, the optical path length $\psi + d\psi$ must be the same for all rays starting from the point $d\xi$ and arriving at the point $d\xi'$. From this we obtain the condition:

$$\alpha_\xi n'_\xi - n_\xi = \text{const.} \tag{57.1}$$

This is the condition we have been seeking, which the paths of the rays in the optical system must satisfy in order to have image formation for a small line segment using broad beams. The relation (57.1) must be fulfilled for all rays starting from the point O.

Let us apply this condition to image formation by means of an axially-symmetric optical system. We start with the image of a line segment coinciding with the optical axis (x axis); clearly the image also coincides with the axis. A ray moving along the optical axis ($n_x = 1$), because of the axial symmetry of the system, does not change its direction after passing through it, that is, n_x' is also 1. From this it follows that const in (57.1) is equal in this case to $\alpha_x - 1$, and we can rewrite (57.1) in the form

$$\frac{1 - n_x}{1 - n_x'} = \alpha_x.$$

Denoting by θ and θ' the angles subtended by the rays with the optical axis at points of the object and image, we have

$$1 - n_x = 1 - \cos\theta = 2\sin^2\frac{\theta}{2}, \quad 1 - n_x' = 1 - \cos\theta' = 2\sin^2\frac{\theta'}{2}.$$

Thus we obtain the condition for image formation in the form

$$\frac{\sin\frac{\theta}{2}}{\sin\frac{\theta'}{2}} = \text{const} = \sqrt{\alpha_x}. \tag{57.2}$$

Next, let us consider the imaging of a small portion of a plane perpendicular to the optical axis of an axially symmetric system; the image will obviously also be perpendicular to this axis. Applying (57.1) to an arbitrary segment lying in the plane which is to be imaged, we get:

$$\alpha_r \sin\theta' - \sin\theta = \text{const},$$

where θ and θ' are again the angles made by the beam with the optical axis. For rays emerging from the point of intersection of the object plane with the optical axis, and directed along this axis ($\theta = 0$), we must have $\theta' = 0$, because of symmetry. Therefore const is zero, and we obtain the condition for imaging in the form

$$\frac{\sin\theta}{\sin\theta'} = \text{const} = \alpha_r. \tag{57.3}$$

As for the formation of an image of a three-dimensional object using broad beams, it is easy to see that this is impossible even for a small volume, since the conditions (57.2) and (57.3) are incompatible.

§ 58. The limits of geometrical optics

From the definition of a monochromatic plane wave, its amplitude is the same everywhere and at all times. Such a wave is infinite in extent in all directions in space, and exists over the whole range of time from $-\infty$ to $+\infty$. Any wave whose amplitude is not constant everywhere at all times can only be more or less monochromatic. We now take up the question of the "degree of non-monochromaticity" of a wave.

Let us consider an electromagnetic wave whose amplitude at each point is a function of the time. Let ω_0 be some average frequency of the wave. Then the field of the wave, for example the electric field, at a given point has the form $\mathbf{E}_0(t)e^{-i\omega_0 t}$. This field, although it is of course not monochromatic, can be expanded in monochromatic waves, that is, in a Fourier integral. The amplitude of the component in the expansion, with frequency ω, is proportional to the integral

$$\int_{-\infty}^{+\infty} \mathbf{E}_0(t)e^{i(\omega-\omega_0)t}\, dt.$$

The factor $e^{i(\omega-\omega_0)t}$ is a periodic function whose average value is zero. If $\mathbf{E}_0$ were exactly constant, then the integral would be exactly zero, for $\omega \neq \omega_0$. If, however, $\mathbf{E}_0(t)$ is variable, but hardly changes over a time interval of order $1/|\omega - \omega_0|$, then the integral is almost equal to zero, the more exactly the slower the variation of $\mathbf{E}_0$. In order for the integral to be significantly different from zero, it is necessary that $\mathbf{E}_0(t)$ vary significantly over a time interval of the order of $1/|\omega - \omega_0|$.

We denote by Δt the order of magnitude of the time interval during which the amplitude of the wave at a given point in space changes significantly. From these considerations, it now follows that the frequencies deviating most from ω_0, which appear with reasonable intensity in the spectral resolution of this wave, are determined by the condition $1/|\omega - \omega_0| \sim \Delta t$. If we denote by $\Delta\omega$ the frequency interval (around the average frequency ω_0) which enters in the spectral resolution of the wave, then we have the relation

$$\Delta\omega\Delta t \sim 1. \tag{58.1}$$

We see that a wave is the more monochromatic (i.e. the smaller $\Delta\omega$) the larger Δt, i.e. the slower the variation of the amplitude at a given point in space.

Relations similar to (58.1) are easily derived for the wave vector. Let Δx, Δy, Δz be the orders of magnitude of distances along the X, Y, Z axes, in which the wave amplitude changes significantly. At a given time, the field of the wave as a function of the coordinates has the form

$$\mathbf{E}_0(\mathbf{r})e^{i\mathbf{k}_0 \cdot \mathbf{r}},$$

where $\mathbf{k}_0$ is some average value of the wave vector. By a completely analogous derivation to that for (58.1) we can obtain the interval $\Delta\mathbf{k}$ of values contained in the expansion of the wave into a Fourier integral:

$$\Delta k_x \Delta x \sim 1, \quad \Delta k_y \Delta y \sim 1, \quad \Delta k_z \Delta z \sim 1. \tag{58.2}$$

Let us consider, in particular, a wave which is radiated during a finite time interval. We denote by Δt the order of magnitude of this interval. The amplitude at a given point in space changes significantly during the time Δt in the course of which the wave travels completely past the point. Because of the relations (58.1) we can now say that the "lack of monochromaticity" of such a wave, $\Delta\omega$, cannot be smaller than $1/\Delta t$ (it can of course be larger):

$$\Delta\omega \gtrsim \frac{1}{\Delta t}. \tag{58.3}$$

Similarly, if Δx, Δy, Δz are the orders of magnitude of the extension of the wave in space, then for the spread in the values of components of the wave vector, entering in the resolution of the wave, we obtain

$$\Delta k_x \gtrsim \frac{1}{\Delta x}, \quad \Delta k_y \gtrsim \frac{1}{\Delta y}, \quad \Delta k_z \lesssim \frac{1}{\Delta z}. \tag{58.4}$$

From these formulas it follows that if we have a beam of light of finite width, then the direction of propagation of the light in such a beam cannot be strictly constant. Taking the X axis along the (average) direction of light in the beam, we obtain

$$\theta_y \gtrsim \frac{1}{k\Delta y} \sim \frac{\lambda}{\Delta y}, \tag{58.5}$$

where θ_y is the order of magnitude of the deviation of the beam from its average direction in the XY plane and λ is the wavelength.

On the other hand, the formula (58.5) answers the question of the limit of sharpness of optical image formation. A beam of light whose rays, according to geometrical optics, would all intersect in a point, actually gives an image not in the form of a point but in the form of a spot. For the width Δ of this spot, we obtain, according to (58.5),

$$\Delta \sim \frac{1}{k\theta} \sim \frac{\lambda}{\theta}, \tag{58.6}$$

where θ is the opening angle of the beam. This formula can be applied not only to the image but also to the object. Namely, we can state that in observing a beam of light emerging from a luminous point, this point cannot be distinguished from a body of dimensions λ/θ. In this way formula (58.6) determines the limiting *resolving power* of a microscope. The minimum value of Δ, which is reached for $\theta \sim 1$, is λ, in complete agreement with the fact that the limit of geometrical optics is determined by the wavelength of the light.

PROBLEM

Determine the order of magnitude of the smallest width of a light beam produced from a parallel beam at a distance l from a diaphragm.

Solution: Denoting the size of the aperture in the diaphragm by d, we have from (58.5) for the angle of deflection of the beam (the "diffraction angle"), λ/d, so that the width of the beam is of order $d + (\lambda/d)l$. The smallest value of this quantity $\sim \sqrt{\lambda l}$.

§ 59. Diffraction

The laws of geometrical optics are strictly correct only in the ideal case when the wavelength can be considered to be infinitely small. The more poorly this condition is fulfilled, the greater are the deviations from geometrical optics. Phenomenon which are the consequence of such deviations are called *diffraction phenomena*.

Diffraction phenomena can be observed, for example, if along the path of propagation of the light† there is an obstacle—an opaque body (we call it a *screen*) of arbitrary form or, for example, if the light passes through holes in opaque screens. If the laws of geometrical optics were strictly satisfied, there would be beyond the screen regions of "shadow" sharply delineated from regions where light falls. The diffraction has the consequence that, instead of a sharp boundary between light and shadow, there is a quite complex distribution of the

† In what follows, in discussing diffraction we shall talk of the diffraction of light; all these same considerations also apply, of course, to any electromagnetic wave.

intensity of the light. These diffraction phenomena appear the more strongly the smaller the dimensions of the screens and the apertures in them, or the greater the wavelength.

The problems of the theory of diffraction consists in determining, for given positions and shapes of the objects (and locations of the light sources), the distribution of the light, that is, the electromagnetic field over all space. The exact solution of this problem is possible only through solution of the wave equation with suitable boundary conditions at the surface of the body, these conditions being determined also by the optical properties of the material. Such a solution usually presents great mathematical difficulties.

However, there is an approximate method which for many cases is a satisfactory solution of the problem of the distribution of light near the boundary between light and shadow. This method is applicable to cases of small deviation from geometrical optics, i.e. when firstly, the dimensions of all bodies are large compared with the wavelength (this requirement applies both to the dimensions of screens and apertures and also to the distances from the bodies to the points of emission and observation of the light); and secondly when there are only small deviations of the light from the directions of the rays given by geometrical optics.

Let us consider a screen with an aperture through which the light passes from given sources. Figure 9 shows the screen in profile (the heavy line); the light travels from left to right. We denote by u some one of the components of $\mathbf{E}$ or $\mathbf{H}$. Here we shall understand u to mean a function only of the coordinates, i.e. without the factor $e^{-i\omega t}$ determining the time dependence. Our problem is to determine the light intensity, that is, the field u, at any point of observation P beyond the screen. For an approximate solution of this problem in cases where the deviations from geometrical optics are small, we may assume that at the points of the aperture the field is the same as it would have been in the absence of the screen. In other words, the values of the field here are those which follow directly from geometrical optics. At all points immediately behind the screen, the field can be set equal to zero. In this the properties of the screen (i.e. of the screen material) obviously play no part. It is also obvious that in the cases we are considering, what is important for the diffraction is only the shape of the edge of the aperture, while the shape of the opaque screen is unimportant.

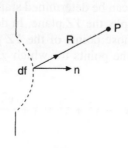

Fig. 9.

We introduce some surface which covers the aperture in the screen and is bounded by its edges (a profile of such a surface is shown in Fig. 9 as a dashed line). We break up this surface into sections with area df, whose dimensions are small compared with the size of the aperture, but large compared with the wavelength of the light. We can then consider each of these sections through which the light passes as if it were itself a source of light waves spreading out on all sides from this section. We shall consider the field at the point P to be the result of superposition of the fields produced by all the sections df of the surface covering the aperture. (This is called *Huygens' principle*.)

The field produced at the point P by the section df is obviously proportional to the value u of the field at the section df itself (we recall that the field at df is assumed to be the same as it would have been in the absence of the screen). In addition, it is proportional to the projection df_n of the area df on the plane perpendicular to the direction n of the ray coming from the light source to df. This follows from the fact that no matter what shape the element df has, the same rays will pass through it provided its projection df_n remain fixed, and therefore its effect on the field at P will be the same.

Thus the field produced at the point P by the section df is proportional to $u\, df_n$. Furthermore, we must still take into account the change in the amplitude and phase of the wave during its propagation from df to P. The law of this change is determined by formula (54.3). Therefore $u\, df_n$ must be multiplied by $(1/R)e^{ikR}$ (where R is the distance from df to P, and k is the absolute value of the wave vector of the light), and we find that the required field is

$$au\,\frac{e^{ikR}}{R}\,df_n,$$

where a is an as yet unknown constant. The field at the point P, being the result of the addition of the fields produced by all the elements df, is consequently equal to

$$u_p = a \int u\,\frac{e^{ikR}}{R}\,df_n, \tag{59.1}$$

where the integral extends over the surface bounded by the edge of the aperture. In the approximation we are considering, this integral cannot, of course, depend on the form of this surface. Formula (59.1) is, obviously, applicable not only to diffraction by an aperture in a screen, but also to diffraction by a screen around which the light passes freely. In that case the surface of integration in (59.1) extends on all sides from the edge of the screen.

To determine the constant a, we consider a plane wave propagating along the X axis; the wave surfaces are parallel to the plane YZ. Let u be the value of the field in the YZ plane. Then at the point P, which we choose on the X axis, the field is equal to $u_p = ue^{ikx}$. On the other hand, the field at the point P can be determined starting from formula (59.1), choosing as surface of integration, for example, the YZ plane. In doing this, because of the smallness of the angle of diffraction, only those points of the YZ plane are important in the integral which lie close to the origin, i.e. the points for which $y, z \ll x$ (x is the coordinate of the point P). Then

$$R = \sqrt{x^2 + y^2 + z^2} \approx x + \frac{y^2 + z^2}{2x},$$

and (59.1) gives

$$u_p = au\,\frac{e^{ikx}}{x}\int\limits_{-\infty}^{+\infty} e^{i\frac{ky^2}{2x}}\,dy \int\limits_{-\infty}^{+\infty} e^{i\frac{kz^2}{2x}}\,dz,$$

where u is a constant (the field in the YZ plane); in the factor $1/R$, we can put $R \simeq x = \text{const.}$ By the substitution $y = \xi\sqrt{2x/k}$ these two integrals can be transformed to the integral

$$\int\limits_{-\infty}^{+\infty} e^{i\xi^2}\,d\xi = \int\limits_{-\infty}^{+\infty} \cos\xi^2\,d\xi + i\int\limits_{-\infty}^{+\infty} \sin\xi^2\,d\xi = \sqrt{\frac{\pi}{2}}\,(1 + i),$$

and we get

$$u_p = aue^{ikx}\frac{2i\pi}{k}.$$

On the other hand, $u_p = ue^{ikx}$, and consequently

$$a = \frac{k}{2\pi i}$$

Substituting in (59.1), we obtain the solution to our problem in the form

$$u_p = \int \frac{ku}{2\pi i R}e^{ikR}\,df_n.\tag{59.2}$$

In deriving formula (59.2), the light source was assumed to be essentially a point, and the light was assumed to be strictly monochromatic. The case of a real, extended source, which emits non-monochromatic light, does not, however, require special treatment. Because of the complete independence (incoherence) of the light emitted by different points of the source, and the incoherence of the different spectral components of the emitted light, the total diffraction pattern is simply the sum of the intensity distributions obtained from the diffraction of the independent components of the light.

Let us apply formula (59.2) to the solution of the problem of the change in phase of a ray on passing through its point of tangency to the caustic (see the end of § 54). We choose as our surface of integration in (59.2) any wave surface, and determine the field u_p at a point P, lying on some given ray at a distance x from its point of intersection with the wave surface we have chosen (we choose this point as coordinate origin O, and as YZ plane the plane tangent to the wave surface at the point O). In the integration of (59.2) only a small area of the wave surface in the neighbourhood of O is important. If the XY and XZ planes are chosen to coincide with the principal planes of curvature of the wave surface at the point O, then near this point the equation of the surface is

$$X = \frac{y^2}{2R_1} + \frac{z^2}{2R_2},$$

where R_1 and R_2 are the radii of curvature. The distance R from the point on the wave surface with coordinates X, y, z, to the point P with coordinates x, 0, 0, is

$$R = \sqrt{(x-X)^2 + y^2 + z^2} \simeq x + \frac{y^2}{2}\left(\frac{1}{x} - \frac{1}{R_1}\right) + \frac{z^2}{2}\left(\frac{1}{x} - \frac{1}{R_2}\right).$$

On the wave surface, the field u can be considered constant; the same applies to the factor $1/R$. Since we are interested only in changes in the place of the wave, we drop coefficients and write simply

$$u_p \sim \frac{1}{i}\int e^{ikR}\,df_n \simeq \frac{e^{ikx}}{i}\int\limits_{-\infty}^{+\infty} dy\, e^{ik\frac{y^2}{2}\left(\frac{1}{x} - \frac{1}{R_1}\right)} \int\limits_{-\infty}^{+\infty} dz\, e^{ik\frac{z^2}{2}\left(\frac{1}{x} - \frac{1}{R_2}\right)}.\tag{59.3}$$

The centres of curvature of the wave surface lie on the ray we are considering, at the points $x = R_1$ and $x = R_2$; these are the points where the ray is tangent to the caustic. Suppose $R_2 < R_1$. For $x < R_2$, the coefficients of i in the exponentials appearing in the two integrands are positive, and each of these integrals is proportional to $(1 + i)$. Therefore on the part of the ray before its first tangency to the caustic, we have $u_p \sim e^{ikx}$. For $R_2 < x < R_1$, that is, on the segment of the ray between its two points of tangency, the integral over y is proportional to $1 + i$, but the integral over z is proportional to $1 - i$, so that their product does not contain i. Thus we have here $u_p \sim -ie^{ikx} = e^{i(kx - (\pi/2))}$, that is, as the ray passes in the neighbourhood of the first caustic, its phase undergoes an additional change of $-\pi/2$. Finally, for $x > R_1$, we have $u_p \sim - e^{ikx} = e^{i(kx-\pi)}$, that is, on passing in the neighbourhood of the second caustic, the phase once more changes by $- \pi/2$.

PROBLEM

Determine the distribution of the light intensity in the neighbourhood of the point where the ray is tangent to the caustic.

Solution: To solve the problem, we use formula (59.2), taking the integral in it over any wave surface which is sufficiently far from the point of tangency of the ray to the caustic. In Fig. 10, ab is a section of this wave surface, and $a'b'$ is a section of the caustic; $\alpha'b'$ is the evolute of the curve ab. We are interested in the intensity distribution in the neighbourhood of the point O where the ray QO is tangent to the caustic; we assume the length D of the segment QO of the ray to be large. We denote by x the distance from the point O along the normal to the caustic, and assume positive values x for points on the normal in the direction of the centre of curvature.

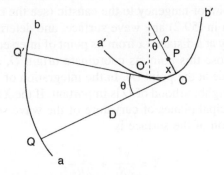

FIG. 10.

The integrand in (59.2) is a function of the distance R from the arbitrary point Q' on the wave surface to the point P. From a well-known property of the evolute, the sum of the length of the segment $Q'O'$ of the tangent at the point O' and the length of the arc OO' is equal to the length QO of the tangent at the point O. For points O and O' which are near to each other we have $OO' = \theta\varrho$ (ϱ is the radius of curvature of the caustic at the point O). Therefore the length $Q'O' = D - \theta\varrho$. The distance $Q'O$ (along a straight line) is approximately (the angle θ is assumed to be small)

$$Q'O \cong Q'O' + \varrho \sin \theta = D - \theta\varrho + \varrho \sin \theta \cong D - \varrho \frac{\theta^3}{6}.$$

Finally, the distance $R = Q'P$ is equal to $R \cong Q'O - x \sin \theta \cong Q'O - x\theta$, that is,

$$R \cong D - x\theta - \tfrac{1}{6}\varrho\theta^3.$$

Substituting this expression in (59.2), we obtain

$$u_p \sim \int_{-\infty}^{+\infty} e^{-ikr\theta - i\frac{k\varrho}{6}\theta^3} \, d\theta = 2 \int_{0}^{\infty} \cos\left(kx\theta + \frac{k\varrho}{6}\theta^3 \right) d\theta$$

(the slowly varying factor $1/D$ in the integrand is unimportant compared with the exponential factor, so we assume it constant). Introducing the new integration variable $\xi = (k\varrho/2)^{1/3}\,\theta$, we get

$$u_p \sim \Phi\left(x\left(\frac{2k^2}{\varrho}\right)^{1/3} \right),$$

where $\Phi(t)$ is the Airy function.†

For the intensity $I \sim |u_p|^2$, we write:

$$I = 2A\left(\frac{2k^2}{\varrho}\right)^{1/6} \Phi^2\left(x\left(\frac{2k^2}{\varrho}\right)^{1/3} \right)$$

(concerning the choice of the constant factor, cf, below).

For large positive values of x, we have from this the asymptotic formula

$$I \approx \frac{A}{2\sqrt{x}}\exp\left\{ -\frac{4x^{3/2}}{3}\sqrt{\frac{2k^2}{\varrho}} \right\},$$

that is, the intensity drops exponentially (shadow region). For large negative values of x, we have

$$I \approx \frac{2A}{\sqrt{-x}}\sin^2\left\{ \frac{2(-x)^{3/2}}{3}\sqrt{\frac{2k^2}{\rho}} + \frac{\pi}{4} \right\},$$

that is, the intensity oscillates rapidly; its average value over these oscillations is

$$\bar{I} = \frac{A}{\sqrt{-x}}.$$

From this meaning of the constant A is clear—it is the intensity far from the caustic which would be obtained from geometrical optics neglecting diffraction effects.

† The Airy function $\Phi(t)$ is defined as

$$\Phi(t) = \frac{1}{\sqrt{\pi}}\int_0^\infty \cos\left(\frac{\xi^3}{3} + \xi t\right)d\xi. \qquad (1)$$

(see *Quantum Mechanics*, Mathematical Appendices, § b). For large positive values of the argument, the asymptotic expression for $\Phi(t)$ is

$$\Phi(t) \approx \frac{1}{2t^{1/4}}\exp\left(-\frac{2}{3}t^{3/2}\right), \qquad (2)$$

that is, $\Phi(t)$ goes exponentially to zero. For large negative values of t, the function $\Phi(t)$ oscillates with decreasing amplitude according to the law:

$$\Phi(t) \approx \frac{1}{(-t)^{1/4}}\sin\left(\frac{2}{3}(-t)^{3/2} + \frac{\pi}{4}\right). \qquad (3)$$

The Airy function is related to the MacDonald function (modified Hankel function) of order $1/3$:

$$\Phi(t) = \sqrt{t/3\pi}\,K_{1/3}(\tfrac{2}{3}t^{3/2}). \qquad (4)$$

Formula (2) corresponds to the asymptotic expansion of $K_\nu(t)$:

$$K\nu(t) \approx \sqrt{\frac{\pi}{2t}}e^{-t}.$$

The function $\Phi(t)$ attains its largest value, 0.949, for $t = -1.02$; correspondingly, the maximum intensity is reached at $x(2k^2/\varrho)^{1/3} = -1.02$, where

$$I = 2.03\ Ak^{1/3}\varrho^{-1/6}.$$

At the point where the ray is tangent to the caustic ($x = 0$), we have $I = 0.89\ Ak^{1/3}\varrho^{-1/6}$ [since $\Phi(0) = 0.629$].

Thus near the caustic the intensity is proportional to $k^{1/3}$, that is, to $\lambda^{-1/3}$ (λ is the wavelength). For $\lambda \to 0$, the intensity goes to infinity, as it should (see § 54).

§ 60. Fresnel diffraction

If the light source and the point P at which we determine the intensity of the light are located at finite distances from the screen, then in determining the intensity at the point P, only those points are important which lie in a small region of the wave surface over which we integrate in (59.2)—the region which lies near the line joining the source and the point P. In fact, since the deviations from geometrical optics are small, the intensity of the light arriving at P from various points of the wave surface decreases very rapidly as we move away from this line. Diffraction phenomena in which only a small portion of the wave surface plays a role are called *Fresnel diffraction* phenomena.

Let us consider the Fresnel diffraction by a screen. From what we have just said, for a given point P only a small region at the edge of the screen is important for this diffraction. But over sufficiently small regions, the edge of the screen can always be considered to be a straight line. We shall therefore, from now on, understand the edge of the screen to mean just such a small straight line segment.

We choose as the XY plane a plane passing through the light source Q (Fig. 11) and through the line of the edge of the screen. Perpendicular to this, we choose the plane XZ so that it passes through the point Q and the point of observation P, at which we try to determine the light intensity. Finally, we choose the origin of coordinates O on the line of the edge of the screen, after which the positions of all three axes are completely determined.

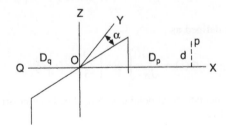

FIG. 11.

Let the distance from the light source Q to the origin be D_q. We denote the x-coordinate of the point of observation P by D_p, and its z-coordinate, i.e. its distance from the XY plane, by d. According to geometrical optics, the light should pass only through points lying above the XY plane; the region below the XY plane is the region which according to geometrical optics should be in shadow (region of geometrical shadow).

We now determine the distribution of light intensity on the screen near the edge of the geometrical shadow, i.e. for values of d small compared with D_p and D_q. A negative d means that the point P is located within the geometrical shadow.

As the surface of integration in (59.2) we choose the half-plane passing through the line of the edge of the screen and perpendicular to the XY plane. The coordinates x and y of points on this surface are related by the equation $x = y \tan \alpha$ (α is the angle between the line of the edge of the screen and the Y axis), and the z-coordinate is positive. The field of the wave produced by the source Q, at the distance R_q from it, is proportional to the factor e^{ikR_q}. Therefore the field u on the surface of integration is proportional to

$$u \sim \exp\{ik\sqrt{y^2 + z^2 + (D_q + y \tan \alpha)^2}\}.$$

In the integral (59.2) we must now substitute for R,

$$R = y^2 + (z - d)^2 + (D_p - y \tan \alpha)^2.$$

The slowly varying factors in the integrand are unimportant compared with the exponential. Therefore we may consider $1/R$ constant, and write $dy\, dz$ in place of df_n. We then find that the field at the point P is

$$u_p \sim \int_{-\infty}^{+\infty}\int_0^{\infty} \exp\{ik\sqrt{(D_q + y \tan \alpha)^2 + y^2 + z^2}$$

$$+ \sqrt{(D_p - y \tan \alpha)^2 + (z - d)^2 + y^2}\}\, dy\, dz. \tag{60.1}$$

As we have already said, the light passing through the point P comes mainly from points of the plane of integration which are in the neighbourhood of O. Therefore in the integral (60.1) only values of y and z which are small (compared with D_q and D_p) are important. For this reason we can write

$$\sqrt{(D_q + y \tan \alpha)^2 + y^2 + z^2} \simeq D_q + \frac{y^2 + z^2}{2D_q} + y \tan \alpha,$$

$$\sqrt{(D_p - y \tan \alpha)^2 + (z - d)^2 + y^2} \simeq D_p + \frac{(z - d)^2 + y^2}{2D_p} - y \tan \alpha.$$

We substitute this in (60.1). Since we are interested only in the field as a function of the distance d, the constant factor $\exp\{ik(D_p + D_q)\}$ can be omitted; the integral over y also gives an expression not containing d, so we omit it also. We then find

$$u_p \sim \int_0^{\infty} \exp\left\{ik\left(\frac{1}{2D_q}z^2 + \frac{1}{2D_p}(z - d)^2\right)\right\}dz.$$

This expression can also be written in the form

$$u_p \sim \exp\left\{ik\frac{d^2}{2(D_p + D_q)}\right\}\int_0^{\infty} \exp\left\{ik\frac{\frac{1}{2}\left[\left(\frac{1}{D_p} + \frac{1}{D_q}\right)z - \frac{d}{D_p}\right]^2}{\frac{1}{D_p} + \frac{1}{D_q}}\right\}dz. \tag{60.2}$$

The light intensity is determined by the square of the field, that is, by the square modulus $|u_p|^2$. Therefore, when calculating the intensity, the factor standing in front of the integral is irrelevant, since when multiplied by the complex conjugate expression it gives unity. An obvious substitution reduces the integral to

$$u_p \sim \int_{-w}^{\infty} e^{i\eta^2} \, d\eta, \tag{60.3}$$

where

$$w = d \sqrt{\frac{kD_q}{2D_p(D_q + D_p)}}. \tag{60.4}$$

Thus, the intensity I at the point P is :

$$I = \frac{I_0}{2} \left| \sqrt{\frac{2}{\pi}} \int_{-w}^{\infty} e^{i\eta^2} \, d\eta \right|^2 = \frac{I_0}{2} \left\{ \left(C(w^2) + \frac{1}{2} \right)^2 + \left(S(w^2) + \frac{1}{2} \right)^2 \right\}, \tag{60.5}$$

where

$$C(z) = \sqrt{\frac{2}{\pi}} \int_0^{\sqrt{z}} \cos \eta^2 \, d\eta, \quad S(z) = \sqrt{\frac{2}{\pi}} \int_0^{\sqrt{z}} \sin \eta^2 \, d\eta$$

are called the *Fresnel integrals*. Formula (60.5) solves our problem of determining the light intensity as a function of d. The quantity I_0 is the intensity in the illuminated region at points not too near the edge of the shadow; more precisely, at those points with $w \gg 1$ ($C(\infty) = S(\infty) = \frac{1}{2}$ in the limit $w \to \infty$).

The region of geometrical shadow corresponds to negative w. It is easy to find the asymptotic form of the function $I(w)$ for large negative values of w. To do this we proceed as follows. Integrating by parts, we have

$$\int_{|w|}^{\infty} e^{i\eta^2} \, d\eta = -\frac{1}{2i|w|} e^{iw^2} + \frac{1}{2i} \int_{|w|}^{\infty} e^{i\eta^2} \frac{d\eta}{\eta^2}.$$

Integrating by parts once more on the right side of the equation and repeating this process, we obtain an expansion in powers of $1/|w|$:

$$\int_{|w|}^{\infty} e^{i\eta^2} \, d\eta = e^{iw^2} \left[-\frac{1}{2i|w|} + \frac{1}{4|w|^3} - \cdots \right]. \tag{60.6}$$

Although an infinite series of this type does not converge, nevertheless, because the sucessive terms decrease very rapidly for large values of $|w|$, the first term already gives a good representation of the function on the left for sufficiently large $|w|$ (such a series is said to be *asymptotic*). Thus, for the intensity $I(w)$, (60.5), we obtain the following asymptotic formula, valid for large negative values of w:

$$I = \frac{I_0}{4\pi w^2}. \tag{60.7}$$

We see that in the region of geometric shadow, far from its edge, the intensity goes to zero as the inverse square of the distance from the edge of the shadow.

We now consider positive values of w, that is, the region above the XY plane. We write

$$\int_{-w}^{\infty} e^{i\eta^2}\, d\eta = \int_{-\infty}^{+\infty} e^{i\eta^2}\, d\eta - \int_{-\infty}^{-w} e^{i\eta^2}\, d\eta = (1 + i)\sqrt{\frac{\pi}{2}} - \int_{w}^{\infty} e^{i\eta^2}\, d\eta .$$

For sufficiently large w, we can use an asymptotic representation for the integral standing on the right side of the equation, and we have

$$\int_{-w}^{\infty} e^{i\eta^2}\, d\eta \cong (1 + i)\sqrt{\frac{\pi}{2}} + \frac{1}{2iw} e^{iw^2} . \tag{60.8}$$

Substituting this expression in (60.5), we obtain

$$I = I_0 \left(1 + \sqrt{\frac{1}{\pi}} \; \frac{\sin\left(w^2 - \dfrac{\pi}{4} \right)}{w} \right). \tag{60.9}$$

Thus in the illuminated region, far from the edge of the shadow, the intensity has an infinite sequence of maxima and minima, so that the ratio I/I_0 oscillates on both sides of unity. With increasing w, the amplitude of these oscillations decreases inversely with the distance from the edge of the geometric shadow, and the positions of the maxima and minima steadily approach one another.

For small w, the function $I(w)$ has qualitatively this same character (Fig. 12). In the region of the geometric shadow, the intensity decreases monotonically as we move away from the boundary of the shadow. (On the boundary itself, $I/I_0 = \frac{1}{4}$.) For positive w, the intensity has alternating maxima and minima. At the first (largest) maximum, $I/I_0 = 1.37$.

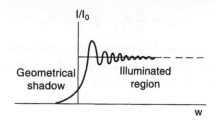

FIG. 12.

§ 61. Fraunhofer diffraction

Of special interest for physical applications are those diffraction phenomena which occur when a plane parallel bundle of rays is incident on a screen. As a result of the diffraction, the beam ceases to be parallel, and there is light propagation along directions other than the initial one. Let us consider the problem of determining the distribution over direction of the intensity of the diffracted light at large distances beyond the screen (this formulation of the problem corresponds to *Fraunhofer diffraction*). Here we shall again restrict ourselves to the case of small deviations from geometrical optics, i.e. we shall assume that the angles of deviation of the rays from the initial direction (the diffraction angles) are small.

This problem can be solved by starting from the general formula (59.2) and passing to the limit where the light source and the point of observation are at infinite distances from the screen. A characteristic feature of the case we are considering is that, in the integral which determines the intensity of the diffracted light, the whole wave surface over which the integral is taken is important (in contrast to the case of Fresnel diffraction, where only the portions of the wave surface near the edge of the screens are important).†

However, it is simpler to treat this problem anew, without recourse to the general formula (59.2).

Let us denote by u_0 the field which would exist beyond the screens if geometrical optics were rigorously valid. This field is a plane wave, but its cross-section has certain regions (corresponding to the "shadows" of opaque screens) in which the field is zero. We denote by S the part of the plane cross-section on which the field u_0 is different from zero; since each such plane is a wave surface of the plane wave, $u_0 = $ const over the whole surface S.

Actually, however, a wave with a limited cross-sectional area cannot be strictly plane (see § 58). In its spatial Fourier expansion there appear components with wave vectors having different directions, and this is precisely the origin of the diffraction.

Let us expand the field u_0 into a two-dimensional Fourier integral with respect to the coordinates y, z in the plane of the transverse cross-section of the wave. For the Fourier components, we have:

$$u_{\mathbf{q}} = \iint u_0 e^{-i\mathbf{q} \cdot \mathbf{r}} dy \, dz, \tag{61.1}$$

where the vectors $\mathbf{q}$ are constant vectors in the y, z plane; the integration actually extends only over that portion S of the y, z plane on which u_0 is different from zero. If $\mathbf{k}$ is the wave vector of the incident wave, the field component $u_{\mathbf{q}} e^{i\mathbf{q} \cdot \mathbf{r}}$ gives the wave vector $\mathbf{k}' = \mathbf{k} + \mathbf{q}$. Thus the vector $\mathbf{q} = \mathbf{k}' - \mathbf{k}$ determines the change in the wave vector of the light in the diffraction. Since the absolute values $k = k' = \omega/c$, the small diffraction angles θ_y, θ_z in the xy- and xz-planes are related to the components of the vector q by the equations

$$q_y = \frac{\omega}{c}\theta_y, \quad q_z = \frac{\omega}{c}\theta_z. \tag{61.2}$$

For small deviations from geometrical optics, the components in the expansion of the field u_0 can be assumed to be identical with the components of the actual field of the diffracted light, so that formula (61.1) solves our problem.

---

† The criteria for Fresnel and Fraunhofer diffraction are easily found by returning to formula (60.2) and applying it, for example, to a slit of width a (instead of to the edge of an isolated screen). The integration over z in (60.2) should then be taken between the limits from 0 to a. Fresnel diffraction corresponds to the case when the term containing z^2 in the exponent of the integrand is important, and the upper limit of the integral can be replaced by ∞. For this to be the case, we must have

$$ka^2\left(\frac{1}{D_p} + \frac{1}{D_q}\right) \gg 1.$$

On the other hand, if this inequality is reversed, the term in z^2 can be dropped; this corresponds to the case of Fraunhofer diffraction.

The intensity distribution of the diffracted light is given by the square $|u_{\mathbf{q}}|^2$ as a function of the vector $\mathbf{q}$. The quantitative connection with the intensity of the incident light is established by the formula

$$\iint u_0^2 \, dy \, dz = \iint |u_{\mathbf{q}}|^2 \frac{dq_y \, dq_z}{(2\pi)^2} \tag{61.3}$$

[compare (49.8)]. From this we see that the relative intensity diffracted into the solid angle $do = d\theta_y \, d\theta_z$ is given by

$$\frac{|u_{\mathbf{q}}|^2}{u_0^2} \frac{dq_y \, dq_z}{(2\pi)^2} = \left(\frac{\omega}{2\pi c}\right)^2 \left|\frac{u_{\mathbf{q}}}{u_0}\right|^2 do. \tag{61.4}$$

Let us consider the Fraunhofer diffraction from two screens which are "complementary": the first screen has holes where the second is opaque and conversely. We denote by $u^{(1)}$ and $u^{(2)}$ the field of the light diffracted by these screens (when the same light is incident in both cases). Since $u_{\mathbf{q}}^{(1)}$ and $u_{\mathbf{q}}^{(2)}$ are expressed by integrals (61.1) taken over the surfaces of the apertures in the screens, and since the apertures in the two screens complement one another to give the whole plane, the sum $u_{\mathbf{q}}^{(1)} + u_{\mathbf{q}}^{(2)}$ is the Fourier component of the field obtained in the absence of the screens, i.e. it is simply the incident light. But the incident light is a rigorously plane wave with definite direction of propagation, so that $u_{\mathbf{q}}^{(1)} + u_{\mathbf{q}}^{(2)} = 0$ for all nonzero values of $\mathbf{q}$. Thus we have $u_{\mathbf{q}}^{(1)} = -u_{\mathbf{q}}^{(2)}$, or for the corresponding intensities,

$$|u_{\mathbf{q}}^{(1)}|^2 = |u_{\mathbf{q}}^{(2)}|^2 \quad \text{for } \mathbf{q} \neq 0. \tag{61.5}$$

This means that complementary screens give the same distribution of intensity of the diffracted light (this is called *Babinet's principle*).

We call attention here to one interesting consequence of the Babinet principle. Let us consider a black body, i.e. one which absorbs completely all the light falling on it. According to geometrical optics, when such a body is illuminated, there is produced behind it a region of geometrical shadow, whose cross-sectional area is equal to the area of the body in the direction perpendicular to the direction of incidence of the light. However, the presence of diffraction causes the light passing by the body to be partially deflected from its initial direction. As a result, at large distances behind the body there will not be complete shadow but, in addition to the light propagating in the original direction, there will also be a certain amount of light propagating at small angles to the original direction. It is easy to determine the intensity of this scattered light. To do this, we point out that according to Babinet's principle, the amount of light deviated because of diffraction by the body under consideration is equal to the amount of light which would be deviated by diffraction from an aperture cut in an opaque screen, the shape and size of the aperture being the same as that of the transverse section of the body. But in Fraunhofer diffraction from an aperture *all* the light passing through the aperture is deflected. From this it follows that the total amount of light scattered by a black body is equal to the amount of light falling on its surface and absorbed by it.

PROBLEMS

1. Calculate the Fraunhofer diffraction of a plane wave normally incident on an infinite slit (of width $2a$) with parallel sides cut in an opaque screen.

Solution: We choose the plane of the slit as the *yz* plane, with the *z* axis along the slit (Fig. 13 shows a section of the screen). For normally incident light, the plane of the slit is one of the wave surfaces, and we choose it as the surface of integration in (61.1). Since the slit is infinitely long, the light is deflected only in the *xy* plane [since the integral (61.1)] becomes zero for $q_z \neq 0$}.

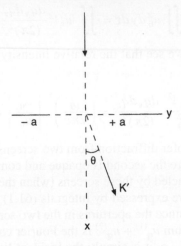

Therefore the field should be expanded only in the *y* coordinate:

$$u_q = u_0 \int_{-a}^{a} e^{-iqy} dy = \frac{2u_0}{q} \sin qa.$$

The intensity of the diffracted light in the angular range $d\theta$ is

$$u dI = \frac{I_0}{2a} \left| \frac{u_q}{u_0} \right|^2 \frac{dq}{2\pi} = \frac{I_0}{\pi ak} \frac{\sin^2 ka\theta}{\theta^2} d\theta,$$

where $k = \omega/c$, and I_0 is the total intensity of the light incident on the slit.

$dI/d\theta$ as a function of diffraction angle has the form shown in Fig. 14. As θ increases toward either side from $\theta = 0$, the intensity goes through a series of maxima with rapidly decreasing height. The successive maxima are separated by minima at the points $\theta = n\pi/ka$ (where *n* is an integer); at the minima, the intensity falls to zero.

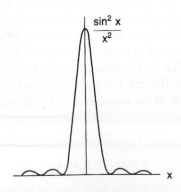

FIG. 14.

2. Calculate the Fraunhofer diffraction by a diffraction grating—a plane screen in which are cut a series of identical parallel slits (the width of the slits is $2a$, the width of opaque screen between neighbouring slits is $2b$, and the number of slits is N).

Solution: We choose the plane of the grating as the yz plane, with the z axis parallel to the slits. Diffraction occurs only in the xy plane, and integration of (61.1) gives:

$$u_y = u_y' \sum_{n=0}^{N-1} e^{-2inqd} = u_q' \frac{1 - e^{-2iNqd}}{1 - e^{-2iqd}},$$

where $d = a + b$, and u_q' is the result of the integration over a single slit. Using the results of problem 1, we get:

$$dI = \frac{I_0 a}{N\pi} \left(\frac{\sin Nqd}{\sin qd} \right)^2 \left(\frac{\sin qa}{qa} \right)^2 dq = \frac{I_0}{N\pi ak} \left(\frac{\sin Nk\theta d}{\sin k\theta d} \right)^2 \frac{\sin^2 ka\theta}{\theta^2} d\theta$$

(I_0 is the total intensity of the light passing through all the slits).

For the case of a large number of slits ($N \to \infty$), this formula can be written in another form. For values $q = \pi n/d$, where n is an integer, dI/dq has a maximum; near such a maximum (i.e. for $qd = n\pi + \varepsilon$, with ε small)

$$dI = I_0 a \left(\frac{\sin qa}{qa} \right)^2 \frac{\sin^2 N\varepsilon}{\pi N \varepsilon^2} dq.$$

But for $N \to \infty$, we have the formula[†]

$$\lim_{N \to \infty} \frac{\sin^2 Nx}{\pi N x^2} = \delta(x).$$

We therefore have, in the neighbourhood of each maximum:

$$dI = I_0 \frac{a}{d} \left(\frac{\sin qa}{qa} \right)^2 \delta(\varepsilon) d\varepsilon,$$

i.e., in the limit the widths of the maxima are infinitely narrow and the total light intensity in the n'th maximum is

$$I^{(n)} = I_0 \frac{d}{\pi^2 a} \frac{\sin^2 (n\pi a/d)}{n^2}.$$

3. Find the distribution of intensity over direction for the diffraction of light which is incident normal to the plane of a circular aperture of radius a.

Solution: We introduce cylindrical coordinates z, r, ϕ with the z axis passing through the centre of the aperture and perpendicular to its plane. It is obvious that the diffraction is symmetric about the z axis, so that the vector $\mathbf{q}$ has only a radial component $q_r \quad \gamma = k\theta$. Measuring the angle ϕ from the direction $\mathbf{q}$, and integrating in (61.1) over the plane of the aperture, we find:

[†] For $x \neq 0$ the function on the left side of the equation is zero, while according to a well-known formula of the theory of Fourier series,

$$\lim_{N \to \infty} \left(\frac{1}{\pi} \int_{-a}^{a} f(x) \frac{\sin^2 Nx}{Nx^2} dx \right) = f(0).$$

From this we see that the properties of this function actually coincide with those of the δ-function (see the footnote on p. 74).

$$u_q = u_0 \int\limits_0^a \int\limits_0^{2\pi} e^{-iq'r\cos\phi}\, r\, d\phi\, dr = 2\pi u_0 \int\limits_0^a J_0(qr) r\, dr,$$

where J_0 is the zero'th order Bessel function. Using the well-known formula

$$\int\limits_0^a J_0(qr) r\, dr = \frac{a}{q} J_1(aq),$$

we then have

$$u_q = 2\pi \frac{u_0 a}{q} J_1(aq),$$

and according to (61.4) we obtain for the intensity of the light diffracted into the element of solid angle do:

$$dI = I_0 \frac{J_1^2(ak\theta)}{\pi\theta^2}\, do,$$

where I_0 is the total intensity of the light incident on the aperture.

CHAPTER 8

THE FIELD OF MOVING CHARGES

§ 62. The retarded potentials

In Chapter 5 we studied the constant field, produced by charges at rest, and in Chapter 6, the variable field in the absence of charges. Now we take up the study of varying fields in the presence of arbitrarily moving charges.

We derive equations determining the potentials for arbitrarily moving charges. This derivation is most conveniently done in four-dimensional form, repeating the derivation at the end of § 46, with the one change that we use the second pair of Maxwell equations in the form (30.2)

$$\frac{\partial F^{ik}}{\partial x^k} = -\frac{4\pi}{c} j^i.$$

The same right-hand side also appears in (46.8), and after imposing the Lorentz condition

$$\frac{\partial A^i}{\partial x^i} = 0, \quad \text{i.e.} \quad \frac{1}{c}\frac{\partial \phi}{\partial t} + \operatorname{div} \mathbf{A} = 0, \tag{62.1}$$

on the potentials, we get

$$\frac{\partial^2 A^i}{\partial x_k \partial x^k} = \frac{4\pi}{c} j^i. \tag{62.2}$$

This is the equation which determines the potentials of an arbitrary electromagnetic field. In three-dimensional form it is written as two equations, for $\mathbf{A}$ and for ϕ:

$$\Delta \mathbf{A} - \frac{1}{c^2}\frac{\partial^2 \mathbf{A}}{\partial t^2} = -\frac{4\pi}{c}\mathbf{j}, \tag{62.3}$$

$$\Delta \phi - \frac{1}{c^2}\frac{\partial^2 \phi}{\partial t^2} = -4\pi\varrho. \tag{62.4}$$

For constant fields, these reduce to the already familiar equations (36.4) and (43.4), and for variable fields without charges, to the homogeneous wave equation.

As we know, the solution of the inhomogeneous linear equations (62.3) and (62.4) can be represented as the sum of the solution of these equations without the right-hand side, and a particular integral of these equations with the right-hand side. To find the particular solution, we divide the whole space into infinitely small regions and determine the field produced by

171

the charges located in one of these volume elements. Because of the linearity of the field equations, the actual field will be the sum of the fields produced by all such elements.

The charge de in a given volume element is, generally speaking, a function of the time. If we choose the origin of coordinates in the volume element under consideration, then the charge density is $\varrho = de(t)\,\delta(\mathbf{R})$, where $\mathbf{R}$ is the distance from the origin. Thus we must solve the equation

$$\Delta\phi - \frac{1}{c^2}\frac{\partial^2\phi}{\partial t^2} = -4\pi\,de(t)\,\delta(\mathbf{R}). \tag{62.5}$$

Everywhere, except at the origin, $\delta(\mathbf{R}) = 0$, and we have the equation

$$\Delta\phi - \frac{1}{c^2}\frac{\partial^2\phi}{\partial t^2} = 0. \tag{62.6}$$

It is clear that in the case we are considering ϕ has central symmetry, i.e. ϕ is a function only of R. Therefore if we write the Laplace operator in spherical coordinates, (62.6) reduces to

$$\frac{1}{R^2}\frac{\partial}{\partial R}\left(R^2\frac{\partial\phi}{\partial R}\right) - \frac{1}{c^2}\frac{\partial^2\phi}{\partial t^2} = 0.$$

To solve this equation, we make the substitution $\phi = \chi(R, t)/R$. Then, we find for χ

$$\frac{\partial^2\chi}{\partial R^2} - \frac{1}{c^2}\frac{\partial^2\chi}{\partial t^2} = 0.$$

But this is the equation of plane waves, whose solution has the form (see § 47):

$$\chi = f_1\left(t - \frac{R}{c}\right) + f_2\left(t + \frac{R}{c}\right).$$

Since we only want a particular solution of the equation, it is sufficient to choose only one of the functions f_1 and f_2. Usually it turns out to be convenient to take $f_2 = 0$ (concerning this, see below). Then, everywhere except at the origin, ϕ has the form

$$\phi = \frac{\chi\left(t - \dfrac{R}{c}\right)}{R}. \tag{62.7}$$

So far the function χ is arbitrary; we now choose it so that we also obtain the correct value for the potential at the origin. In other words, we must select χ so that at the origin equation (62.5) is satisfied. This is easily done noting that as $R \to 0$, the potential increases to infinity, and therefore its derivatives with respect to the coordinates increase more rapidly than its time derivative. Consequently as $R \to 0$, we can, in equation (62.5), neglect the term $(1/c^2)/(\partial^2\phi/\partial t^2)$ compared with $\Delta\phi$. Then (62.5) goes over into the familiar equation (36.9) leading to the Coulomb law. Thus, near the origin, (62.7) must go over into the Coulomb law, from which it follows that $\chi(t) = de(t)$, that is,

$$\phi = \frac{de\left(t - \dfrac{R}{c}\right)}{R}.$$

From this it is easy to get to the solution of equation (62.4) for an arbitrary distribution of charges $\varrho(x, y, z, t)$. To do this, it is sufficient to write $de = \varrho dV$ (dV is the volume element) and integrate over the whole space. To this solution of the inhomogeneous equation (62.4) we can still add the solution ϕ_0 of the same equation without the right-hand side. Thus, the general solution has the form:

$$\phi(\mathbf{r}, t) = \int \frac{1}{R} \varrho \left(\mathbf{r}', t - \frac{R}{c} \right) dV' + \psi_0, \tag{62.8}$$

$$\mathbf{R} = \mathbf{r} - \mathbf{r}', \qquad dV' = dx' \, dy' \, dz'$$

where

$$\mathbf{r} = (x, y, z), \qquad \mathbf{r}' = (x', y', z');$$

R is the distance from the volume element dV to the "field point" at which we determine the potential. We shall write this expression briefly as

$$\phi = \int \frac{\varrho_{t-(R/c)}}{R} \, dV + \phi_0, \tag{62.9}$$

where the subscript means that the quantity ϱ is to be taken at the time $t - (R/c)$, and the prime on dV has been omitted.

Similarly we have for the vector potential:

$$\mathbf{A} = \frac{1}{c} \int \frac{\mathbf{j}_{t-(R/c)}}{R} \, dV + \mathbf{A}_0, \tag{62.10}$$

where $\mathbf{A}_0$ is the solution of equation (62.3) without the right-hand term.

The potentials (62.9) and (62.10) (without ϕ_0 and $\mathbf{A}_0$) are called the retarded potentials.

In case the charges are at rest (i.e. density ρ independent of the time), formula (62.9) goes over into the well-known formula (36.8) for the electrostatic field; for the case of stationary motion of the charges, formula (62.10), after averaging, goes over into formula (43.5) for the vector potential of a constant magnetic field.

The quantities $\mathbf{A}_0$ and ϕ_0 in (62.9) and (62.10) are to be determined so that the conditions of the problem are fulfilled. To do this it is clearly sufficient to impose initial conditions, that is, to fix the values of the field at the initial time. However we do not usually have to deal with such initial conditions. Instead we are usually given conditions at large distances form the system of charges throughout all of time. Thus, we may be told that radiation is incident on the system from outside. Corresponding to this, the field which is developed as a result of the interaction of this radiation with the system can differ from the external field only by the radiation originating from the system. This radiation emitted by the system must, at large distances, have the form of waves spreading out from the system, that is, in the direction of increasing R. But precisely this condition is satisfied by the retarded potentials. Thus these solutions represent the field produced by the system, while ϕ_0 and $\mathbf{A}_0$ must be set equal to the external field acting on the system.

§ 63. The Lienard–Wiechert potentials

Let us determine the potentials for the field produced by a charge carrying out an assigned motion along a trajectory $\mathbf{r} = \mathbf{r}_0(t)$.

According to the formulas for the retarded potentials, the field at the point of observation $P(x, y, z)$ at time t is determined by the state of motion of the charge at the earlier time t', for which the time of propagation of the light signal from the point $\mathbf{r}_0(t')$, where the charge was located, to the field point P just coincides with the difference $t - t'$. Let $\mathbf{R}(t) = \mathbf{r} - \mathbf{r}_0(t)$ be the radius vector from the charge e to the point P; like $\mathbf{r}_0(t)$ it is a given function of the time. Then the time t' is determined by the equation

$$t' + \frac{R(t')}{c} = t. \tag{63.1}$$

For each value of t this equation has just one root t'.†

In the system of reference in which the particle is at rest at time t', the potential at the point of observation at time t is just the Coulomb potential,

$$\phi = \frac{e}{R(t')}, \qquad \mathbf{A} = 0. \tag{63.2}$$

The expressions for the potentials in an arbitrary reference system can be found directly by finding a four-vector which for $\mathbf{v} = 0$ coincides with the expressions just given for ϕ and $\mathbf{A}$. Noting that, according to (63.1), ϕ in (63.2) can also be written in the form

$$\phi = \frac{e}{c(t - t')},$$

we find that the required four-vector is:

$$A^i = e \frac{u^i}{R_k u^k}, \tag{63.3}$$

where u^k is the four-velocity of the charge, $R^k = [c(t - t'), \mathbf{r} - \mathbf{r}']$, where x', y', z', t' are related by the equation (63.1), which in four-dimensional form is

$$R_k R^k = 0. \tag{63.4}$$

Now once more transforming to three-dimensional notation, we obtain, for the potentials of the field produced by an arbitrarily moving point charge, the following expressions:

$$\phi = \frac{e}{\left(R - \dfrac{\mathbf{v} \cdot \mathbf{R}}{c} \right)}, \qquad \mathbf{A} = \frac{e\mathbf{v}}{c\left(R - \dfrac{\mathbf{v} \cdot \mathbf{R}}{c} \right)}, \tag{63.5}$$

where $\mathbf{R}$ is the radius vector, taken from the point where the charge is located to the point of observation P, and all the quantities on the right sides of the equations must be evaluated at the time t', determined from (63.1). The potentials of the field, in the form (63.5), are called the *Lienard–Wiechert potentials*.

† This point is obvious but it can be verified directly. To do this we choose the field point P and the time of observation t as the origin O of the four-dimensional coordinate system and construct the light cone (§ 2) with its vertex at O. The lower half of the cone, containing the absolute past (with respect to the event O), is the geometrical locus of world points such that signals sent from them reach O. The points in which this hypersurface intersects the world line of the charge are precisely the roots of (63.1). But since the velocity of a particle is always less than the velocity of light, the inclination of its world line relative to the time axis is everywhere less than the slope of the light cone. It then follows that the world line of the particle can intersect the lower half of the light cone in only one point.

To calculate the intensities of the electric and magnetic fields from the formulas

$$\mathbf{E} = -\frac{1}{c}\frac{\partial \mathbf{A}}{\partial t} - \text{grad } \phi, \qquad \mathbf{H} = \text{curl } \mathbf{A},$$

we must differentiate ϕ and $\mathbf{A}$ with respect to the coordinates x, y, z of the point, and the time t of observation. But the formulas (63.5) express the potentials as functions of t', and only through the relation (63.1) as implicit functions of r, y, z, t. Therefore to calculate the required derivatives we must first calculate the derivatives of t'. Differentiating the relation $R(t') = c(t - t')$ with respect to t, we get

$$\frac{\partial R}{\partial t} = \frac{\partial R}{\partial t'}\frac{\partial t'}{\partial t} = -\frac{\mathbf{R}\cdot\mathbf{v}}{R}\frac{\partial t'}{\partial t} = c\left(1 - \frac{\partial t'}{\partial t}\right).$$

(The value of $\partial R/\partial t'$ is obtained by differentiating the identity $R^2 = \mathbf{R}^2$ and substituting $\partial\mathbf{R}(t')/\partial t' = -\mathbf{v}(t')$. The minus sign is present because $\mathbf{R}$ is the radius vector from the charge e to the point P, and not the reverse.)
   Thus,

$$\frac{\partial t'}{\partial t} = \frac{1}{1 - \dfrac{\mathbf{v}\cdot\mathbf{R}}{Rc}}. \tag{63.6}$$

Similarly differentiating the same relation with respect to the coordinates, we find

$$\text{grad } t' = -\frac{1}{c}\text{ grad } R(t') = -\frac{1}{c}\left(\frac{\partial R}{\partial t'}\text{ grad } t' + \frac{\mathbf{R}}{R}\right),$$

so that

$$\text{grad } t' = -\frac{\mathbf{R}}{c\left(R - \dfrac{\mathbf{R}\cdot\mathbf{v}}{c}\right)}. \tag{63.7}$$

With the aid of these formulas, there is no difficulty in carrying out the calculation of the fields $\mathbf{E}$ and $\mathbf{H}$. Omitting the intermediate calculations, we give the final results:

$$\mathbf{E} = e\frac{1 - \dfrac{v^2}{c^2}}{\left(R - \dfrac{\mathbf{R}\cdot\mathbf{v}}{c}\right)^3}\left(\mathbf{R} - \frac{\mathbf{v}}{c}R\right) + \frac{e}{c^2\left(R - \dfrac{\mathbf{R}\cdot\mathbf{v}}{c}\right)^3}\mathbf{R}\times\left\{\left(\mathbf{R} - \frac{\mathbf{v}}{c}R\right)\times\dot{\mathbf{v}}\right\}, \tag{63.8}$$

$$\mathbf{H} = \frac{1}{R}\mathbf{R}\times\mathbf{E}. \tag{63.9}$$

Here, $\dot{\mathbf{v}} = \partial\mathbf{v}/\partial t'$; all quantities on the right sides of the equations refer to the time t'. It is interesting to note that the magnetic field turns out to be everywhere perpendicular to the electric.
   The electric field (63.8) consists of two parts of different type. The first term depends only on the velocity of the particle (and not on its acceleration) and varies at large distances like $1/R^2$. The second term depends on the acceleration, and for large R it varies like $1/R$. Later

(§ 66) we shall see that this latter term is related to the electromagnetic waves radiated by the particle.

As for the first term, since it is independent of the acceleration it must correspond to the field produced by a uniformly moving charge. In fact, for constant velocity the difference

$$\mathbf{R}_{t'} - \frac{\mathbf{v}}{c} R_{t'} = \mathbf{R}_{t'} - \mathbf{v}(t - t')$$

is the distance $\mathbf{R}_t$ from the charge to the point of observation at precisely the moment of observation. It is also easy to show directly that

$$R_{t'} - \frac{1}{c} \mathbf{R}_{t'} \cdot \mathbf{v} = \sqrt{R_t^2 - \frac{1}{c^2}(\mathbf{v} \times \mathbf{R}_t)^2} = R_t \sqrt{1 - \frac{v^2}{c^2} \sin^2 \theta_t},$$

where θ_t is the angle between $\mathbf{R}_t$ and $\mathbf{v}$. Consequently the first term in (63.8) is identical with the expression (38.8).

PROBLEM

Derive the Lienard–Wiechert potentials by integrating (62.9)–(62.10).

Solution: We write formula (62.8) in the form:

$$\phi(\mathbf{r}, t) = \iint \frac{\varrho(\mathbf{r}', \tau)}{|\mathbf{r} - \mathbf{r}'|} \delta\left(\tau - t + \frac{1}{c}|\mathbf{r} - \mathbf{r}'|\right) d\tau dV'$$

(and similarly for $\mathbf{A}(\mathbf{r}, t)$), introducing the additional delta function and thus eliminating the implicit arguments in the function ϱ. For a point charge, moving in a trajectory $\mathbf{r} = \mathbf{r}_0(t)$; we have:

$$\varrho(\mathbf{r}', \tau) = e\delta[\mathbf{r}' - \mathbf{r}_0(\tau)].$$

Substituting this expression and integrating over dV', we get:

$$\phi(\mathbf{r}, t) = e \int \frac{d\tau}{|\mathbf{r} - \mathbf{r}_0(\tau)|} \delta\left[\tau - t + \frac{1}{c}|\mathbf{r} - \mathbf{r}_0(\tau)|\right],$$

The τ integration is done using the formula

$$\delta[F(\tau)] = \frac{\delta(\tau - t')}{F'(t')}$$

[where t' is the root of $F(t') = 0$], and gives formula (63.5).

§ 64. Spectral resolution of the retarded potentials

The field produced by moving charges can be expanded into monochromatic waves. The potentials of the different monochromatic components of the field have the form $\phi_\omega e^{-i\omega t}$, $\mathbf{A}_\omega e^{-i\omega t}$. The charge and current densities of the system of charges producing the field can also be expanded in a Fourier series or integral. It is clear that each Fourier component of ϱ and $\mathbf{j}$ is responsible for the creation of the corresponding monochromatic component of the field.

In order to express the Fourier components of the field in terms of the Fourier components of the charge density and current, we substitute in (62.9) for ϕ and ϱ respectively, $\phi_\omega e^{-i\omega t}$, and $\varrho_\omega e^{-i\omega t}$. We then obtain

$$\phi_\omega e^{-i\omega t} = \int \varrho_\omega \, \frac{e^{-i\omega\left(t-\frac{R}{c}\right)}}{R} \, dV.$$

Factoring $e^{-i\omega t}$ and introducing the absolute value of the wave vector $k = \omega/c$, we have:

$$\phi_\omega = \int \varrho_\omega \, \frac{e^{ikR}}{R} \, dV. \tag{64.1}$$

Similarly, for $\mathbf{A}_\omega$ we get

$$\mathbf{A}_\omega = \int \mathbf{j}_\omega \, \frac{e^{ikR}}{cR} \, dV. \tag{64.2}$$

We note that formula (64.1) represents a generalization of the solution of the Poisson equation to a more general equation of the form

$$\Delta\phi_\omega + k^2\phi_\omega = -4\pi\varrho_\omega \tag{64.3}$$

(obtained from equations (62.4) for ϱ, ϕ depending on the time through the factor $e^{-i\omega t}$).

If we were dealing with expansion into a Fourier integral, then the Fourier components of the charge density would be

$$\varrho_\omega = \int\limits_{-\infty}^{+\infty} \varrho e^{i\omega t} \, dt.$$

Substituting this expression in (64.1), we get

$$\phi_\omega = \int\limits_{-\infty}^{+\infty}\!\!\int \frac{\varrho}{R} \, e^{i(\omega t + kR)} \, dV \, dt. \tag{64.4}$$

We must still go over from the continuous distribution of charge density to the point charges whose motion we are actually considering. Thus, if there is just one point charge, we set

$$\varrho = e\delta[\mathbf{r} - \mathbf{r}_0(t)],$$

where $\mathbf{r}_0(t)$ is the radius vector of the charge, and is a given function of the time. Substituting this expression in (64.4) and carrying out the space integration [which reduces to replacing $\mathbf{r}$ by $\mathbf{r}_0(t)$], we get:

$$\phi_\omega = e \int\limits_{-\infty}^{\infty} \frac{1}{R(t)} \, e^{i\omega[t+R(t)/c]} \, dt, \tag{64.5}$$

where now $R(t)$ is the distance from the moving particle to the point of observation. Similarly we find for the vector potential:

$$\mathbf{A}_\omega = \frac{e}{c} \int\limits_{-\infty}^{\infty} \frac{\mathbf{v}(t)}{R(t)} \, e^{i\omega[t+R(t)/c]} \, dt, \tag{64.6}$$

where $\mathbf{v} = \dot{\mathbf{r}}_0(t)$ is the velocity of the particle.

Formulas analogous to (64.5), (64.6) can also be written for the case where the spectral resolution of the charge and current densities contains a discrete series of frequencies. Thus, for a periodic motion of a point charge (with period $T = 2\pi/\omega_0$) the spectral resolution of the field contains only frequencies of the form $n\omega_0$, and the corresponding components of the vector potential are

$$\mathbf{A}_n = \frac{e}{cT} \int\limits_0^T \frac{\mathbf{v}(t)}{R(t)} \, e^{in\omega_0 [t+R(t)/c]} \, dt \tag{64.7}$$

(and similarly for ϕ_n). In both (64.6) and (64.7) the Fourier components are defined in accordance with § 49.

PROBLEM

Find the expansion in plane waves of the field of a charge in uniform rectilinear motion.

Solution: We proceed in similar fashion to that used in § 51. We write the charge density in the form $\varrho = e\delta(\mathbf{r} - \mathbf{v}t)$, where $\mathbf{v}$ is the velocity of the particle. Taking Fourier components of the equation $\Box \phi = -4\pi e \, \delta(\mathbf{r} - \mathbf{v}t)$, we find $(\Box \phi)_\mathbf{k} = -4\pi e \, e^{-i(\mathbf{v}\cdot\mathbf{k})t}$.

On the other hand, from

$$\phi = \int e^{i\mathbf{k}\cdot\mathbf{r}} \phi_\mathbf{k} \frac{d^3 k}{(2\pi)^3}$$

we have

$$(\Box\phi)_\mathbf{k} = -k^2\phi_\mathbf{k} \frac{1}{c^2} \frac{\partial^2 \phi_\mathbf{k}}{\partial t^2}.$$

Thus,

$$\frac{1}{c^2} \frac{\partial^2 \phi_\mathbf{k}}{\partial t^2} + k^2\phi_\mathbf{k} = 4\pi e e^{-i(\mathbf{k}\cdot\mathbf{v})t},$$

from which, finally

$$\phi_\mathbf{k} = 4\pi e \frac{e^{-i(\mathbf{k}\cdot\mathbf{v})t}}{k^2 - \left(\dfrac{\mathbf{k}\cdot\mathbf{v}}{c}\right)^2}.$$

From this it follows that the wave with wave vector $\mathbf{k}$ has the frequency $\omega = \mathbf{k}\cdot\mathbf{v}$. Similarly, we obtain for the vector potential,

$$\mathbf{A}_\mathbf{k} = 4\pi e - \frac{\mathbf{v}e^{-i(\mathbf{k}\cdot\mathbf{v})t}}{k^2 - \left(\dfrac{\mathbf{k}\cdot\mathbf{v}}{c}\right)^2}.$$

Finally, we have for the fields,

$$\mathbf{E}_\mathbf{k} = -i\mathbf{k}\phi_\mathbf{k} + i\frac{\mathbf{k}\cdot\mathbf{v}}{c}\mathbf{A}_\mathbf{k} = 4\pi e \, i \frac{-\mathbf{k} + \dfrac{(\mathbf{k}\cdot\mathbf{v})}{c^2}\mathbf{v}}{k^2 - \left(\dfrac{\mathbf{k}\cdot\mathbf{v}}{c}\right)^2} e^{-i(\mathbf{k}\cdot\mathbf{v})t},$$

$$\mathbf{H_k} = i\mathbf{k} \times \mathbf{A_k} = \frac{4\pi e}{c} \, i \, \frac{\mathbf{k} \times \mathbf{v}}{k^2 - \left(\dfrac{\mathbf{k} \cdot \mathbf{v}}{c}\right)^2} \, e^{-i(\mathbf{k} \cdot \mathbf{v})t}.$$

§ 65. The Lagrangian to terms of second order

In ordinary classical mechanics, we can describe a system of particles interacting with each other with the aid of a Lagrangian which depends only on the coordinates and velocities of these particles (at one and the same time). The possibility of doing this is, in the last analysis, dependent on the fact that in mechanics the velocity of propagation of interactions is assumed to be infinite.

We already know that because of the finite velocity of propagation, the field must be considered as an independent system with its own "degrees of freedom". From this it follows that if we have a system of interacting particles (charges), then to describe it we must consider the system consisting of these particles and the field. Therefore, when we take into account the finite velocity of propagation of interactions, it is impossible to describe the system of interacting particles rigorously with the aid of a Lagrangian, depending only on the coordinates and velocities of the particles and containing no quantities related to the internal "degrees of freedom" of the field.

However, if the velocity v of all the particles is small compared with the velocity of light, then the system can be described by a certain approximate Lagrangian. It turns out to be possible to introduce a Lagrangian describing the system, not only when all powers of v/c are neglected (classical Lagrangian), but also to terms of second order, v^2/c^2. This last remark is related to the fact that the radiation of electromagnetic waves by moving charges (and consequently, the appearance of a "self"-field) occurs only in the third approximation in v/c (see later, in § 67).†

As a preliminary, we note that in zero'th approximation, that is, when we completely neglect the retardation of the potentials, the Lagrangian for a system of charges has the form

$$L^{(0)} = \sum_a \frac{1}{2} m_a \mathbf{v}_a^2 - \sum_{a>b} \frac{e_a e_b}{R_{ab}} \tag{65.1}$$

(the summation extends over the charges which make up the system). The second term is the potential energy of interaction as it would be for charges at rest.

To get the next approximation, we proceed in the following fashion. The Lagrangian for a charge e_a in an external field is

$$L_a = -m_a c^2 \sqrt{1 - \frac{v_a^2}{c^2}} - e_a \phi + \frac{e_a}{c} \mathbf{A} \cdot \mathbf{v}_a. \tag{65.2}$$

Choosing any one of the charges of the system, we determine the potentials of the field produced by all the other charges at the position of the first, and express them in terms of the coordinates and velocities of the charges which produce this field (this can be done only approximately—for ϕ, to terms of order v^2/c^2, and for $\mathbf{A}$, to terms in v/c). Substituting the expressions for the potentials obtained in this way in (65.2), we get the Lagrangian for one

---

† For systems consisting of particles with the same charge-to-mass ratio, the appearance of radiation is put off to the fifth approximation in v/c; in such a case there is a Lagrangian to terms of fourth order in v/c. [See B.M. Barker and R.F. O'Connel, *Can. J. Phys.* **58**, 1659 (1980).]

of the charges of the system (for a given motion of the other charges). From this, one can then easily find the Lagrangian for the whole system.

We start from the expressions for the retarded potentials

$$\phi = \int \frac{\varrho_{t-R/c}}{R} \, dV, \quad \mathbf{A} = \frac{1}{c} \int \frac{\mathbf{j}_{t-R/c}}{R} \, dV.$$

If the velocities of all the charges are small compared with the velocity of light, then the charge distribution does not change significantly during the time R/c. Therefore we can expand $\rho_{t-R/c}$ and $\mathbf{j}_{t-R/c}$ in series of powers of R/c. For the scalar potential we thus find, to terms of second order:

$$\phi = \int \frac{\varrho dV}{R} - \frac{1}{c} \frac{\partial}{\partial t} \int \varrho dV + \frac{1}{2c^2} \frac{\partial^2}{\partial t^2} \int R\varrho dV$$

(ϱ without indices is the value of ϱ at time t; the time differentiations can clearly be taken out from under the integral sign). But $\int \varrho dV$ is the constant total charge of the system. Therefore the second term in our expression is zero, so that

$$\phi = \int \frac{\varrho dV}{R} + \frac{1}{2c^2} \frac{\partial^2}{\partial t^2} \int R\varrho dV. \tag{65.3}$$

We can proceed similarly with $\mathbf{A}$. But the expression for the vector potential in terms of the current density already contains $1/c$, and when substituted in the Lagrangian is multiplied once more by $1/c$. Since we are looking for a Lagrangian which is correct only to terms of second order, we can limit ourselves to the first term in the expansion of $\mathbf{A}$, that is,

$$\mathbf{A} = \frac{1}{c} \int \frac{\varrho \mathbf{v}}{R} \, dV \tag{65.4}$$

(we have substituted $\mathbf{j} = \varrho\mathbf{v}$).

Let us first assume that there is only a single point charge e. Then we obtain from (65.3) and (65.4),

$$\phi = \frac{e}{R} + \frac{e}{2c^2} \frac{\partial^2 R}{\partial t^2}, \quad \mathbf{A} = \frac{e\mathbf{v}}{cR}, \tag{65.5}$$

where R is the distance from the charge.

We choose in place of ϕ and $\mathbf{A}$ other potentials ϕ' and $\mathbf{A}'$, making the transformation (see § 18):

$$\phi' = \phi - \frac{1}{c} \frac{\partial f}{\partial t}, \quad \mathbf{A}' = \mathbf{A} + \mathrm{grad}\, f,$$

in which we choose for f the function

$$f = \frac{e}{2c} \frac{\partial R}{\partial t}.$$

Then we get†

---

† These potentials no longer satisfy the Lorentz condition (62.1), nor the equations (62.3)–(62.4).

$$\phi' = \frac{e}{R}, \qquad \mathbf{A}' = \frac{e\mathbf{v}}{cR} + \frac{e}{2c} \nabla \frac{\partial R}{\partial t}.$$

To calculate $\mathbf{A}'$ we note first of all that $\nabla(\partial R/\partial t) = (\partial/\partial t)\nabla R$. The grad operator here means differentiation with respect to the coordinates of the field point at which we seek the value of $\mathbf{A}'$. Therefore ∇R is the unit vector $\mathbf{n}$ directed from the charge e to the field point, so that

$$\mathbf{A}' = \frac{e\mathbf{v}}{cR} + \frac{e}{2c} \dot{\mathbf{n}}.$$

We also write:

$$\dot{\mathbf{n}} = \frac{\partial}{\partial t}\left(\frac{\mathbf{R}}{R}\right) = \frac{\dot{\mathbf{R}}}{R} - \frac{\mathbf{R}\dot{R}}{R^2}.$$

But the derivative $-\dot{\mathbf{R}}$ for a given field point is the velocity $\mathbf{v}$ of the charge, and the derivative $\dot{R}$ is easily determined by differentiating $R^2 = \mathbf{R}^2$, that is, by writing

$$R\dot{R} = \mathbf{R} \cdot \dot{\mathbf{R}} = - \mathbf{R} \cdot \mathbf{v}.$$

Thus,

$$\dot{\mathbf{n}} = \frac{-\mathbf{v} + \mathbf{n}(\mathbf{n} \cdot \mathbf{v})}{R}.$$

Substituting this in the expression for $\mathbf{A}'$, we get finally:

$$\phi' = \frac{e}{R}, \qquad \mathbf{A}' = \frac{e[\mathbf{v} + (\mathbf{v} \cdot \mathbf{n})\mathbf{n}]}{2cR}. \tag{65.6}$$

If there are several charges then we must, clearly, sum these expressions over all the charges.

Substituting these expressions in (65.2), we obtain the Lagrangian L_a for the charge e_a (for a fixed motion of the other charges). In doing this we must also expand the first term in (65.2) in powers of v_a/c, retaining terms up to the second order. Thus we find:

$$L_a = \frac{m_a v_a^2}{2} + \frac{1}{8}\frac{m_a v_a^4}{c^2} - e_a \sum_b{}' \frac{e_b}{R_{ab}} + \frac{e_a}{2c^2} \sum_b{}' \frac{e_b}{R_{ab}} [\mathbf{v}_a \cdot \mathbf{v}_b + (\mathbf{v}_a \cdot \mathbf{n}_{ab})(\mathbf{v}_b \cdot \mathbf{n}_{ab})]$$

(the summation goes over all the charges except e_a; $\mathbf{n}_{ab}$ is the unit vector from e_b to e_a).

From this, it is no longer difficult to get the Lagrangian for the whole system. It is easy to convince oneself that this function is not the sum of the L_a for all the charges, but has the form

$$L = \sum_a \frac{m_a v_a^2}{2} + \sum_a \frac{m_a v_a^4}{8c^2} - \sum_{a>b} \frac{e_a e_b}{R_{ab}} + \sum_{a>b} \frac{e_a e_b}{2c^2 R_{ab}} [\mathbf{v}_a \cdot \mathbf{v}_b + (\mathbf{v}_a \cdot \mathbf{n}_{ab})(\mathbf{v}_b \cdot \mathbf{n}_{ab})].$$

$$\tag{65.7}$$

Actually, for each of the charges under a given motion of all the others, this function L goes over into L_a as given above. The expression (65.7) determines the Lagrangian of a system of charges correctly to terms of second order. (It was first obtained by C. G. Darwin, 1922.)

Finally we find the Hamiltonian of a system of charges in this same approximation. This could be done by the general rule for calculating $\mathcal{H}$ from L; however it is simpler to proceed as follows. The second and fourth terms in (65.7) are small corrections to $L^{(0)}$(65.1). On the other hand, we know from mechanics that for small changes of L and $\mathscr{H}$, the additions to them are equal in magnitude and opposite in sign (here the variations of L are considered for constant coordinates and velocities, while the changes in $\mathscr{H}$ refer to constant coordinates and momenta).†

Therefore we can at once write $\mathscr{H}$, subtracting from

$$\mathscr{H}^{(0)} = \sum_a \frac{p_a^2}{2m_a} + \sum_{a>b} \frac{e_a e_b}{R_{ab}}$$

the second and fourth terms of (65.7), replacing the velocities in them by the first approximation $\mathbf{v}_a = \mathbf{p}_a/m_a$. Thus,

$$\mathscr{H} = \sum_a \frac{p_a^2}{2m_a} - \sum_a \frac{p_a^4}{8c^2 m_a^3} + \sum_{a>b} \frac{e_a e_b}{R_{ab}} -$$

$$- \sum_{a>b} \frac{e_a e_b}{2c^2 m_a m_b R_{ab}} [\mathbf{p}_a \cdot \mathbf{p}_b + (\mathbf{p}_a \cdot \mathbf{n}_{ab})(\mathbf{p}_b \cdot \mathbf{n}_{ab})]. \tag{65.8}$$

PROBLEMS

1. Determine (correctly to terms of second order) the centre of inertia of a system of interacting particles.

Solution: The problem is solved most simply by using the formula

$$\mathbf{R} = \frac{\sum_a \mathscr{E}_a \mathbf{r}_a + \int W \mathbf{r} \, dV}{\sum_a \mathscr{E}_a + \int W \, dV}$$

[see (14.6)], where $\mathscr{E}_a$ is the kinetic energy of the particle (including its rest energy), and W is the energy density of the field produced by the particles. Since the $\mathscr{E}_a$ contain the large quantities $m_a c^2$, it is sufficient, in obtaining the next approximation, to consider only those terms in $\mathscr{E}_a$ and W which do not contain c, i.e. we need consider only the nonrelativistic kinetic energy of the particles and the energy of the electrostatic field. We then have:

$$\int W \mathbf{r} \, dV = \frac{1}{8\pi} \int E^2 \mathbf{r} \, dV$$

$$= \frac{1}{8\pi} \int (\nabla \varphi)^2 \mathbf{r} \, dV$$

$$= \frac{1}{8\pi} \int \left(d\mathbf{f} \cdot \nabla \frac{\varphi^2}{2} \right) \mathbf{r} - \frac{1}{8\pi} \int \nabla \frac{\varphi^2}{2} \, dV - \frac{1}{8\pi} \int \varphi \Delta \varphi \cdot \mathbf{r} \, dV;$$

the integral over the infinitely distant surface vanishes; the second integral also is transformed into a surface integral and vanishes, while we substitute $\Delta \varphi = -4\pi \varrho$ in the third integral and obtain:

$$\int W \mathbf{r} \, dV = \frac{1}{2} \int \varrho \varphi \mathbf{r} \, dV = \frac{1}{2} \sum_a e_a \varphi_a \mathbf{r}_a,$$

† See *Mechanics*, § 40.

where φ_a is the potential produced at the point $\mathbf{r}_a$ by all the charges other than e_a.†

Finally, we get:

$$\mathbf{R} = \frac{1}{\mathscr{E}} \sum_a \mathbf{r}_a \left(m_a c^2 + \frac{p_a{}^2}{2m_a} + \frac{e_a}{2} \sum_b{}' \frac{e_b}{R_{ab}} \right)$$

(with a summation over all b except $b = a$), where

$$\mathscr{E} = \sum_a \left(m_a c^2 + \frac{p_a^2}{2m_a} + \sum_{a>b} \frac{e_a e_b}{R_{ab}} \right)$$

is the total energy of the system. Thus in this approximation the coordinates of the centre of inertia can actually be expressed in terms of quantities referring only to the particles.

2. Write the Hamiltonian in second approximation for a system of two particles, omitting the motion of the system as a whole.

Solution: We choose a system of reference in which the total momentum of the two particles is zero. Expressing the momenta as derivatives of the action, we have

$$\mathbf{p}_1 + \mathbf{p}_2 = \partial S/\partial \mathbf{r}_1 + \partial S/\partial \mathbf{r}_2 = 0.$$

From this it is clear that in the reference system chosen the action is a function of $\mathbf{r} = \mathbf{r}_2 - \mathbf{r}_1$, the difference of the radius vectors of the two particles. Therefore we have $\mathbf{p}_2 = -\mathbf{p}_1 = \mathbf{p}$, where $\mathbf{p} = \partial S/\partial \mathbf{r}$ is the momentum of the relative motion of the particles. The Hamiltonian is

$$\mathscr{H} = \frac{1}{2} \left(\frac{1}{m_1} + \frac{1}{m_2} \right) p^2 - \frac{1}{8c^2} \left(\frac{1}{m_1{}^3} + \frac{1}{m_2{}^3} \right) p^4 + \frac{e_1 e_2}{r} + \frac{e_1 e_2}{2m_1 m_2 c^2 r} [p^2 + (\mathbf{p} \cdot \mathbf{n})^2].$$

---

† The elimination of the self-field of the particles corresponds to the mass "renormalization" mentioned in the footnote on p. 97).

CHAPTER 9

RADIATION OF ELECTROMAGNETIC WAVES

§ 66. The field of a system of charges at large distances

We consider the field produced by a system of moving charges at distances large compared with the dimensions of the system.

We choose the origin of coordinates O anywhere in the interior of the system of charges. The radius vector from O to the point P, where we determine the field, we denote by $\mathbf{R}_0$, and the unit vector in this direction by $\mathbf{n}$. Let the radius vector of the charge element $de = \varrho \, dV$ be $\mathbf{r}$, and the radius vector from de to the point P be $\mathbf{R}$. Obviously $\mathbf{R} = \mathbf{R}_0 - \mathbf{r}$.

At large distances from the system of charges, $R_0 \gg r$, and we have approximately,

$$R = |\mathbf{R}_0 - \mathbf{r}| \cong R_0 - \mathbf{r} \cdot \mathbf{n}.$$

We substitute this in formulas (62.9), (62.10) for the retarded potentials. In the denominator of the integrands we can neglect $\mathbf{r} \cdot \mathbf{n}$ compared with R_0. In $t - (R/c)$, however, this is generally not possible; whether it is possible to neglect these terms is determined not by the relative values of R_0/c and $\mathbf{r} \cdot (\mathbf{n}/c)$, but by how much the quantities ϱ and $\mathbf{j}$ change during the time $\mathbf{r} \cdot (\mathbf{n}/c)$. Since R_0 is constant in the integration and can be taken out from under the integral sign, we get for the potentials of the field at large distances from the system of charges the expressions:

$$\phi = \frac{1}{R_0} \int \varrho_{t - \frac{R_0}{c} + \mathbf{r} \cdot \frac{\mathbf{n}}{c}} \, dV, \tag{66.1}$$

$$\mathbf{A} = \frac{1}{cR_0} \int \mathbf{j}_{t - \frac{R_0}{c} + \mathbf{r} \cdot \frac{\mathbf{n}}{c}} \, dV. \tag{66.2}$$

At sufficiently large distances from the system of charges, the field over small regions of space can be considered to be a plane wave. For this it is necessary that the distance be large compared not only with the dimensions of the system, but also with the wavelength of the electromagnetic waves radiated by the system. We refer to this region of space as the *wave zone* of the radiation.

In a plane wave, the fields $\mathbf{E}$ and $\mathbf{H}$ are related to each other by (47.4), $\mathbf{E} = \mathbf{H} \times \mathbf{n}$. Since $\mathbf{H} = \text{curl } \mathbf{A}$, it is sufficient for a complete determination of the field in the wave zone to calculate only the vector potential. In a plane wave we have $\mathbf{H} = (1/c)\dot{\mathbf{A}} \times \mathbf{n}$ [see (47.3)], where the dot indicates differentiation with respect to time.† Thus, knowing $\mathbf{A}$, we find $\mathbf{H}$

---

† In the present case, this formula is easily verified also by direct computation of the curl of the expression (66.2), and dropping terms in $1/R_0^2$ in comparison with terms $\sim 1/R_0$.

and $\mathbf{E}$ from the formulas:†

$$\mathbf{H} = \frac{1}{c}\,\dot{\mathbf{A}} \times \mathbf{n}, \qquad \mathbf{E} = \frac{1}{c}\,(\dot{\mathbf{A}} \times \mathbf{n}) \times \mathbf{n}. \tag{66.3}$$

We note that the field at large distances is inversely proportional to the first power of the distance R_0 from the radiating system. We also note that the time t enters into the expressions (66.1) to (66.3) always in the combination $t - (R_0/c)$.

For the radiation produced by a single arbitrarily moving point charge, it turns out to be convenient to use the Lienard–Wiechert potentials. At large distances, we can replace the radius vector $\mathbf{R}$ in formula (63.5) by the constant vector $\mathbf{R}_0$, and in the condition (63.1) determining t', we must set $R = R_0 - \mathbf{r}_0 \cdot \mathbf{n}$ ($\mathbf{r}_0(t)$ is the radius vector of the charge). Thus,‡

$$\mathbf{A} = \frac{e\mathbf{v}(t')}{cR_0\left(1 - \dfrac{\mathbf{n} \cdot \mathbf{v}(t')}{c}\right)}, \tag{66.4}$$

where t' is determined from the equality

$$t' - \frac{\mathbf{r}_0(t')}{c} \cdot \mathbf{n} = t - \frac{R_0}{c}. \tag{66.5}$$

The radiated electromagnetic waves carry off energy. The energy flux is given by the Poynting vector which, for a plane wave, is

$$\mathbf{S} = c\,\frac{H^2}{4\pi}\,\mathbf{n}.$$

The intensity dI of radiation into the element of solid angle do is defined as the amount of energy passing in unit time through the element $df = R_0^2\,do$ of the spherical surface with centre at the origin and radius R_0. This quantity is clearly equal to the energy flux density S multiplied by df, i.e.

$$dI = c\,\frac{H^2}{4\pi}\,R_0^2\,do. \tag{66.6}$$

Since the field H is inversely proportional to R_0, we see that the amount of energy radiated by the system in unit time into the element of solid angle do is the same for all distances (if the values of $t - (R_0/c)$ are the same for them). This is, of course, as it should be, since the energy radiated from the system spreads out with velocity c into the surrounding space, not accumulating or disappearing anywhere.

We derive the formulas for the spectral resolution of the field of the waves radiated by the system. These formulas can be obtained directly from those in § 64. Substituting in (64.2) $R = R_0 - \mathbf{r} \cdot \mathbf{n}$ (in which we can set $R = R_0$ in the denominator of the integrand), we get for the Fourier components of the vector potential:

$$\mathbf{A}_\omega = \frac{e^{ikR_0}}{cR_0} \int \mathbf{j}_\omega e^{-i\mathbf{k}\cdot\mathbf{r}}\,dV \tag{66.7}$$

---

† The formula $\mathbf{E} = -(1/c)\dot{\mathbf{A}}$ [see (47.3)] is here not applicable to the potentials ϕ, $\mathbf{A}$, since they do not satisfy the same auxiliary condition as was imposed on them in § 47.

‡ In formula (63.8) for the electric field, the present approximation corresponds to dropping the first term in comparison with the second.

(where $\mathbf{k} = k\mathbf{n}$). The components $\mathbf{H}_\omega$ and $\mathbf{E}_\omega$ are determined using formula (66.3). Substituting in it for $\mathbf{H}$, $\mathbf{E}$, $\mathbf{A}$, respectively, $\mathbf{H}_\omega e^{-i\omega t}$, $\mathbf{E}_\omega e^{-i\omega t}$, $\mathbf{A}_\omega e^{-i\omega t}$, and then dividing by $e^{-i\omega t}$, we find

$$\mathbf{H}_\omega = i\mathbf{k} \times \mathbf{A}_\omega, \qquad \mathbf{E}_\omega = \frac{ic}{\omega}(\mathbf{k} \times \mathbf{A}_\omega) \times \mathbf{k}. \qquad (66.8)$$

When speaking of the spectral distribution of the intensity of radiation, we must distinguish between expansions in Fourier series and Fourier integrals. We deal with the expansion into a Fourier integral in the case of the radiation accompanying the collision of charged particles. In this case the quantity of interest is the total amount of energy radiated during the time of the collision (and correspondingly lost by the colliding particles). Suppose $d\,\mathcal{E}_{n\omega}$ is the energy radiated into the element of solid angle do in the form of waves with frequencies in the interval $d\omega$. According to the general formula (49.8), the part of the total radiation lying in the frequency interval $d\omega/2\pi$ is obtained from the usual formula for the intensity by replacing the square of the field by the square modulus of its Fourier component and multiplying by 2. Therefore we have in place of (66.6):

$$d\mathcal{E}_{n\omega} = \frac{c}{2\pi}|\mathbf{H}_\omega|^2 R_0^2\, do\, \frac{d\omega}{2\pi}. \qquad (66.9)$$

If the charges carry out a periodic motion, then the radiation field must be expanded in a Fourier series. According to the general formula (49.4) the intensities of the various components of the Fourier resolution are obtained from the usual formula for the intensity by replacing the field by the Fourier components and then multiplying by two. Thus the intensity of the radiation into the element of solid angle do, with frequency $\omega = n\omega_0$ equals

$$dI_n = \frac{c}{2\pi}|\mathbf{H}_n|^2 R_0^2\, do. \qquad (66.10)$$

Finally, we give the formulas for determining the Fourier components of the radiation field directly from the given motion of the radiating charges. For the Fourier integral expansion, we have:

$$\mathbf{j}_\omega = \int_{-\infty}^{\infty} \mathbf{j}e^{i\omega t}\, dt.$$

Substituting this in (66.7) and changing from the continuous distribution of currents to a point charge moving along a trajectory $\mathbf{r}_0 = \mathbf{r}_0(t)$ (see § 64), we obtain:

$$\mathbf{A}_\omega = \frac{e^{ikR_0}}{cR_0}\int_{-\infty}^{+\infty} e\mathbf{v}(t)e^{i[\omega t - \mathbf{k}\cdot\mathbf{r}_0(t)]}\, dt, \qquad (66.11)$$

Since $\mathbf{v} = d\mathbf{r}_0/dt$, $\mathbf{v}\, dt = d\mathbf{r}_0$ and this formula can also be written in the form of a line integral taken along the trajectory of the charge:

$$\mathbf{A}_\omega = e\frac{e^{ikR_0}}{cR_0}\int e^{i(\omega t - \mathbf{k}\cdot\mathbf{r}_0)}\, d\mathbf{r}_0. \qquad (66.12)$$

According to (66.8), the Fourier components of the magnetic field have the form:

$$\mathbf{H}_\omega = e \frac{i\omega e^{ikR_0}}{c^2 R_0} \int e^{i(\omega t - \mathbf{k}\cdot\mathbf{r}_0)} \mathbf{n} \times d\mathbf{r}_0. \tag{66.13}$$

If the charge carries out a periodic motion in a closed trajectory, then the field must be expanded in a Fourier series. The components of the Fourier series expansion are obtained by replacing the integration over all times in formulas (66.11) to (66.13) by an average over the period T of the motion (see § 49). For the Fourier component of the magnetic field with frequency $\omega = n\omega_0 = n(2\pi/T)$, we have

$$\mathbf{H}_n = e \frac{2\pi i n e^{ikR_0}}{c^2 T^2 R_0} \int_0^T e^{i[n\omega_0 t - \mathbf{k}\cdot\mathbf{r}_0(t)]} \mathbf{n} \times \mathbf{v}(t)\, dt$$

$$= e \frac{2\pi i n e^{ikR_0}}{c^2 T^2 R_0} \oint e^{i(n\omega_0 t - \mathbf{k}\cdot\mathbf{r}_0)} \mathbf{n} \times d\mathbf{r}_0 \tag{66.14}$$

In the second integral, the integration goes over the closed orbit of the particle.

PROBLEM

Find the four-dimensional expression for the spectral resolution of the four-momentum radiated by a charge moving along a given trajectory.

Solution: Substituting (66.8) in (66.9), and using the fact that, because of the condition (62.1), $k\phi_\omega = \mathbf{k}\cdot\mathbf{A}_\omega$, we find:

$$d\mathcal{E}_{n\omega} = \frac{c}{2\pi}(k^2|\mathbf{A}_\omega|^2 - |\mathbf{k}\cdot\mathbf{A}_\omega|^2) R_0^2 \, do \, \frac{d\omega}{2\pi}$$

$$= \frac{ck^2}{2\pi}(|\mathbf{A}_\omega|^2 - |\phi_\omega|^2) R_0^2 \, do \, \frac{d\omega}{2\pi} = -\frac{ck^2}{2\pi} A_{i\omega} A_\omega^{i*} R_0^2 \, do \, \frac{d\omega}{2\pi}.$$

Representing the four-potential $A_{i\omega}$ in a form analogous to (66.12), we get:

$$d\mathcal{E}_{n\omega} = -\frac{k^2 e^2}{4\pi^2} \chi_i \chi^{i*} \, do \, dk,$$

where χ^i denotes the four-vector

$$\chi^i = \int \exp(-ik_l x^l)\, dx^i$$

and the integration is performed along the world line of the trajectory of the particle. Finally, changing to four-dimensional notation [including the four-dimensional "volume element" in k-space, as in (10.1a)], we find for the radiated four-momentum:

$$dP^i = \frac{e^2 k^i}{2\pi^2 c} \chi_i \chi^{i*} \delta(k_m k^m)\, d^4k.$$

§ 67. Dipole radiation

The time $\mathbf{r}\cdot(\mathbf{n}/c)$ in the integrands of the expressions (66.1) and (66.2) for the retarded potentials can be neglected in cases where the distribution of charge changes little during this time. It is easy to find the conditions for satisfying this requirement. Let T denote the order of magnitude of the time during which the distribution of the charges in the system changes significantly. The radiation of the system will obviously contain periods of order T (i.e. frequencies of order $1/T$). We further denote by a the order of magnitude of the dimensions of the system. Then the time $\mathbf{r} \cdot (\mathbf{n}/c) \sim a/c$. In order that the distribution of the charges in the system shall not undergo a significant change during this time, it is necessary that $a/c \ll T$. But cT is just the wavelength λ of the radiation. Thus the condition $a \ll cT$ can be written in the form

$$a \ll \lambda, \tag{67.1}$$

that is, the dimensions of the system must be small compared with the radiated wavelength.

We note that this same condition (67.1) can also be obtained from (66.7). In the integrand, $\mathbf{r}$ goes through values in an interval of the order of the dimensions of the system, since outside the system $\mathbf{j}$ is zero. Therefore the exponent $i\mathbf{k} \cdot \mathbf{r}$ is small, and can be neglected for those waves in which $ka \ll 1$, which is equivalent to (67.1).

This condition can be written in still another form by noting that $T \sim a/v$, so that $\lambda \sim ca/v$, if v is of the order of magnitude of the velocities of the charges. From $a \ll \lambda$, we then find

$$v \ll c, \tag{67.2}$$

that is, the velocities of the charges must be small compared with the velocity of light.

We shall assume that this condition is fulfilled, and take up the study of the radiation at distances from the radiating system large compared with the wavelength (and consequently, in any case, large compared with the dimensions of the system). As was pointed out in § 66, at such distances the field can be considered as a plane wave, and therefore in determining the field it is sufficient to calculate only the vector potential.

The vector potential (66.2) of the field now has the form

$$\mathbf{A} = \frac{1}{cR_0} \int \mathbf{j}_{t'}\, dV, \tag{67.3}$$

where the time $t' = t - (R_0/c)$ now no longer depends on the variable of integration. Substituting $\mathbf{j} = \varrho\mathbf{v}$, we rewrite (67.3) in the form

$$\mathbf{A} = \frac{1}{cR_0} (\Sigma\, e\mathbf{v})$$

(the summation goes over all the charges of the system; for brevity, we omit the index t'— all quantities on the right side of the equation refer to time t'). But

$$\Sigma\, e\mathbf{v} = \frac{d}{dt} \Sigma\, e\mathbf{r} = \dot{\mathbf{d}},$$

where $\mathbf{d}$ is the dipole moment of the system. Thus,

$$\mathbf{A} = \frac{1}{cR_0}\, \dot{\mathbf{d}}. \tag{67.4}$$

With the aid of formula (66.3) we find that the magnetic field is equal to

$$\mathbf{H} = \frac{1}{c^2 R_0}\, \ddot{\mathbf{d}} \times \mathbf{n}, \tag{67.5}$$

and the electric field to

$$\mathbf{E} = \frac{1}{c^2 R_0}\, (\ddot{\mathbf{d}} \times \mathbf{n}) \times \mathbf{n}. \tag{67.6}$$

We note that in the approximation considered here, the radiation is determined by the second derivative of the dipole moment of the system. Radiation of this kind is called *dipole radiation*.

Since $\mathbf{d} = \Sigma\, er$, $\ddot{\mathbf{d}} = \Sigma\, e\dot{\mathbf{v}}$. Thus the charges can radiate only if they move with acceleration. Charges in uniform motion do not radiate. This also follows directly from the principle of relativity, since a charge in uniform motion can be considered in the inertial system in which it is at rest, and a charge at rest does not radiate.

Substituting (67.5) in (66.6), we get the intensity of the dipole radiation:

$$dI = \frac{1}{4\pi c^3}\, (\ddot{\mathbf{d}} \times \mathbf{n})^2\, do = \frac{\ddot{\mathbf{d}}^2}{4\pi c^3} \sin^2 \theta\, do\,, \tag{67.7}$$

where θ is the angle between $\ddot{\mathbf{d}}$ and $\mathbf{n}$. This is the amount of energy radiated by the system in unit time into the element of solid angle do. We note that the angular distribution of the radiation is given by the factor $\sin^2 \theta$.

Substituting $do = 2\pi \sin \theta\, d\theta$ and integrating over θ from θ to π, we find for the total radiation

$$I = \frac{2}{3c^3}\, \ddot{\mathbf{d}}^2\,. \tag{67.8}$$

If we have just one charge moving in the external field, then $\mathbf{d} = e\mathbf{r}$ and $\ddot{\mathbf{d}} = e\mathbf{w}$, where $\mathbf{w}$ is the acceleration of the charge. Thus the total radiation of the moving charge is

$$I = \frac{2e^2 w^2}{3c^3}\,, \tag{67.9}$$

We note that a closed system of particles, for all of which the ratio of charge to mass is the same, cannot radiate (by dipole radiation). In fact, for such a system, the dipole moment

$$\mathbf{d} = \Sigma\, er = \Sigma \frac{e}{m}\, m\mathbf{r} = \text{const } \Sigma\, m\mathbf{r}\,,$$

where const is the charge-to-mass ratio common to all the charges. But $\Sigma\, m\mathbf{r} = \mathbf{R} \Sigma\, m$, where $\mathbf{R}$ is the radius vector of the centre of inertia of the system (remember that all of the velocities are small, $v \ll c$, so that non-relativistic mechanics is applicable). Therefore $\ddot{\mathbf{d}}$ is proportional to the acceleration of the centre of inertia, which is zero, since the centre of inertia moves uniformly.

Finally, we give the formula for the spectral resolution of the intensity of dipole radiation. For radiation accompanying a collision, we introduce the quantity $d\mathscr{E}_\omega$ of energy radiated throughout the time of the collision in the form of waves with frequencies in the interval $d\omega/2\pi$ (see § 66). It is obtained by replacing the vector $\ddot{\mathbf{d}}$ in (67.8) by its Fourier component $\ddot{\mathbf{d}}_\omega$ and multiplying by 2:

$$d\mathscr{E}_\omega = \frac{4}{3c^3}\, (\ddot{\mathbf{d}}_\omega)^2\, \frac{d\omega}{2\pi}\,.$$

For determining the Fourier components, we have

$$\ddot{\mathbf{d}}_\omega e^{-i\omega t} = \frac{d^2}{dt^2}\, (\mathbf{d}_\omega e^{-i\omega t}) = -\omega^2 \mathbf{d}_\omega e^{-i\omega t}\,,$$

from which $\ddot{\mathbf{d}}_\omega = -\omega^2 \mathbf{d}_\omega$. Thus, we get

$$d\mathscr{E}_\omega = \frac{4\omega^4}{3c^3} |\mathbf{d}_\omega|^2 \frac{d\omega}{2\pi}. \tag{67.10}$$

For periodic motion of the particles, we obtain in similar fashion the intensity of radiation with frequency $\omega = n\omega_0$ in the form

$$I_n = \frac{4\omega_0^4 n^4}{3c^3} |\mathbf{d}_n|^2. \tag{67.11}$$

PROBLEMS

1. Find the radiation from a dipole $\mathbf{d}$, rotating in a plane with constant angular velocity Ω.†

Solution: Choosing the plane of the rotation as the x, y plane, we have:

$$d_x = d_0 \cos \Omega t, \quad d_y = d_0 \sin \Omega t.$$

Since these functions are monochromatic, the radiation is also monochromatic, with frequency $\omega = \Omega$. From formula (67.7) we find for the angular distribution of the radiation (averaged over the period of the rotation):

$$\bar{d}I = \frac{d_0^2 \Omega^4}{8\pi c^3} (1 + \cos^2 \theta) \, do,$$

where θ is the angle between the direction $\mathbf{n}$ of the radiation and the z axis. The total radiation is

$$\bar{I} = \frac{2d_0^2 \Omega^4}{3c^3}.$$

The polarization of the radiation is along the vector $\ddot{\mathbf{d}} \times \mathbf{n} = \omega^2 \mathbf{n} \times \mathbf{d}$. Resolving it into components in the $\mathbf{n}, z$ plane and perpendicular to it, we find that the radiation is elliptically polarized, and that the ratio of the axes of the ellipse is equal to $n_z = \cos \theta$; in particular, the radiation along the z axis is circularly polarized.

2. Determine the angular distribution of the radiation from a system of charges, moving as a whole (with velocity $\mathbf{v}$), if the distribution of the radiation is known in the reference system in which the system is at rest as a whole.

Solution: Let

$$dI' = f(\cos \theta', \phi') \, do', \quad do' = d(\cos \theta') \, d\phi'$$

be the intensity of the radiation in the K' frame which is attached to the moving charge system (θ', ϕ' are the polar coordinates; the polar axis is along the direction of motion of the system). The energy $d\mathscr{E}$ radiated during a time interval dt in the fixed (laboratory) reference frame K, is related to the energy $d\mathscr{E}'$ radiated in the K' system by the transformation formula

$$d\mathscr{E}' = \frac{d\mathscr{E} - \mathbf{V} \cdot d\mathbf{P}}{\sqrt{1 - \dfrac{V^2}{c^2}}} = d\mathscr{E} \frac{1 - \dfrac{V}{c} \cos \theta}{\sqrt{1 - \dfrac{V^2}{c^2}}}$$

(the momentum of radiation propagating in a given direction is related to its energy by the equation

† The radiation from a rotator or a symmetric top which has a dipole moment is of this type. In the first case, $\mathbf{d}$ is the total dipole moment of the rotator; in the second case $\mathbf{d}$ is the projection of the dipole moment of the top on a plane perpendicular to its axis of precession (i.e. the direction of the total angular momentum).

$|d\mathbf{P}| = d\,\mathcal{E}/c$). The polar angles, θ, θ' of the direction of the radiation in the K and K' frames are related by formulas (5.6), and the azimuths ϕ and ϕ' are equal. Finally, the time interval dt' in the K' system corresponds to the time

$$dt = \frac{dt'}{\sqrt{1 - \dfrac{V^2}{c^2}}}$$

in the K system.

As a result, we find for the intensity $dI = (d\,\mathcal{E}/dt)\,do$ in the K system:

$$dI = \frac{\left(1 - \dfrac{V^2}{c^2}\right)^2}{\left(1 - \dfrac{V}{c}\cos\theta\right)^3}\,f\left(\frac{\cos\theta - \dfrac{V}{c}}{1 - \dfrac{V}{c}\cos\theta},\,\phi\right)do\,.$$

Thus, for a dipole moving along the direction of its own axis, $f = \mathrm{const}\cdot\sin^2\theta'$, and by using the formula just obtained, we find:

$$dI = \mathrm{const}\cdot\frac{\left(1 - \dfrac{V^2}{c^2}\right)^3\sin^2\theta}{\left(1 - \dfrac{V}{c}\cos\theta\right)^5}\,do\,.$$

§ 68. Dipole radiation during collisions

In problems of radiation during collisions, one is seldom interested in the radiation accompanying the collision of two particles moving along definite trajectories. Usually we have to consider the scattering of a whole beam of particles moving parallel to each other, and the problem consists in determining the total radiation per unit current density of particles.

If the current density is unity, i.e. if one particle passes per unit time across unit area of the cross-section of the beam, then the number of particles in the flux which have "impact parameters" between ϱ and $\varrho + d\varrho$ is $2\pi\varrho\,d\varrho$ (the area of the ring bounded by the circles of radius ϱ and $\varrho + d\varrho$). Therefore the required total radiation is gotten by multiplying the total radiation $\Delta\mathcal{E}$ from a single particle (with given impact parameter) by $2\pi\varrho\,d\varrho$ and integrating over ϱ from 0 to ∞. The quantity determined in this way has the dimensions of energy times area. We call it the *effective radiation* (in analogy to the effective cross-section for scattering) and denote it by $\varkappa$:[†]

$$\varkappa = \int_0^\infty \Delta\mathcal{E}\cdot 2\pi\varrho\,d\varrho. \tag{68.1}$$

We can determine in completely analogous manner the effective radiation in a given solid angle element do, in a given frequency interval $d\omega$, etc.[‡]

[†] The ratio of $\varkappa$ to the energy of the radiating system is called the cross section for energy loss by radiation.

[‡] If the expression to be integrated depends on the angle of orientation of the projection of the dipole moment of the particle on the plane transverse to the beam, then we must first average over all directions in this plane and only then multiply by $2\pi\varrho\,d\varrho$ and integrate.

We derive the general formula for the angular distribution of radiation emitted in the scattering of a beam of particles by a centrally symmetric field, assuming dipole radiation.

The intensity of the radiation (at a given time) from each of the particles of the beam under consideration is determined by formula (67.7), in which $\mathbf{d}$ is the dipole moment of the particle relative to the scattering centre.† First of all we average this expression over all directions of the vectors $\ddot{\mathbf{d}}$ in the plane perpendicular to the beam direction. Since $(\ddot{\mathbf{d}} \times \mathbf{n})^2 = \ddot{\mathbf{d}}^2 - (\mathbf{n} \cdot \ddot{\mathbf{d}})^2$, the averaging affects only $(\mathbf{n} \cdot \ddot{\mathbf{d}})^2$. Because the scattering field is centrally symmetric and the incident beam is parallel, the scattering, and also the radiation, has axial symmetry around an axis passing through the centre. We choose this axis as x axis. From symmetry, it is obvious that the first powers $\ddot{d}_y$, $\ddot{d}_z$ give zero on averaging, and since $\ddot{d}_x$ is not subjected to the averaging process,

$$\overline{\ddot{d}_x \ddot{d}_y} = \overline{\ddot{d}_x \ddot{d}_z} = 0.$$

The average values of $\ddot{d}_y^2$ and $\ddot{d}_z^2$ are equal to each other, so that

$$\overline{\ddot{d}_y^2} = \overline{\ddot{d}_z^2} = \tfrac{1}{2} [(\ddot{\mathbf{d}})^2 - \ddot{d}_x^2].$$

Keeping all this in mind, we find without difficulty:

$$\overline{(\ddot{\mathbf{d}} \times \mathbf{n})^2} = \tfrac{1}{2} (\ddot{\mathbf{d}}^2 + \ddot{d}_x^2) + \tfrac{1}{2} (\ddot{\mathbf{d}}^2 - 3\ddot{d}_x^2) \cos^2 \theta,$$

where θ is the angle between the direction $\mathbf{n}$ of the radiation and the x axis.

Integrating the intensity over the time and over all impact parameters, we obtain the following final expression giving the effective radiation as a function of the direction of radiation:

$$d\varkappa_\mathbf{n} = \frac{do}{4\pi c^3} \left[A + B \frac{3 \cos^2 \theta - 1}{2} \right], \tag{68.2}$$

where

$$A = \frac{2}{3} \int_0^\infty \int_{-\infty}^{+\infty} \ddot{\mathbf{d}}^2 \, dt \, 2\pi \varrho \, d\varrho, \qquad B = \frac{1}{3} \int_0^\infty \int_{-\infty}^{+\infty} (\ddot{\mathbf{d}}^2 - 3\ddot{d}_x^2) \, dt \, 2\pi \varrho \, d\varrho. \tag{68.3}$$

The second term in (68.2) is written in such a form that it gives zero when averaged over all directions, so that the total effective radiation is $\varkappa = A/c^3$. We call attention to the fact that the angular distribution of the radiation is symmetric with respect to the plane passing through the scattering centre and perpendicular to the beam, since the expression (68.2) is unchanged if we replace θ by $\pi - \theta$. This property is specific to dipole radiation, and is no longer true for higher approximations in v/c.

The intensity of the radiation accompanying the scattering can be separated into two parts—radiation polarized in the plane passing through the x axis and the direction $\mathbf{n}$ (we choose this plane as the xy plane), and radiation polarized in the perpendicular plane xz.

The vector of the electric field has the direction of the vector

† Actually one usually deals with the dipole moment of two particles—the scatterer and the scattered particle—relative to their common centre of inertia.

$$\mathbf{n} \times (\ddot{\mathbf{d}} \times \mathbf{n}) = \mathbf{n}(\mathbf{n} \cdot \ddot{\mathbf{d}}) - \ddot{\mathbf{d}}$$

[see (67.6)]. The component of this vector in the direction perpendicular to the xy plane is—$\ddot{d}_z$, and its projection on the xy plane is $|\sin \theta \ddot{d}_x - \cos \theta \ddot{d}_y|$. This latter quantity is most conveniently determined from the z-component of the magnetic field which has the direction $\ddot{\mathbf{d}} \times \mathbf{n}$.

Squaring $\mathbf{E}$ and averaging over all directions of the vector $\ddot{\mathbf{d}}$ in the yz plane, we see first of all that the product of the projections of the field on the xy plane and perpendicular to it, vanishes. This means that the intensity can actually be represented as the sum of two independent parts—the intensities of the radiation polarized in the two mutually perpendicular planes.

The intensity of the radiation with its electric vector perpendicular to the xy plane is determined by the mean square of $\ddot{d}_z^2 = \frac{1}{2}(\ddot{\mathbf{d}}^2 - \ddot{d}_x^2)$. For the corresponding part of the effective radiation, we obtain the expression

$$d\varkappa_{\mathbf{n}}^{\perp} = \frac{do}{4\pi c^3} \frac{1}{2} \int_0^\infty \int_{-\infty}^{+\infty} (\ddot{\mathbf{d}}^2 - \ddot{d}_x^2) \, dt \, 2\pi\varrho \, d\varrho. \tag{68.4}$$

We note that this part of the radiation is isotropic. It is unnecessary to give the expression for the effective radiation with electric vector in the xy plane since it is clear that

$$d\varkappa_{\mathbf{n}}^{\parallel} + d\varkappa_{\mathbf{n}}^{\perp} = d\varkappa_{\mathbf{n}}.$$

In a similar way we can get the expression for the angular distribution of the effective radiation in a given frequency interval $d\omega$:

$$d\varkappa_{\mathbf{n},\omega} = \left[A(\omega) + B(\omega) \frac{3\cos^2\theta - 1}{2} \right] \frac{do}{2\pi c^3} \frac{d\omega}{2\pi} \tag{68.5}$$

where

$$A(\omega) = \frac{2\omega^4}{3} \int_0^\infty \mathbf{d}_\omega^2 \, 2\pi\varrho \, d\varrho, \qquad B(\omega) = \frac{\omega^4}{3} \int_0^\infty (\mathbf{d}_\omega^2 - 3d_{x\omega}^2) \, 2\pi\varrho \, d\varrho \tag{68.6}$$

§ 69. Radiation of low frequency in collisions

Let us consider the low-frequency "tail" of the spectral distribution of the *bremsstrahlung:* the range of frequencies that is low compared to the frequency ω_0 around which the main part of the radiation is concentrated:

$$\omega \ll \omega_0. \tag{69.1}$$

We shall not assume that the velocities of the colliding particles are small compared to the light velocity, as was done in the preceding paragraph; the following formulas are valid for arbitrary velocities. In the nonrelativistic case, $\omega_0 \sim 1/\tau$, where τ is the order of magnitude of the duration of the collision; in the ultrarelativistic case, ω_0 is proportional to the square of the energy of the radiating particle (cf. § 77).

In the integral

$$\mathbf{H}_\omega = \int\limits_{-\infty}^{\infty} \mathbf{H} e^{i\omega t}\, dt\,,$$

the field $\mathbf{H}$ of the radiation is significantly different from zero only during a time interval of the order of $1/\omega_0$. Therefore, in accord with condition (69.1), we can assume that $\omega t \ll 1$ in the integral, so that we can replace $e^{i\omega t}$ by unity; then

$$\mathbf{H}_\omega = \int\limits_{-\infty}^{\infty} \mathbf{H}\, dt\,.$$

Substituting $\mathbf{H} = \mathring{\mathbf{A}} \times \mathbf{n}/c$ and carrying out the time integration, we get:

$$\mathbf{H}_\omega = \frac{1}{c}(\mathbf{A}_2 - \mathbf{A}_1) \times \mathbf{n},\tag{69.2}$$

where $\mathbf{A}_2 - \mathbf{A}_1$ is the change in the vector poential produced by the colliding particles during the time of the collision.

The total radiation (with frequency ω) during the time of the collision is found by substituting (69.2) in (66.9):

$$d\mathscr{E}_{\mathbf{n}\omega} = \frac{R_0^2}{4c\pi^2}[(\mathbf{A}_2 - \mathbf{A}_1) \times \mathbf{n}]^2\, do\, d\omega\,.\tag{69.3}$$

We can use the Lienard–Wiechert expression (66.4) for the vector potential, and obtain:

$$d\mathscr{E}_{\mathbf{n}\omega} = \frac{1}{4\pi^2 c^3}\left[\sum e\left\{\frac{\mathbf{v}_2 \times \mathbf{n}}{1 - (1/c)\mathbf{n}\cdot\mathbf{v}_2} - \frac{\mathbf{v}_1 \times \mathbf{n}}{1 - (1/c)\mathbf{n}\cdot\mathbf{v}_1}\right\}\right]^2\, do\, d\omega,\tag{69.4}$$

where $\mathbf{v}_1$ and $\mathbf{v}_2$ are the velocities of the particle before and after the collision, and the sum is taken over the two colliding particles. We note that the coefficient of $d\omega$ is independent of frequency. In other words, at low frequencies [condition (69.1)], the spectral distribution is independent of frequency, i.e. $d\mathscr{E}_{\mathbf{n}\omega}/d\omega$ tends toward a constant limit as $\omega \to 0$.†

If the velocities of the colliding particles are small compared with the velocity of light, then (69.4) becomes

$$d\mathscr{E}_{\mathbf{n}\omega} = \frac{1}{4\pi^2 c^3}[\sum e(\mathbf{v}_2 - \mathbf{v}_1) \times \mathbf{n}]^2\, do\, d\omega\,.\tag{69.5}$$

This expression corresponds to the case of dipole radiation, with the vector potential given by formula (67.4).

An interesting application of these formulas is to the radiation produced in the emission of a new charged particle (e.g. the emergence of a β-particle from a nucleus). This process is to be treated as an instantaneous change in the velocity of the particle from zero to its

---

† By integrating over the impact parameters, we can obtain an analogous result for the effective radiation in the scattering of a beam of particles. However it must be remembered that this result is not valid for the effective radiation when there is a Coulomb interaction of the colliding particles, because then the integral over ρ is divergent (logarithmically) for large ρ. We shall see in the next section that in this case the effective radiation at low frequencies depends logarithmically on frequency and does not remain constant.

actual value. [Because of the symmetry of formula (69.5) with respect to interchange of $\mathbf{v}_1$ and $\mathbf{v}_2$, the radiation originating in this process is identical with the radiation which would be produced in the inverse process—the instantaneous stopping of the particle.] The essential point is that, since the "time" for the process is $\tau \to 0$, condition (69.1) is actually satisfied for all frequencies.†

PROBLEM

Find the spectral distribution of the total radiation produced when a charged particle is emitted which moves with velocity v.

Solution: According to formula (69.4) (in which we set $\mathbf{v}_2 = \mathbf{v}$, $\mathbf{v}_1 = 0$), we have:

$$d\mathscr{E}_\omega = d\omega \frac{e^2 v^2}{4\pi^2 c^3} \int_0^\pi \frac{\sin^2\theta}{\left(1 - \frac{v}{c}\cos\theta\right)^2} \, 2\pi \sin\theta \, d\theta.$$

Evaluation of the integral gives:‡

$$d\mathscr{E}_\omega = \frac{e^2}{\pi c}\left(\frac{c}{v}\ln\frac{c+v}{c-v} - 2\right) d\omega. \tag{1}$$

For $v \ll c$, this formula goes over into

$$d\mathscr{E}_\omega = \frac{2e^2 v^2}{3\pi c^3} \, d\omega,$$

which can also be obtained directly from (69.5).

§ 70. Radiation in the case of Coulomb interaction

In this section we present, for reference purposes, a series of formulas relating to the dipole radiation of a system of two charged particles; it is assumed that the velocities of the particles are small compared with the velocity of light.

Uniform motion of the system as a whole, i.e. motion of its centre of mass, is not of interest, since it does not lead to radiation, therefore we need only consider the relative motion of the particles. We choose the origin of coordinates at the centre of mass. Then the dipole moment of the system $\mathbf{d} = e_1\mathbf{r}_1 + e_2\mathbf{r}_2$ has the form

$$\mathbf{d} = \frac{e_1 m_2 - e_2 m_1}{m_1 + m_2}\,\mathbf{r} = \mu\left(\frac{e_1}{m_1} - \frac{e_2}{m_2}\right)\mathbf{r} \tag{70.1}$$

where the indices 1 and 2 refer to the two particles, and $\mathbf{r} = \mathbf{r}_1 - \mathbf{r}_2$ is the radius vector between them, and

---

† However, the applicability of these formulas is limited by the quantum condition that $\hbar\omega$ be small compared with the total kinetic energy of the particle.

‡ Even though, as we have already pointed out, condition (69.1) is satisfied for all frequencies, because the process is "instantaneous" we cannot find the total radiated energy by integrating (1) over ω—the integral diverges at high frequencies. We mention that, aside from the violation of the conditions for classical behaviour at high frequencies, in the present case the cause of the divergence lies in the incorrect formulation of the classical problem, in which the particle has an infinite acceleration at the initial time.

$$\mu = \frac{m_1 m_2}{m_1 + m_2}$$

is the reduced mass.

We start with the radiation accompanying the elliptical motion of two particles attracting each other according to the Coulomb law. As we know from mechanics†, this motion can be expressed as the motion of a particle with mass μ in the ellipse whose equation in polar coordinates is

$$1 + \varepsilon \cos \phi = \frac{a(1 - \varepsilon^2)}{r}, \tag{70.2}$$

where the semimajor axis a and the eccentricity ε are

$$a = \frac{\alpha}{2|\mathscr{E}|}, \qquad \varepsilon = \sqrt{1 - \frac{2|\mathscr{E}|M^2}{\mu\alpha^2}}. \tag{70.3}$$

Here $\mathscr{E}$ is the total energy of the particles (omitting their rest energy!) and is negative for a finite motion; $M = \mu r^2 \dot{\phi}$ is the angular momentum, and α is the constant in the Coulomb law:

$$\alpha = |e_1 e_2|.$$

The time dependence of the coordinates can be expressed in terms of the parametric equations

$$r = a(1 - \varepsilon \cos \xi), \qquad t = \sqrt{\frac{\mu a^3}{\alpha}} (\xi - \varepsilon \sin \xi). \tag{70.4}$$

One full revolution in the ellipse corresponds to a change of the parameter ξ from 0 to 2π; the period of the motion is

$$T = 2\pi \sqrt{\frac{\mu a^3}{\alpha}}.$$

We calculate the Fourier components of the dipole moment. Since the motion is periodic we are dealing with an expansion in Fourier series. Since the dipole moment is proportional to the radius vector $\mathbf{r}$, the problem reduces to the calculation of the Fourier components of the coordinates $x = r \cos \phi$, $y = r \sin \phi$. The time dependence of x and y is given by the parametric equations

$$x = a (\cos \xi - \varepsilon), \qquad y = a\sqrt{1 - \varepsilon^2} \sin \xi,$$

$$\omega_0 t = \xi - \varepsilon \sin \xi. \tag{70.5}$$

Here we have introduced the frequency

$$\omega_0 = 2\pi/T = \sqrt{\alpha/\mu a^3} = \frac{(2|\mathscr{E}|)^{\frac{3}{2}}}{\alpha \mu^{\frac{1}{2}}}.$$

† See *Mechanics*, § 15.

Instead of the Fourier components of the coordinates, it is more convenient to calculate the Fourier components of the velocities, using the fact that $\dot{x}_n = -i\omega_0 n x_n$; $\dot{y}_n = -i\omega_0 n y_n$. We have

$$x_n = \frac{\dot{x}_n}{-i\omega_0 n} = \frac{i}{\omega_0 nT} \int_0^T e^{i\omega_0 nt} \dot{x}\, dt .$$

But $\dot{x}\, dt = dx = -a \sin \xi\, d\xi$; transforming from an integral over t to one over ξ, we have

$$x_n = -\frac{ia}{2\pi n} \int_0^{2\pi} e^{in(\xi - \varepsilon \sin \xi)} \sin \xi\, d\xi .$$

Similarly, we find

$$y_n = \frac{ia\sqrt{1-\varepsilon^2}}{2\pi n} \int_0^{2\pi} e^{in(\xi - \varepsilon \sin \xi)} \cos \xi\, d\xi = \frac{ia\sqrt{1-\varepsilon^2}}{2\pi n\varepsilon} \int_0^{2\pi} e^{in(\xi - \varepsilon \sin \xi)} d\xi$$

(in going from the first to the second integral, we write the integrand as $\cos \xi \equiv (\cos \xi - 1/\varepsilon) + 1/\varepsilon$; then the integral with $\cos \xi - 1/\varepsilon$ can be done, and gives identically zero). Finally, we use a formula of the theory of Bessel functions,

$$\frac{1}{2\pi} \int_0^{2\pi} e^{i(n\xi - x\sin\xi)}\, d\xi = \frac{1}{\pi} \int_0^{\pi} \cos (n\xi - x \sin \xi)\, d\xi = J_n(x), \qquad (70.6)$$

where $J_n(x)$ is the Bessel function of integral order n. As a final result, we obtain the following expression for the required Fourier components:

$$x_n = \frac{a}{n} J_n'(n\varepsilon), \qquad y_n = \frac{ia\sqrt{1-\varepsilon^2}}{n\varepsilon} J_n(n\varepsilon) \qquad (70.7)$$

(the prime on the Bessel function means differentiation with respect to its argument).

The expression for the intensity of the monochromatic components of the radiation is obtained by substituting x_n and y_n into the formula

$$I_n = \frac{4\omega_0^4 n^4}{3c^3} \mu^2 \left(\frac{e_1}{m_1} - \frac{e_2}{m_2} \right)^2 (|x_n|^2 + |y_n|^2)$$

[see (67.11)]. Expressing a and ω_0 in terms of the characteristics of the particles, we obtain finally:

$$I_n = \frac{64n^4 e^4}{3c^3\alpha^2} \left(\frac{e_1}{m_1} - \frac{e_2}{m_2} \right)^2 \left[J_n'^2(n\varepsilon) + \frac{1-\varepsilon^2}{\varepsilon^2} J_n^2(n\varepsilon) \right]. \qquad (70.8)$$

In particular, we shall give the asymptotic formula for the intensity of very high harmonics (large n) for motion in an orbit which is close to a parabola (ε close to 1). For this purpose, we use the formula

$$J_n(n\varepsilon) \cong \frac{1}{\sqrt{\pi}} \left(\frac{2}{n}\right)^{1/3} \Phi\left[\left(\frac{n}{2}\right)^{2/3} (1 - \varepsilon^2)\right] \tag{70.9}$$

$$n \gg 1, \qquad 1 - \varepsilon \ll 1,$$

where Φ is the Airy function defined on p. 161.†
Substituting in (70.8) gives:

$$I_n = \frac{64.2^{2/3}}{3\pi} \frac{n^{4/3}\mathscr{E}^4}{c^3\alpha^2} \left(\frac{e_1}{m_1} - \frac{e_2}{m_2}\right)^2 \left\{(1 - \varepsilon^2)\Phi^2\left[\left(\frac{n}{2}\right)^{2/3} (1 - \varepsilon^3)\right]\right.$$

$$\left. + \left(\frac{2}{n}\right)^{2/3} \Phi'^2\left[\left(\frac{n}{2}\right)^{2/3} (1 - \varepsilon^2)\right]\right\}. \tag{70.10}$$

This result can also be expressed in terms of the MacDonald function K_y:

$$I_n = (1 - \varepsilon^2)^2 \frac{64}{9\pi^2} \frac{n^2\mathscr{E}^4}{c^3\alpha^2} \left(\frac{e_1}{m_1} - \frac{e_2}{m_2}\right)^2 \left\{K_{1/3}^2\left[\frac{n}{3}(1 - \varepsilon^2)^{3/2}\right] + K_{2/3}^2\left[\frac{n}{3}(1 - \varepsilon^2)^{3/2}\right]\right\}$$

(the necessary formulas are given in the footnote on p. 218).

Next, we consider the collision of two attracting charged particles. Their relative motion is described as the motion of a particle with mass μ in the hyperbola

$$1 + \varepsilon \cos \phi = \frac{a(\varepsilon^2 - 1)}{r}, \tag{70.11}$$

where

$$a = \frac{\alpha}{2\mathscr{E}}, \quad \varepsilon = \sqrt{1 + \frac{2\mathscr{E}M^2}{\mu\alpha^2}} \tag{70.12}$$

(now $\mathscr{E} > 0$). The time dependence of r is given by the parametric equations

† For $n \gg 1$, the main contributions to the integral

$$J_n(n\varepsilon) = \frac{1}{\pi} \int_0^\pi \cos[n(\xi - \varepsilon \sin \xi)] d\xi$$

come from small values of ξ (for larger values of ξ, the integrand oscillates rapidly). In accordance with this, we expand the argument of the cosine in powers of ξ:

$$J_n(n\varepsilon) = \frac{1}{\pi} \int_0^\infty \cos\left[n\left(\frac{1 - \varepsilon^2}{2}\xi + \frac{\xi^3}{6}\right)\right] d\xi;$$

because of the rapid convergence of the integral, the upper limit has been replaced by ∞; the term in ξ^3 must be kept because the first order term contains the small coefficient $1 - \varepsilon \cong (1 - \varepsilon^2)/2$. The integral above is reduced to the form (70.9) by an obvious substitution.

$$r = a(\varepsilon \cosh \xi - 1), \quad t = \sqrt{\frac{\mu a^3}{\alpha}} \, (\varepsilon \sinh \xi - \xi), \tag{70.13}$$

where the parameter ξ runs through values from $-\infty$ to $+\infty$. For the coordinates x, y, we have

$$x = a(\varepsilon - \cosh \xi), \quad y = a\sqrt{\varepsilon^2 - 1} \, \sinh \xi. \tag{70.14}$$

The calculation of the Fourier components (we are now dealing with expansion in a Fourier integral) proceeds in complete analogy to the preceding case. We find the result:

$$x_\omega = \frac{\pi a}{\omega} H_{iv}^{(1)'}(iv\varepsilon), \quad y_\omega = -\frac{\pi a\sqrt{\varepsilon^2 - 1}}{\omega\varepsilon} H_{iv}^{(1)}(iv\varepsilon). \tag{70.15}$$

where $H_{iv}^{(1)}$ is the Hankel function of the first kind, of order iv, and we have introduced the notation

$$v = \frac{\omega}{\sqrt{\dfrac{\alpha}{\mu a^3}}} = \frac{\omega\alpha}{\mu v_0^3} \tag{70.16}$$

(v_0 is the relative velocity of the particles at infinity; the energy $\mathscr{E} = \mu v_0^2/2$).† In the calculation we have used the formula from the theory of Bessel functions:

$$\int_{-\infty}^{+\infty} e^{p\xi - ix \sinh \xi} \, d\xi = i\pi H_p^{(1)}(ix). \tag{70.17}$$

Substituting (70.15) in the formula

$$d\mathscr{E}_\omega = \frac{4\omega^4\mu^2}{3c^3} \left(\frac{e_1}{m_1} - \frac{e_2}{m_2}\right)^2 (|x_\omega|^2 + |y_\omega|^2) \frac{d\omega}{2\pi}$$

[see (67.10)], we get:

$$d\mathscr{E}_\omega = \frac{\pi\mu^2\alpha^2\omega^3}{6c^3\mathscr{E}^2} \left(\frac{e_1}{m_1} - \frac{e_2}{m_2}\right)^2 \left\{[H_{iv}^{(1)'}(iv\varepsilon)]^2 + \frac{\varepsilon^2 - 1}{\varepsilon^2}[H_{iv}^{(1)}(iv\varepsilon)]^2\right\} d\omega. \tag{70.18}$$

A quantity of greater interest is the "effective radiation" during the scattering of a parallel beam of particles (see § 68). To calculate it, we multiply $d\mathscr{E}_\omega$ by $2\pi\varrho \, d\varrho$ and integrate over all ϱ from zero to infinity. We transform from an integral over ϱ to one over ε (between the limits 1 and ∞) using the fact that $2\pi\varrho \, d\varrho = 2\pi a^2\varepsilon d\varepsilon$; this relation follows from the definition (70.12), in which the angular momentum M and the energy $\mathscr{E}$ are related to the impact parameter ϱ and the velocity v_0 by

$$M = \mu\varrho v_0, \quad \mathscr{E} = \mu \frac{v_0^2}{2}.$$

The resultant integral can be directly integrated with the aid of the formula

† Note that the function $H_{iv}^{(1)}(iv\varepsilon)$ is purely imaginary, while its derivative $H_{iv}^{(1)'}(iv\varepsilon)$ is real.

$$z \left[Z_p'^2 + \left(\frac{p^2}{z^2} - 1 \right) Z_p^2 \right] = \frac{d}{dz} (z Z_p Z_p'),$$

where $Z_p(z)$ is an arbitrary solution of the Bessel equation of order p.† Keeping in mind that for $\varepsilon \to \infty$, the Hankel function $H_{iv}^{(1)}(iv\varepsilon)$ goes to zero, we get as our result the following formula:

$$d\varkappa_\omega = \frac{4\pi^2 \alpha^3 \omega}{3 c^3 \mu v_0^5} \left(\frac{e_1}{m_1} - \frac{e_2}{m_2} \right)^2 |H_{iv}^{(1)}(iv)| H_{iv}^{(1)'}(iv) \, d\omega. \tag{70.19}$$

Let us consider the limiting cases of low and high frequencies. In the integral

$$\int_{-\infty}^{+\infty} e^{iv(\xi - \sinh \xi)} \, d\xi = i\pi H_{iv}^{(1)}(iv) \tag{70.20}$$

defining the Hankel function, the only important range of the integration parameter ξ is that in which the exponent is of order unity. For low frequencies ($v \ll 1$), only the region of large ξ is important. But for large ξ we have $\sinh \xi \gg \xi$. Thus, approximately,

$$H_{iv}^{(1)}(iv) \cong -\frac{i}{\pi} \int_{-\infty}^{+\infty} e^{-iv\sinh \xi} \, d\xi = H_0^{(1)}(iv).$$

Similarly, we find that

$$H_{iv}^{(1)'}(iv) \cong H_0^{(1)'}(iv).$$

Using the approximate expression (for small x) from the theory of Bessel functions:

$$iH_0^{(1)}(ix) \cong \frac{2}{\pi} \ln \frac{2}{\gamma x}$$

($\gamma = e^C$, where C is the Euler constant; $\gamma = 1.781\ldots$), we get the following expression for the effective radiation at low frequencies:

$$d\varkappa_\omega = \frac{16 \alpha^2}{3 v_0^2 c^3} \left(\frac{e_1}{m_1} - \frac{e_2}{m_2} \right)^2 \ln \left(\frac{2 \mu v_0^3}{\gamma \omega \alpha} \right) d\omega \quad \text{for} \quad \omega \ll \frac{\mu v_0^3}{\alpha}. \tag{70.21}$$

It depends logarithmically on the frequency.

For high frequencies ($v \gg 1$), on the other hand, the region of small ξ is important in the integral (70.20). In accordance with this, we expand the exponent of the integrand in powers of ξ and get, approximately,

---

† This formula is a direct consequence of the Bessel equation

$$Z'' + \frac{1}{z} Z' + \left(1 - \frac{p^2}{z^2} \right) Z = 0.$$

$$H_{iv}^{(1)}(iv) \cong -\frac{i}{\pi} \int_{-\infty}^{+\infty} e^{-\frac{iv\xi^3}{6}} d\xi = -\frac{2i}{\pi} \, \mathrm{Re} \left(\int_{0}^{\infty} e^{-\frac{iv\xi^3}{6}} d\xi \right).$$

By the substitution $iv\xi^3/6 = \eta$, the integral goes over into the Γ-function, and we obtain the result:

$$H_{iv}^{(1)}(iv) \cong -\frac{i}{\pi\sqrt{3}} \left(\frac{6}{v}\right)^{1/3} \Gamma\left(\frac{1}{3}\right).$$

Similarly, we find

$$H_{iv}^{(1)'}(iv) \cong \frac{1}{\pi\sqrt{3}} \left(\frac{6}{v}\right)^{2/3} \Gamma\left(\frac{2}{3}\right).$$

Next, using the formula of the theory of the Γ-function,

$$\Gamma(x)\Gamma(1-x) = \frac{\pi}{\sin \pi x},$$

we obtain for the effective radiation at high frequencies:

$$d\varkappa_\omega = \frac{16\pi\alpha^2}{3^{3/2} \, v_0^2 c^3} \left(\frac{e_1}{m_1} - \frac{e_2}{m_2}\right)^2 d\omega, \quad \text{for} \quad \omega \gg \frac{\mu v_0^3}{\alpha}, \tag{70.22}$$

that is, an expression which is independent of the frequency.

We now proceed to the radiation accompanying the collision of two particles repelling each other according to the Coulomb law $U = \alpha/r (\alpha > 0)$. The motion occurs in a hyperbola,

$$-1 + \varepsilon \cos \phi = \frac{a(\varepsilon^2 - 1)}{r}, \tag{70.23}$$

$$x = a(\varepsilon + \cosh \xi), \quad y = a\sqrt{\varepsilon^2 - 1} \, \sinh \xi,$$

$$t = \sqrt{\frac{\mu a^3}{\alpha}} \, (\varepsilon \sinh \xi + \xi) \tag{70.24}$$

[a and ε as in (70.12)]. All the calculations for this case reduce immediately to those given above, so it is not necessary to present them. Namely, the integral

$$x_\omega = \frac{ia}{\omega} \int_{-\infty}^{+\infty} e^{iv(\varepsilon\sinh\xi+\xi)} \sinh \xi \, d\xi$$

for the Fourier component of the coordinate x reduces, by making the substitution $\xi \to i\pi - \xi$, to the integral for the case of attraction, multiplied by $-e^{-xv}$; the same holds for y_∞.

Thus the expressions for the Fourier components x_ω, y_ω in the case of repulsion differ from the corresponding expressions for the case of attraction by the factor $e^{-\pi v}$. So the only change in the formulas for the radiation is an additional factor $e^{-2\pi v}$. In particular, for low frequencies we get the previous formula (70.21) (since for $v \ll 1$, $e^{-2\pi v} \cong 1$). For high frequencies, the effective radiation has the form

$$d\varkappa_\omega = \frac{16\pi\alpha^2}{3^{3/2}\,v_0^2 c^3} \left(\frac{e_1}{m_1} - \frac{e_2}{m_2}\right)^2 \exp\left(-\frac{2\pi\omega\alpha}{\mu v_0^3}\right) d\omega, \quad \text{for} \quad \omega \gg \frac{\mu v_0^3}{\alpha}. \quad (70.25)$$

It drops exponentially with increasing frequency.

PROBLEMS

1. Calculate the average total intensity of the radiation for elliptical motion of two attracting charges.

Solution: From the expression (70.1) for the dipole moment, we have for the total intensity of the radiation:

$$I = \frac{2\mu^2}{3c^3}\left(\frac{e_1}{m_1} - \frac{e_2}{m_2}\right)^2 \ddot{\mathbf{r}}^2 = \frac{2\alpha^2}{3c^3}\left(\frac{e_1}{m_1} - \frac{e_2}{m_2}\right)^2 \frac{1}{r^4},$$

where we have used the equation of motion $\mu\ddot{\mathbf{r}} = -\alpha\mathbf{r}/r^3$. We express the coordinate r in terms of ϕ from the orbit equation (70.2) and, by using the equation $dt = \mu r^2\, d\phi/M$, we replace the time integration by an integration over the angle ϕ (from 0 to 2π). As a result, we find for the average intensity:

$$\bar{I} = \frac{1}{T}\int\limits_0^T I\, dt = \frac{2^{3/2}}{3c^3}\left(\frac{e_1}{m_1} - \frac{e_2}{m_2}\right)^2 \frac{\mu^{5/2}\alpha^3|\mathscr{E}|^{3/2}}{M^5}\left(3 - \frac{2|\mathscr{E}|M^2}{\mu\alpha^2}\right).$$

2. Calculate the total radiation $\Delta\mathscr{E}$ for the collision of two charged particles.

Solution: In the case of attraction the trajectory is the hyperbola (70.11) and in the case of repulsion, (70.23). The angle between the asymptotes of the hyperbola and its axis is ϕ_0, determined from $\pm\cos\phi_0 = 1/\varepsilon$, and the angle of deflection of the particles (in the system of coordinates in which the centre of mass is at rest) is $\chi = |\pi - 2\phi_0|$. The calculation proceeds the same as in Problem 1 (the integral over ϕ is taken between the limits $-\phi_0$ and $+\phi_0$). The result for the case of attraction is

$$\Delta\mathscr{E} = \frac{\mu^3 v_0^5}{3c^3|\alpha|}\tan^3\frac{\chi}{2}\left\{(\pi + \chi)\left(1 + 3\tan^2\frac{\chi}{2}\right) + 6\tan\frac{\chi}{2}\right\}\left(\frac{e_1}{m_1} - \frac{e_2}{m_2}\right)^2,$$

and for the case of repulsion:

$$\Delta\mathscr{E} = \frac{\mu^3 v_0^5}{3c^3\alpha}\tan^3\frac{\chi}{2}\left\{(\pi + \chi)\left(1 + 3\tan^2\frac{\chi}{2}\right) + 6\tan\frac{\chi}{2}\right\}\left(\frac{e_1}{m_1} - \frac{e_2}{m_2}\right)^2.$$

In both, χ is understood to be a positive angle, determined from the relation

$$\cot\frac{\chi}{2} = \frac{\mu v_0^2 \varrho}{\alpha}.$$

Thus for a head-on collision ($\varrho \to 0$, $\chi \to \pi$) of charges repelling each other:

$$\Delta\mathscr{E} = \frac{8\mu^3 v_0^5}{45c^3\alpha}\left(\frac{e_1}{m_1} - \frac{e_2}{m_2}\right)^2.$$

3. Calculate the total effective radiation in the scattering of a beam of particles in a repulsive Coulomb field.

Solution: The required quantity is

$$\varkappa = \int\limits_0^\infty \int\limits_{-\infty}^{+\infty} I\, dt\, 2\pi\varrho\, d\varrho = \frac{2\alpha^2}{3c^3}\left(\frac{e_1}{m_1} - \frac{e_2}{m_2}\right)^2 2\pi \int\limits_0^\infty \int\limits_{-\infty}^{+\infty} \frac{1}{r^4}\, dt\, \varrho\, d\varrho.$$

We replace the time integration by integration over r along the trajectory of the charge, writing $dt = dr/v_r$, where the radial velocity $v_r \equiv \dot{r}$ is expressed in terms of r by the formula

$$v_r = \sqrt{\frac{2}{\mu}\left[\mathscr{E} - \frac{M^2}{2\mu r^2} - U(r)\right]} = \sqrt{v_0^2 - \frac{\varrho^2 v_0^2}{r^2} - \frac{2\alpha}{\mu r}}.$$

The integration over r goes between the limits from ∞ to the distance of closest approach $r_0 = r_0(\varrho)$ (the point at which $v_r = 0$), and then from r_0 once again to infinity; this reduces to twice the integral from r_0 to ∞. The calculation of the double integral is conveniently done by changing the order of integration—integrating first over ϱ and then over r. The result of the calculation is:

$$\varkappa = \frac{8\pi}{9c^3}\,\alpha\mu v_0\left(\frac{e_1}{m_1} - \frac{e_2}{m_2}\right)^2.$$

4. Calculate the angular distribution of the total radiation emitted when one charge passes by another, if the velocity is so large (though still small compared with the velocity of light) that the deviation from straight-line motion can be considered small.

Solution: The angle of deflection is small if the kinetic energy $\mu v^2/2$ is large compared to the potential energy, which is of order $\alpha/\varrho(\mu\alpha^2 \gg \alpha/\varrho)$. We choose the plane of the motion as the x, y plane, with the origin at the centre of inertia and the x axis along the direction of the velocity. In first approximation, the trajectory is the straight line $x = vt$, $y = \varrho$. In the next approximation, the equations of motion give

$$\mu\ddot{x} = \frac{\alpha}{r^2}\frac{x}{r} \cong \frac{\alpha vt}{r^3}, \qquad \mu\ddot{y} = \frac{\alpha y}{r^2 r} \cong \frac{\alpha\varrho}{r^3},$$

with

$$r = \sqrt{x^2 + y^2} \cong \sqrt{\varrho^2 + v^2 t^2}.$$

Using formula (67.7), we have:

$$d\mathscr{E}_\mathbf{n} = do\,\frac{\mu^2}{4\pi c^3}\left(\frac{e_1}{m_1} - \frac{e_2}{m_2}\right)^2\int\limits_{-\infty}^{\infty}[\ddot{x}^2 + \ddot{y}^2 - (\ddot{x}n_x + \ddot{y}n_y)^2]\,dt,$$

where $\mathbf{n}$ is the unit vector in the direction of do. Expressing the integrand in terms of t and performing the integration, we get:

$$d\mathscr{E}_\mathbf{n} = \frac{\alpha^2}{32\,vc^3\varrho^3}\left(\frac{e_1}{m_1} - \frac{e_2}{m_2}\right)^2(4 - n_z^2 - 3n_y^2)\,do$$

§ 71. Quadrupole and magnetic dipole radiation

We now consider the radiation associated with the succeeding terms in the expansion of the vector potential in powers of the ratio a/λ of the dimensions of the system to the wavelength. Since a/λ is assumed to be small, these terms are generally small compared with the first (dipole) term, but they are important in those cases where the dipole moment of the system is zero, so that dipole radiation does not occur.

Expanding the integrand in (66.2),

$$\mathbf{A} = \frac{1}{cR_0}\int \mathbf{j}_{t' + \frac{\mathbf{r}\cdot\mathbf{n}}{c}}\,dV,$$

in powers of $\mathbf{r}\cdot\mathbf{n}/c$, we find, correct to terms of first order:

$$A = \frac{1}{cR_0} \int \mathbf{j}_{t'} \, dV + \frac{1}{c^2 R_0} \frac{\partial}{\partial t'} \int (\mathbf{r} \cdot \mathbf{n}) \, \mathbf{j}_{t'} \, dV.$$

Substituting $\mathbf{j} = \varrho \mathbf{v}$ and changing to point charges, we obtain:

$$A = \frac{\Sigma \, e\mathbf{v}}{cR_0} + \frac{1}{c^2 R_0} \frac{\partial}{\partial t} \Sigma \, e\mathbf{v}(\mathbf{r} \cdot \mathbf{n}). \tag{71.1}$$

(From now on, as in § 67, we drop the index t' in all quantities).

In the second term we write

$$\mathbf{v}(\mathbf{r} \cdot \mathbf{n}) = \frac{1}{2} \frac{\partial}{\partial t} \mathbf{r}(\mathbf{n} \cdot \mathbf{r}) + \frac{1}{2} \mathbf{v}(\mathbf{n} \cdot \mathbf{r}) - \frac{1}{2} \mathbf{r}(\mathbf{n} \cdot \mathbf{v})$$

$$= \frac{1}{2} \frac{\partial}{\partial t} \mathbf{r}(\mathbf{n} \cdot \mathbf{r}) + \frac{1}{2} (\mathbf{r} \times \mathbf{v}) \times \mathbf{n}.$$

We then find for A the expression

$$A = \frac{\dot{\mathbf{d}}}{cR_0} + \frac{1}{2c^2 R_0} \frac{\partial^2}{\partial t^2} \Sigma \, e\mathbf{r}(\mathbf{n} \cdot \mathbf{r}) + \frac{1}{cR_0} (\dot{m} \times \mathbf{n}), \tag{71.2}$$

where $\mathbf{d}$ is the dipole moment of the system, and

$$m = \frac{1}{2c} \Sigma \, e\mathbf{r} \times \mathbf{v}$$

is its magnetic moment. For further transformation, we note that we can, without changing the field, add to A any vector proportional to $\mathbf{n}$, since according to formula (66.3), $\mathbf{H}$ and $\mathbf{E}$ are unchanged by this. For this reason we can replace (71.2) by

$$A = \frac{\dot{\mathbf{d}}}{cR_0} + \frac{1}{6c^2 R_0} \frac{\partial^2}{\partial t^2} \Sigma \, e[3\mathbf{r}(\mathbf{n} \cdot \mathbf{r}) - \mathbf{n}r^2] + \frac{1}{cR_0} \dot{m} \times \mathbf{n}.$$

But the expression under the summation sign is just the product $n_\beta D_{\alpha\beta}$ of the vector $\mathbf{n}$ and the quadrupole moment tensor $D_{\alpha\beta} = \Sigma \, e(3x_\alpha x_\beta - \delta_{\alpha\beta}r^2)$ (see § 41). We introduce the vector $\mathbf{D}$ with components $D_\alpha = D_{\alpha\beta} n_\beta$, and get the final expression for the vector potential:

$$A = \frac{\dot{\mathbf{d}}}{cR_0} + \frac{1}{6c^2 R_0} \ddot{\mathbf{D}} + \frac{1}{cR_0} \dot{m} \times \mathbf{n}. \tag{71.3}$$

Knowing A, we can now determine the fields $\mathbf{H}$ and $\mathbf{E}$ of the radiation, using the general formula (66.3):

$$\mathbf{H} = \frac{1}{c^2 R_0} \left\{ \ddot{\mathbf{d}} \times \mathbf{n} + \frac{1}{6c} \dddot{\mathbf{D}} \times \mathbf{n} + (\ddot{m} \times \mathbf{n}) \times \mathbf{n} \right\},$$

$$\mathbf{E} = \frac{1}{c^2 R_0} \left\{ (\ddot{\mathbf{d}} \times \mathbf{n}) \times \mathbf{n} + \frac{1}{6c} (\dddot{\mathbf{D}} \times \mathbf{n}) \times \mathbf{n} + \mathbf{n} \times \ddot{m} \right\}. \tag{71.4}$$

The intensity dI of the radiation in the solid angle do is given by the general formula (66.6). We calculate here the total radiation, i.e., the energy radiated by the system in unit time in all directions. To do this, we average dI over all directions of $\mathbf{n}$; the total radiation is equal to this average multiplied by 4π. In averaging the square of the magnetic field, all the cross-products of the three terms in **H** vanish, so that there remain only the mean squares of the three. A simple calculation† gives the following result for I:

$$I = \frac{2}{3c^3}\ddot{d}^2 + \frac{1}{180c^5}\dddot{D}^2_{\alpha\beta} + \frac{2}{3c^3}\ddot{m}^2 \tag{71.5}$$

Thus the total radiation consists of three independent parts; they are called, respectively, *dipole*, *quadrupole*, and *magnetic dipole* radiation.

We note that the magnetic dipole radiation is actually not present for many systems. Thus it is not present for a system in which the charge-to-mass ratio is the same for all the moving charges (in this case the dipole radiation also vanishes, as already shown in § 67). Namely, for such a system the magnetic moment is proportional to the angular momentum (see § 44) and therefore, since the latter is conserved, $\dot{m} = 0$. For the same reason, magnetic dipole radiation does not occur for a system consisting of just two particles (cf. the problem in § 44. In this case we cannot draw any conclusion concerning the dipole radiation).

PROBLEMS

1. Calculate the total effective radiation in the scattering of a beam of charged particles by particles identical with them.

Solution: In the collision of identical particles, dipole radiation (and also magnetic dipole radiation) does not occur, so that we must calculate the quadrupole radiation. The quadrupole moment tensor of a system of two identical particles (relative to their centre of mass) is

$$D_{\alpha\beta} = \frac{e}{2}\,(3x_\alpha x_\beta - r^2\delta_{\alpha\beta}),$$

where x_α are the components of the radius vector **r** between the particles. After threefold differentiation of $D_{\alpha\beta}$, we express the first, second, and third derivatives with respect to time of x_α in terms of the relative velocity of the particles v_α as:

$$\dot{x}_\alpha = v_\alpha, \quad \mu\ddot{x}_\alpha = \frac{m}{2}\ddot{x}_\alpha = \frac{e^2 x_\alpha}{r^3}, \quad \frac{m}{2}\dddot{x}_\alpha = e^2\,\frac{v_\alpha r - 3x_\alpha v_r}{r^4},$$

where $v_r = \mathbf{v}\cdot\mathbf{r}/r$ is the radial component of the velocity (the second equality is the equation of motion of the charge, and the third is obtained by differentiating the second). The calculation leads to the following expression for the intensity:

$$I = \frac{\dddot{D}^2_{\alpha\beta}}{180c^5} = \frac{2e^6}{15m^2c^5}\frac{1}{r^4}\,(v^2 + 11v_\phi^2)$$

† We present a convenient method for averaging the products of components of a unit vector. Since the tensor $n_\alpha n_\beta$ is symmetric, it can be expressed in terms of the unit tensor $\delta_{\alpha\beta}$. Also noting that its trace is 1, we have:

$$\overline{n_\alpha n_\beta} = \tfrac{1}{3}\delta_{\alpha\beta}.$$

The average value of the product of four components is:

$$\overline{n_\alpha n_\beta n_\gamma n_\delta} = \tfrac{1}{15}(\delta_{\alpha\beta}\delta_{\gamma\delta} + \delta_{\alpha\gamma}\delta_{\beta\delta} + \delta_{\alpha\delta}\delta_{\beta\gamma}).$$

The right side is constructed from unit tensors to give a fourth-rank tensor that is symmetric in all its indices; the overall coefficient is determined by contracting on two pairs of indices, which must give unity.

$(v^2 = v_r^2 + v_\phi^2)$; v and v_ϕ are expressible in terms of r by using the equalities

$$v^2 = v_0^2 - \frac{4e^2}{mr}, \qquad v_\phi = \frac{\varrho v_0}{r}.$$

We replace the time integration by an integration over r in the same way as was done in Problem 3 of § 70, namely, we write

$$dt = \frac{dr}{v_r} = \frac{dr}{\sqrt{v_0^2 - \frac{\varrho^2 v_0^2}{r^2} - \frac{4e^2}{mr}}}.$$

In the double integral (over ϱ and r), we first carry out the integration over ϱ and then over r. The result of the calculation is:

$$\varkappa = \frac{4\pi}{9} \frac{e^4 v_0^3}{mc^5}.$$

2. Find the force reacting on a radiating system of particles which is carrying out a stationary finite motion.

Solution: The required force **F** is obtained by calculating the loss of momentum of the system per unit time, i.e. it is the momentum flux carried off by the electromagnetic waves radiated by the system:

$$F_\alpha = - \int \sigma_{\alpha\beta} \, df_\beta = - \int \sigma_{\beta\alpha} n_\beta R_0^2 \, do;$$

the integration is over a large sphere of radius R_0. The stress tensor is given by formula (33.3) and the fields **E** and **H** by (71.4). In view of the transversality of these fields, the integral reduces to

$$\mathbf{F} = - \frac{1}{8\pi} \int 2H^2 \mathbf{n} R_0^2 \, do.$$

The average over the direction of **n** is done using the formulas in the footnote on p. 205 (where the product of an odd number of components of **n** gives zero). The result is:†

$$F_\alpha = - \frac{1}{4\pi c^4} \left\{ \frac{1}{15c} \dddot{D}_{\alpha\beta} \ddot{d}_\beta + \frac{2}{3} (\ddot{\mathbf{d}} \times \dddot{m})_\alpha \right\}.$$

§ 72. The field of the radiation at near distances

The formulas for the dipole radiation were derived by us for the field at distances large compared with the wavelength (and, all the more, large compared with the dimensions of the radiating system). In this section we shall assume, as before, that the wavelength is large compared with the dimensions of the system, but shall consider the field at distances which are not large compared with, but *of the same order as,* the wavelength.

The formula (67.4) for the vector potential

$$\mathbf{A} = \frac{1}{cR_0} \dot{\mathbf{d}} \tag{72.1}$$

is still valid, since in deriving it we used only the fact that R_0 was large compared with the dimensions of the system. However, now the field cannot be considered to be a plane wave

---

† We note that this force is of higher order in $1/c$ than the Lorentz frictional forces (§ 75). The latter give no contribution to the total force of recoil: the sum of the forces (75.5) acting on the particles of an electrically neutral system is zero.

even over small regions. Therefore the formulas (67.5) and (67.6) for the electric and magnetic fields are no longer applicable, so that to calculate them, we must first determine both $\mathbf{A}$ and ϕ.

The formula for the scalar potential can be derived directly from that for the vector potential, using the general condition (62.1),

$$\operatorname{div} \mathbf{A} + \frac{1}{c}\frac{\partial \phi}{\partial t} = 0,$$

imposed on the potentials. Substituting (72.1) in this, and integrating over the time, we get

$$\phi = -\operatorname{div} \frac{\mathbf{d}}{R_0}. \tag{72.2}$$

The integration constant (an arbitrary function of the coordinates) is omitted, since we are interested only in the variable part of the potential. We recall that in the formula (72.2) as well as in (72.1) the value of $\mathbf{d}$ must be taken at the time $t' = t - (R_0/c)$.†

Now it is no longer difficult to calculate the electric and magnetic field. From the usual formulas, relating $\mathbf{E}$ and $\mathbf{H}$ to the potentials,

$$\mathbf{H} = \frac{1}{c}\operatorname{curl} \frac{\dot{\mathbf{d}}}{R_0}, \tag{72.3}$$

$$\mathbf{E} = \operatorname{grad} \operatorname{div} \frac{\mathbf{d}}{R_0} - \frac{1}{c^2}\frac{\ddot{\mathbf{d}}}{R_0}. \tag{72.4}$$

The expression for $\mathbf{E}$ can be rewritten in another form, noting that $\mathbf{d}_{t'}/R_0$ [just as any function of coordinates and time of the form $\frac{1}{R_0}f\left(t - \frac{R_0}{c}\right)$] satisfies the wave equation:

$$\frac{1}{c^2}\frac{\partial^2}{\partial t^2}\left(\frac{\mathbf{d}}{R_0}\right) = \Delta\left(\frac{\mathbf{d}}{R_0}\right).$$

Also using the formula

$$\operatorname{curl} \operatorname{curl} \mathbf{a} = \operatorname{grad} \operatorname{div} \mathbf{a} - \Delta \mathbf{a},$$

we find that

$$\mathbf{E} = \operatorname{curl} \operatorname{curl} \frac{\mathbf{d}}{R_0}. \tag{72.5}$$

The results obtained determine the field at distances of the order of the wavelength. It is

† Sometimes one introduces the so-called Hertz vector, defined by

$$\mathbf{Z} = -\frac{1}{R_0}\mathbf{d}\left(t - \frac{R_0}{c}\right).$$

Then

$$\mathbf{A} = -\frac{1}{c}\dot{\mathbf{Z}}, \qquad \phi = \operatorname{div} \mathbf{Z}.$$

understood that in all these formulas it is not permissible to take $1/R_0$ out from under the differentiation sign, since the ratio of terms containing $1/R_0^2$ to terms with $1/R_0$ is just of the same order as λ/R_0.

Finally, we give the formulas for the Fourier components of the field. To determine $\mathbf{H}_\omega$ we substitute in (72.3) for $\mathbf{H}$ and $\mathbf{d}$ their monochromatic components $\mathbf{H}_\omega e^{-i\omega t}$ and $\mathbf{d}_\omega e^{-i\omega t}$ respectively. However, we must remember that the quantities on the right sides of equations (72.1) to (72.5) refer to the time $t' = t - (R_0/c)$. Therefore we must substitute in place of $\mathbf{d}$ the expression

$$\mathbf{d}_\omega e^{-i\omega\left(t-\frac{R_0}{c}\right)} = \mathbf{d}_\omega e^{-i\omega t + ikR_0}.$$

Making the substitution and dividing by $e^{-i\omega t}$, we get

$$\mathbf{H}_\omega = -ik\,\mathrm{curl}\left(\mathbf{d}_\omega \frac{e^{ikR_0}}{R_0}\right) = ik\mathbf{d}_\omega \times \nabla\frac{e^{ikR_0}}{R_0}.$$

or, performing the differentiation,

$$\mathbf{H}_\omega = -ik\mathbf{d}_\omega \times \mathbf{n}\left(\frac{ik}{R_0} - \frac{1}{R_0^2}\right)e^{ikR_0}, \tag{72.6}$$

where $\mathbf{n}$ is a unit vector along $\mathbf{R}_0$.

In similar fashion, we find from (72.4):

$$\mathbf{E}_\omega = k^2\mathbf{d}_\omega\frac{e^{ikR_0}}{R_0} + (\mathbf{d}_\omega \cdot \nabla)\nabla\frac{e^{ikR_0}}{R_0},$$

and differentiation gives

$$\mathbf{E}_\omega = \mathbf{d}_\omega\left(\frac{k^2}{R_0} + \frac{ik}{R_0^2} - \frac{1}{R_0^3}\right)e^{ikR_0} + \mathbf{n}(\mathbf{n}\cdot\mathbf{d}_\omega)\left(-\frac{k^2}{R_0} - \frac{3ik}{R_0^2} + \frac{3}{R_0^3}\right)e^{ikR_0}. \tag{72.7}$$

At distances large compared to the wavelength ($kR_0 \gg 1$), we can neglect the terms in $1/R_0^2$ and $1/R_0^3$ in formulas (72.7) and (72.6), and we arrive at the field in the "wave zone",

$$\mathbf{E}_\omega = \frac{k^2}{R_0}\,\mathbf{n}\times(\mathbf{d}_\omega\times\mathbf{n})\,e^{ikR_0}, \qquad \mathbf{H}_\omega = -\frac{k^2}{R_0}\,\mathbf{d}_\omega\times\mathbf{n}e^{ikR_0}.$$

At distances which are small compared to the wavelength ($kR_0 \ll 1$), we neglect the terms in $1/R_0$ and $1/R_0^2$ and set $e^{ikR_0} \cong 1$; then

$$\mathbf{E}_\omega = \frac{1}{R_0^3}\{3\mathbf{n}(\mathbf{d}_\omega\cdot\mathbf{n}) - \mathbf{d}_\omega\},$$

which corresponds to the static field of an electric dipole (§ 40); in this approximation, the magnetic field vanishes.

PROBLEMS

1. Calculate the quadrupole and magnetic dipole radiation fields at near distances.

Solution: Assuming, for brevity, that dipole radiation is not present, we have (see the calculation carried out in § 71)

$$\mathbf{A} = \frac{1}{c} \int \mathbf{j}_{t-\frac{R}{c}} \frac{dV}{R} \cong -\frac{1}{c} \int (\mathbf{r} \cdot \nabla) \frac{\mathbf{j}_{t-\frac{R_0}{c}}}{R_0} \, dV$$

where we have expanded in powers of $\mathbf{r} = \mathbf{R}_0 - \mathbf{R}$. In contrast to what was done in § 71, the factor $1/R_0$ cannot here be taken out from under the differentiation sign. We take the differential operator out of the integral and rewrite the integral in tensor notation:

$$A_\alpha = -\frac{1}{c} \frac{\partial}{\partial X_\beta} \int \frac{x_\beta j_\alpha}{R_0} \, dV$$

(X_β are the components of the radius vector $\mathbf{R}_0$). Transforming from the integral to a sum over the charges, we find

$$A_\alpha = -\frac{1}{c} \frac{\partial}{\partial X_\beta} \frac{(\Sigma e v_\alpha x_\beta)_{t'}}{R_0}.$$

In the same way as in § 71, this expression breaks up into a quadrupole part and a magnetic dipole part. The corresponding scalar potentials are calculated from the vector potentials in the same way as in the text. As a result, we obtain for the quadrupole radiation:

$$A_\alpha = -\frac{1}{6c} \frac{\partial}{\partial X_\beta} \frac{\dot{D}_{\alpha\beta}}{R_0}, \qquad \phi = \frac{1}{6} \frac{\partial^2}{\partial X_\alpha \partial X_\beta} \frac{D_{\alpha\beta}}{R_0},$$

and for the magnetic dipole radiation:

$$\mathbf{A} = \text{curl} \frac{m}{R_0}, \qquad \phi = 0$$

[all quantities on the right sides of the equations refer as usual to the time $t' = t - (R_0/c)$].

The field intensities for magnetic dipole radiation are:

$$\mathbf{E} = -\frac{1}{c} \text{curl} \frac{\dot{m}}{R_0}, \qquad \mathbf{H} = \text{curl curl} \frac{m}{R_0}.$$

Comparing with (72.3), (72.5), we see that in the magnetic dipole case, $\mathbf{E}$ and $\mathbf{H}$ are expressed in terms of m in the same way as—$\mathbf{H}$ and $\mathbf{E}$ are expressed in terms of $\mathbf{d}$ for the electric dipole case.

The spectral components of the potentials of the quadrupole radiation are:

$$A_\alpha^{(\omega)} = \frac{ik}{6} D_{\alpha\beta}^{(\omega)} \frac{\partial}{\partial X_\beta} \frac{e^{ikR_0}}{R_0}, \qquad \phi^{(\omega)} = \frac{1}{6} D_{\alpha\beta}^{(\omega)} \frac{\partial^2}{\partial X_\alpha \partial X_\beta} \frac{e^{ikR_0}}{R_0}.$$

Bacause of their complexity, we shall not give the expressions for the field.

2. Find the rate of loss of angular momentum of a system of charges through dipole radiation of electromagnetic waves.

Solution: According to (32.9) the density of flux of angular momentum of the electromagnetic field is given by the spatial components of the four-tensor $x^i T^{kl} - x^k T^{il}$. Changing to three-dimensional notation, we introduce the three-dimensional angular momentum vector with components $\frac{1}{2} e_{\sigma\beta\gamma} M^{\beta\gamma}$; the flux density is given by the three-dimensional tensor

$$\frac{1}{2} e_{\alpha\beta\gamma}(x_\beta \sigma_{\gamma\delta} - x_\gamma \sigma_{\beta\delta}) = e_{\alpha\beta} x_\beta \sigma_{\gamma\delta},$$

where $\sigma_{\alpha\beta} \equiv T^{\alpha\beta}$ is the three-dimensional Maxwell stress tensor (and we write all indices as subscripts in accordance with the usual three-dimensional notation). The total angular momentum lost by the system per unit time is equal to the flux of angular momentum of the radiation field through a spherical surface of radius R_0:

$$-\frac{dM_\alpha}{dt} = \oint e_{\alpha\beta\gamma} x_\beta \sigma_{\gamma\delta} n_\delta \, df,$$

where $df = R_0^2 \, do$, and $\mathbf{n}$ is a unit vector in the direction of $\mathbf{R}_0$. Using the tensor $\sigma_{\alpha\beta}$ from (33.3), we find:

$$\frac{d\mathbf{M}}{dt} = \frac{R_0^3}{4\pi} \int \{(\mathbf{n} \times \mathbf{E})(\mathbf{n} \cdot \mathbf{E}) + (\mathbf{n} \times \mathbf{H})(\mathbf{n} \cdot \mathbf{H})\} \, do. \tag{1}$$

In applying this formula to the radiation field at large distances from the system, we must not, however, stop at terms $\sim 1\, R_0$; in this approximation $\mathbf{n} \cdot \mathbf{E} = \mathbf{n} \cdot \mathbf{H} = 0$, so that the integrand vanishes. These terms [given by (67.5)–(67.6)] are sufficient only for calculating the factors $\mathbf{n} \times \mathbf{E}$ and $\mathbf{n} \times \mathbf{H}$; the longitudinal field components $\mathbf{n} \cdot \mathbf{E}$ and $\mathbf{n} \cdot \mathbf{H}$ arise from terms $\sim 1/R_0^2$ (as a result, the integrand in (1) becomes $\sim 1\ 1/R_0^3$, and the distance R_0 drops out of the answer, as it should). In the dipole approximation $\lambda \gg a$, and we must distinguish between terms containing additional factors [relative to (67.5)–(67.6)] $\sim \lambda\ R_0$ or $\sim a\ R_0$; it is sufficient to keep only the former. These terms can be obtained from (72.3) and (72.5); a calculation to second order in $1\ R_0$ gives:†

$$\mathbf{E} \cdot \mathbf{n} = \frac{2}{cR_0^2} \, \mathbf{n} \cdot \dot{\mathbf{d}}, \quad \mathbf{H} \cdot \mathbf{n} = 0. \tag{2}$$

Substituting (2) and (67.6) in (1), we find:

$$\frac{d\mathbf{M}}{dt} = \frac{1}{2\pi c^3} \int (\mathbf{n} \cdot \ddot{\mathbf{d}})(\mathbf{n} \cdot \dot{\mathbf{d}}) \, do.$$

Finally, writing the integrand in the form $e_{\alpha\beta\gamma} n_\beta \dot{d}_\gamma n_\delta \dot{d}_\delta$ and averaging over the direction of $\mathbf{n}$, we obtain:

$$\frac{d\mathbf{M}}{dt} = \frac{2}{3c^3} \, \dot{\mathbf{d}} \cdot \ddot{\mathbf{d}}, \tag{3}$$

We note that for a linear oscillator ($\mathbf{d} + \mathbf{d}_0 \cos \omega t$ with real amplitude $\mathbf{d}_0$) the expression (3) vanishes; there is no loss of angular momentum in the radiation.

§ 73. Radiation from a rapidly moving charge

Now we consider a charged particle moving with a velocity which is not small compared with the velocity of light.

The formulas of § 67, derived under the assumption that $v \ll c$, are not immediately applicable to this case. We can, however, consider the particle in that system of reference in which the particle is at rest at a given moment; in this system of reference the formulas referred to are of course valid (we call attention to the fact that this can be done only for the case of a *single* moving particle; for a system of several particles there is generally no system of reference in which all the particles are at rest simultaneously).

Thus in this particular system of reference the particle radiates, in time dt, the energy

$$d\mathscr{E} = \frac{2e^2}{3c^3} \, w^2 \, dt \tag{73.1}$$

[in accordance with formula (67.9)], where w is the acceleration of the particle in this system of reference. In this system of reference, the total radiated momentum is zero:

$$d\mathbf{P} = 0. \tag{73.2}$$

† A nonzero value of $\mathbf{n} \cdot \mathbf{H}$ would be obtained only if one included terms of higher order in a/R_0.

In fact, the radiated momentum is given by the integral of the momentum flux density in the radiation field over a closed surface surrounding the particle. But because of the symmetry of the dipole radiation, the momenta carried off in opposite directions are equal in magnitude and opposite in direction; therefore the integral is identically zero.

For the transformation to an arbitrary reference system, we rewrite formulas (73.1) and (73.2) in four-dimensional form. It is easy to see that the "radiated four-momentum" dP_i must be written as

$$dP^i = - \frac{2e^2}{3c} \frac{du^k}{ds} \frac{du_k}{ds} dx^i = - \frac{2e^2}{3c} \frac{du^k}{ds} \frac{du_k}{ds} u^i ds. \tag{73.3}$$

In fact, in the reference frame in which the particle is at rest, the space components of the four-velocity u^i are equal to zero, and

$$\frac{du^k}{ds} \frac{du_k}{ds} = - \frac{w^2}{c^4};$$

therefore the space components of dP^i become zero and the time component gives equation (73.1).

The total four-momentum radiated during the time of passage of the particle through a given electromagnetic field is equal to the integral of (73.3), that is,

$$\Delta P^i = - \frac{2e^2}{3c} \int \frac{du^k}{ds} \frac{du_k}{ds} dx^i. \tag{73.4}$$

We rewrite this formula in another form, expressing the four-acceleration du^i/ds in terms of the electromagnetic field tensor, using the equation of motion (23.4):

$$mc \frac{du_k}{ds} = \frac{e}{c} F_{kl} u^l.$$

We then obtain

$$\Delta P^i = - \frac{2e^4}{3m^2 c^5} \int (F_{kl} u^l)(F^{km} u_m) \, dx^i. \tag{73.5}$$

The time component of (73.4) or (73.5) gives the total radiated energy $\Delta \mathcal{E}$. Substituting for all the four-dimensional quantities their expressions in terms of three-dimensional quantities, we find

$$\Delta \mathcal{E} = \frac{2e^2}{3c^3} \int_{-\infty}^{\infty} \frac{w^2 - \dfrac{(\mathbf{v} \times \mathbf{w})^2}{c^2}}{\left(1 - \dfrac{v^2}{c^2}\right)^3} \, dt \tag{73.6}$$

($\mathbf{w} = \dot{\mathbf{v}}$ is the acceleration of the particle), or, in terms of the external electric and magnetic fields:

$$\Delta \mathcal{E} = \frac{2e^4}{3m^2 c^3} \int_{-\infty}^{\infty} \frac{\left\{\mathbf{E} + \dfrac{1}{c} \mathbf{v} \times \mathbf{H}\right\}^2 - \dfrac{1}{c^2}(\mathbf{E} \cdot \mathbf{v})^2}{1 - \dfrac{v^2}{c^2}} \, dt. \tag{73.7}$$

The expressions for the total radiated momentum differ by having an extra factor **v** in the integrand.

It is clear from formula (73.7) that for velocities close to the velocity of light, the total energy radiated per unit time varies with the velocity essentially like $[1 - (v^2/c^2)]^{-1}$, that is, proportionally to the square of the energy of the moving particle. The only exception is motion in an electric field, along the direction of the field. In this case the factor $[1 - (v^2/c^2)]$ standing in the denominator is cancelled by an identical factor in the numerator, and the radiation does not depend on the energy of the particle.

Finally there is the question of the angular distribution of the radiation from a rapidly moving charge. To solve this problem, it is convenient to use the Lienard–Wiechert expressions for the fields (63.8) and (63.9). At large distances we must retain only the term of lowest order in $1/R$ [the second term in (63.8)]. Introducing the unit vector **n** in the direction of the radiation ($\mathbf{R} = \mathbf{n}R$), we get the formulas

$$\mathbf{E} = \frac{e}{c^2 R} \frac{\mathbf{n} \times \left\{ \left(\mathbf{n} - \dfrac{\mathbf{v}}{c} \right) \times \mathbf{w} \right\}}{\left(1 - \dfrac{\mathbf{n} \cdot \mathbf{v}}{c} \right)^3}, \qquad \mathbf{H} = \mathbf{n} \times \mathbf{E}, \qquad (73.8)$$

where all the quantities on the right sides of the equations refer to the retarded time $t' = t - (R/c)$.

The intensity radiated into the solid angle do is $dI = (c/4\pi)E^2 R^2\, do$. Expanding E^2, we get

$$dI = \frac{e^2}{4\pi c^3} \left\{ \frac{2(\mathbf{n} \cdot \mathbf{w})(\mathbf{v} \cdot \mathbf{w})}{c\left(1 - \dfrac{\mathbf{v} \cdot \mathbf{n}}{c} \right)^5} + \frac{\mathbf{w}^2}{\left(1 - \dfrac{\mathbf{v} \cdot \mathbf{n}}{c} \right)^4} - \frac{\left(1 - \dfrac{v^2}{c^2} \right)(\mathbf{n} \cdot \mathbf{w})^2}{\left(1 - \dfrac{\mathbf{v} \cdot \mathbf{n}}{c} \right)^6} \right\} do. \quad (73.9)$$

If we want to determine the angular distribution of the total radiation throughout the whole motion of the particle, we must integrate the intensity over the time. In doing this, it is important to remember that the integrand is a function of t'; therefore we must write

$$dt = \frac{\partial t}{\partial t'} dt' = \left(1 - \frac{\mathbf{n} \cdot \mathbf{v}}{c} \right) dt' \qquad (73.10)$$

[see (63.6)], after which the integration over t' is immediately done. Thus we have the following expression for the total radiation into the solid angle do:

$$d\mathcal{E}_\mathbf{n} = \frac{e^2}{4\pi c^3} do \int \left\{ \frac{2(\mathbf{n} \cdot \mathbf{w})(\mathbf{v} \cdot \mathbf{w})}{c\left(1 - \dfrac{\mathbf{v} \cdot \mathbf{n}}{c} \right)^4} + \frac{\mathbf{w}^2}{\left(1 - \dfrac{\mathbf{v} \cdot \mathbf{n}}{c} \right)^3} - \frac{\left(1 - \dfrac{v^2}{c^2} \right)(\mathbf{n} \cdot \mathbf{w})^2}{\left(1 - \dfrac{\mathbf{n} \cdot \mathbf{v}}{c} \right)^5} \right\} dt'. \quad (73.11)$$

As we see from (73.9), in the general case the angular distribution of the radiation is quite complicated. In the ultrarelativistic case, $(1 - (v/c) \ll 1)$ it has a characteristic appearance, which is related to the presence of high powers of the difference $1 - (\mathbf{v} \cdot \mathbf{n}/c)$ in the denominators of the various terms in this expression. Thus, the intensity is large within the narrow range of angles in which the difference $1 - (\mathbf{v} \cdot \mathbf{n}/c)$ is small. Denoting by θ the small angle between $\mathbf{n}$ and $\mathbf{v}$, we have:

$$1 - \frac{v}{c} \cos \theta \cong 1 - \frac{v}{c} + \frac{\theta^2}{2} \cong \frac{1}{2}\left(1 - \frac{v^2}{c^2} + \theta^2\right);$$

this difference is small for

$$\theta \sim \sqrt{1 - \frac{v^2}{c^2}}. \tag{73.12}$$

Thus an ultrarelativistic particle radiates mainly along the direction of its own motion, within the small range (73.12) of angles around the direction of its velocity.

We also point out that, for arbitrary velocity and acceleration of the particle, there are always two directions for which the radiated intensity is zero. These are the directions for which the vector $\mathbf{n} - (\mathbf{v}/c)$ is parallel to the vector $\mathbf{w}$, so that the field (73.8) becomes zero. (See also problem 2 of this section.)

Finally, we give the simpler formulas to which (73.9) reduces in two special cases.

If the velocity and acceleration of the particle are parallel,

$$\mathbf{H} = \frac{e}{c^2 R} \frac{\mathbf{w} \times \mathbf{n}}{\left(1 - \dfrac{\mathbf{n} \cdot \mathbf{v}}{c}\right)^3}$$

and the intensity is

$$dI = \frac{e^2}{4\pi c^3} \frac{w^2 \sin^2 \theta}{\left(1 - \dfrac{v}{c} \cos \theta\right)^6} \, do. \tag{73.13}$$

It is naturally, symmetric around the common direction of $\mathbf{v}$ and $\mathbf{w}$, and vanishes along $(\theta = 0)$ and opposite to $(\theta = \pi)$ the direction of the velocity. In the ultrarelativistic case, the intensity as a function of θ has a sharp double maximum in the region (73.12), with a steep drop to zero for $\theta = 0$.

If the velocity and acceleration are perpendicular to one another, we have from (73.9):

$$dI = \frac{e^2 w^2}{4\pi c^3}\left[\frac{1}{\left(1 - \dfrac{v}{c}\cos\theta\right)^4} - \frac{\left(1 - \dfrac{v^2}{c^2}\right)\sin^2\theta\cos^2\phi}{\left(1 - \dfrac{v}{c}\cos\theta\right)^6}\right] do, \tag{73.14}$$

where θ is again the angle between $\mathbf{n}$ and $\mathbf{v}$, and ϕ is the azimuthal angle of the vector $\mathbf{n}$ relative to the plane passing through $\mathbf{v}$ and $\mathbf{w}$. This intensity is symmetric only with respect to the plane of $\mathbf{v}$ and $\mathbf{w}$, and vanishes along the two directions in this plane which form the angle $\theta = \cos^{-1}(v/c)$ with the velocity.

PROBLEMS

1. Find the total radiation from a relativistic particle with charge e_1, which passes with impact parameter ϱ through the Coulomb field of a fixed centre (with potential $\phi = e_2/r$).

Solution: In passing through the field, the relativistic particle is hardly deflected at all.† We may therefore regard the velocity **v** in (73.7) as constant, so that the field at the position of the particle is

$$\mathbf{E} = \frac{e_2\mathbf{r}}{r^3} \cong \frac{e_2\mathbf{r}}{(\varrho^2 + v^2t^2)^{3/2}},$$

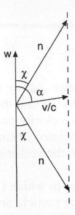

FIG. 15.

with $x = vt$, $y = \varrho$. Performing the time integration in (73.7), we obtain:

$$\Delta\mathscr{E} = \frac{\pi e_1^4 e_2^2}{12m^2c^3\varrho^3 v}\frac{4c^2 - v^2}{c^2 - v^2}.$$

2. Find the directions along which the intensity of the radiation from a moving particle vanishes.

Solution: From the geometrical construction (Fig. 15) we find that the required directions **n** lie in the plane passing through **v** and **w**, and form an angle χ with the direction of **w** where

$$\sin\chi = \frac{v}{c}\sin\alpha,$$

and α is the angle between **v** and **w**.

3. Find the intensity of the radiation from a particle which is carrying out a stationary motion in the field of a circularly polarized plane electromagnetic wave.

Solution: According to the results of problem 3 of § 48, the particle moves in a circle, and its velocity at each moment is parallel to **H** and perpendicular to **E**. Its kinetic energy is

$$\frac{mc^2}{\sqrt{1 - v^2c^2}} = c\sqrt{p^2 - m^2c^2} = c\gamma$$

(where we use the notation of the problem cited). From formula (73.7) we find the intensity of the radiation:

$$I = \frac{2e^4}{3m^2c^3}\frac{\mathbf{E}^2}{1 - v^2c^2} = \frac{2e^4E_0^2}{3m^2c^3}\left[1 - \left(\frac{eE_0}{mc\omega}\right)^2\right].$$

4. The same problem in the field of a linearly polarized wave.

Solution: According to the results of problem 2 of § 48, the motion occurs in the plane xy, passing through

† For $v \sim c$, deviations through sizable angles can occur only for impact parameters $\varrho \sim e^2/mc^2$, which cannot in general be treated classically.

the direction of propagation of the wave (the x axis) and the direction of $\mathbf{E}$ (the y axis); the field $\mathbf{H}$ is along the z direction (and $H_z = E_y$). Form (73.7) we find:

$$I = \frac{2e^4 \mathbf{E}^2}{3m^2 c^3} \frac{(1 - v_x c)^2}{1 - v^2 c^2}.$$

The average over the period of the motion, which was done using a parametric representation in the problem cited, gives the result

$$\bar{I} = \frac{e^4 E_0^2}{3m^2 c^3} \left[1 - \frac{3}{8} \left(\frac{eE_0}{mc\omega} \right)^2 \right].$$

§ 74. Synchrotron radiation (magnetic bremsstrahlung)

We consider the radiation from a charge moving with arbitrary velocity in a circle in a uniform constant magnetic field; such radiation is called magnetic bremsstrahlung. The radius of the orbit r and the cyclic frequency of the motion ω_H are expressible in terms of the field intensity H and the velocity of the particle v, by the formulas (see § 21):

$$r = \frac{mcv}{eH \sqrt{1 - \frac{v^2}{c^2}}}, \qquad \omega_H = \frac{v}{r} = \frac{eH}{mc} \sqrt{1 - \frac{v^2}{c^2}}. \qquad (74.1)$$

The total intensity of the radiation over all directions is given directly by (73.7), omitting the time integration, in which we must set $\mathbf{E} = 0$ and $\mathbf{H} \perp \mathbf{v}$:

$$I = \frac{2e^4 H^2 v^2}{3m^2 c^5 \left(1 - \frac{v^2}{c^2} \right)}. \qquad (74.2)$$

We see that the total intensity is proportional to the square of the momentum of the particle.

If we are interested in the angular distribution of the radiation, then we must use formula (73.11). One quantity of interest is the average intensity during a period of the motion. For this we integrate (73.11) over the time of revolution of the particle in the circle and divide the result by the period $T = 2\pi/\omega_H$.

We choose the plane of the orbit as the XY plane (the origin is at the centre of the circle), and we draw the YZ plane to pass through the direction $\mathbf{n}$ of the radiation (Fig. 16). The

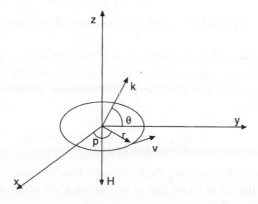

FIG. 16.

magnetic field is along the negative Z axis (the direction of motion of the particle in Fig. 16 corresponds to a positive charge e). Further, let θ be the angle between the direction $\mathbf{k}$ of the radiation and the Y axis, and $\phi = \omega_H t$ be the angle between the radius vector of the particle and the X axis. Then the cosine of the angle between $\mathbf{k}$ and the velocity $\mathbf{v}$ is $\cos\theta \cos\phi$ (the vector $\mathbf{v}$ lies in the XY plane, and at each moment is perpendicular to the radius vector of the particle). We express the acceleration $\mathbf{w}$ of the particle in terms of the field $\mathbf{H}$ and the velocity $\mathbf{v}$ by means of the equation of motion [see (21.1)]:

$$\mathbf{w} = \frac{e}{mc}\sqrt{1 - \frac{v^2}{c^2}}\,\mathbf{v} \times \mathbf{H}.$$

After a simple calculation, we get:

$$\overline{dI} = do\,\frac{e^4 H^2 v^2}{8\pi^2 m^2 c^5}\left(1 - \frac{v^2}{c^2}\right)\int_0^{2\pi}\frac{\left(1 - \frac{v^2}{c^2}\right)\sin^2\theta + \left(\frac{v}{c} - \cos\theta\cos\phi\right)^2}{\left(1 - \frac{v}{c}\cos\theta\cos\phi\right)^5}\,d\phi \quad (74.3)$$

(the time integration has been converted into integration over $\phi = \omega_H t$). The integration is elementary, though rather lengthy. As a result one finds the following formula:

$$dI = do\,\frac{e^4 H^2 v^2\left(1 - \frac{v^2}{c^2}\right)}{8\pi m^2 c^5}\left[\frac{2 + \frac{v^2}{c^2}\cos^2\theta}{\left(1 - \frac{v^2}{c^2}\cos^2\theta\right)^{5/2}} - \frac{\left(1 - \frac{v^2}{c^2}\right)\left(4 + \frac{v^2}{c^2}\cos^2\theta\right)\cos^2\theta}{4\left(1 - \frac{v^2}{c^2}\cos^2\theta\right)^{7/2}}\right]$$

$$(74.4)$$

The ratio of the intensity of radiation for $\theta = \pi/2$ (perpendicular to the plane of the orbit) to the intensity for $\theta = 0$ (in the plane of the orbit) is

$$\frac{\left(\dfrac{dI}{do}\right)_0}{\left(\dfrac{dI}{do}\right)_{\pi/2}} = \frac{4 + 3\dfrac{v^2}{c^2}}{8\left(1 - \dfrac{v^2}{c^2}\right)^{5/2}}. \quad (74.5)$$

As $v \to 0$, this ratio approaches $\frac{1}{2}$, but for velocities close to the velocity of light, it becomes very large.

Next we consider the spectral distribution of the radiation. Since the motion of the charge is periodic, we are dealing with expansion in a Fourier series. The calculation starts conveniently with the vector potential. For the Fourier components of the vector potential we have the formula [see (66.12)]:

$$\mathbf{A}_n = e\,\frac{e^{ikR_0}}{cR_0 T}\oint e^{i(\omega_H nt - \mathbf{k}\cdot\mathbf{r})}\,d\mathbf{r},$$

where the integration is taken along the trajectory of the particle (the circle). For the coordinates of the particle we have $x = r\cos\omega_H t$, $y = r\sin\omega_H t$. As integration variable we choose the angle $\phi = \omega_H t$. Noting that

$$\mathbf{k} \cdot \mathbf{r} = kr \cos \theta \sin \phi = (nv/c) \cos \theta \sin \phi$$

($k = n\omega_H/c = nv/cr$), we find for the Fourier components of the x-component of the vector potential:

$$A_{xn} = -\frac{eV}{2\pi c R_0} e^{ikR_0} \int_0^{2\pi} e^{in\left(\phi - \frac{v}{c}\cos\theta\sin\phi\right)} \sin\phi \, d\phi \, .$$

We have already had to deal with such an integral in § 70. It can be expressed in terms of the derivative of a Bessel function:

$$A_{xn} = -\frac{ieV}{cR_0} e^{ikR_0} J_n'\left(\frac{nv}{c}\cos\theta\right) \tag{74.6}$$

Similarly, one calculates A_{yn}:

$$A_{yn} = -\frac{e}{R_0 \cos \theta} e^{ikR_0} J_n\left(\frac{nv}{c}\cos\theta\right). \tag{74.7}$$

The component along the Z axis obviously vanishes.

From the formulas of § 66 we have for the intensity of radiation with frequency $\omega = n\omega_H$, in the element of solid angle do:

$$dI_n = \frac{c}{2\pi} |\mathbf{H}_n|^2 R_0^2 \, do = \frac{c}{2\pi} |\mathbf{k} \times \mathbf{A}_n|^2 R_0^2 \, do.$$

Noting that

$$|\mathbf{A} \times \mathbf{k}|^2 = A_x^2 k^2 + A_y^2 k^2 \sin^2\theta,$$

and substituting (74.6) and (74.7), we get for the intensity of radiation the following formula (G.A. Schott, 1912):

$$dI_n = \frac{n^2 e^4 H^2}{2\pi c^3 m^2}\left(1 - \frac{v^2}{c^2}\right)\left[\tan^2\theta \cdot J_n^2\left(\frac{nv}{c}\cos\theta\right) + \frac{v^2}{c^2} J_n'^2\left(\frac{nv}{c}\cos\theta\right)\right] do. \tag{74.8}$$

To determine the total intensity over all directions of the radiation with frequency $\omega = n\omega_H$, this expression must be integrated over all angles. However, the integration cannot be carried out in finite form. By a series of transformations, making use of certain relations from the theory of Bessel functions, the required integral can be written in the following form:

$$I_n = \frac{2e^4 H^2\left(1 - \frac{v^2}{c^2}\right)}{m^2 c^2 v}\left[\frac{nv^2}{c^2} J_{2n}'\left(\frac{2nv}{c}\right) - n^2\left(1 - \frac{v^2}{c^2}\right)\int_0^{v/c} J_{2n}(2n\zeta)\,d\zeta\right]. \tag{74.9}$$

We consider in more detail the ultrarelativistic case where the velocity of motion of the particle is close to the velocity of light.

Setting $v = c$ in the numerator of (74.2), we find that in the ultrarelativistic case the total intensity of the synchrotron radiation is proportional to the square of the particle energy $\mathscr{E}$:

$$I = \frac{2e^4 H^2}{3m^2 c^3} \left(\frac{\mathscr{E}}{mc^2}\right)^2. \tag{74.10}$$

The angular distribution of the radiation is highly anisotropic. The radiation is concentrated mainly in the plane of the orbit. The "width" $\Delta\theta$ of the angular range within which most of the radiation is included is easily evaluated from the condition $1 - (v^2/c^2) \cos^2 \theta \sim 1 - (v^2/c^2)$, writing $\theta = \pi/2 \pm \Delta\theta$, $\sin \theta \cong 1 - (\Delta\theta)^2/2$. It is clear that†

$$\Delta\theta \sim \sqrt{1 - \frac{v^2}{c^2}} = \frac{mc^2}{\mathscr{E}}. \tag{74.11}$$

We shall see below that in the ultrarelativistic case the main role in the radiation is played by frequencies with large n (Arzimovich and Pomeranchuk, 1945). We can therefore use the asymptotic formula (70.9), according to which:

$$J_{2n}(2n\xi) \cong \frac{1}{\sqrt{\pi n^{\frac{1}{3}}}} \Phi[n^{\frac{2}{3}}(1 - \xi^2)] \tag{74.12}$$

Substituting in (74.9), we get the following formula for the spectral distribution of the radiation for large values of n:‡

$$I_n = - \frac{2e^4 H^2 \frac{mc^2}{\mathscr{E}} u^{\frac{1}{2}}}{\sqrt{\pi m^2 c^3}} \left\{ \Phi'(u) + \frac{u}{2} \int_n^\infty \Phi(u)\, du \right\}, \tag{74.13}$$

$$u = n^{\frac{2}{3}} \left(\frac{mc^2}{\mathscr{E}}\right)^2.$$

For $u \to 0$ the function in the curly brackets approaches the constant limit $\Phi'(0) = -0.4587$ ... § Therefore for $u \ll 1$, we have

$$I_n = 0.52 \frac{e^4 H^2}{m^2 c^3} \left(\frac{mc^2}{\mathscr{E}}\right)^2 u^{\frac{1}{3}}, \qquad 1 \ll n \ll \left(\frac{\mathscr{E}}{mc^2}\right)^3. \tag{74.14}$$

For $u \gg 1$, we can use the asymptotic expression for the Airy function (see the footnote on p. 161, and obtain):

† This result is, of course, in agreement with the angular distribution of the instantaneous intensity which we found in the preceding section [see (73.12)]; however, the reader should not confuse the angle θ of this section with the angle θ between **n** and **v** in § 73!

‡ In making the substitution, the limit $n^{2/3}$ of the integral can be changed to infinity, to within the required accuracy; we have also set $v = c$ wherever possible. Even though values of ξ close to 1 are important in the integral (74.9), the use of formula (74.12) is still permissible, since the integral converges rapidly at the lower limit.

§ From the definition of the Airy function, we have:

$$\Phi'(0) = -\frac{1}{\sqrt{\pi}} \int_0^\infty \xi \sin \frac{\xi^3}{3} d\xi = -\frac{1}{\sqrt{\pi} \cdot 3^{1/3}} \int_0^\infty x^{-1/3} \sin x\, dx = -\frac{3^{1/6} \Gamma(\frac{2}{3})}{2\sqrt{\pi}}.$$

$$I_n = \frac{e^4 H^2 \left(\frac{mc^2}{\mathscr{E}}\right)^{\frac{5}{2}} n^{\frac{1}{2}}}{2\sqrt{\pi}\, m^2 c^3} \exp\left\{-\frac{2}{3} n \left(\frac{mc^2}{\mathscr{E}}\right)^3\right\}, \qquad n \gg \left(\frac{\mathscr{E}}{mc^2}\right)^3 \qquad (74.15)$$

that is, the intensity drops exponentially for large n.

Consequently the spectrum has a maximum for

$$n \sim \left(\frac{\mathscr{E}}{mc^2}\right)^3,$$

and the main part of the radiation is concentrated in the region of frequencies for which

$$\omega \sim \omega_H \left(\frac{\mathscr{E}}{mc^2}\right)^3 = \frac{eH}{mc}\left(\frac{\mathscr{E}}{mc^2}\right)^2. \qquad (74.16)$$

These values of ω are very large, compared to the distance ω_H between neighbouring frequencies. We may say that the spectrum has a "quasicontinuous" character, consisting of a large number of closely spaced lines.

In place of the distribution function I_n we can therefore introduce a distribution over the continuous series of frequencies $\omega = n\omega_H$, writing

$$dI = I_n dn = I_n \frac{d\omega}{\omega_H}.$$

For numerical computations it is convenient to express this distribution in terms of the MacDonald function K_ν.† After some simple transformations of formula (74.13), it can be written as

$$dI = d\omega \frac{\sqrt{3} e^3 H}{2\pi mc^2} F\left(\frac{\omega}{\omega_c}\right), \qquad F(\xi) = \xi \int_\xi^\infty K_{\frac{5}{3}}(\xi)\, d\xi, \qquad (74.17)$$

where we use the notation

$$\omega_c = \frac{3eH}{2mc}\left(\frac{\mathscr{E}}{mc^2}\right)^2. \qquad (74.18)$$

Figure 17 shows a graph of the function $F(\xi)$.

Finally, a few comments on the case when the particle moves, not in a plane orbit, but in a helical trajectory, i.e. has a longitudinal velocity (along the field) $v_\parallel = v \cos \chi$ (where χ is the angle between **H** and **v**). The frequency of the rotational motion is given by the same

---

† The connection between the Airy function and the function $K_{1/3}$ is given by formula (4) of the footnote on p. 161. In the further transformations one uses the recursion relations

$$K_{\nu-1}(x) - K_{\nu+1}(x) = -\frac{2\nu}{x} K_\nu, \qquad 2K'_\nu(x) = -K_{\nu-1}(x) - K_{\nu+1}(x),$$

where $K_{-\nu}(x) = K_\nu(x)$. In particular, it is easy to show that

$$\Phi'(t) = -\frac{t}{\sqrt{3\pi}} K_{2/3}\left(\frac{2}{3} t^{3/2}\right).$$

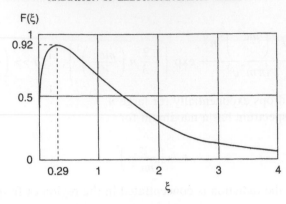

FIG. 17.

formula (74.1), but now the vector **v** moves, not in a circle, but on the surface of a cone with its axis along **H** and with vertex angle 2χ. The total intensity of the radiation (defined as the total energy loss per sec from the particle) will differ from (74.2) in having H replaced by $H_\perp = H \sin \chi$.

In the ultrarelativistic case the radiation is concentrated in directions near the generators of the "velocity cone". The spectral distribution and the total intensity (in the same sense as above) are obtained from (74.17) and (74.10) by the substitution $H \to H_\perp$. If we are talking about the intensity as seen in these directions by an observer at rest, then we must introduce an additional factor which takes into account the approach or moving away of the radiator (a particle moving in a circle). This factor is given by the ratio dt/dt_{obs}, where dt_{obs} is the time interval between arrival at the observer of signals emitted by the source at an interval dt. It is obvious that

$$dt_{obs} = dt \left(1 - \frac{1}{c}\, v_\| \cos \vartheta \right),$$

where ϑ is the angle between the directions of **k** and **H** (the latter being taken as the positive direction of the velocity $v_\|$). In the ultrarelativistic case, when the direction of **k** is close to the direction of **v**, we have $\vartheta \approx \chi$, so that

$$\frac{dt}{dt_{obs}} = \left(1 - \frac{v_\|}{c} \cos \chi \right)^{-1} \approx \frac{1}{\sin^2 \chi}. \tag{74.19}$$

PROBLEMS

1. Find the law of variation of energy with time for a charge moving in a circular orbit in a constant uniform magnetic field, and losing energy by radiation.

Solution: According to (74.2), we have for the energy loss per unit time:

$$-\frac{d\mathscr{E}}{dt} = \frac{2e^4 H^2}{3m^4 c^7} (\mathscr{E}^2 - m^2 c^4)$$

($\mathscr{E}$ is the energy of the particle). From this we find:

$$\frac{\mathscr{E}}{mc^2} = \coth \left(\frac{2e^4 H^2}{3m^3 c^5}\, t + \text{const} \right).$$

As t increases, the energy decreases monotonically, approaching the value $\mathscr{E} - mc^2$ (for complete stopping of the particle) asymptotically as $t \to \infty$.

2. Find the asymptotic formula for the spectral distribution of the radiation at large values of n for a particle moving in a circle with a velocity which is not close to the velocity of light.

Solution: We use the well-known asymptotic formula of the theory of Bessel functions

$$J_n(n\varepsilon) = \frac{1}{\sqrt{2\pi n}(1 - \varepsilon^2)^{1/4}} \left[\frac{\varepsilon'}{1 + \sqrt{1 - \varepsilon^2}} \, e^{\sqrt{1 - \varepsilon^2}} \right]^n,$$

which is valid for $n(1 - \varepsilon^2)^{2/3} \gg 1$. Using this formula, we find from (74.9):

$$I_n = \frac{e^4 H^2 n^{1/2}}{2\sqrt{\pi} m^2 c^3} \left(1 - \frac{v^2}{c^2}\right)^{5/4} \left[\frac{\frac{v}{c}}{1 + \sqrt{1 - \frac{v^2}{c^2}}} \, e^{\sqrt{1 - \frac{v^2}{c^2}}} \right]^{2n}$$

This formula is applicable for $n[1 - (v^2/c^2)]^{3/2} \gg 1$; if in addition $1 - (v^2/c^2)$ is small, the formula goes over into (74.15).

3. Find the polarization of the synchrotron radiation.

Solution: The electric field $\mathbf{E}_n$ is calculated from the vector potential $\mathbf{A}_n$ (74.6–74.7) according to the formula

$$\mathbf{E}_n - \frac{i}{k}(\mathbf{k} \times \mathbf{A}_n) \times \mathbf{k} - \frac{i}{k}\mathbf{k}(\mathbf{k} \cdot \mathbf{A}_n) + ik\mathbf{A}_n.$$

Let $\mathbf{e}_1$, $\mathbf{e}_2$ be unit vectors in the plane perpendicular to $\mathbf{k}$, where $\mathbf{e}_1$ is parallel to the x axis and $\mathbf{e}_2$ lies in the yz plane (their components are $\mathbf{e}_1 = (1, 0, 0)$, $\mathbf{e}_2 = (0, \sin\theta, -\cos\theta)$; the vectors $\mathbf{e}_1$, $\mathbf{e}_2$ and $\mathbf{k}$ form a right-hand system). Then the electric field will be:

$$\mathbf{E}_n = ikA_{xn}\mathbf{e}_1 + ik\sin\theta \, A_{yn}\mathbf{e}_2,$$

or, dropping the unimportant common factors:

$$\mathbf{E}_n \sim \frac{v}{c} J_n'\left(\frac{nv}{c}\cos\theta\right) \mathbf{e}_1 + \tan\theta \, J_n\left(\frac{nv}{c}\cos\theta\right) i\mathbf{e}_2,$$

The wave is elliptically polarized (see § 48).

In the ultrarelativistic case, for large n and small angles θ, the functions J_n and J_n' are expressed in terms of $K_{1/3}$ and $K_{2/3}$, where we set

$$1 - \frac{v^2}{c^2}\cos^2\theta \approx 1 - \frac{v^2}{c^2} + \theta^2 - \left(\frac{mc^2}{\mathscr{E}}\right)^2 + \theta^2$$

in the arguments. As a result we get:

$$\mathbf{E}_n = \mathbf{e}_1 \psi K_{2/3}\left(\frac{n}{3}\psi^3\right) + i\mathbf{e}_2\theta K_{1/3}\left(\frac{n}{3}\psi^3\right), \quad \psi - \sqrt{\left(\frac{mc^2}{\mathscr{E}}\right)^2 + \theta^2}.$$

For $\theta = 0$ the elliptical polarization degenerates into linear polarization along $\mathbf{e}_1$. For large θ ($|\theta| \gg mc^2/\mathscr{E}$, $n\theta^3 \gg 1$), we have $K_{1/3}(x) \approx K_{2/3}(x) \approx \sqrt{\pi/2x}\, e^{-x}$, and the polarization tends to become circular: $\mathbf{E}_n \sim \mathbf{e}_1 + i\mathbf{e}_2$; the intensity of the radiation, however, also becomes exponentially small. In the intermediate range of angles the minor axis of the ellipse lies along $\mathbf{e}_2$ and the major axis along $\mathbf{e}_1$. The direction of rotation depends on the sign of the angle θ ($\theta > 0$ if the directions of $\mathbf{H}$ and $\mathbf{k}$ lie on opposite sides of the orbit plane, as shown in Fig. 16).

§ 75. Radiation damping

In § 65 we showed that the expansion of the potentials of the field of a system of charges in a series of powers of v/c leads in the second approximation to a Lagrangian completely describing (in this approximation) the motion of the charges. We now continue the expansion of the field to terms of higher order and discuss the effects to which these terms lead.

In the expansion of the scalar potential

$$\phi = \int \frac{1}{R} \, \varrho_{t-\frac{R}{c}} \, dV,$$

the term of third order in $1/c$ is

$$\phi^{(3)} = -\frac{1}{6c^3} \frac{\partial^3}{\partial t^3} \int R^3 \varrho \, dV, \tag{75.1}$$

For the same reason as in the derivation following (65.3), in the expansion of the vector potential we need only take the term of second order in $1/c$, that is,

$$\mathbf{A}^{(2)} = -\frac{1}{c^2} \frac{\partial}{\partial t} \int \mathbf{j} \, dV. \tag{75.2}$$

We make a transformation of the potentials:

$$\phi' = \phi - \frac{1}{c} \frac{\partial f}{\partial f}, \qquad \mathbf{A}' = \mathbf{A} + \operatorname{grad} f,$$

choosing the function f so that the scalar potential $\phi'^{(3)}$ becomes zero:

$$f = -\frac{1}{6c^2} \frac{\partial^2}{\partial t^2} \int R^2 \varrho \, dV.$$

Then the new vector potential is equal to

$$\mathbf{A}'^{(2)} = -\frac{1}{c^2} \frac{\partial}{\partial t} \int \mathbf{j} \, dV - \frac{1}{6c^2} \frac{\partial^2}{\partial t^2} \nabla \int R^2 \varrho \, dV$$

$$= -\frac{1}{c^2} \frac{\partial}{\partial t} \int \mathbf{j} \, dV - \frac{1}{3c^2} \frac{\partial^2}{\partial t^2} \int \mathbf{R} \varrho \, dV.$$

Making the transition from the integral to a sum over individual charges, we get for the first term on the right the expression

$$-\frac{1}{c^2} \sum e\dot{\mathbf{v}}.$$

In the second term, we write $\mathbf{R} = \mathbf{R}_0 - \mathbf{r}$, where $\mathbf{R}_0$ and $\mathbf{r}$ have their usual meaning (see § 66); then $\dot{\mathbf{R}} = -\dot{\mathbf{r}} = -\mathbf{v}$ and the second term takes the form

$$\frac{1}{3c^2} \sum e\dot{\mathbf{v}}.$$

Thus,

$$\mathbf{A}'^{(2)} = -\frac{2}{3c^2} \sum e\dot{\mathbf{v}}, \tag{75.3}$$

The magnetic field corresponding to this potential is zero ($\mathbf{H} = \operatorname{curl} \mathbf{A}'^{(2)} = 0$), since $\mathbf{A}'^{(2)}$ does not contain the coordinates explicitly. The electric field $\mathbf{E} = - (1/c)\dot{\mathbf{A}}'^{(2)}$ is

$$\mathbf{E} = \frac{2}{3c^3}\, \dddot{\mathbf{d}}, \tag{75.4}$$

where $\mathbf{d}$ is the dipole moment of the system.

Thus the third order terms in the expansion of the field lead to certain additional forces acting on the charges, not contained in the Lagrangian (65.7); these forces depend on the time derivatives of the accelerations of the charges.

Let us consider a system of charges carrying out a stationary motion† and calculate the average work done by the field (75.4) per unit time. The force acting on each charge e is $\mathbf{f} = e\mathbf{E}$, that is,

$$\mathbf{f} = \frac{2e}{3c^2}\, \dddot{\mathbf{d}}. \tag{75.5}$$

The work done by this force in unit time is $\mathbf{f} \cdot \mathbf{v}$, so that the total work performed on all the charges is equal to the sum, taken over all the charges:

$$\Sigma \mathbf{f} \cdot \mathbf{v} = \frac{2}{3c^3}\, \dddot{\mathbf{d}} \cdot \Sigma e\mathbf{v} = \frac{2}{3c^3}\, \dddot{\mathbf{d}} \cdot \dot{\mathbf{d}} = \frac{2}{3c^3}\, \frac{d}{dt}(\dot{\mathbf{d}} \cdot \ddot{\mathbf{d}}) - \frac{2}{3c^3}\, (\ddot{\mathbf{d}})^2.$$

When we average over the time, the first term vanishes, so that the average work is equal to

$$\overline{\Sigma \mathbf{f} \cdot \mathbf{v}} = - \frac{2}{3c^3}\, \overline{\ddot{\mathbf{d}}^2}. \tag{75.6}$$

The expression standing on the right is (except for a sign reversal) just the average energy radiated by the system in unit time [see (67.8)]. Thus, the forces (75.5) appearing in third approximation, describe the reaction of the radiation on the charges. These forces are called *radiation damping* or *Lorentz frictional forces*.

Simultaneously with the energy loss from a radiating system of charges, there also occurs a certain loss of angular momentum. The decrease in angular momentum per unit time, $d\mathbf{M}/dt$, is easily calculated with the aid of the expression for the damping forces. Taking the time derivative of the angular momentum $\mathbf{M} = \Sigma\, \mathbf{r} \times \mathbf{p}$, we have $\dot{\mathbf{M}} = \Sigma\, \mathbf{r} \times \dot{\mathbf{p}}$, since $\Sigma\, \dot{\mathbf{r}} \times \mathbf{p} = \Sigma\, m(\mathbf{v} \times \mathbf{v}) \equiv 0$. We replace the time derivative of the momentum of the particle by the friction force (75.5) acting on it, and find

$$\dot{\mathbf{M}} = \Sigma\, \mathbf{r} \times \mathbf{f} = \frac{2}{3c^3}\, \Sigma\, e\mathbf{r} \times \dddot{\mathbf{d}} = \frac{2}{3c^3}\, \mathbf{d} \times \dddot{\mathbf{d}}.$$

We are interested in the time average of the loss of angular momentum for a stationary motion, just as before, we considered the time average of the energy loss. Writing

$$\mathbf{d} \times \dddot{\mathbf{d}} = \frac{d}{dt}(\mathbf{d} \times \ddot{\mathbf{d}}) - \dot{\mathbf{d}} \times \ddot{\mathbf{d}}$$

and noting that the time derivative (first term) vanishes on averaging, we finally obtain the following expression for the average loss of angular momentum of a radiating system:‡

---

† More precisely, a motion which, although it would have been stationary if radiation were neglected, proceeds with continual slowing down.

‡ In agreement with the result (3) obtained in problem 2 of § 72.

$$\frac{\overline{d\mathbf{M}}}{dt} = -\frac{2}{3c^3}\,\overline{\mathbf{d} \times \ddot{\mathbf{d}}}. \tag{75.7}$$

Radiation damping occurs also for a single charge moving in an external field. It is equal to

$$\mathbf{f} = \frac{2e^2}{3c^3}\,\dddot{\mathbf{v}}. \tag{75.8}$$

For a single charge, we can always choose such a system of reference that the charge at the given moment is at rest in it. If, in this reference frame, we calculate the higher terms in the expansion of the field produced by the charge, it turns out that they have the following property. As the radius vector R from the charge to the field point approaches zero, all these terms become zero. Thus in the case of a single charge, formula (75.8) is an exact formula for the reaction of the radiation, in the system of reference in which the charge is at rest.

Nevertheless, we must keep in mind that the description of the action of the charge "on itself" with the aid of the damping force is unsatisfactory in general, and contains contradictions. The equation of motion of a charge, in the absence of an external field, on which only the force (75.8) acts, has the form

$$m\dot{\mathbf{v}} = \frac{2e^2}{3c^3}\,\dddot{\mathbf{v}}.$$

This equation has, in addition to the trivial solution $\mathbf{v} = \text{const}$, another solution in which the acceleration $\dot{\mathbf{v}}$ is proportional to $\exp(3mc^3t/2e^2)$, that is, increases indefinitely with the time. This means, for example, that a charge passing through any field, upon emergence from the field, would have to be infinitely "self-accelerated". The absurdity of this result is evidence for the limited applicability of formula (75.8).

One can raise the question of how electrodynamics, which satisfies the law of conservation of energy, can lead to the absurd result that a free charge increases its energy without limit. Actually the root of this difficulty lies in the earlier remarks (§ 37) concerning the infinite electromagnetic "intrinsic mass" of elementary particles. When in the equation of motion we write a finite mass for the charge, then in doing this we essentially assign to it formally an infinite negative "intrinsic mass" of nonelectromagnetic origin, which together with the electromagnetic mass should result in a finite mass for the particle. Since, however, the subtraction of one infinity from another is not an entirely correct mathematical operation, this leads to a series of further difficulties, among which is the one mentioned here.

In a system of coordinates in which the velocity of the particle is small, the equation of motion when we include the radiation damping has the form

$$m\dot{\mathbf{v}} = e\mathbf{E} + \frac{e}{c}\,\mathbf{v} \times \mathbf{H} + \frac{2}{3}\frac{e^2}{c^3}\,\dddot{\mathbf{v}}. \tag{75.9}$$

From our discussion, this equation is applicable only to the extent that the damping force is small compared with the force exerted on the charge by the external field.

To clarify the physical meaning of this condition, we proceed as follows. In the system of reference in which the charge is at rest at a given moment, the second time derivative of the velocity is equal, neglecting the damping force, to

$$\ddot{\mathbf{v}} = \frac{e}{m}\,\dot{\mathbf{E}} + \frac{e}{mc}\,\dot{\mathbf{v}} \times \mathbf{H}.$$

In the second term we substitute (to the same order of accuracy) $\dot{\mathbf{v}} = (e/m)\mathbf{E}$, and obtain

$$\ddot{\mathbf{v}} = \frac{e}{m}\,\dot{\mathbf{E}} + \frac{e^2}{m^2 c}\,\mathbf{E} \times \mathbf{H}.$$

Corresponding to this, the damping force consists of two terms:

$$\mathbf{f} = \frac{2e^3}{3mc^3}\,\dot{\mathbf{E}} + \frac{2e^4}{3m^2 c^4}\,\mathbf{E} \times \mathbf{H}. \qquad (75.10)$$

If ω is the frequency of the motion, then $\dot{\mathbf{E}}$ is proportional to ωE and, consequently, the first term is of order $(e^2\omega/mc^3)E$; the second is of order $(e^4/m^2 c^4)EH$. Therefore the condition for the damping force to be small compared with the force eE exerted by the external field on the charge gives, first of all,

$$\frac{e^2}{mc^3}\,\omega \ll 1,$$

or, introducing the wavelength $\lambda \sim c/\omega$,

$$\lambda \gg \frac{e^2}{mc^2}. \qquad (75.11)$$

Thus formula (75.8) for the radiation damping is applicable only if the wavelength of the radiation incident on the charge is large compared with the "*radius*" of the charge e^2/mc^2. We see that once more a distance of order e^2/mc^2 appears as the limit at which electrodynamics leads to internal contradictions (see § 37).

Secondly, comparing the second term in the damping force to the force eE, we find the condition

$$H \ll \frac{m^2 c^4}{e^3}. \qquad (75.12)$$

Thus it is also necessary that the field itself be not too large. A field of order $m^2 c^4/e^3$ also represents a limit at which classical electrodynamics leads to internal contradictions. Also we must remember here that actually, because of quantum effects, electrodynamics is already not applicable for considerably smaller fields.†

To avoid misunderstanding, we remind the reader that the wavelength in (75.11) and the field value in (75.12) refer to the system of reference in which the particle is at rest at the given moment.

PROBLEM

Calculate the time in which two attracting charges, performing an elliptic motion (with velocity small compared with the velocity of light) and losing energy due to radiation, "fall in" toward each other.

Solution: Assuming that the relative energy loss in one revolution is small, we can equate the time derivative of the energy to the average intensity of the radiation (which was determined in problem 1 of § 70):

---

† For fields of order $m^2 c^3/\hbar e$, i.e., when $\hbar\omega_H \sim mc^2$. This limit is $\hbar c/e^2 = 137$ times smaller than the limit set by the condition (75.12). The ratio of these distances to R_0 is of order $\hbar c/e^2 \sim 137$.

$$\frac{d|\mathscr{E}|}{dt} = \frac{(2|\mathscr{E}|)^{3/2} \mu^{5/2} \alpha^3}{3c^3 M^5} \left(\frac{e_1}{m_1} - \frac{e_2}{m_2}\right)^2 \left(3 - \frac{2|\mathscr{E}|M^2}{\mu\alpha^2}\right), \tag{1}$$

where $\alpha = |e_1 e_2|$. Together with the energy, the particles lose angular momentum. The loss of angular momentum per unit time is given by formula (75.7); substituting the expression (70.1) for $\mathbf{d}$, and noting that $\mu\ddot{\mathbf{r}} = -\alpha \mathbf{r}/r^3$ and $\mathbf{M} = \mu\mathbf{r} \times \mathbf{v}$, we find:

$$\frac{d\mathbf{M}}{dt} = -\frac{2\alpha}{3c^3}\left(\frac{e_1}{m_1} - \frac{e_2}{m_2}\right)^2 \frac{\mathbf{M}}{r^3}.$$

We average this expression over a period of the motion. Because of the slowness of the changes in $\mathbf{M}$, it is sufficient to average on the right only over r^{-3}; this average value is computed in precisely the same way as the average of r^{-4} was found in problem 1 of § 70. As a result we find for the average loss of angular momentum per unit time the following expression:

$$\frac{dM}{dt} = -\frac{2\alpha(2\mu|\mathscr{E}|)^{3/2}}{3c^3 M^2}\left(\frac{e_1}{m_1} - \frac{e_2}{m_2}\right)^2 \tag{2}$$

[as in equation (1), we omit the average sign]. Dividing (1) by (2), we get the differential equation

$$\frac{d|\mathscr{E}|}{dM} = -\frac{\mu\alpha^2}{2M^3}\left(3 - 2\frac{|\mathscr{E}|M^2}{\mu\alpha^2}\right),$$

which, on integration, gives:

$$|\mathscr{E}| = \frac{\mu\alpha^2}{2M^2}\left(1 - \frac{M^3}{M_0^3}\right) + \frac{|\mathscr{E}_0|}{M_0}M. \tag{3}$$

The constant of integration is chosen so that for $M = M_0$, we have $\mathscr{E} = \mathscr{E}_0$, where M_0 and $\mathscr{E}_0$ are the initial angular momentum and energy of the particles.

The "falling in" of the particles toward one another corresponds to $M \to 0$. From (3) we see that then $\mathscr{E} \to -\infty$.

We note that the product $|\mathscr{E}|M^2$ tends toward $\mu\alpha^2/2$, and from formula (70.3) it is clear that the eccentricity $\varepsilon \to 0$, i.e. as the particles approach one another, the orbit approaches a circle. Substituting (3) in (2), we determine the derivative dt/dM expressed as a function of M, after which integration with respect to $\mathbf{M}$ between the limits M_0 and 0 gives the time of fall:

$$t_{\text{fall}} = \frac{c^3 M_0^5}{\alpha\sqrt{2|\mathscr{E}_0|\mu^3}}\left(\frac{e_1}{m_1} - \frac{e_2}{m_2}\right)^{-2} (\sqrt{\mu\alpha^2} + \sqrt{2M_0^2|\mathscr{E}_0|})^{-2}.$$

§ 76. Radiation damping in the relativistic case

We derive the relativistic expression for the radiation damping (for a single charge), which is applicable also to motion with velocity comparable to that of light. This force is now a four-vector g^i, which must be included in the equation of motion of the charge, written in four-dimensional form:

$$mc\frac{du^i}{ds} = \frac{e}{c}F^{ik}u_k + g^i. \tag{76.1}$$

To determine g^i we note that for $v \ll c$, its three space components must go over into the components of the vector $\mathbf{f}/c$ (75.8). It is easy to see that the vector $(2e^2/3c)/(d^2u^i/ds^2)$ has this property. However, it does not satisfy the identity $g^i u_i = 0$, which is valid for any force four-vector. In order to satisfy this condition, we must add to the expression given a certain auxiliary four-vector, made up from the four-velocity u^i and its derivatives. The three space components of this vector must become zero in the limiting case $\mathbf{v} = 0$, in order not to change the correct values of $\mathbf{f}$ which are already given by $(2e^2/3c)/(d^2u^i/ds^2)$. The four-vector u^i has this property, and therefore the required auxiliary term has the form αu^i. The scalar α must be chosen so that we satisfy the auxiliary relation $g^i u_i = 0$. As a result we find

$$g^i = \frac{2e^2}{3c} \left(\frac{d^2 u^i}{ds^2} - u^i u^k \frac{d^2 u_k}{ds^2} \right). \tag{76.2}$$

In accordance with the equations of motion, this expression can be written in another form, by expressing $d^2 u^i/ds^2$ directly in terms of the field tensor of the external field acting on the particle:

$$\frac{du^i}{ds} = \frac{e}{mc^2} F^{ik} u_k,$$

$$\frac{d^2 u^i}{ds^2} = \frac{e}{mc^2} \frac{\partial F^{ik}}{\partial x^i} u_k u^l + \frac{e^2}{m^2 c^4} F^{ik} F_{kl} u^l.$$

In making substitutions, we must keep in mind that the product of the tensor $\partial F^{ik}/\partial x^l$, which is antisymmetric in the indices i, k, and the symmetric tensor $u_i u_k$ gives identically zero. So,

$$g^i = \frac{2e^3}{3mc^3} \frac{\partial F^{ik}}{\partial x^l} u_k u^l - \frac{2e^4}{3m^2 c^5} F^{il} F_{kl} u^k + \frac{2e^4}{3m^2 c^5} (F_{kl} u^l)(F^{km} u_m) u^i. \tag{76.3}$$

The integral of the four-force g^i over the world line of the motion of a charge, passing through a given field, must coincide (except for opposite sign) with the total four-momentum ΔP^i of the radiation from the charge [just as the average value of the work of the force $\mathbf{f}$ in the nonrelativistic case coincides with the intensity of dipole radiation; see (75.6)]. It is easy to check that this is actually so. The first term in (76.2) goes to zero on performing the integration, since at infinity the particle has no acceleration, i.e. $(du^i/ds) = 0$. We integrate the second term by parts and get:

$$-\int g^i \, ds = \frac{2e^2}{3c} \int u^k u^i \frac{d^2 u_k}{ds^2} \, ds = -\frac{2e^2}{3c} \int \frac{du_k}{ds} \frac{du^k}{ds} \, dx^i$$

which coincides exactly with (73.4).

When the velocity of the particle approaches the velocity of light, those terms in the space components of the four-vector (76.3) increase most rapidly which contain the components of the four-velocity in the third degree. Therefore, keeping only these terms in (76.3) and using the relation (9.18) between the space components of the four vector g^i and the three-dimensional force $\mathbf{f}$, we find for the latter:

$$\mathbf{f} = \frac{2e^4}{3m^2 c^5} (F_{kl} u^l)(F^{km} u_m) \mathbf{n},$$

where $\mathbf{n}$ is a unit vector in the direction of $\mathbf{v}$. Consequently, in this case the force $\mathbf{f}$ is opposite to the velocity of the particle; choosing the latter as the X axis, and writing out the four-dimensional expressions, we obtain:

$$f_x = -\frac{2e^4}{3m^2c^4}\frac{(E_y - H_z)^2 + (E_z + H_y)^2}{1 - \dfrac{v^2}{c^2}} \tag{76.4}$$

(where we have set $v = c$ everywhere except in the denominator).

We see that for an ultrarelativistic particle, the radiation damping is proportional to the square of its energy.

Let us call attention to the following interesting situation. Earlier we pointed out that the expression obtained for the radiation damping is applicable only to fields which (in the reference system K_0, in which the particle is at rest) are small compared with m^2c^4/e^3. Let F be the order of magnitude of the external field in the reference system K, in which the particle moves with velocity v. Then in the K_0 frame, the field has the order of magnitude $F/\sqrt{1 - v^2/c^2}$ (see the transformation formulas in § 24). Therefore F must satisfy the condition

$$\frac{e^3 F}{m^2c^4\sqrt{1 - \dfrac{v^2}{c^2}}} \ll 1. \tag{76.5}$$

At the same time, the ratio of the damping force (76.4) to the external force ($\sim eF$) is of the order of

$$\frac{e^3 F}{m^2c^4\left(1 - \dfrac{v^2}{c^2}\right)},$$

and we see that, even though the condition (76.5) is satisfied, it may happen (for sufficiently high energy of the particle) that the damping force is large compared with the ordinary Lorentz force acting on the particle in the electromagnetic field.† Thus for an ultrarelativistic particle we can have the case where the radiation damping is the main force acting on the particle.

In this case the loss of (kinetic) energy of the particle per unit length of path can be equated to the damping force f_x alone; keeping in mind that the latter is proportional to the square of the energy of the particle, we write

$$\frac{d\mathscr{E}_{\text{kin}}}{dx} = -k(x)\,\mathscr{E}_{\text{kin}}^2,$$

where we denote by $k(x)$ the coefficient, depending on the x coordinate and expressed in terms of the transverse components of the field in accordance with (76.4). Integrating this differential equation, we find

$$\frac{1}{\mathscr{E}_{\text{kin}}} = \frac{1}{\mathscr{E}_0} + \int_{-\infty}^{x} k(x)\,dx,$$

---

† We should emphasize that this result does not in any way contradict the derivation given earlier of the relativistic expression for the four-force g^i, in which it was assumed to be "small" compared with the four-force $(e/c)F^{ik}u_k$. It is sufficient to satisfy the requirement that the components of one vector be small compared to those of another in just one frame of reference; by virtue of relativistic invariance, the four-dimensional formulas obtained on the basis of such an assumption will be valid in any other reference frame.

where $\mathscr{E}_0$ represents the initial energy of the particle (its energy for $x \to -\infty$). In particular, the final energy $\mathscr{E}_1$ of the particle (after passage of the particle through the field) is given by the formula

$$\frac{1}{\mathscr{E}_1} = \frac{1}{\mathscr{E}_0} + \int_{-\infty}^{+\infty} k(x)\, dx.$$

We see that for $\mathscr{E}_0 \to \infty$, the final $\mathscr{E}_1$ approaches a constant limit independent of $\mathscr{E}_0$ (I. Pomeranchuk, 1939). In other words, after passing through the field, the energy of the particle cannot exceed the energy $\mathscr{E}_{\text{crit}}$, defined by the equation

$$\frac{1}{\mathscr{E}_{\text{crit}}} = \int_{-\infty}^{+\infty} k(x)\, dx,$$

or, substituting the expression for $k(x)$,

$$\mathscr{E}_{\text{crit}}^{-1} = \frac{2}{3m^2 c^4} \left(\frac{e^2}{mc^2} \right)^2 \int_{-\infty}^{+\infty} [(E_y - H_z)^2 + (E_z + H_y)^2]\, dx. \tag{76.6}$$

PROBLEMS

1. Calculate the limiting energy which a particle can have after passing through the field of a magnetic dipole m; the vector m and the direction of motion lie in a plane.

Solution: We choose the plane passing through the vector m and the direction of motion as the XZ plane, where the particle moves parallel to the X axis at a distance ϱ from it. For the transverse components of the field of the magnetic dipole we have (see 44.4):

$$H_y = 0,$$

$$H_z = \frac{3(m \cdot r)_z - m_z r^2}{r^2} = \frac{m}{(\varrho^2 + x^2)^{5/2}} \{3(\varrho \cos \phi + x \sin \phi)\varrho - (\varrho^2 + x^2) \cos \phi\}$$

(ϕ is the angle between m and the Z axis). Substituting in (76.6) and performing the integration, we obtain

$$\frac{1}{\mathscr{E}_{\text{crit}}} = \frac{m^2 \pi}{64 m^2 c^4 \varrho^5} \left(\frac{e^2}{mc^2} \right)^2 (15 + 26 \cos^2 \phi).$$

2. Write the three-dimensional expression for the damping force in the relativistic case.

Solution: Calculating the space components of the four-vector (76.3), we find

$$\mathbf{f} = \frac{2e^3}{3mc^3} \left(1 - \frac{v^2}{c^2} \right)^{-1/2} \left\{ \left(\frac{\partial}{\partial t} + \mathbf{v} \cdot \nabla \right) \mathbf{E} + \frac{1}{c} \mathbf{v} \times \left(\frac{\partial}{\partial t} + \mathbf{v} \cdot \nabla \right) \mathbf{H} \right\}$$

$$+ \frac{2e^4}{3m^2 c^4} \left\{ \mathbf{E} \times \mathbf{H} + \frac{1}{c} \mathbf{H} \times (\mathbf{H} \times \mathbf{v}) + \frac{1}{c} \mathbf{E}(\mathbf{v} \cdot \mathbf{E}) \right\}$$

$$- \frac{2e^4}{3m^2 c^5 \left(1 - \dfrac{v^2}{c^2} \right)} \mathbf{v} \left\{ \left(\mathbf{E} + \frac{1}{c} \mathbf{v} \times \mathbf{H} \right)^2 - \frac{1}{c^2} (\mathbf{E} \cdot \mathbf{v})^2 \right\}.$$

§ 77. Spectral resolution of the radiation in the ultrarelativistic case

Earlier (in § 73) it was shown that the radiation from an ultrarelativistic particle is directed mainly in the forward direction, along the velocity of the particle: it is contained almost entirely within the small range of angles

$$\Delta\theta \sim \sqrt{1 - \frac{v^2}{c^2}}$$

around the direction of **v**.

In evaluating the spectral resolution of the radiation, the relation between the magnitude of the angular range $\Delta\theta$ and the angle of deflection α of the particle in passing through the external electromagnetic field is essential.

The angle α can be calculated as follows. The change in the transverse (to the direction of motion) momentum of the particle is of the order of the product of the transverse force eF† and the time of passage through the field, $t \sim a/v \cong a/c$ (where a is the distance within which the field is significantly different from 0). The ratio of this quantity to the momentum

$$p = \frac{mv}{\sqrt{1 - v^2/c^2}} \cong \frac{mc}{\sqrt{1 - v^2/c^2}}$$

determines the order of magnitude of the small angle α:

$$\alpha \sim \frac{eFa}{mc^2} \sqrt{1 - \frac{v^2}{c^2}} \, .$$

Dividing by $\Delta\theta$, we find:

$$\frac{\alpha}{\Delta\theta} \sim \frac{eFa}{mc^2} \, . \tag{77.1}$$

We call attention to the fact that the ratio does not depend on the velocity of the particle, and is completely determined by the properties of the external field itself.

We assume first that

$$eFa \gg mc^2, \tag{77.2}$$

that is, the total deflection of the particle is large compared with $\Delta\theta$. Then we can say that radiation in a given direction occurs mainly from that portion of the trajectory in which the velocity of the particle is almost parallel to that direction (subtending with it an angle in the interval $\Delta\theta$) and the length of this segment is small compared with a. The field F can be considered constant within this segment, and since a small segment of a curve can be considered as an arc of a circle, we can apply the results obtained in § 74 for radiation during uniform motion in a circle (replacing H by F). In particular, we may state that the main part of the radiation is concentrated in the frequency range

$$\omega \sim \frac{eF}{mc\left(1 - \frac{v^2}{c^2}\right)} \tag{77.3}$$

---

† If we choose the X axis along the direction of motion of the particle, then $(eF)^2$ is the sum of the squares of the y and z components of the Lorentz force, $e\mathbf{E} + e\mathbf{v}/c \times \mathbf{H}$, in which we can here set $v \cong c$:

$$F^2 = (E_y - H_z)^2 + (E_z + H_y)^2 \, .$$

[see (74.16)].

In the opposite limiting case,

$$eFa \ll mc^2, \tag{77.4}$$

the total angle of deflection of the particle is small compared with $\Delta\theta$. In this case the radiation is directed mainly into the narrow angular range $\Delta\theta$ around the direction of motion, while radiation arrives at a given point from the whole trajectory.

To compute the spectral distribution of the intensity, it is convenient to start in this case from the Lienard–Wiechert expressions (73.8) for the field in the wave zone. Let us compute the Fourier component

$$\mathbf{E}_\omega = \frac{1}{2\pi} \int\limits_{-\infty}^{\infty} \mathbf{E} e^{i\omega t} dt.$$

The expression on the right of formula (73.8) is a function of the retarded time t', which is determined by the condition $t' = t - R(t')/c$. At large distances from a particle which is moving with an almost constant velocity $\mathbf{v}$, we have:

$$t' \cong t - \frac{R_0}{c} + \frac{1}{c}\mathbf{n}\cdot\mathbf{r}(t') \cong t - \frac{R_0}{c} + \frac{1}{c}\mathbf{n}\cdot\mathbf{v}t'$$

$(\mathbf{r} = \mathbf{r}(t') \cong \mathbf{v}t'$ is the radius vector of the particle), or

$$t = t'\left(1 - \frac{\mathbf{n}\cdot\mathbf{v}}{c}\right) + \frac{R_0}{c}.$$

We replace the t integration by an integration over t', by setting

$$dt = dt'\left(1 - \frac{\mathbf{n}\cdot\mathbf{v}}{c}\right),$$

and obtain:

$$\mathbf{E}_\omega = \frac{e}{c^2} \frac{e^{ikR_0}}{R_0\left(1 - \dfrac{\mathbf{n}\cdot\mathbf{v}}{c}\right)^2} \int\limits_{-\infty}^{\infty} \mathbf{n}\times\left\{\left(\mathbf{n} - \frac{\mathbf{v}}{c}\right)\times\mathbf{w}(t')\right\} e^{i\omega t'\left(1 - \frac{\mathbf{n}\cdot\mathbf{v}}{c}\right)} dt'.$$

We treat the velocity $\mathbf{v}$ as constant; only the acceleration $\mathbf{w}(t')$ is variable. Introducing the notation

$$\omega' = \omega\left(1 - \frac{\mathbf{n}\cdot\mathbf{v}}{c}\right), \tag{77.5}$$

and the corresponding frequency component of the acceleration, we write $\mathbf{E}_\omega$ in the form

$$\mathbf{E}_\omega = \frac{e}{c^2} \frac{e^{ikR_0}}{R_0}\left(\frac{\omega}{\omega'}\right)^2 \mathbf{n}\times\left\{\left(\mathbf{n} - \frac{\mathbf{v}}{c}\right)\times\mathbf{w}_{\omega'}\right\}.$$

Finally from (66.9) we get for the energy radiated into solid angle do, with frequency in $d\omega$:

$$d\mathcal{E}_{n\omega} = \frac{e^2}{2\pi c^3} \left(\frac{\omega}{\omega'}\right)^4 \left|\mathbf{n} \times \left\{\left(\mathbf{n} - \frac{\mathbf{v}}{c}\right) \times \mathbf{w}_{\omega'}\right\}\right|^2 do \frac{d\omega}{2\pi}. \tag{77.6}$$

An estimate of the order of magnitude of the frequencies in which the radiation is mainly concentrated in the case of (77.4) is easily made by noting that the Fourier component $\mathbf{w}_{\omega'}$ is significantly different from zero only if the time $1/\omega'$, or

$$\frac{1}{\omega\left(1 - \frac{v^2}{c^2}\right)}$$

is of the same order as the time $a/v \sim a/c$ during which the acceleration of the particle changes significantly. Therefore we find:

$$\omega \sim \frac{c}{a\left(1 - \frac{v^2}{c^2}\right)}. \tag{77.7}$$

The energy dependence of the frequency is the same as in (77.3), but the coefficient is different.

In the treatment of both cases (77.2) and (77.4) it was assumed that the total loss of energy by the particle during its passage through the field was relatively small. We shall now show that the first of these cases also covers the problem of the radiation by an ultrarelativistic particle, whose total loss of energy is comparable with its initial energy.

The total loss of energy by the particle in the field can be determined from the work of the Lorentz frictional force. The work done by the force (76.4) over the path $\sim a$ is of order

$$af \sim \frac{e^4 F^2 a}{m^2 c^4 \left(1 - \frac{v^2}{c^2}\right)}.$$

In order for this to be comparable with the total energy of the particle,

$$mc^2 \bigg/ \sqrt{1 - \frac{v^2}{c^2}},$$

the field must exist at distances

$$a \sim \frac{m^3 c^6}{e^4 F^2} \sqrt{1 - \frac{v^2}{c^2}}.$$

But then condition (77.2) is satisfied automatically:

$$aeF \sim \frac{m^3 c^6}{e^3 F} \sqrt{1 - \frac{v^2}{c^2}} \gg mc^2,$$

since the field F must necessarily satisfy condition (76.5)

$$\frac{F}{\sqrt{1 - \frac{v^2}{c^2}}} \ll \frac{m^2 c^4}{e^3},$$

since otherwise we could not even apply ordinary electrodynamics.

PROBLEMS

1. Determine the spectral distribution of the total (over all directions) radiation intensity for the condition (77.2).

Solution: For each element of length of the trajectory, the radiation is determined by (74.13), where we must replace H by the value of the transverse force F at the given point and, in addition, we must go over from a discrete to a continuous frequency spectrum. This transformation is accomplished by formally multiplying by dn and the replacement

$$I_n\, dn = I_n \frac{dn}{d\omega}\, d\omega = I_n \frac{d\omega}{\omega_0}.$$

Next, integrating over all time, we obtain the spectral distribution of the total radiation in the following form:

$$d\mathscr{E}_\omega = - d\omega\, \frac{2e^2\omega}{\sqrt{\pi}c}\left(1 - \frac{v^2}{c^2}\right) \int_{-\infty}^{+\infty}\left[\frac{1}{u}\,\Phi'(u) + \frac{1}{2}\int_u^\infty \Phi(u)\,du\right]dt,$$

where $\Phi(u)$ is the Airy function of the argument

$$u = \left[\frac{mc\omega}{eF}\left(1 - \frac{v^2}{c^2}\right)\right]^{2/3}.$$

The integrand depends on the integration variable t implicitly through the quantity u (F and with it u, varies along the trajectory of the particle; for a given motion this variation can be considered as a time dependence).

2. Determine the spectral distribution of the total (over all directions) radiated energy for the condition (77.4).

Solution: Keeping in mind that the main role is played by the radiation at small angles to the direction of motion, we write:

$$\omega' = \omega\left(1 - \frac{v}{c}\cos\theta\right) \cong \omega\left(1 - \frac{v}{c} + \frac{\theta^2}{2}\right) \cong \frac{\omega}{2}\left(1 - \frac{v^2}{c^2} + \theta^2\right).$$

We replace the integration over angles $do = \sin\theta\,d\theta\,d\phi \cong \theta\,d\theta\,d\phi$ in (77.6) by an integration over $d\phi\,d\omega'/\omega$. In writing out the square of the vector triple product in (77.6) it must be remembered that in the ultrarelativistic case the longitudinal component of the acceleration is small compared with the transverse component [in the ratio $1 - (v^2/c^2)$], and that in the present case we can, to sufficient accuracy, consider $\mathbf{w}$ and $\mathbf{v}$ to be mutually perpendicular. As a result, we find for the spectral distribution of the total radiation the following formula:

$$d\mathscr{E}_\omega = \frac{e^2\omega d\omega}{2\pi c^3} \int_{\frac{\omega}{2}\left(1 - \frac{v^2}{c^2}\right)}^\infty \frac{|\mathbf{w}_{\omega'}|^2}{\omega'^2}\left[1 - \frac{\omega}{\omega'}\left(1 - \frac{v^2}{c^2}\right) + \frac{\omega^2}{2\omega'^2}\left(1 - \frac{v^2}{c^2}\right)^2\right]d\omega'.$$

§ 78. Scattering by free charges

If an electromagnetic wave falls on a system of charges, then under its action the charges are set in motion. This motion in turn produces radiation in all directions; there occurs, we say, a *scattering* of the original wave.

The scattering is most conveniently characterized by the ratio of the amount of energy emitted by the scattering system in a given direction per unit time, to the energy flux density of the incident radiation. This ratio clearly has dimensions of area, and is called the *effective scattering cross-section* (or simply the *cross-section*).

Let dI be the energy radiated by the system into solid angle do per second for an incident wave with Poynting vector $\mathbf{S}$. Then the effective cross-section for scattering (into the solid angle do) is

$$d\sigma = \frac{\overline{dI}}{\overline{S}} \tag{78.1}$$

(the dash over a symbol means a time average). The integral σ of $d\sigma$ over all directions is the *total* scattering cross-section.

Let us consider the scattering produced by a free charge at rest. Suppose there is incident on this charge a plane monochromatic linearly polarized wave. Its electric field can be written in the form

$$\mathbf{E} = \mathbf{E}_0 \cos (\mathbf{k} \cdot \mathbf{r} - \omega t + \alpha).$$

We shall assume that the velocity acquired by the charge under the influence of the incident wave is small compared with the velocity of light (which is usually the case). Then we can consider the force acting on the charge to be $e\mathbf{E}$, while the force $(e/c)\mathbf{v} \times \mathbf{H}$ due to the magnetic field can be neglected. In this case we can also neglect the effect of the displacement of the charge during its vibrations under the influence of the field. If the charge carries out vibrations around the coordinate origin, then we can assume that the field which acts on the charge at all times is the same as that at the origin, that is,

$$\mathbf{E} = \mathbf{E}_0 \cos (\omega t - \alpha).$$

Since the equation of motion of the charge is

$$m\ddot{\mathbf{r}} = e\mathbf{E}$$

and its dipole moment $\mathbf{d} = e\mathbf{r}$, then

$$\ddot{\mathbf{d}} = \frac{e^2}{m} \mathbf{E}. \tag{78.2}$$

To calculate the scattered radiation, we use formula (67.7) for dipole radiation (this is justified, since the velocity acquired by the charge is assumed to be small). We also note that the frequency of the wave radiated by the charge (i.e., scattered by it) is clearly the same as the frequency of the incident wave.

Substituting (78.2) in (67.7), we find

$$dI = \frac{e^4}{4\pi m^2 c^3} (\mathbf{E} \times \mathbf{n}')^2 \, do,$$

where $\mathbf{n}'$ is a unit vector in the scattering direction. On the other hand, the Poynting vector of the incident wave is

$$S = \frac{c}{4\pi} E^2. \tag{78.3}$$

From this we find, for the cross-section for scattering into the solid angle do,

$$d\sigma = \left(\frac{e^2}{mc^2}\right)^2 \sin^2 \theta \, do, \qquad (78.4)$$

where θ is the angle between the direction of scattering (the vector $\mathbf{n}$), and the direction of the electric field $\mathbf{E}$ of the incident wave. We see that the effective scattering cross-section of a free charge is independent of frequency.

We determine the total cross-section σ. To do this, we choose the polar axis along $\mathbf{E}$. Then $do = \sin \theta \, d\theta \, d\phi$; substituting this and integrating with respect to θ from 0 to π, and over ϕ from 0 to 2π, we find

$$\sigma = \frac{8\pi}{3} \left(\frac{e^2}{mc^2}\right)^2 \qquad (78.5)$$

(This is the *Thomson formula*.)

Finally, we calculate the differential cross-section $d\sigma$ in the case where the incident wave is unpolarized (ordinary light). To do this we must average (78.4) over all directions of the vector $\mathbf{E}$ in a plane perpendicular to the direction of propagation of the incident wave (direction of the wave vector $\mathbf{k}$). Denoting by $\mathbf{e}$ the unit vector along the direction of $\mathbf{E}$, we write:

$$\overline{\sin^2 \theta} = 1 - \overline{(\mathbf{n} \cdot \mathbf{e})^2} = 1 - n_\alpha n_\beta \overline{e_\alpha e_\beta}.$$

The averaging is done using the formula†

$$\overline{e_\alpha e_\beta} = \frac{1}{2}\left(\delta_{\alpha\beta} - \frac{k_\alpha k_\beta}{k^2}\right), \qquad (78.6)$$

and gives

$$\overline{\sin^2 \theta} = \frac{1}{2}\left(1 + \frac{(\mathbf{n} \cdot \mathbf{k})^2}{k^2}\right) = \frac{1}{2}(1 + \cos^2 \Theta)$$

where Θ is the angle between the directions of the incident and scattered waves (the scattering angle). Thus the effective cross-section for scattering of an unpolarized wave by a free charge is

$$d\sigma = \frac{1}{2}\left(\frac{e^2}{mc^2}\right)^2 (1 + \cos^2 \Theta) \, do, \qquad (78.7)$$

The occurrence of scattering leads, in particular, to the appearance of a certain force acting on the scattering particle. One can verify this by the following considerations. On the average, in unit time, the wave incident on the particle loses energy $c\overline{W}\sigma$, where $\overline{W}$ is the average energy density, and σ is the total effective scattering cross-section. Since the momentum of the field is equal to its energy divided by the velocity of light, the incident wave loses momentum equal in magnitude to $\overline{W}\sigma$. On the other hand, in a system of reference in which the charge carries out only small vibrations under the action of the force $e\mathbf{E}$, and its velocity

† In fact, $e_\alpha e_\beta$ is a symmetric tensor with trace equal to 1, which gives zero when multiplied by k_α, because $\mathbf{e}$ and $\mathbf{k}$ are perpendicular. The expression given here satisfies these conditions.

v is small, the total flux of momentum in the scattered wave is zero, to terms of higher order in v/c (in § 73 it was shown that in a reference system in which $v = 0$, radiation of momentum by the particle does not occur). Therefore all the momentum lost by the incident wave is "absorbed" by the scattering particle. The average force $\bar{\mathbf{f}}$ acting on the particle is equal to the average momentum absorbed per unit time, i.e.

$$\bar{\mathbf{f}} = \sigma \overline{W} \, \mathbf{n}_0 \tag{78.8}$$

($\mathbf{n}_0$ is a unit vector in the direction of propagation of the incident wave). We note that the average force appears as a second order quantity in the field of the incident wave, while the "instantaneous" force (the main part of which is $e\mathbf{E}$) is of first order in the field.

Formula (78.8) can also be obtained directly by averaging the damping force (75.10). The first term, proportional to $\dot{\mathbf{E}}$, goes to zero on averaging, as does the average of the main part of the force, $e\mathbf{E}$. The second term gives

$$\bar{\mathbf{f}} = \frac{2e^4}{3m^2c^4} \, \overline{E^2} \mathbf{n}_0 = \frac{8\pi}{3} \left(\frac{e^2}{mc^2} \right)^2 \frac{\overline{E^2}}{4\pi} \, \mathbf{n}_0,$$

which, using (78.5), coincides with (78.8).

PROBLEMS

1. Determine the effective cross-section for scattering of an elliptically polarized wave by a free charge.

Solution: The field of the wave has the form $\mathbf{E} = \mathbf{A} \cos(\omega t + \alpha) + \mathbf{B} \sin(\omega t + \alpha)$, where $\mathbf{A}$ and $\mathbf{B}$ are mutually perpendicular vectors (see § 48). By a derivation similar to the one in the text, we find

$$d\sigma = \left(\frac{e^2}{mc^2} \right)^2 \frac{(\mathbf{A} \times \mathbf{n})^2 + (\mathbf{B} \times \mathbf{n})^2}{A^2 \times B^2} \, do.$$

2. Determine the effective cross-section for scattering of a linearly polarized wave by a charge carrying out small vibrations under the influence of an elastic force (oscillator).

Solution: The equation of motion of the charge in the incident field $\mathbf{E} = \mathbf{E}_0 \cos(\omega t + \alpha)$ is

$$\ddot{\mathbf{r}} + \omega_0^2 \mathbf{r} = \frac{e}{m} \mathbf{E}_0 \cos(\omega t + \alpha),$$

where ω_0 is the frequency of its free vibrations. For the forced vibrations, we then have

$$\mathbf{r} = \frac{e\mathbf{E}_0 \cos(\omega t + \alpha)}{m(\omega_0^2 - \omega^2)}.$$

Calculating $\mathbf{d}$ from this, we find

$$d\sigma = \left(\frac{e^2}{mc^2} \right)^2 \frac{\omega^4}{(\omega_0^2 - \omega^2)^2} \sin^2 \theta \, do$$

(θ is the angle between $\mathbf{E}$ and $\mathbf{n}'$).

3. Determine the total effective cross-section for scattering of light by an electric dipole which, mechanically, is a rotator. The frequency ω of the wave is assumed to be large compared with the frequency Ω_0 of free rotation of the rotator.

Solution: Because of the condition $\omega \gg \Omega_0$, we can neglect the free rotation of the rotator, and consider only the forced rotation under the action of the moment of the forces $\mathbf{d} \times \mathbf{E}$ exerted on it by the scattered wave. The equation for this motion is: $J\dot{\mathbf{\Omega}} = \mathbf{d} \times \mathbf{E}$, where J is the moment of inertia of the rotator and $\mathbf{\Omega}$ is the angular velocity of rotation. The change in the dipole moment vector, as it rotates without changing its absolute value, is given by the formula $\dot{\mathbf{d}} = \mathbf{\Omega} \times \mathbf{d}$. From these two equations, we find (omitting the quadratic term in the small quantity $\mathbf{\Omega}$):

$$\ddot{\mathbf{d}} = \frac{1}{J}(\mathbf{d} \times \mathbf{E}) \times \mathbf{d} = \frac{1}{J}[E d^2 - (\mathbf{E} \cdot \mathbf{d})\,\mathbf{d}].$$

Assuming that all orientations of the dipole in space are equally probable, and averaging $\ddot{\mathbf{d}}^2$ over them, we find for the total effective cross-section,

$$\sigma = \frac{16\pi d^4}{9c^4 J^2}.$$

4. Determine the degree of depolarization in the scattering of ordinary light by a free charge.

Solution: From symmetry considerations, it is clear that the two incoherent polarized components of the scattered light (see § 50) will be linearly polarized: one in the plane of scattering (the plane passing through the incident and scattered waves) and the other perpendicular to this plane. The intensities of these components are determined by the components of the field of the incident wave in the plane of scattering ($\mathbf{E}_{\parallel}$) and perpendicular to it ($\mathbf{E}_{\perp}$), and, according to (78.4), are proportional respectively to

$$(\mathbf{E}_{\parallel} \times \mathbf{n})^2 = E_{\parallel}^2 \cos^2\Theta \quad \text{and} \quad (\mathbf{E}_{\perp} \times \mathbf{n}')^2 = E_{\perp}^2$$

(where Θ is the angle of scattering). Since for the ordinary incident light, $\overline{E_{\parallel}^2} = \overline{E_{\perp}^2}$, the degree of depolarization [see the definition in (50.9)] is:

$$\varrho = \cos^2 \Theta.$$

5. Determine the frequency ω' of the light scattered by a moving charge.

Solution: In a system of coordinates in which the charge is at rest, the frequency of the light does not change on scattering ($\omega = \omega'$). This relation can be written in invariant form as

$$k'_i u'^i = k_i u^i,$$

where u^i is the four-velocity of the charge. From this we find without difficulty

$$\omega'\left(1 - \frac{v}{c}\cos\theta'\right) = \omega\left(1 - \frac{v}{c}\cos\theta\right),$$

where θ and θ' are the angles made by the incident and scattered waves with the direction of motion (v is the velocity of the charge).

6. Determine the angular distribution of the scattering of a linearly polarized wave by a charge moving with velocity v in the direction of propagation of the wave.

Solution: The velocity of the particle is perpendicular to the fields $\mathbf{E}$ and $\mathbf{H}$ of the incident wave, and is therefore also perpendicular to the acceleration $\mathbf{w}$ given to the particle. The scattered intensity is given by (73.14), where the acceleration $\mathbf{w}$ of the particle must be expressed in terms of the fields $\mathbf{E}$ and $\mathbf{H}$ of the incident wave by the formulas obtained in the problem in § 17. Dividing the intensity dI by the Poynting vector of the incident wave, we get the following expression for the scattering cross-section:

$$d\sigma = \left(\frac{e^2}{mc^2}\right)^2 \frac{\left(1 - \frac{v^2}{c^2}\right)\left(1 - \frac{v}{c}\right)^2}{\left(1 - \frac{v}{c}\sin\theta\cos\phi\right)^6}\left[\left(1 - \frac{v}{c}\sin\theta\cos\phi\right)^2 - \left(1 - \frac{v^2}{c^2}\right)\cos^2\theta\right]do,$$

where θ and ϕ are the polar angle and azimuth of the direction $\mathbf{n}'$ relative to a system of coordinates with Z axis along $\mathbf{E}$, and X axis along $\mathbf{v}$ ($\cos(\mathbf{n}', \mathbf{E}) = \cos\theta$; $\cos(\mathbf{n}', \mathbf{v}) = \sin\theta\cos\phi$).

7. Calculate the motion of a charge under the action of the average force exerted upon it by the wave scattered by it.

Solution: The force (78.8), and therefore the velocity of the motion under consideration, is along the direction of propagation of the incident wave (X axis). In the auxiliary reference system K_0, in which the particle is at rest (we recall that we are dealing with the motion averaged over the period of the small vibrations), the force acting on it is σW_0, and the acceleration acquired by it under the action of this force is

$$w_0 = \frac{\sigma}{m}\,\overline{W}_0$$

(the index zero refers to the reference system K_0). The transformation to the original reference system K (in which the charge moves with velocity v) is given by the formulas obtained in the problem of § 7 and by formula (47.7), and gives:

$$\frac{d}{dt}\frac{v}{\sqrt{1-\dfrac{v^2}{c^2}}} = \frac{1}{\left(1-\dfrac{v^2}{c^2}\right)^{3/2}}\frac{dv}{dt} = \frac{\overline{W}\sigma}{m}\frac{1-\dfrac{v}{c}}{1+\dfrac{v}{c}}.$$

Integrating this expression, we find

$$\frac{\overline{W}\sigma}{mc}\,t = \frac{1}{3}\sqrt{\frac{1+\dfrac{v}{c}}{1-\dfrac{v}{c}}}\cdot\frac{2-\dfrac{v}{c}}{1-\dfrac{v}{c}} - \frac{2}{3},$$

which determines the velocity $v = dx/dt$ as an implicit function of the time (the integration constant has been chosen so that $v = 0$ at $t = 0$).

8. Determine the effective cross-section for scattering of a linearly polarized wave by an oscillator, taking into account the radiation damping.

Solution: We write the equation of motion of the charge in the incident field in the form

$$\ddot{\mathbf{r}} + \omega_0^2\mathbf{r} = \frac{e}{m}\,\mathbf{E}_0 e^{-i\omega t} + \frac{2e^2}{3mc^3}\,\dddot{\mathbf{r}}.$$

In the damping force, we can substitute approximately $\dddot{\mathbf{r}} = -\omega_0^2\dot{\mathbf{r}}$; then we find

$$\ddot{\mathbf{r}} + \gamma\dot{\mathbf{r}} + \omega_0^2\mathbf{r} = \frac{e}{m}\,\mathbf{E}_0 e^{-i\omega t},$$

where $\gamma = (2e^2/3mc^3)\,\omega_0^2$. From this we obtain

$$\mathbf{r} = \frac{e}{m}\,\mathbf{E}_0\frac{e^{-i\omega t}}{\omega_0^2 - \omega^2 - i\omega\gamma}.$$

The effective cross-section is

$$\sigma = \frac{8\pi}{3}\left(\frac{e^2}{mc^2}\right)^2\frac{\omega^4}{(\omega_0^2 - \omega^2)^2 + \omega^2\gamma^2}.$$

§ 79. Scattering of low-frequency waves

The scattering of a wave by a system of charges differs from the scattering by a single charge (at rest), first of all in the fact that because of the presence of internal motion of the charges of the system, the frequency of the scattered radiation can be different from the frequency of the incident wave. Namely, in the spectral resolution of the scattered wave there appear, in addition to the frequency ω of the incident wave, frequencies ω' differing from ω by one of the internal frequencies of motion of the scattering system. The scattering with changed frequency is called *incoherent* (or *combinational*), in contrast to the *coherent* scattering without change in frequency.

Assuming that the field of the incident wave is weak, we can represent the current density in the form $\mathbf{j} = \mathbf{j}_0 + \mathbf{j}'$, where $\mathbf{j}_0$ is the current density in the absence of the external field, and $\mathbf{j}'$ is the change in the current under the action of the incident wave. Correspondingly, the vector potential (and other quantities) of the field of the system also has the form $\mathbf{A} = \mathbf{A}_0 + \mathbf{A}'$, where $\mathbf{A}_0$ and $\mathbf{A}'$ are determined by the currents $\mathbf{j}_0$ and $\mathbf{j}'$. Clearly, $\mathbf{A}'$ describes the wave scattered by the system.

Let us consider the scattering of a wave whose frequency ω is small compared with all the internal frequencies of the system. The scattering will consist of an incoherent as well as a coherent part, but we shall here consider only the coherent scattering.

In calculating the field of the scattered wave, for sufficiently low frequency ω, we can use the expansion of the retarded potentials which was presented in §§ 67 and 71, even if the velocities of the particles of the system are not small compared with the velocity of light. Namely, for the validity of the expansion of the integral

$$\mathbf{A}' = \frac{1}{cR_0} \int \mathbf{j}'_{t - \frac{R_0}{c} + \frac{\mathbf{r} \cdot \mathbf{n}}{c}} \, dV,$$

it is necessary only that the time $\mathbf{r} \cdot \mathbf{n}'/c \sim a/c$ be small compared with the time $1/\omega$; for sufficiently low frequencies ($\omega \ll c/a$), this condition is fulfilled independently of the velocities of the particles of the system.

The first terms in the expansion give

$$\mathbf{H}' = \frac{1}{c^2 R_0} \{ \ddot{\mathbf{d}}' \times \mathbf{n}' + (\dddot{m}' \times \mathbf{n}') \times \mathbf{n}' \},$$

where $\mathbf{d}'$, m' are the parts of the dipole and magnetic moments of the system which are produced by the radiation falling on the system. The succeeding terms contain higher time derivatives than the second, and we drop them.

The component $\mathbf{H}'_\omega$ of the spectral resolution of the field of the scattered wave, with frequency equal to that of the incident wave, is given by this same formula, when we substitute for all quantities their Fourier components: $\ddot{\mathbf{d}}'_\omega = -\omega^2 \mathbf{d}'_\omega$, $\dddot{m}'_\omega = -\omega^2 m'_\omega$. Then we obtain

$$\mathbf{H}'_\omega = \frac{\omega^2}{c^2 R_0} \{ \mathbf{n}' + \mathbf{d}'_\omega + \mathbf{n}' \times (m'_\omega \times \mathbf{n}') \}. \tag{79.1}$$

The later terms in the expansion of the field would give quantities proportional to higher powers of the small frequency. If the velocities of all the particles of the system are small ($v \ll c$), then in (79.1) we can neglect the second term in comparison to the first, since the magnetic moment contains the ratio v/c. Then

$$\mathbf{H}'_\omega = \frac{1}{c^2 R_0} \omega^2 \mathbf{n}' \times \mathbf{d}'_\omega. \tag{79.2}$$

If the total charge of the system is zero, then for $\omega \to 0$, $\mathbf{d}'_\omega$ and m'_ω approach constant limits (if the sum of the charges were different from zero, then for $\omega = 0$, i.e. for a constant field, the system would begin to move as a whole). Therefore for low frequencies ($\omega \ll v/a$) we can consider $\mathbf{d}'_\omega$ and m'_ω as independent of frequency, so that the field of the scattered wave is proportional to the square of the frequency. Its intensity is consequently proportional to ω^4. Thus for the scattering of a low-frequency wave, the effective cross-section for (coherent) scattering is proportional to the fourth power of the frequency of the incident radiation.†

§ 80. Scattering of high-frequency waves

We consider the scattering of a wave by a system of charges in the opposite limit, when the frequency ω of the wave is large compared with the fundamental internal frequencies of the system. The latter have the order of magnitude $\omega_0 \sim v/a$, so that ω must satisfy the condition

$$\omega \gg \omega_0 \sim \frac{v}{a}. \tag{80.1}$$

In addition, we assume that the velocities of the charges of the system are small ($v \ll c$).

According to condition (80.1), the periods of the motion of the charges of the system are large compared with the period of the wave. Therefore during a time interval of the order of the period of the wave, the motion of the charges of the system can be considered uniform. This means that in considering the scattering of short waves, we need not take into account the interaction of the charges of the system with each other, that is, we can consider them as free.

Thus in calculating the velocity $\mathbf{v}'$, acquired by a charge in the field of the incident wave, we can consider each of the charges in the system separately, and write for it an equation of motion of the form

$$m \frac{d\mathbf{v}'}{dt} = e\mathbf{E} = e\mathbf{E}_0 e^{-i(\omega t - \mathbf{k} \cdot \mathbf{r})},$$

where $\mathbf{k} = (\omega/c)\mathbf{n}$ is the wave vector of the incident wave. The radius vector of the charge is, of course, a function of the time. In the exponent on the right side of this equation the time rate of change of the first term is large compared with that of the second (the first is ω, while the second is of order $kv \sim v(\omega/c) \ll \omega$). Therefore in integrating the equation of motion, we can consider the term $\mathbf{r}$ on the right side as constant. Then

$$\mathbf{v}' = -\frac{e}{i\omega m} \mathbf{E}_0 e^{-i(\omega t - \mathbf{k} \cdot \mathbf{r})} \tag{80.2}$$

For the vector potential of the scattered wave (at large distances from the system), we have from (79.1):

$$\mathbf{A}' = \frac{1}{cR_0} \Sigma \, (e\mathbf{v}')_{t - \frac{R_0}{c} + \frac{\mathbf{r} \cdot \mathbf{n}'}{c}} \,,$$

where the sum goes over all the charges of the system. Substituting (80.2), we find

---

† This also applies to the scattering of light by ions as well as by neutral atoms. Because of the large mass of the nucleus, the scattering resulting from the motion of the ion as a whole can be neglected.

$$\mathbf{A}' = -\frac{1}{icR_0\omega} e^{-i\omega\left(t-\frac{R_0}{c}\right)} \mathbf{E}_0 \sum \frac{e^2}{m} e^{-i\mathbf{q}\cdot\mathbf{r}},\tag{80.3}$$

where $\mathbf{q} = \mathbf{k}' - \mathbf{k}$ is the difference between the wave vector $\mathbf{k} = (\omega/c)\,\mathbf{n}$ of the incident wave, and the wave vector $\mathbf{k}' = (\omega/c)\mathbf{n}'$ of the scattered wave.† The value of the sum in (80.3) must be taken at the time $t' = t - (R_0/c)$ (for brevity as usual, we omit the index t' on $\mathbf{r}$); the change of $\mathbf{r}$ in the time $\mathbf{r}\cdot\mathbf{n}'/c$ can be neglected in view of our assumption that the velocities of the particles are small. The absolute value of the vector $\mathbf{q}$ is

$$q = 2\frac{\omega}{c}\sin\frac{\Theta}{2},\tag{80.4}$$

where Θ is the scattering angle.

For scattering by an atom (or molecule), we can neglect the terms in the sum in (80.3) which come from the nuclei, because their masses are large compared with the electron mass. Later we shall be looking at just this case, so that we remove the factor e^2/m from the summation sign, and understand by e and m the charge and mass of the electron.

For the field $\mathbf{H}'$ of the scattered wave we find from (66.3):

$$\mathbf{H}' = \frac{\mathbf{E}_0\times\mathbf{n}'}{c^2 R_0} e^{-i\omega\left(t-\frac{R_0}{c}\right)} \frac{e^2}{m} \sum e^{-i\mathbf{q}\cdot\mathbf{r}},\tag{80.5}$$

The energy flux into an element of solid angle in the direction $\mathbf{n}'$ is

$$\frac{c|\mathbf{H}'|^2}{8\pi} R_0^2\,do = \frac{e^4}{8\pi c^3 m^2}(\mathbf{n}'\times\mathbf{E}_0)^2 \left|\sum e^{-i\mathbf{q}\cdot\mathbf{r}}\right|^2 do.$$

Dividing this by the energy flux $c|\mathbf{E}_0|^2/8\pi$ of the incident wave, and introducing the angle θ between the direction of the field $\mathbf{E}$ of the incident wave and the direction of scattering, we finally obtain the effective scattering cross-section in the form

$$d\sigma = \left(\frac{e^2}{mc^2}\right)^2 \overline{\left|\sum e^{-i\mathbf{q}\cdot\mathbf{r}}\right|^2} \sin^2\theta\,do.\tag{80.6}$$

The dash means a time average, i.e. an average over the motion of the charges of the system; it appears because the scattering is observed over a time interval large compared with the periods of motion of the charges of the system.

For the wavelength of the incident radiation, there follows from the condition (80.1) the inequality $\lambda \ll ac/v$. As for the relative values of λ and a, both the limiting cases $\lambda \gg a$ and $\lambda \ll a$ are possible. In both these cases the general formula (80.6) simplifies considerably.

In the case of $\lambda \gg a$, in the expression (80.6) $\mathbf{q}\cdot\mathbf{r} \ll 1$, since $q \sim 1/\lambda$, and r is of order of a. Replacing $e^{-i\mathbf{q}\cdot\mathbf{r}}$ by unity in accordance with this, we have:

$$d\sigma = \left(\frac{Ze^2}{mc^2}\right)^2 \sin^2\theta\,do.\tag{80.7}$$

that is, the scattering is proportional to the square of the atomic number Z.

---

† Strictly speaking, the wave vector $\mathbf{k}' = \omega'\mathbf{n}'/c$, where the frequency ω' of the scattered wave may differ from ω. However, in the present case of high frequencies the difference $\omega' - \omega \sim \omega_0$ can be neglected.

We now go over to the case of $\lambda \ll a$. In the square of the sum which appears in (80.6), in addition to the square modulus of each term, there appear products of the form $e^{-i\mathbf{q} \cdot (\mathbf{r}_1 - \mathbf{r}_2)}$.

In averaging over the motion of the charges, i.e. over their mutual separations, $\mathbf{r}_1 - \mathbf{r}_2$ takes on values in an interval of order a. Since $q \sim 1/\lambda$, $\lambda \ll a$, the exponential factor $e^{-i\mathbf{q} \cdot (\mathbf{r}_1 - \mathbf{r}_2)}$ is a rapidly oscillating function in this interval, and its average value vanishes. Thus for $\lambda \ll a$, the effective scattering cross-section is

$$d\sigma = Z \left(\frac{e^2}{mc^2} \right)^2 \sin^2 \theta \, do, \tag{80.8}$$

that is, the scattering is proportional to the first power of the atomic number. We note that this formula is not applicable for small angles of scattering ($\Theta \sim \lambda/a$), since in this case $q \sim \Theta/\lambda \sim 1/a$ and the exponent $\mathbf{q} \cdot \mathbf{r}$ is not large compared to unity.

To determine the effective coherent scattering cross-section, we must separate out that part of the field of the scattered wave which has the frequency ω. The expression (80.5) depends on the time through the factor $e^{-i\omega t}$, and also involves the time in the sum $\sum e^{-i\mathbf{q} \cdot \mathbf{r}}$. This latter dependence leads to the result that in the field of the scattered wave there are contained, along with the frequency ω, other (though close to ω) frequencies. That part of the field which has the frequency ω (i.e. depends on the time only through the factor $e^{-i\omega t}$), is obtained if we average the sum $\sum e^{-i\mathbf{q} \cdot \mathbf{r}}$ over time. In accordance with this, the expression for the effective coherent scattering cross-section $d\sigma_{\text{coh}}$, differs from the total cross-section $d\sigma$ in that it contains, in place of the average value of the square modulus of the sum, the square modulus of the average value of the sum,

$$d\sigma_{\text{coh}} = \left(\frac{e^2}{mc^2} \right)^2 \left| \overline{\sum e^{-i\mathbf{q} \cdot \mathbf{r}}} \right|^2 \sin^2 \theta \, do. \tag{80.9}$$

It is useful to note that this average value of the sum is (except for a factor) just the space Fourier component of the average distribution $\varrho(\mathbf{r})$ of the electric charge density in the atom:

$$e \, \overline{\sum e^{-i\mathbf{q} \cdot \mathbf{r}}} = \int \varrho(\mathbf{r}) e^{-i\mathbf{q} \cdot \mathbf{r}} \, dV = \varrho_{\mathbf{q}}. \tag{80.10}$$

In case $\lambda \gg a$, we can again replace $e^{i\mathbf{q} \cdot \mathbf{r}}$ by unity, so that

$$d\sigma_{\text{coh}} = \left(Z \frac{e^2}{mc^2} \right)^2 \sin^2 \theta \, do. \tag{80.11}$$

Comparing this with the total effective cross-sction (80.7), we see that $d\sigma_{\text{coh}} = d\sigma$, that is, all the scattering is coherent.

If $\lambda \ll a$, then when we average in (80.9) all the terms of the sum (being rapidly oscillating functions of the time) vanish, so that $d\sigma_{\text{coh}} = 0$. Thus in this case the scattering is completely incoherent.

CHAPTER 10

PARTICLE IN A GRAVITATIONAL FIELD

§ 81. Gravitational fields in nonrelativistic mechanics

Gravitational fields, or fields of gravity, have the basic property that all bodies move in them in the same manner, independently of mass, provided the initial conditions are the same.

For example, the laws of free fall in the gravity field of the earth are the same for all bodies; whatever their mass, all acquire one and the same acceleration.

This property of gravitational fields provides the possibility of establishing an analogy between the motion of a body in a gravitational field and the motion of a body not located in any external field, but which is considered from the point of view of a noninertial system of reference. Namely, in an inertial reference system, the free motion of all bodies is uniform and rectilinear, and if, say, at the initial time their velocities are the same, they will be the same for all times. Clearly, therefore, if we consider this motion in a given noninertial system, then relative to this system all the bodies will move in the same way.

Thus the properties of the motion in a noninertial system are the same as those in an inertial system in the presence of a gravitational field. In other words, a noninertial reference system is equivalent to a certain gravitational field. This is called the *principle of equivalence.*

Let us consider, for example, motion in a uniformly accelerated reference system. A body of arbitrary mass, freely moving in such a system of reference, clearly has relative to this system a constant acceleration, equal and opposite to the acceleration of the system itself. The same applies to motion in a uniform constant gravitational field, e.g. the field of gravity of the earth (over small regions, where the field can be considered uniform). Thus a uniformly accelerated system of reference is equivalent to a constant, uniform external field. In the same way, nonuniformity accelerated linear motion of the reference system is clearly equivalent to a uniform but gravitational field.

However, the fields to which noninertial reference systems are equivalent are not completely identical with "actual" gravitational fields which occur also in inertial frames. For there is a very essential difference with respect to their behaviour at infinity. At infinite distances from the bodies producing the field, "actual" gravitational fields always go to zero. Contrary to this, the fields to which noninertial frames are equivalent increase without limit at infinity, or, in any event, remain finite in value. Thus, for example, the centrifugal force which appears in a rotating reference system increases without limit as we move away from the axis of rotation; the field to which a reference system in accelerated linear motion is equivalent is the same over all space and also at infinity.

The fields to which noninertial systems are equivalent vanish as soon as we transform to an inertial system. In contrast to this, "actual" gravitational fields (existing also in an inertial reference frame) cannot be eliminated by any choice of reference system. This is already clear from what has been said above concerning the difference in conditions at infinity between "actual" gravitational fields and fields to which noninertial systems are equivalent; since the latter do not approach zero at infinity, it is clear that it is impossible, by any choice of reference frame, to eliminate an "actual" field, since it vanishes at infinity.

All that can be done by a suitable choice of reference system is to eliminate the gravitational field in a given region of space, sufficiently small so that the field can be considered uniform over it. This can be done by choosing a system in accelerated motion, the acceleration of which is equal to that which would be acquired by a particle placed in the region of the field which we are considering.

The motion of a particle in a gravitational field is determined, in nonrelativistic mechanics, by a Lagrangian having (in an inertial reference frame) the form

$$L = \frac{mv^2}{2} - m\phi,$$ (81.1)

where ϕ is a certain function of the coordinates and time which characterizes the field and is called the *gravitational potential*.† Correspondingly, the equation of motion of the particle is

$$\dot{\mathbf{v}} = -\text{grad } \phi.$$ (81.2)

It does not contain the mass or any other constant characterizing the properties of the particle; this is the mathematical expression of the basic property of gravitational fields.

§ 82. The gravitational field in relativistic mechanics

The fundamental property of gravitational fields that all bodies move in them in the same way, remains valid also in relativistic mechanics. Consequently there remains also the analogy between gravitational fields and noninertial reference systems. Therefore in studying the properties of gravitational fields in relativistic mechanics, we naturally also start from this analogy.

In an inertial reference system, in cartesian coordinates, the interval ds is given by the relation:

$$ds^2 = c^2 dt^2 - dx^2 - dy^2 - dz^2.$$

Upon transforming to any other inertial reference system (i.e. under Lorentz transformation), the interval, as we know, retains the same form. However, if we transform to a noninertial system of reference ds^2 will no longer be a sum of squares of the four coordinate differentials.

So, for example, when we transform to a uniformly rotating system of coordinates,

$$x = x' \cos \Omega t - y' \sin \Omega t, \quad y = x' \sin \Omega t + y' \cos \Omega t, \quad z = z'$$

(Ω is the angular velocity of the rotation, directed along the Z axis), the interval takes on the form

$$ds^2 = [c^2 - \Omega^2(x'^2 + y'^2)]\, dt^2 - dx'^2 - dy'^2 - dz'^2 + 2\Omega y'\, dx'\, dt - 2\Omega x'\, dy'\, dt.$$

---

† In what follows we shall seldom have to use the electromagnetic potential ϕ, so that the designation of the gravitational potential by the same symbol cannot lead to misunderstanding.

No matter what the law of transformation of the time coordinate, this expression cannot be represented as a sum of squares of the coordinate differentials.

Thus in a noninertial system of reference the square of an interval appears as a quadratic form of general type in the coordinate differentials, that is, it has the form

$$ds^2 = g_{ik}\, dx^i dx^k, \tag{82.1}$$

where the g_{ik} are certain functions of the space coordinates x^1, x^2, x^3 and the time coordinate x^0. Thus, when we use a noninertial system, the four-dimensional coordinate system x^0, x^1, x^2, x^3 is a curvilinear. The quantities g_{ik}, determining all the geometric properties in each curvilinear system of coordinates, represent, we say, the *space-time metric*.

The quantities g_{ik} can clearly always be considered symmetric in the indices i and k ($g_{ki} = g_{ik}$), since they are determined from the symmetric form (82.1), where g_{ik} and g_{ki} enter as factors of one and the same product $dx^i\, dx^k$. In the general case, there are ten different quantities g_{ik} —four with equal, and 4.3/2 = 6 with different indices. In an inertial reference system, when we use cartesian space coordinates $x^{1,\,2,\,3} = x, y, z$, and the time, $x^0 = ct$, the quantities g_{ik} are

$$g_{00} = 1, \quad g_{11} = g_{22} = g_{33} = -1, \quad g_{ik} = 0 \quad \text{for} \quad i \neq k. \tag{82.2}$$

We call a four-dimensional system of coordinates with these values of g_{ik} *galilean*.

In the previous section it was shown that a noninertial system of reference is equivalent to a certain field of force. We now see that in relativistic mechanics, these fields are determined by the quantities g_{ik}.

The same applies also to "actual" gravitational fields. Any gravitational field is just a change in the metric of space-time, as determined by the quantities g_{ik}. This important fact means that the geometrical properties of space-time (its metric) are determined by physical phenomena, and are not fixed properties of space and time.

The theory of gravitational fields, constructed on the basis of the theory of relativity, is called the *general theory of relativity*. It was established by Einstein (and finally formulated by him in 1915), and represents probably the most beautiful of all existing physical theories. It is remarkable that it was developed by Einstein in a purely deductive manner and only later was substantiated by astronomical observations.

As in nonrelativistic mechanics, there is a fundamental difference between "actual" gravitational fields and fields to which noninertial reference systems are equivalent. Upon transforming to a noninertial reference system, the quadratic form (82.1), i.e. the quantities g_{ik}, are obtained from their galilean values (82.2) by a simple transformation of coordinates, and can be reduced over all space to their galilean values by the inverse coordinate transformation. That such forms for g_{ik} are very special is clear from the fact that it is impossible by a mere transformation of the four coordinates to bring the ten quantities g_{ik} to a preassigned form.

An "actual" gravitational field cannot be eliminated by any transformation of coordinates. In other words, in the presence of a gravitational field space-time is such that the quantities g_{ik} determining its metric cannot, by any coordinate transformation, be brought to their galilean values over all space. Such a space-time is said to be *curved*, in contrast to *flat* space-time, where such a reduction is possible.

By an appropriate choice of coordinates, we can, however, bring the quantities g_{ik} to galilean form at any individual point of the non-galilean space-time: this amounts to the reduction to diagonal form of a quadratic form with constant coefficients (the values of g_{ik} at the given point). Such a coordinate system is said to be *galilean for the given point.*†

We note that, after reduction to diagonal form at a given point, the matrix of the quantities g_{ik} has one positive and three negative principal values.‡ From this it follows, in particular, that the determinant g, formed from the quantities g_{ik}, is always negative for a real space-time:

$$g < 0 . \tag{82.3}$$

A change in the metric of space-time also means a change in the purely spatial metric. To a galilean g_{ik} in flat space-time, there corresponds a euclidean geometry of space. In a gravitational field, the geometry of space becomes non-euclidean. This applies both to "true" gravitational fields, in which space-time is "curved", as well as to fields resulting from the fact that the reference system is non-inertial, which leave the space-time flat.

The problem of spatial geometry in a gravitational field will be considered in more detail in § 84. It is useful to give here a simple argument which shows pictorially that space will become non-euclidean when we change to a non-inertial system of reference. Let us consider two reference frames, of which one (K) is inertial, while the other (K') rotates uniformly with respect to K around their common z axis. A circle in the x, y plane of the K system (with its centre at the origin) can also be regarded as a circle in the x', y' plane of the K' system. Measuring the length of the circle and its diameter with a yardstick in the K system, we obtain values whose ratio is equal to π, in accordance with the euclidean character of the geometry in the inertial reference system. Now let the measurement be carried out with a yardstick at rest relative to K'. Observing this process from the K system, we find that the yardstick laid along the circumference suffers a Lorentz contraction, whereas the yardstick placed radially is not changed. It is therefore clear that the ratio of the circumference to the diameter, obtained from such a measurement, will be greater than π.

In the general case of an arbitrary, varying gravitational field, the metric of space is not only non-euclidean, but also varies with the time. This means that the relations between different geometrical distances change with time. As a result, the relative position of "test bodies" introduced into the field cannot remain unchanged in any coordinate system.§ Thus if the particles are placed around the circumference of a circle and along a diameter, since the ratio of the circumference to the diameter is not equal to π and changes with time, it is clear that if the separations of the particles along the diameter remain unchanged the separations around the circumference must change, and conversely. Thus in the general theory of relativity it is impossible in general to have a system of bodies which are fixed relative to one another.

This result essentially changes the very concept of a system of reference in the general theory of relativity, as compared to its meaning in the special theory. In the latter we meant

---

† To avoid misunderstanding, we state immediately that the choice of such a coordinate system does not mean that the gravitational field has been eliminated over the corresponding infinitesimal volume of four-space. Such an elimination is also always possible, by virtue of the principle of equivalence, and has a greater significance (see § 87).

‡ This set of signs is called the *signature* of the matrix.

§ Strictly speaking, the number of particles should be greater than four. Since we can construct a tetrahedron from any six line segments, we can always, by a suitable definition of the reference system, make a system of four particles form an invariant tetrahedron. *A fortiori,* we can fix the particles relative to one another in system of three or two particles.

by a reference system a set of bodies at rest relative to one another in unchanging relative positions. Such systems of bodies do not exist in the presence of a variable gravitational field, and for the exact determination of the position of a particle in space we must, strictly speaking, have an infinite number of bodies which fill all the space like some sort of "medium". Such a system of bodies with arbitrarily running clocks fixed on them constitutes a reference system in the general theory of relativity.

In connection with the arbitrariness of the choice of a reference system, the laws of nature must be written in the general theory of relativity in a form which is appropriate to any four-dimensional system of coordinates (or, as one says, in "*covariant*" form). This, of course, does not imply the physical equivalence of all these reference systems (like the physical equivalence of all inertial reference systems in the special theory). On the contrary, the specific appearances of physical phenomena, including the properties of the motion of bodies, become different in all systems of reference.

§ 83. Curvilinear coordinates

Since, in studying gravitational fields we are confronted with the necessity of considering phenomena in an arbitrary reference frame, it is necessary to develop four-dimensional geometry in arbitrary curvilinear coordinates. Sections 83, 85 and 86 are devoted to this.

Let us consider the transformation from one coordinate system, x^0, x^1, x^2, x^3, to another x'^0, x'^1, x'^2, x'^3:

$$x^i = f^i(x'^0, x'^1, x'^2, x'^3),$$

where the f^i are certain functions. When we transform the coordinates, their differentials transform according to the relation

$$dx^i = \frac{\partial x^i}{\partial x'^k} dx'^k.$$ (83.1)

Every aggregate of four quantities A^i ($i = 0, 1, 2, 3$), which under a transformation of coordinates transform like the coordinate differentials, is called a *contravariant* four-vector:

$$A^i = \frac{\partial x^i}{\partial x'^k} A'^k.$$ (83.2)

Let ϕ be some scalar. Under a coordinate transformation, the four quantities $\partial \phi / \partial x^i$ transform according to the formula

$$\frac{\partial \phi}{\partial x^i} = \frac{\partial \phi}{\partial x'^k} \frac{\partial x'^k}{\partial x^i},$$ (83.3)

which is different from formula (83.2). Every aggregate of four quantities A_i which, under a coordinate transformation, transform like the derivatives of a scalar, is called a *covariant* four-vector:

$$A_i = \frac{\partial x'^k}{\partial x^i} A'_k.$$ (83.4)

Because two types of vectors appear in curvilinear coordinates, there are three types of tensors of the second rank. We call a *contravariant tensor* of the second rank, A^{ik}, an aggregate of sixteen quantities which transform like the products of the components of two contravariant vectors, i.e. according to the law

$$A^{ik} = \frac{\partial x^i}{\partial x'^l} \frac{\partial x^k}{\partial x'^m} A'^{lm}. \tag{83.5}$$

A *covariant tensor* of rank two, transforms according to the formula

$$A_{ik} = \frac{\partial x'^l}{\partial x^i} \frac{\partial x'^m}{\partial x^k} A'_{lm}, \tag{83.6}$$

and a *mixed tensor* transforms as follows:

$$A^i_k = \frac{\partial x^i}{\partial x'^l} \frac{\partial x'^m}{\partial x^k} A'^l_m. \tag{83.7}$$

The definitions given here are the natural generalization of the definitions of four-vectors and four-tensors in galilean coordinates (§ 6), according to which the differentials dx^i constitute a contravariant four-vector and the derivatives $\partial\phi/\partial x^i$ form a covariant four-vector.†

The rules for forming four-tensor by multiplication or contraction of products of other four-tensors remain the same in curvilinear coordinates as they were in galilean coordinates. For example, it is easy to see that, by virtue of the transformation laws (83.2) and (83.4), the scalar product of two four-vectors $A^i B_i$ is invariant:

$$A^i B_i = \frac{\partial x^i}{\partial x'^l} \frac{\partial x'^m}{\partial x^i} A'^l B'_m = \frac{\partial x'^m}{\partial x'^l} A'^l B'_m = A'^l B'_l.$$

The unit four-tensor δ^i_k is defined the same as before in curvilinear coordinates: its components are again $\delta^i_k = 0$ for $i \neq k$, and are equal to 1 for $i = k$. If A^k is a four-vector, then multiplying by δ^i_k we get:

$$A^k \delta^i_k = A^i,$$

i.e. another four-vector; this proves that δ^i_k is a tensor.

The square of the line element in curvilinear coordinates is a quadratic form in the differentials dx^i:

$$ds^2 = g_{ik} dx^i dx^k, \tag{83.8}$$

where the g_{ik} are functions of the coordinates; g_{ik} is symmetric in the indices i and k:

$$g_{ik} = g_{ki}. \tag{83.9}$$

Since the (contracted) product of g_{ik} and the contravariant tensor $dx^i dx^k$ is a scalar, the g_{ik} form a covariant tensor; it is called the *metric tensor*.

Two tensors A_{ik} and B^{ik} are said to be *reciprocal* to each other if

$$A_{ik} B^{kl} = \delta^l_i.$$

In particular the contravariant metric tensor is the tensor g^{ik} reciprocal to the tensor g_{ik}, that is,

† Nevertheless, while in a galilean system the coordinates x^i themselves (and not just their differentials) also form a four-vector, this is, of course, not the case in curvilinear coordinates.

$$g_{ik}g^{kl} = \delta_i^l.$$ (83.10)

The same physical quantity can be represented in contra- or co-variant components. It is obvious that the only quantities that can determine the connection between the different forms are the components of the metric tensor. This connection is given by the formulas:

$$A^i = g^{ik}A_k, \quad A_i = g_{ik}A^k.$$ (83.11)

In a galilean coordinate system the metric tensor has components:

$$g_{ik}^{(0)} = g^{ik(0)} = \begin{pmatrix} 1 & 0 & 0 & 0 \\ 0 & -1 & 0 & 0 \\ 0 & 0 & -1 & 0 \\ 0 & 0 & 0 & -1 \end{pmatrix}.$$ (83.12)

Then formula (83.11) gives the familiar relation $A^0 = A_0$, $A^{1,2,3} = -A_{1,2,3}$, etc.†

These remarks also apply to tensors. The transition between the different forms of a given physical tensor is accomplished by using the metric tensor according to the formulas:

$$A^i{}_k = g^{il}A_{ik}, \quad A^{ik} = g^{il}g^{km}A_{lm},$$

etc.

In § 6 we defined (in galilean coordinates) the completely antisymmetric unit pseudo-tensor e^{iklm}. Let us transform it to an arbitrary system of coordinates, and now denote it by E^{iklm}. We keep the notation e^{iklm} for the quantities defined as before by $e^{0123} = 1$ (or $e_{0123} = -1$).

Let the x'^i be galilean, and the x^i be arbitrary curvilinear coordinates. According to the general rules for transformation of tensors, we have:

$$E^{iklm} = \frac{\partial x^i}{\partial x'^p} \frac{\partial x^k}{\partial x'^r} \frac{\partial x^l}{\partial x'^s} \frac{\partial x^m}{\partial x'^t} e^{prst},$$

or

$$E^{iklm} = J e^{iklm},$$

where J is the determinant formed from the derivatives $\partial x^i/\partial x'^p$, i.e. it is just the Jacobian of the transformation from the galilean to the curvilinear coordinates:

$$J = \frac{\partial(x^0, x^1, x^2, x^3)}{\partial(x'^0, x'^1, x'^2, x'^3)}.$$

This Jacobian can be expressed in terms of the determinant of the metric tensor g_{ik} (in the system x^i). To do this we write the formula for the transformation of the metric tensor:

$$g^{ik} = \frac{\partial x^i}{\partial x'^l} \frac{\partial x^k}{\partial x'^m} g^{lm(0)},$$

and equate the determinants of the two sides of this equation. The determinant of the

---

† Whenever, in giving analogies, we use galilean coordinate systems, one should realize that such a system can be selected only in a flat space. In the case of a curved four-space, one should speak of a coordinate system that is galilean over a given infinitesimal element of four-volume, which can always be found. None of the derivations are affected by this change.

reciprocal tensor $|g^{ik}| = 1/g$. The determinant $|g^{lm(0)}| = -1$. Thus we have $1/g = -J^2$, and so $J = 1/\sqrt{-g}$.

Thus, in curvilinear coordinates the antisymmetric unit tensor of rank four must be defined as

$$E^{iklm} = \frac{1}{\sqrt{-g}} e^{iklm}. \qquad (83.13)$$

The indices of this tensor are lowered by using the formula

$$e^{prst} g_{ip} g_{kr} g_{ls} g_{mt} = -g\, e_{iklm},$$

so that its covariant components are

$$E_{iklm} = \sqrt{-g}\, e_{iklm}. \qquad (83.14)$$

In a galilean coordinate system x'^i the integral of a scalar with respect to $d\Omega' = dx'^0 dx'^1 dx'^2 dx'^3$ is also a scalar, i.e. the element $d\Omega'$ behaves like a scalar in the integration (§ 6). On transforming to curvilinear coordinates x^i, the element of integration $d\Omega'$ goes over into

$$d\Omega' \to \frac{1}{J} d\Omega = \sqrt{-g}\, d\Omega.$$

Thus, in curvilinear coordinates, when integrating over a four-volume the quantity $\sqrt{-g}\, d\Omega$ behaves like an invariant.†

All the remarks at the end of § 6 concerning elements of integration over hypersurfaces, surfaces and lines remain valid for curvilinear coordinates, with the one difference that the definition of dual tensors changes. The element of "area" of the hypersurface spanned by three infinitesimal displacements is the contravariant antisymmetric tensor dS^{ikl}; the vector dual to it is gotten by multiplying by the tensor $\sqrt{-g}\, e_{iklm}$, so it is equal to

$$\sqrt{-g}\, dS_i = -\tfrac{1}{6} e_{iklm} dS^{klm} \sqrt{-g}. \qquad (83.15)$$

Similarly, if df^{ik} is the element of (two-dimensional) surface spanned by two infinitesimal displacements, the dual tensor is defined as‡

$$\sqrt{-g}\, df_{ik}^* = \tfrac{1}{2}\sqrt{-g}\, e_{iklm} df^{lm}. \qquad (83.16)$$

We keep the designations dS_i and df_{ki}^* as before for $\tfrac{1}{6} e_{iklm} dS^{klm}$ and $\tfrac{1}{2} e_{iklm} df^{lm}$ (and not

† If ϕ is a scalar, the quantity $\sqrt{-g}\,\phi$, which gives an invariant when integrated over $d\Omega$, is called a *scalar density*. Similarly, we speak of *vector* and *tensor densities* $\sqrt{-g}\, A^i$, $\sqrt{-g}\, A^{ik}$, etc. These quantities give a vector or tensor on multiplication by the four-volume element $d\Omega$ (the integral $\int A^i \sqrt{-g}\, d\Omega$ over a finite region cannot, generally speaking, be a vector, since the laws of transformation of the vector A^i are different at different points).

‡ It is understood that the elements dS^{klm} and df^{ik} are constructed on the infinitesimal displacements dx^i, dx'^i, dx''^i in the same way as in § 6, no matter what the geometrical significance of the coordinates x^i. Then the formal significance of the elements dS_i and df_{ik}^* is the same as before. In particular, as before $dS_0 = dx_1 dx_2 dx_3 \equiv dV$. We keep the earlier definition of dV for the product of differentials of the three space coordinates; we must, however, remember that the element of geometrical spatial volume is given in curvilinear coordinates not by dV, but by $\sqrt{\gamma}\, dV$, where γ is the determinant of the spatial metric tensor (which will be defined in the next section).

their products by $\sqrt{-g}$); the rules (6.14)–(6.19) for transforming the various integrals into one another remain the same, since their derivation was formal in character and not related to the tensor properties of the different quantities. Of particular importance is the rule for transforming the integral over a hypersurface into an integral over a four-volume (Gauss' theorem), which is accomplished by the substitution:

$$dS_i \to d\Omega \frac{\partial}{\partial x^i}. \tag{83.17}$$

§ 84. Distances and time intervals

We have already said that in the general theory of relativity the choice of a coordinate system is not limited in any way; the triplet of space coordinates x^1, x^2, x^3, can be any quantities defining the position of bodies in space, and the time coordinate x^0 can be defined by an arbitrarily running clock. The question arises of how, in terms of the values of the quantities x^1, x^2, x^3, x^0, we can determine actual distances and time intervals.

First we find the relation of the proper time, which from now on we shall denote by τ, to the coordinate x^0. To do this we consider two infinitesimally separated events, occurring at one and the same point in space. Then the interval ds between the two events is, as we know, just $c\,d\tau$, where $d\tau$ is the (proper) time interval between the two events. Setting $dx^1 = dx^2 = dx^3 = 0$ in the general expression $ds^2 = g_{ik}\,dx^i\,dx^k$, we consequently find

$$ds^2 = c^2\,d\tau^2 = g_{00}(dx^0)^2,$$

from which

$$d\tau = \frac{1}{c}\sqrt{g_{00}}\,dx^0, \tag{84.1}$$

or else, for the time between any two events occurring at the same point in space.

$$\tau = \frac{1}{c}\int \sqrt{g_{00}}\,dx^0. \tag{84.2}$$

This relation determines the actual time interval (or as it is also called, the *proper time* for the given point in space) for a change of the coordinate x^0. We note in passing that the quantity g_{00}, as we see from these formulas, is positive:

$$g_{00} > 0. \tag{84.3}$$

It is necessary to emphasize the difference between the meaning of (84.3) and the meaning of the signature [the signs of three principal values of the tensor g_{ik} (§ 82)]. A tensor g_{ik} which does not satisfy the second of these conditions cannot correspond to any real gravitational field, i.e. cannot be the metric of a real space-time. Nonfulfilment of the condition (84.3) would mean only that the corresponding system of reference cannot be realized with real bodies; if the condition on the principal values is fulfilled, then a suitable transformation of the coordinates can make g_{00} positive (an example of such a system is given by the rotating system of coordinates, see § 89).

We now determine the element dl of *spatial distance*. In the special theory of relativity we can define dl as the interval between two infinitesimally separated events occurring at one and the same time. In the general theory of relativity, it is usually impossible to do this, i.e. it is impossible to determine dl by simply setting $dx^0 = 0$ in ds. This is related to the fact that in a gravitational field the proper time at different points in space has a different dependence on the coordinate x^0.

To find dl, we now proceed as follows.

Suppose a light signal is directed from some point B in space (with coordinates $x^\alpha + dx^\alpha$) to a point A infinitely near to it (and having coordinates x^α) and then back over the same path. Obviously, the time (as observed from the one point B) required for this, when multiplied by c, is twice the distance between the two points.

Let us write the interval, separating the space and time coordinates:

$$ds^2 = g_{\alpha\beta}\, dx^\alpha\, dx^\beta + 2g_{0\alpha}\, dx^0\, dx^\alpha + g_{00}\, (dx^0)^2 \tag{84.4}$$

where it is understood that we sum over repeated Greek indices from 1 to 3. The interval between the events corresponding to the departure and arrival of the signal from one point to the other is equal to zero. Solving the equation $ds^2 = 0$ with respect to dx^0, we find two roots:

$$dx^{0(1)} = \frac{1}{g_{00}} \left\{ -g_{0\alpha}\, dx^\alpha - \sqrt{(g_{0\alpha}g_{0\beta} - g_{\alpha\beta}g_{00})\, dx^\alpha dx^\beta} \right\},$$

$$dx^{0(2)} = \frac{1}{g_{00}} \left\{ -g_{0\alpha}\, dx^\alpha + \sqrt{(g_{0\alpha}g_{0\beta} - g_{\alpha\beta}g_{00})\, dx^\alpha dx^\beta} \right\}, \tag{84.5}$$

corresponding to the propagation of the signal in the two directions between A and B. If x^0 is the moment of arrival of the signal at A, the times when it left B and when it will return to B are, respectively, $x^0 + dx_0^{(1)}$ and $x^0 + dx_0^{(2)}$. In the schematic diagram of Fig. 18 the solid lines are the world lines corresponding to the given coordinates x^α and $x^\alpha + dx^\alpha$, while the dashed lines are the world lines of the signals.† It is clear that the total interval of "time" between the departure of the signal and its return to the original point is equal to

$$dx^{0(2)} - dx^{0(1)} = \frac{2}{g_{00}} \sqrt{(g_{0\alpha}g_{0\beta} - g_{\alpha\beta}g_{00})\, dx^\alpha dx^\beta}.$$

The corresponding interval of proper time is obtained, according to (84.1), by multiplying by $\sqrt{g_{00}}/c$, and the distance dl between the two points by multiplying once more by $c/2$. As a result, we obtain

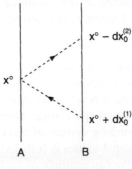

FIG. 18.

† In Fig. 18, it is assumed that $dx_0^{(2)} > 0$, $dx_0^{(1)} < 0$, but this is not necessary: $dx^{0(1)}$ and $dx^{0(2)}$ may have the same sign. The fact that in this case the value x^0(A) at the moment of arrival of the signal at A might be less than the value x^0(B) at the moment of its departure from B contains no contradiction, since the rates of clocks at different points in space are not assumed to be synchronized in any way.

$$dl^2 = \left(-g_{\alpha\beta} + \frac{g_{0\alpha}g_{0\beta}}{g_{00}}\right)dx^{\alpha}dx^{\beta}.$$

This is the required expression, defining the distance in terms of the space coordinate elements. We rewrite it in the form

$$dl^2 = \gamma_{\alpha\beta}\,dx^{\alpha}dx^{\beta}, \tag{84.6}$$

where

$$\gamma_{\alpha\beta} = \left(-g_{\alpha\beta} + \frac{g_{0\alpha}g_{0\beta}}{g_{00}}\right) \tag{84.7}$$

is the three-dimensional metric tensor, determining the metric, i.e., the geometric properties of the space. The relations (84.7) give the connection between the metric of real space and the metric of the four-dimensional space-time.†

However, we must remember that the g_{ik} generally depend on x^0, so that the space metric (84.6) also changes with time. For this reason, it is meaningless to integrate dl; such an integral would depend on the world line chosen between the two given space points. Thus, generally speaking, in the general theory of relativity the concept of a definite distance between bodies loses its meaning, remaining valid only for infinitesimal distances. The only case where the distance can be defined also over a finite domain is that in which the g_{ik} do not depend on the time, so that the integral $\int dl$ along a space curve has a definite meaning.

It is worth noting that the tensor $-\gamma_{\alpha\beta}$ is the reciprocal of the contravariant three-dimensional tensor $g^{\alpha\beta}$. In fact, from $g^{ik}g_{kl} = \delta^i_l$, we have, in particular,

$$g^{\alpha\beta}g_{\beta\gamma} + g^{\alpha 0}g_{0\gamma} = \delta^\alpha_\gamma, \quad g^{\alpha\beta}g_{\beta 0} + g^{\alpha 0}g_{00} = 0, \quad g^{0\beta}g_{\beta 0} + g^{00}g_{00} = 1. \tag{84.8}$$

Determining $g^{\alpha 0}$ from the second equation and substituting in the first, we obtain:

$$-g^{\alpha\beta}\gamma_{\beta\gamma} = \delta^\alpha_\gamma.$$

This result can be formulated differently, by the statement that the quantities $-g^{\alpha\beta}$ form the contravariant three-dimensional metric tensor corresponding to the metric (84.6):

$$\gamma^{\alpha\beta} = -g^{\alpha\beta}. \tag{84.9}$$

† The quadratic form (84.6) must clearly be positive definite. For this, its coefficients must, as we know from the theory of forms, satisfy the conditions

$$\gamma_{11} > 0, \quad \begin{vmatrix} \gamma_{11} & \gamma_{12} \\ \gamma_{21} & \gamma_{22} \end{vmatrix} > 0, \quad \begin{vmatrix} \gamma_{11} & \gamma_{12} & \gamma_{13} \\ \gamma_{21} & \gamma_{22} & \gamma_{33} \\ \gamma_{31} & \gamma_{32} & \gamma_{33} \end{vmatrix} > 0.$$

Expressing g_{ik} in terms of g_{ik}, it is easy to show that these conditions take the form

$$\begin{vmatrix} g_{00} & g_{01} \\ g_{10} & g_{11} \end{vmatrix} < 0, \quad \begin{vmatrix} g_{00} & g_{01} & g_{02} \\ g_{10} & g_{11} & g_{12} \\ g_{20} & g_{21} & g_{22} \end{vmatrix} > 0, \quad g < 0.$$

These conditions, together with the condition (84.3), must be satisfied by the components of the metric tensor in every system of reference which can be realized with the aid of real bodies.

We also state that the determinants g and γ, formed respectively from the quantities g_{ik} and $\gamma_{\alpha\beta}$, are related to one another by

$$-g = g_{00}\gamma. \tag{84.10}$$

In some of the later applications it will be convenient to introduce the three-dimensional vector **g**, whose covariant components are defined as

$$g_\alpha = -\frac{g_{0\alpha}}{g_{00}}. \tag{84.11}$$

Considering **g** as a vector in the space with metric (84.6), we must define its contravariant components as $g^\alpha = \gamma^{\alpha\beta}g_\beta$. Using (84.9) and the second of equations (84.8), it is easy to see that

$$g^\alpha = \gamma^{\alpha\beta}g_\beta = -g^{0\alpha}. \tag{84.12}$$

We also note the formula

$$g^{00} = \frac{1}{g_{00}} - g_\alpha g^\alpha, \tag{84.13}$$

which follows from the third of equations (84.8).

We now turn to the definition of the concept of simultaneity in the general theory of relativity. In other words, we discuss the question of the possibility of synchronizing clocks located at different points in space, i.e. the setting up of a correspondence between the readings of these clocks.

Such a synchronization must obviously be achieved by means of an exchange of light signals between the two points. We again consider the process of propagation of signals between two infinitely near points A and B, as shown in Fig. 18. We should regard as simultaneous with the moment x^0 at the point A that reading of the clock at point B which is half-way between the moments of departure and return of the signal to that point, i.e. the moment

$$x^0 + \Delta x^0 = x^0 + \tfrac{1}{2}(dx^{0(2)} + dx^{0(1)}).$$

Substituting (84.5), we thus find that the difference in the values of the "time" x^0 for two simultaneous events occurring at infinitely near points is given by

$$\Delta x^0 = -\frac{g_{0\alpha}dx^\alpha}{g_{00}} \equiv g_\alpha dx^\alpha. \tag{84.14}$$

This relation enables us to synchronize clocks in any infinitesimal region of space. Carrying out a similar synchronization from the point A, we can synchronize clocks, i.e. we can define simultaneity of events, along any open curve.†

However, synchronization of clocks along a closed contour turns out to be impossible in general. In fact, starting out along the contour and returning to the initial point, we would obtain for Δx^0 a value different from zero. Thus it is, *a fortiori*, impossible to synchronize

---

† Multiplying (84.14) by g_{00} and bringing both terms to one side, we can state the condition for synchronization in the form $dx_0 = g_{0i}dx^i = 0$: the "covariant differential" dx_0 between two infinitely near simultaneous events must be equal to zero.

clocks over all space. The exceptional cases are those reference systems in which all the components $g_{0\alpha}$ are equal to zero.†

It should be emphasized that the impossibility of synchronization of all clocks is a property of the arbitrary reference system, and not of the space-time itself. In any gravitational field, it is always possible (in infinitely many ways) to choose the reference system so that the three quantities $g_{0\alpha}$ become identically equal to zero, and thus make possible a complete synchronization of clocks (see § 97).

Even in the special theory of relativity, proper time elapses differently for clocks moving relative to one another. In the general theory of relativity, proper time elapses differently even at different points of space in the same reference system. This means that the interval of proper time between two events occurring at some point in space, and the interval of time between two events simultaneous with these at another point in space, are in general different from one another.

§ 85. Covariant differentiation

In galilean coordinates‡ the differentials dA_i of a vector A_i form a vector, and the derivatives $\partial A_i / \partial x^k$ of the components of a vector with respect to the coordinates form a tensor. In curvilinear coordinates this is not so; dA_i is not a vector, and $\partial A_i / \partial x^k$ is not a tensor. This is due to the fact that dA_i is the difference of vectors located at different (infinitesimally separated) points of space; at different points in space vectors transform differently, since the coefficients in the transformation formulas (83.2), (83.4) are functions of the coordinates.

It is also easy to verify these statements directly. To do this we determine the transformation formulas for the differentials dA_i in curvilinear coordinates. A covariant vector is transformed according to the formula

$$A_i = \frac{\partial x'^k}{\partial x^i} A'_k;$$

therefore

$$dA_i = \frac{\partial x'^k}{\partial x^i} dA'_k + A'_k d\frac{\partial x'^k}{\partial x^i} = \frac{\partial x'^k}{\partial x^i} dA'_k + A'_k \frac{\partial^2 x'^k}{\partial x^i \partial x^l} dx^l.$$

Thus dA_i does not transform at all like a vector (the same also applies, of course, to the differential of a contravariant vector). Only if the second derivatives $\partial^2 x'^k / \partial x^i \partial x^l = 0$, i.e. if the x'^k are linear functions of the x^k, do the transformation formulas have the form

$$dA_i = \frac{\partial x'^k}{\partial x^i} dA'_k,$$

that is, dA_i transforms like a vector.

We now undertake the definition of a tensor which in curvilinear coordinates plays the same role as $\partial A_i / \partial x^k$ in galilean coordinates. In other words, we must transform $\partial A_i / \partial x^k$ from galilean to curvilinear coordinates.

---

† We should also assign to this class those cases where the $g_{0\alpha}$ can be made equal to zero by a simple transformation of the time coordinate, which does not involve any choice of the system of objects serving for the definition of the space coordinates.

‡ In general, whenever the quantities g_{ik} are constant.

In curvilinear coordinates, in order to obtain a differential of a vector which behaves like a vector, it is necessary that the two vectors to be subtracted from each other be located at the same point in space. In other words, we must somehow "translate" one of the vectors (which are separated infinitesimally from each other) to the point where the second is located, after which we determine the difference of two vectors which now refer to one and the same point in space. The operation of translation itself must be defined so that in galilean coordinates the difference shall coincide with the ordinary differential dA_i. Since dA_i is just the difference of the components of two infinitesimally separated vectors, this means that when we use galilean coordinates the components of the vector should not change as a result of the translation operation. But such a translation is precisely the translation of a vector parallel to itself. Under a *parallel translation* of a vector, its components in galilean coordinates do not change. If, on the other hand, we use curvilinear coordinates, then in general the components of the vector will change under such a translation. Therefore in curvilinear coordinates, the difference in the components of the two vectors after translating one of them to the point where the other is located will not coincide with their difference before the translation (i.e. with the differential dA_i).

Thus to compare two infinitesimally separated vectors we must subject one of them to a parallel translation to the point where the second is located. Let us consider an arbitrary contravariant vector; if its value at the point x^i is A^i, then at the neighbouring point $x^i + dx^i$ it is equal to $A^i + dA^i$. We subject the vector A^i to an infinitesimal parallel displacement to the point $x^i + dx^i$; the change in the vector which results from this we denote by δA^i. Then the difference DA^i between the two vectors which are now located at the same point is

$$DA^i = dA^i - \delta A^i. \tag{85.1}$$

The change δA^i in the components of a vector under an infinitesimal parallel displacement depends on the values of the components themselves, where the dependence must clearly be linear. This follows directly from the fact that the sum of two vectors must transform according to the same law as each of the constituents. Thus δA^i has the form

$$\delta A^i = -\Gamma^i_{\lambda l} A^k dx^l, \tag{85.2}$$

where the $\Gamma^i_{\lambda l}$ are certain functions of the coordinates. Their form depends, of course, on the coordinate system; for a galilean coordinate system $\Gamma^i_{\lambda l} = 0$.

From this it is already clear that the quantities Γ^i_{kl} do not form a tensor, since a tensor which is equal to zero in one coordinate system is equal to zero in every other one. In a curvilinear space it is, of course, impossible to make all the Γ^i_{kl} vanish over all of space.

But the principle of equivalence requires that by a suitable choice of coordinate system we can eliminate the gravitational field over a given infinitesimal region of space, i.e. we can make the quantities Γ^i_{kl} vanish in it. We shall see later in §87 that the Γ^i_{kl} play the role of field strengths.†

The quantities Γ^i_{kl} are called "connection coefficients" or "Christoffel symbols".

In addition to the quantities Γ^i_{kl} we shall later also use quantities $\Gamma_{i,kl}$ ‡ defined as follows:

† This is precisely the coordinate system which we have in mind in arguments where we, for brevity's sake, speak of a "galilean" system; still all the proofs remain applicable not only to flat, but also to curved 4-space.

‡ In place of Γ^i_{kl} and $\Gamma_{i,kl}$, the symbols $\begin{Bmatrix} kl \\ i \end{Bmatrix}$ and $\begin{bmatrix} kl \\ i \end{bmatrix}$ are sometimes used.

$$\Gamma_{i,\,kl} = g_{im}\Gamma_{kl}^{m}.\tag{85.3}$$

Conversely,

$$\Gamma_{kl}^{i} = g^{im}\Gamma_{m,\,kl}.\tag{85.4}$$

It is also easy to relate the change in the components of a covariant vector under a parallel displacement to the Christoffel symbols. To do this we note that under a parallel displacement, a scalar is unchanged. In particular, the scalar product of two vectors does not change under a parallel displacement.

Let A_i and B^i be any covariant and contravariant vectors. Then from $\delta(A_iB^i) = 0$, we have

$$B^i\delta A_i = -A_i\delta B^i = \Gamma_{kl}^{i}B^k A_i dx^l$$

or, changing the indices,

$$B^i\delta A_i = \Gamma_{il}^{k}A_k B^i dx^l.$$

From this, in view of the arbitrariness of the B^i,

$$\delta A_i = \Gamma_{il}^{k}A_k dx^l,\tag{85.5}$$

which determines the change in a covariant vector under a parallel displacement.

Substituting (85.2) and $dA^i = (\partial A^i/\partial x^l)\, dx^l$ in (85.1), we have

$$DA^i = \left(\frac{\partial A^i}{\partial x^l} + \Gamma_{kl}^{i}A^k\right)dx^l.\tag{85.6}$$

Similarly, we find for a covariant vector,

$$DA_i = \left(\frac{\partial A_i}{\partial x^l} - \Gamma_{il}^{k}A_k\right)dx^l.\tag{85.7}$$

The expressions in parentheses in (85.6) and (85.7) are tensors, since when multiplied by the vector dx^l they give a vector. Clearly, these are the tensors which give the desired generalization of the concept of a derivative to curvilinear coordinates. These tensors are called the *covariant derivatives* of the vectors A^i and A_i respectively. We shall denote them by $A^i_{;k}$ and $A_{i;\,k}$. Thus,

$$DA^i = A^i_{;\,l}\, dx^l; \quad DA_i = A_{i;\,l}\, dx^l,\tag{85.8}$$

while the covariant derivatives themselves are:

$$A^i_{;\,l} = \frac{\partial A^i}{\partial x^l} + \Gamma_{kl}^{i}A^k,\tag{85.9}$$

$$A_{i;\,l} = \frac{\partial A_i}{\partial x^l} - \Gamma_{il}^{k}A_k.\tag{85.10}$$

In galilean coordinates, $\Gamma_{kl}^{i} = 0$, and covariant differentiation reduces to ordinary differentiation.

It is also easy to calculate the covariant derivative of a tensor. To do this we must determine the change in the tensor under an infinitesimal parallel displacement. For example, let us consider any contravariant tensor, expressible as a product of two contravariant vectors $A^i B^k$. Under parallel displacement,

$$\delta(A^i B^k) = A^i \delta B^k + B^k \delta A^i = -A^i \Gamma^k_{lm} B^l dx^m - B^k \Gamma^i_{lm} A^l dx^m.$$

By virtue of the linearity of this transformation we must also have, for an arbitrary tensor A^{ik},

$$\delta A^{ik} = -(A^{im} \Gamma^k_{ml} + A^{mk} \Gamma^i_{ml}) dx^l. \tag{85.11}$$

Substituting this in

$$DA^{ik} = dA^{ik} - \delta A^{ik} \equiv A^{ik}_{\;;\,l} dx^l,$$

we get the covariant derivative of the tensor A^{ik} in the form

$$A^{ik}_{\;;\,l} = \frac{\partial A^{ik}}{\partial x^l} + \Gamma^i_{ml} A^{mk} + \Gamma^k_{ml} A^{im}. \tag{85.12}$$

In completely similar fashion we obtain the covariant derivative of the mixed tensor A^i_k and the covariant tensor A_{ik} in the form

$$A^i_{k;\,l} = \frac{\partial A^i_k}{\partial x^l} - \Gamma^m_{kl} A^i_m + \Gamma^i_{ml} A^m_k, \tag{85.13}$$

$$A_{ik;\,l} = \frac{\partial A_{ik}}{\partial x^l} - \Gamma^m_{il} A_{mk} - \Gamma^m_{kl} A_{im}. \tag{85.14}$$

One can similarly determine the covariant derivative of a tensor of arbitrary rank. In doing this one finds the following rule of covariant differentiation. To obtain the covariant derivative of the tensor $A^{...}_{...}$ with respect to x^l, we add to the ordinary derivative $\partial A^{...}_{...}/\partial x^l$ for each covariant index $i(A_{..i..})$ a term $-\Gamma^k_{il} A_{..k..}$, and for each contravariant index $i(A^{..i..})$ a term $+\Gamma^i_{kl} A^{..k..}$.

One can easily verify that the covariant derivative of a product is found by the same rule as for ordinary differentiation of products. In doing this we must consider the covariant derivative of a scalar ϕ as an ordinary derivative, that is, as the covariant vector $\phi_k = \partial \phi/\partial x^k$, in accordance with the fact that for a scalar $\delta\phi = 0$, and therefore $D\phi = d\phi$. For example, the covariant derivative of the product $A_i B_k$ is

$$(A_i B_k)_{;\,l} = A_{i;\,l} B_k + A_i B_{k;\,l}.$$

If in a covariant derivative we raise the index signifying the differentiation, we obtain the so-called *contravariant derivative*. Thus,

$$A_i^{\;;\,k} = g^{kl} A_{i;\,l}, \qquad A^{i;\,k} = g^{kl} A^i_{\;;\,l}.$$

We now derive formulas for the transformation of the Christoffel symbols from one coordinate system to another.

These formulas can be got by comparing the two equations that determine the covariant derivatives and requiring that these laws be the same for both. A simple calculation gives

$$\Gamma^i_{kl} = \Gamma'^m_{np} \frac{\partial x^i}{\partial x'^m} \frac{\partial x'^n}{\partial x^k} \frac{\partial x'^p}{\partial x^l} + \frac{\partial^2 x'^m}{\partial x^k \partial x^l} \frac{\partial x^i}{\partial x'^m}. \tag{85.15}$$

From this formula we see that the Γ^i_{kl} transforms like a tensor only for linear coordinate transformations (when the second term in 85.15 drops out).

However we note that this term is symmetric in k and l, and therefore drops out for the transformation of $S^i_{kl} = \Gamma^i_{kl} - \Gamma^i_{lk}$. It therefore transforms like a tensor:

$$S^i_{kl} = S'^m_{np} \frac{\partial x^i}{\partial x'^m} \frac{\partial x'^n}{\partial x^k} \frac{\partial x'^p}{\partial x^l},$$

and is called the "curvature tensor" of the space.

We now show that in this theory, based on the equivalence principle, the curvature tensor must be zero. In fact, as already stated, by virtue of the equivalence principle there must be a "galilean" coordinate system in which the Γ^i_{kl}, and consequently also the S^i_{kl}, vanish at a given point. Since S^i_{kl} is a tensor, if it vanishes in one coordinate system it must vanish in all frames. This means that the Christoffel symbols must be symmetric in their lower indices:

$$\Gamma^i_{kl} = \Gamma^i_{lk}, \tag{85.16}$$

Clearly, also

$$\Gamma_{i,kl} = \Gamma_{i,lk}. \tag{85.17}$$

In general, there are altogether forty different quantities Γ^i_{kl}; for each of the four values of the index i there are ten different pairs of values of the indices k and l (counting pairs obtained by interchanging k and l as the time).

Formula (85.15) enables us to prove easily the assertion made above that it is always possible under condition (85.16) to choose a coordinate system in which all the Γ^i_{kl} become zero at a previously assigned point (such a system is said to be *locally-inertial* or *locally-geodesic* (see § 87)).†

In fact, let the given point be chosen as the origin of coordinates, and let the values of the Γ^i_{kl} at it be initially (in the coordinates x^i) equal to $(\Gamma^i_{kl})_0$. In the neighbourhood of this point, we now make the transformation

$$x'^i = x^i + \tfrac{1}{2}(\Gamma^i_{kl})_0 x^k x^l. \tag{85.18}$$

Then

$$\left(\frac{\partial^2 x'^m}{\partial x^k \partial x^l} \frac{\partial x^i}{\partial x'^m} \right)_0 = (\Gamma^i_{kl})_0 \tag{85.19}$$

and according to (85.15), all the Γ'^m_{np} become equal to zero.

We emphasize that condition (85.16) is essential: the expression on the left side of (85.19) is symmetric in k and l, and so too must be the right side of the equation.

We note that for the transformation (85.18)

$$\left(\frac{\partial x'^i}{\partial x^k} \right)_0 = \delta^i_k,$$

† It can also be shown that, by a suitable choice of the coordinate system, one can make all the Γ^i_{kl} go to zero not just at a point but all along a given world line. (The proof of this statement can be found in the book by P. K. Rashevskii, *Riemannian Geometry and Tensor Analysis*, Nauka, 1964, § 91.)

so that it does not change the value of any tensor (including the tensor g_{ik}) at the given point, so that we can make the Christoffel symbols vanish at the same time as we bring the g_{ik} to galilean form.

§ 86. The relation of the Christoffel symbols to the metric tensor

Let us show that the covariant derivative of the metric tensor g_{ik} is zero. To do this we note that the relation

$$DA_i = g_{ik}DA^k$$

is valid for the vector DA_i, as for any vector. On the other hand, $A_i = g_{ik}A^k$, so that

$$DA_i = D(g_{ik}A^k) = g_{ik}DA^k + A^k Dg_{ik}.$$

Comparing with $DA_i = g_{ik} DA^k$, and remembering that the vector A^k is arbitrary;

$$Dg_{ik} = 0.$$

Therefore the covariant derivative

$$g_{ik;\,l} = 0. \tag{86.1}$$

Thus g_{ik} may be considered as a constant during covariant differentiation.

The equation $g_{ik;\,l} = 0$ can be used to express the Christoffel symbols Γ^i_{kl} in terms of the metric tensor g_{ik}. To do this we write in accordance with the general definition (85.14):

$$g_{ik;\,l} = \frac{\partial g_{ik}}{\partial x^l} - g_{mk}\Gamma^m_{il} - g_{im}\Gamma^m_{kl} = \frac{\partial g_{ik}}{\partial x^l} - \Gamma_{k,\,il} - \Gamma_{i,\,kl} = 0 \, .$$

Thus the derivatives of g_{ik} are expressed in terms of the Christoffel symbols.† We write the values of the derivatives of g_{ik}, permuting the indices i, k, l:

$$\frac{\partial g_{ik}}{\partial x^l} = \Gamma_{k,\,il} + \Gamma_{i,\,kl} \, ,$$

$$\frac{\partial g_{li}}{\partial x^k} = \Gamma_{i,\,kl} + \Gamma_{l,\,ik} \, ,$$

$$-\frac{\partial g_{kl}}{\partial x^i} = -\Gamma_{l,\,ki} - \Gamma_{k,\,li} \, .$$

Taking half the sum of these equations, we find (remembering that $\Gamma_{i,\,kl} = \Gamma_{i,\,lk}$)

$$\Gamma_{i,\,kl} = \frac{1}{2}\left(\frac{\partial g_{ik}}{\partial x^l} + \frac{\partial g_{il}}{\partial x^k} - \frac{\partial g_{kl}}{\partial x^i} \right). \tag{86.2}$$

From this we have for the symbols $\Gamma^i_{kl} = g^{im}\Gamma_{m,\,kl}$,

$$\Gamma^i_{kl} = \frac{1}{2}g^{im}\left(\frac{\partial g_{mk}}{\partial x^l} + \frac{\partial g_{ml}}{\partial x^k} - \frac{\partial g_{kl}}{\partial x^m} \right). \tag{86.3}$$

† Choosing a locally-geodesic system of coordinates therefore means that at the given point all the first derivatives of the components of the metric tensor vanish.

These formulas give the required expressions for the Christoffel symbols in terms of the metric tensor.

We now derive an expression for the contracted Christoffel symbol Γ^i_{ki} which will be important later on. To do this we calculate the differential dg of the determinant g made up from the components of the tensor g_{ik}; dg can be obtained by taking the differential of each component of the tensor g_{ik} and multiplying it by its coefficient in the determinant, i.e. by the corresponding minor. On the other hand, the components of the tensor g^{ik} reciprocal to g_{ik} are equal to the minors of the determinant of the g_{ik}, divided by the determinant. Therefore the minors of the determinant g are equal to gg^{ik}. Thus,

$$dg = gg^{ik}dg_{ik} = -gg_{ik}dg^{ik} \tag{86.4}$$

(since $g_{ik}g^{ik} = \delta^i_i = 4$, $g^{ik}dg_{ik} = -g_{ik}dg^{ik}$).

From (86.3), we have

$$\Gamma^i_{ki} = \frac{1}{2}g^{im}\left(\frac{\partial g_{mk}}{\partial x^i} + \frac{\partial g_{mi}}{\partial x^k} - \frac{\partial g_{ki}}{\partial x^m}\right).$$

Changing the positions of the indices m and i in the third and first terms in parentheses, we see that these two terms cancel each other, so that

$$\Gamma^i_{ki} = \frac{1}{2}g^{im}\frac{\partial g_{im}}{\partial x^k},$$

or, according to (86.4),

$$\Gamma^i_{ki} = \frac{1}{2g}\frac{\partial g}{\partial x^k} = \frac{\partial \ln\sqrt{-g}}{\partial x^k}. \tag{86.5}$$

It is useful to note also the expression for the quantity $g^{kl}\Gamma^i_{kl}$; we have

$$g^{kl}\Gamma^i_{kl} = \frac{1}{2}g^{kl}g^{im}\left(\frac{\partial g_{mk}}{\partial x^l} + \frac{\partial g_{lm}}{\partial x^k} - \frac{\partial g_{kl}}{\partial x^m}\right) = g^{kl}g^{im}\left(\frac{\partial g_{mk}}{\partial x^l} - \frac{1}{2}\frac{\partial g_{kl}}{\partial x^m}\right).$$

With the help of (86.4) this can be transformed to

$$g^{kl}\Gamma^i_{kl} = -\frac{1}{\sqrt{-g}}\frac{\partial(\sqrt{-g}g^{ik})}{\partial x^k}. \tag{86.6}$$

For various calculations it is important to remember that the derivatives of the contravariant tensor g^{ik} are related to the derivatives of g_{ik} by the relations

$$g_{il}\frac{\partial g^{lk}}{\partial x^m} = -g^{ik}\frac{\partial g_{il}}{\partial x^m} \tag{86.7}$$

(which are obtained by differentiating the equality $g_{il}g^{lk} = \delta^k_i$). Finally we point out that the derivatives of g^{ik} can also be expressed in terms of the quantities Γ^i_{kl}. Namely, from the identity $g^{ik}_{;l} = 0$ it follows directly that

$$\frac{\partial g^{ik}}{\partial x^l} = -\Gamma^i_{ml}g^{mk} - \Gamma^k_{ml}g^{im}. \tag{86.8}$$

With the aid of the formulas which we have obtained we can put the expression for $A^i_{;i}$, the generalized divergence of a vector in curvilinear coordinates, in convenient form. Using (86.5), we have

$$A^i_{;i} = \frac{\partial A^i}{\partial x^i} + \Gamma^i_{li} A^l = \frac{\partial A^i}{\partial x^i} + A^l \frac{\partial \ln \sqrt{-g}}{\partial x^l}$$

or, finally,

$$A^i_{;i} = \frac{1}{\sqrt{-g}} \frac{\partial(\sqrt{-g}\, A^i)}{\partial x^i}. \qquad (86.9)$$

We can derive an analogous expression for the divergence of an antisymmetric tensor A^{ik}. From (85.12), we have

$$A^{ik}_{;k} = \frac{\partial A^{ik}}{\partial x^k} + \Gamma^i_{mk} A^{mk} + \Gamma^k_{mk} A^{im}.$$

But, since $A^{mk} = -A^{km}$,

$$\Gamma^i_{mk} A^{mk} = -\Gamma^i_{km} A^{km} = 0.$$

Substituting the expression (86.5) for Γ^k_{mk}, we obtain

$$A^{ik}_{;k} = \frac{1}{\sqrt{-g}} \frac{\partial(\sqrt{-g}\, A^{ik})}{\partial x^k}. \qquad (86.10)$$

Now suppose A_{ik} is a symmetric tensor; we calculate the expression $A^k_{i;k}$ for its mixed components. We have

$$A^k_{i;k} = \frac{\partial A^k_i}{\partial x^k} + \Gamma^k_{lk} A^l_i - \Gamma^l_{ik} A^k_l = \frac{1}{\sqrt{-g}} \frac{\partial(A^k_i \sqrt{-g})}{\partial x^k} - \Gamma^l_{ki} A^k_l.$$

The last term here is equal to

$$-\frac{1}{2}\left(\frac{\partial g_{il}}{\partial x^k} + \frac{\partial g_{kl}}{\partial x^i} - \frac{\partial g_{ik}}{\partial x^l}\right) A^{kl}.$$

Because of the symmetry of the tensor A^{kl}, two of the terms in parentheses cancel each other, leaving

$$A^k_{i;k} = \frac{1}{\sqrt{-g}} \frac{\partial(\sqrt{-g}\, A^k_i)}{\partial x^k} - \frac{1}{2} \frac{\partial g_{kl}}{\partial x^i} A^{kl}. \qquad (86.11)$$

In cartesian coordinates, $\partial A_i/\partial x^k - \partial A_k/\partial x^i$ is an antisymmetric tensor. In curvilinear coordinates this tensor is $A_{i;k} - A_{k;i}$. However, with the help of the expression for $A_{i;k}$ and since $\Gamma^i_{kl} = \Gamma^i_{lk}$, we have

$$A_{i;k} - A_{k;i} = \frac{\partial A_i}{\partial x^k} - \frac{\partial A_k}{\partial x^i}. \qquad (86.12)$$

Finally, we transform to curvilinear coordinates the sum $\partial^2 \phi / \partial x_i \partial x^i$ of the second derivatives of a scalar ϕ. It is clear that in curvilinear coordinates this sum goes over into $\phi_{;\;i}^{;\;i}$. But $\phi_{;i} = \partial \phi / \partial x^i$, since covariant differentiation of a scalar reduces to ordinary differentiation. Raising the index i, we have

$$\phi^{;i} = g^{ik} \frac{\partial \phi}{\partial x^k},$$

and using formula (86.9), we find

$$\phi_{;\;i}^{;\;i} = -\frac{1}{\sqrt{-g}} \frac{\partial}{\partial x^i} \left(\sqrt{-g} \, g^{ik} \frac{\partial \phi}{\partial x^k} \right). \tag{86.13}$$

It is important to note that Gauss' theorem (83.17) for the transformation of the integral of a vector over a hypersurface into an integral over a four-volume can, in view of (86.9), be written as

$$\oint A^i \sqrt{-g} \, dS_i = \int A^i_{;i} \sqrt{-g} \, d\Omega. \tag{86.14}$$

§ 87. Motion of a particle in a gravitational field

The motion of a free material particle is determined in the special theory of relativity from the principle of least action,

$$\delta S = -mc\delta \int ds = 0, \tag{87.1}$$

according to which the particle moves so that its world line is an extremal between a given pair of world points, in our case a straight line (in ordinary three-dimensional space this corresponds to uniform rectilinear motion).

The motion of a particle in a gravitational field is determined by the principle of least action in this same form (87.1), since the gravitational field is nothing but a change in the metric of space-time, manifesting itself only in a change in the expression for ds in terms of the dx^i. Thus, in a gravitational field the particle moves so that its world point moves along an extremal or, as it is called, a *geodesic line* in the four-space x^0, x^1, x^2, x^3; however, since in the presence of the gravitational field space-time is not galilean, this line is not a "straight line", and the real spatial motion of the particle is neither uniform nor rectilinear.

Instead of starting once again directly from the principle of least action (see the problem at the end of this section), it is simpler to obtain the equations of motion of a particle in a gravitational field by an appropriate generalization of the differential equations for the free motion of a particle in the special theory of relativity, i.e. in a galilean four-dimensional coordinate system. These equations are $du^i/ds = 0$ or $du^i = 0$, where $u^i = dx^i/ds$ is the four-velocity. Clearly, in curvilinear coordinates this equation is generalized to the equation

$$Du^i = 0. \tag{87.2}$$

From the expression (85.6) for the covariant differential of a vector, we have

$$du^i + \Gamma^i_{kl} u^k dx^l = 0.$$

Dividing this equation by ds, we have

$$\frac{d^2 x^i}{ds^2} + \Gamma_{kl}^i \frac{dx^k}{ds} \frac{dx^l}{ds} = 0. \tag{87.3}$$

This is the required equation of motion. We see that the motion of a particle in a gravitational field is determined by the quantities Γ_{kl}^i. The derivative $d^2 x^i/ds^2$ is the four-acceleration of the particle. Therefore we may call the quantity $-m\Gamma_{kl}^i u^k u^l$ the "four-force", acting on the particle in the gravitational field. Here, the tensor g_{ik} plays the role of the "potential" of the gravitational field—its derivatives determine the field "intensity" Γ_{kl}^i.†

In § 85 it was shown that by a suitable choice of the coordinate system one can always make all the Γ_{kl}^i zero at an arbitrary point of space-time. We now see that the choice of such a locally-inertial system of reference means the elimination of the gravitational field in the given infinitesimal element of space-time, and the possibility of making such a choice is an expression of the principle of equivalence in the relativistic theory of gravitation.‡

As before, we define the four-momentum of a particle in a gravitational field as

$$p^i = mcu^i. \tag{87.4}$$

Its square is

$$p_i p^i = m^2 c^2. \tag{87.5}$$

Substituting $-\partial S/\partial x^i$ for p_i, we find the Hamilton–Jacobi equation for a particle in a gravitational field:

$$g^{ik} \frac{\partial S}{\partial x^i} \frac{\partial S}{\partial x^k} - m^2 c^2 = 0. \tag{87.6}$$

The equation of a geodesic in the form (87.3) is not applicable to the propagation of a light signal, since along the world line of the propagation of a light ray the interval ds, as we know, is zero, so that all the terms in equation (87.3) become infinite. To get the equations of motion in the form needed for this case, we use the fact that the direction of propagation of a light ray in geometrical optics is determined by the wave vector tangent to the ray. We can therefore write the four-dimensional wave vector in the form $k^i = dx^i/d\lambda$, where λ is some parameter varying along the ray. In the special theory of relativity, in the propagation of light in vacuum the wave vector does not vary along the path, that is, $dk^i = 0$ (see § 53). In a gravitational field this equation clearly goes over into $Dk^i = 0$ or

---

† We also give the form of the equations of motion expressed in terms of covariant components of the four-acceleration. From the condition $Du_i = 0$, we find

$$\frac{du_i}{ds} - \Gamma_{k,il} u^k u^l = 0.$$

Substituting for $\Gamma_{k,il}$ from (86.2), two of the terms cancel and we are left with

$$\frac{du_i}{ds} - \frac{1}{2} \frac{\partial g_{kl}}{\partial x^i} u^k u^l = 0.$$

‡ In the footnote on p. 259 we also noted the possibility of choosing a reference system which is "inertial along a given world line." In particular, if this line is the time axis (along which $x^1, x^2, x^3 = $ const), then the gravitational field will be eliminated for all times in the given spatial element.

$$\frac{dk^i}{d\lambda} + \Gamma^i_{kl}k^k k^l = 0 \tag{87.7}$$

(these equations also determine the parameter λ).†

The absolute square of the wave four-vector (see § 48) is zero, that is,

$$k_i k^i = 0, \tag{87 8}$$

Substituting $\partial\psi/\partial x^i$ in place of k_i (ψ is the eikonal), we find the eikonal equation in a gravitational field

$$g^{ik}\frac{\partial\psi}{\partial x^i}\frac{\partial\psi}{\partial x^k} = 0. \tag{87.9}$$

In the limiting case of small velocities, the relativistic equations of motion of a particle in a gravitational field must go over into the corresponding non-relativistic equations. In this we must keep in mind that the assumption of small velocity implies the requirement that the gravitational field itself be weak; if this were not so a particle located in it would acquire a high velocity.

Let us examine how, in this limiting case, the metric tensor g_{ik} determining the field is related to the nonrelativistic potential ϕ of the gravitational field.

In nonrelativistic mechanics the motion of a particle in a gravitational field is determined by the Lagrangian (81.1). We now write it in the form

$$L = -mc^2 + \frac{mv^2}{2} - m\phi, \tag{87.10}$$

adding the constant $-mc^2$.‡ This must be done so that the nonrelativistic Lagrangian in the absence of the field, $L = -mc^2 + mv^2/2$, shall be the same exactly as that to which the corresponding relativistic function $L = -mc^2\sqrt{1 - v^2/c^2}$ reduces in the limit as $v/c \to 0$.

Consequently, the nonrelativistic action function S for a particle in a gravitational field has the form

$$S = \int L\, dt = -mc\int\left(c - \frac{v^2}{2c} + \frac{\phi}{c}\right) dt.$$

Comparing this with the expression $S = -mc\int ds$, we see that in the limiting case under consideration

$$ds = \left(c - \frac{v^2}{2c} + \frac{\phi}{c}\right) dt.$$

Squaring and dropping terms which vanish for $c \to \infty$, we find

$$ds^2 = (c^2 + 2\phi)\, dt^2 - d\mathbf{r}^2. \tag{87.11}$$

where we have used the fact $\mathbf{v}\, dt = d\mathbf{r}$.

---

† Geodesics, along which $ds \equiv 0$, are said to be *null* or *isotropic*.

‡ The potential ϕ is, of course, defined only to within an arbitrary additive constant. We assume throughout that one makes the natural choice of this constant so that the potential vanishes far from the bodies producing the field.

Thus in the limiting case the component g_{00} of the metric tensor is

$$g_{00} = 1 + \frac{2\phi}{c^2}. \tag{87.12}$$

As for the other components, from (87.11) it would follow that $g_{\alpha\beta} = \delta_{\alpha\beta}$, $g_{0\alpha} = 0$. Actually, however, the corrections to them are, generally speaking, of the same order of magnitude as the corrections to g_{00} (for more detail, see § 106). The impossibility of determining these corrections by the method given above is related to the fact that the corrections to the $g_{\alpha\beta}$, though of the same order of magnitude as the correction to g_{00}, would give rise to terms in the Lagrangian of a higher order of smallness (because in the expression for ds^2 the components $g_{\alpha\beta}$ are not multiplied by c^2, while this is the case for g_{00}).

PROBLEM

Derive the equation of motion (87.3) from the principle of least action (87.1).
Solution: We have:

$$\delta ds^2 = 2ds\,\delta ds = \delta(g_{ik}dx^i dx^k) = dx^i dx^k \frac{\partial g_{ik}}{\partial x^l}\delta x^l + 2g_{ik}dx^i d\delta x^k.$$

Therefore

$$\delta S = -mc \int \left\{ \frac{1}{2}\frac{dx^i}{ds}\frac{dx^k}{ds}\frac{\partial g_{ik}}{\partial x^l}\delta x^l + g_{ik}\frac{dx^i}{ds}\frac{d\delta x^k}{ds}\right\}ds$$

$$= -mc \int \left\{ \frac{1}{2}\frac{dx^i}{ds}\frac{dx^k}{ds}\frac{\partial g_{ik}}{\partial x^l}\delta x^l - \frac{d}{ds}\left(g_{ik}\frac{dx^i}{ds}\right)\delta x^k\right\}ds$$

(in integrating by parts, we use the fact that $\delta x^k = 0$ at the limits). In the second term in the integral, we replace the index k by the index l. We then find, by equating to zero the coefficient of the arbitrary variation δx^l:

$$\frac{1}{2}u^i u^k \frac{\partial g_{ik}}{\partial x^l} - \frac{d}{ds}(g_{il}u^i) = \frac{1}{2}u^i u^k\frac{\partial g_{ik}}{\partial x^l} - g_{il}\frac{du^i}{ds} - u^i u^k\frac{\partial g_{il}}{\partial x^k} = 0.$$

Noting that the third term can be written as

$$-\frac{1}{2}u^i u^k\left(\frac{\partial g_{il}}{\partial x^k} + \frac{\partial g_{kl}}{\partial x^i}\right),$$

and introducing the Christoffel symbols $\Gamma_{l,ik}$ in accordance with (86.2), we have

$$g_{il}\frac{du^i}{ds} + \Gamma_{l,ik}u^i u^k = 0.$$

Equation (87.3) is obtained form this by raising the index l.

§ 88. The constant gravitational field

A gravitational field is said to be *constant* if one can choose a system of reference in which all the components of the metric tensor are independent of the time coordinate x^0; the latter is then called the *world time*.

The choice of a world time is not completely unique. Thus, if we add to x^0 an arbitrary function of the space coordinates, the g_{ik} will still not contain x^0; this transformation corresponds to the arbitrariness in the choice of the time origin at each point in space.† In addition, of course, the world time can be multiplied by an arbitrary constant, i.e. the units for measuring it are arbitrary.

Strictly speaking, only the field produced by a single body can be constant. In a system of several bodies, their mutual gravitational attraction will give rise to motion, as a result of which the field produced by them cannot be constant.

If the body producing the field is fixed (in the reference system in which the g_{ik} do not depend on x^0), then both directions of time are equivalent . For a suitable choice of the time origin at all the points in space, the interval ds should in this case not be changed when we change the sign of x^0, and therefore all the components $g_{0\alpha}$ of the metric tensor must be identically equal to zero. Such constant gravitational fields are said to be *static*.

However, for the field produced by a body to be constant, it is not necessary for the body to be at rest. Thus the field of an axially symmetric body rotating uniformly about its axis will also be constant. However in this case the two time directions are no longer equivalent by any means—if the sign of the time is changed, the sign of the angular velocity is changed. Therefore in such constant gravitational fields (we shall call them *stationary* fields) the components $g_{0\alpha}$ of the metric tensor are in general different from zero.

The meaning of the world time in a constant gravitational field is that an interval of world time between events at a certain point in space coincides with the interval of world time between any other two events at any other point in space, if these events are respectively simultaneous (in the sense explained in § 84) with the first pair of events. But to the same interval of world time x^0 there correspond, at different points of space, different intervals of proper time τ.

The relation between world time and proper time, formula (84.1), can now be written in the form

$$\tau = \frac{1}{c}\sqrt{g_{00}}\,x^0 , \tag{88.1}$$

applicable to any finite time interval.

If the gravitational field is weak, then we may use the approximate expression (87.12), and (88.1) gives

$$\tau = \frac{x^0}{c}\left(1 + \frac{\phi}{c^2}\right) . \tag{88.2}$$

Thus proper time elapses the more slowly the smaller the gravitational potential at a given point in space, i.e., the larger its absolute value (later, in § 96, it will be shown that the

† It is easy to see that under such a transformation the spatial metric, as expected, does not change. In fact, under the substitution

$$x^0 \to x^0 + f(x^1, x^2, x^3)$$

with an arbitrary function $f(x^1, x^2, x^3)$, the components g_{ik} change to

$$g_{\alpha\beta} \to g_{\alpha\beta} + g_{00}f_{,\alpha}f_{,\beta} + g_{0\alpha}f_{,\beta} + g_{0\beta}f_{,\alpha}$$

$$g_{0\alpha} \to g_{0\alpha} + g_{00}f_{,\alpha},\ g_{00} \to g_{00} ,$$

where $f_{,\alpha} \equiv \partial f/\partial x^\alpha$. This obviously does not change the tensor (84.7).

potential ϕ is negative). If one of two idential clocks is placed in a gravitational field for some time, the clock which has been in the field will thereafter appear to be slow.

As was already indicated above, in a static gravitational field the components $g_{0\alpha}$ of the metric tensor are zero. According to the results of § 84, this means that in such a field synchronization of clocks is possible over all space. We note also that the element of spatial distance in a static field is simply:

$$dl^2 = - g_{\alpha\beta} \, dx^\alpha dx^\beta. \tag{88.3}$$

In a stationary field the $g_{0\alpha}$ are different from zero and the synchronization of clocks over all space is impossible. Since the g_{ik} do not depend on x^0, formula (84.14) for the difference between the values of world time for two simultaneous events occurring at different points in space can be written in the form

$$\Delta x^0 = - \int \frac{g_{0\alpha} dx^\alpha}{g_{00}} \tag{88.4}$$

for any two points on the line along which the synchronization of clocks is carried out. In the synchronization of clocks along a closed contour, the difference in the value of the world time which would be recorded upon returning to the starting point is equal to the integral

$$\Delta x^0 = - \oint \frac{g_{0\alpha} dx^\alpha}{g_{00}} \tag{88.5}$$

taken along the closed contour.†

Let us consider the propagation of a light ray in a constant gravitational field. We have seen in § 53 that the frequency of the light is the time derivative of the eikonal ψ (with opposite sign). The frequency expressed in terms of the world time x^0/c is therefore $\omega_0 = -c(\partial\psi/\partial x^0)$. Since the eikonal equation (87.9) in a constant field does not contain x^0 explicitly, the frequency ω_0 remains constant during the propagation of the light ray. The frequency measured in terms of the proper time is $\omega = -(\partial\psi/\partial\tau)$; this frequency is different at different points of space.

From the relation

$$\frac{\partial\psi}{\partial\tau} = \frac{\partial\psi}{\partial x^0} \frac{\partial x^0}{\partial\tau} = \frac{\partial\psi}{\partial x^0} \frac{c}{\sqrt{g_{00}}},$$

we have

$$\omega = \frac{\omega_0}{\sqrt{g_{00}}}. \tag{88.6}$$

In a weak gravitational field we obtain from this, approximately,

$$\omega = \omega_0 \left(1 - \frac{\phi}{c^2}\right). \tag{88.7}$$

We see that the light frequency increases with increasing absolute value of the potential of

† The integral (88.5) is identically zero if the sum $g_{0\alpha} dx^\alpha/g_{00}$ is an exact differential of some function of the space coordinates. However, such a case would simply mean that we are actually dealing with a static field, and that all the $g_{0\alpha}$ could be made equal to zero by a transformation of the form $x^0 \to x^0 + f(x^\alpha)$.

the gravitational field, i.e. as we approach the bodies producing the field; conversely, as the light recedes from these bodies the frequency decreases. If a ray of light, emitted at a point where the gravitational potential is ϕ_1, has (at that point) the frequency ω, then upon arriving at a point where the potential is ϕ_2, it will have a frequency (measured in units of the proper time at that point) equal to

$$\frac{\omega}{1 - \frac{\phi_1}{c^2}}\left(1 - \frac{\phi_2}{c^2}\right) = \omega\left(1 + \frac{\phi_1 - \phi_2}{c^2}\right).$$

A line spectrum emitted by some atoms located, for example, on the sun, looks the same there as the spectrum emitted by the same atoms located on the earth would appear on it. If, however, we observe on the earth the spectrum emitted by the atoms located on the sun, then, as follows from what has been said above, its lines appear to be shifted with respect to the lines of the same spectrum emitted on the earth. Namely, each line with frequency ω will be shifted through the interval $\Delta\omega$ given by the formula

$$\Delta\omega = \frac{\phi_1 - \phi_2}{c^2}\,\omega, \tag{88.8}$$

where ϕ_1 and ϕ_2 are the potentials of the gravitational field at the points of emission and observation of the spectrum respectively. If we observe on the earth a spectrum emitted on the sun or the stars, then $|\phi_1| > |\phi_2|$, and from (88.8) it follows that $\Delta\omega < 0$, i.e. the shift occurs in the direction of lower frequency. The phenomenon we have described is called the "red shift".

The occurrence of this phenomenon can be explained directly on the basis of what has been said above about world time. Because the field is constant, the interval of world time during which a certain vibration in the light wave propagates from one given point of space to another is independent of x^0. Therefore it is clear that the number of vibrations occurring in a unit interval of world time will be the same at all points along the ray. But to one and the same interval of world time there corresponds a larger and larger interval of proper time, the further away we are from the bodies producing the field. Consequently, the frequency, i.e. the number of vibrations per unit proper time, will decrease as the light recedes from these masses.

During the motion of a particle in a constant field, its energy, defined as

$$-c\frac{\partial S}{\partial x^0},$$

the derivative of the action with respect to the world time, is conserved; this follows, for example, from the fact that x^0 does not appear explicitly in the Hamilton–Jacobi equation. The energy defined in this way is the time component of the covariant four-vector of momentum $p_k = mcu_k = mcg_{ki}u^i$. In a static field, $ds^2 = g_{00}(dx^0)^2 - dl^2$, and we have for the energy, which we here denote by $\mathscr{E}_0$,

$$\mathscr{E}_0 = mc^2 g_{00}\frac{dx^0}{ds} = mc^2 g_{00}\frac{dx^0}{\sqrt{g_{00}(dx^0)^2 - dl^2}}.$$

We introduce the velocity

$$v = \frac{dl}{d\tau} = \frac{c\,dl}{\sqrt{g_{00}}\,dx^0}$$

of the particle, measured in terms of the proper time, that is, by an observer located at the given point. Then we obtain for the energy

$$\mathscr{E}_0 = \frac{mc^2 \sqrt{g_{00}}}{\sqrt{1 - \dfrac{v^2}{c^2}}}. \tag{88.9}$$

This is the quantity which is conserved during the motion of the particle.

It is easy to show that the expression (88.9) remains valid also for a stationary field, if only the velocity v is measured in terms of the proper time, as determined by clocks synchronized along the trajectory of the particle. If the particle departs from point A at the moment of world time x^0 and arrives at the infinitesimally distant point B at the moment $x^0 + dx^0$, then to determine the velocity we must now take, not the time interval $(x^0 + dx^0) - x^0 = dx^0$, but rather the difference between $x^0 + dx^0$ and the moment $x^0 - (g_{0\alpha}/g_{00})dx^\alpha$ which is simultaneous at the point B with the moment x^0 at the point A:

$$(x^0 + dx^0) - \left(x^0 - \frac{g_{0\alpha}}{g_{00}} dx^\alpha \right) = dx^0 + \frac{g_{0\alpha}}{g_{00}} dx^\alpha.$$

Multiplying by $\sqrt{g_{00}}/c$, we obtain the corresponding interval of proper time, so that the velocity is

$$v^\alpha = \frac{c\, dx^\alpha}{\sqrt{h}(dx^0 - g_\alpha dx^\alpha)}, \tag{88.10}$$

where we have introduced the notation

$$h = g_{00}, \quad g_\alpha = -\frac{g_{0\alpha}}{g_{00}} \tag{88.11}$$

for the three-dimensional vector **g** (which was already mentioned in § 84) and for the three-dimensional scalar g_{00}. The covariant components of the velocity **v** form a three-dimensional vector in the space with metric $\gamma_{\alpha\beta}$, and correspondingly the square of this vector is to be taken as†

$$v_\alpha = \gamma_{\alpha\beta} v^\beta, \quad v^2 = v_\alpha v^\alpha. \tag{88.12}$$

We note that with such a definition, the interval ds is expressed in terms of the velocity in the usual fashion:

$$ds^2 = g_{00}(dx^0)^2 + 2g_{0\alpha}\, dx^0\, dx^\alpha + g_{\alpha\beta}\, dx^\alpha\, dx^\beta$$

$$= h(dx^0 - g_\alpha dx^\alpha)^2 - dl^2$$

$$= h(dx^0 - g_\alpha\, dx^\alpha)^2 \left(1 - \frac{v^2}{c^2}\right), \tag{88.13}$$

---

† In our further work we shall repeatedly introduce, in addition to four-vectors and four-tensors, three-dimensional vectors and tensors defined in the space with metric $\gamma_{\alpha\beta}$; in particular the vectors **g** and **v**, which we have already used, are of this type. Just as in four dimensions the tensor operations (in particular, raising and lowering of indices) are done using the metric tensor g_{ik}, so, in three dimensions these are done using the tensor $\gamma_{\alpha\beta}$. To avoid misunderstandings that may arise, we shall denote three-dimensional quantities by symbols other than those used for four-dimensional quantities.

The components of the four-velocity

$$u^i = \frac{dx^i}{ds}$$

are

$$u^\alpha = \frac{v^\mu}{c\sqrt{1 - \dfrac{v^2}{c^2}}} , \quad u^0 = \frac{1}{\sqrt{h}\sqrt{1 - \dfrac{v^2}{c^2}}} + \frac{g_\alpha v^\alpha}{c\sqrt{1 - \dfrac{v^2}{c^2}}} \tag{88.14}$$

The energy is

$$\mathcal{E}_0 = mc^2 g_{0i} u^i = mc^2 h(u^0 - g_\alpha u^\alpha) ,$$

and after substituting (88.14), takes the form (88.9).

In the limiting case of a weak gravitational field and low velocities, by substituting $g_{00} = 1 + (2\phi/c^2)$ in (88.9), we get approximately:

$$\mathcal{E}_0 = mc^2 + \frac{mv^2}{2} + m\phi, \tag{88.15}$$

where $m\phi$ is the potential energy of the particle in the gravitational field, which is in agreement with the Lagrangian (87.10).

PROBLEMS

1. Determine the force acting on a particle in a constant gravitational field.

Solution: For the components of Γ_{kl} which we need, we find the following expressions:

$$\Gamma^\alpha_{00} = \frac{1}{2} h^{;\alpha},$$

$$\Gamma^\alpha_{0\beta} = \frac{h}{2}(g^{\alpha}_{;\beta} - g^{;\alpha}_{\beta}) - \frac{1}{2} g_\beta h^{;\alpha}, \tag{1}$$

$$\Gamma^\alpha_{\beta\gamma} = \lambda^\alpha_{\beta\gamma} + \frac{h}{2}[g_\beta(g^{;\alpha}_\gamma - g^\alpha_{;\gamma}) + g_\gamma(g^{;\alpha}_\beta - g^\alpha_{;\beta})] + \frac{1}{2} g_\beta g_\gamma h^{;\alpha}.$$

In these expressions all the tensor operations (covariant differentiation, raising and lowering of indices) are carried out in the three-dimensional space with metric $\gamma_{\alpha\beta}$, on the three-dimensional vector g^α and the three-dimensional scalar h (88.11); $\lambda^\alpha_{\beta\gamma}$ is the three-dimensional Christoffel symbol, constructed from the components of the tensor $\gamma_{\alpha\beta}$ in just the same way as Γ^i_{kl} is constructed from the components of g_{ik}; in the computations we use (84.9)–(84.12).

Substituting (1) in the equation of motion

$$\frac{du^\alpha}{ds} = -\Gamma^\alpha_{00}(u^0)^2 - 2\Gamma^\alpha_{0\beta} u^0 u^\beta - \Gamma^\alpha_{\beta\gamma} u^\beta u^\gamma$$

and using the expression (88.14) for the components of the four-velocity, we find after some simple transformations:

$$\frac{d}{ds} \frac{v^\alpha}{c\sqrt{1 - \dfrac{v^2}{c^2}}} = -\frac{h^{;\alpha}}{2h\left(1 - \dfrac{v^2}{c^2}\right)} - \frac{\sqrt{h}(g^{;\alpha}_{;\beta} - g^{;\alpha}_\beta)v^\beta}{c\left(1 - \dfrac{v^2}{c^2}\right)} - \frac{\lambda^\alpha_{\beta\gamma} v^\beta v^\gamma}{c^2\left(1 - \dfrac{v^2}{c^2}\right)}. \tag{2}$$

The force **f** acting on the particle is the derivative of its momentum **p** with respect to the (synchronized) proper time, as defined by the three-dimensional covariant differential:

$$f^\alpha = c\sqrt{1 - \frac{v^2}{c^2}}\frac{Dp^\alpha}{ds} = c\sqrt{1 - \frac{v^2}{c^2}}\frac{d}{ds}\frac{mv^\alpha}{\sqrt{1 - \frac{v^2}{c^2}}} + \lambda^\alpha_{\beta\gamma}\frac{mv^\beta v^\gamma}{\sqrt{1 - \frac{v^2}{c^2}}}.$$

From (2) we therefore have (for convenience we lower the index α):

$$f_\alpha = \frac{mc^2}{\sqrt{1 - \frac{v^2}{c^2}}}\left\{-\frac{\partial}{\partial x^\alpha}\ln\sqrt{h} + \sqrt{h}\left(\frac{\partial g_\beta}{\partial x^\alpha} - \frac{\partial g_\alpha}{\partial x^\beta}\right)\frac{v^\beta}{c}\right\},$$

or, in the usual three-dimensional notation,†

$$\mathbf{f} = \frac{mc^2}{\sqrt{1 - \frac{v^2}{c^2}}}\left\{-\nabla\ln\sqrt{h} + \sqrt{h}\,\frac{\mathbf{v}}{c}\times(\text{curl }\mathbf{g})\right\}. \tag{3}$$

We note that if the body is at rest, then the force acting on it [the first term in (3)] has a potential. For low velocities of motion the second term in (3) has the form $mc\sqrt{h}\mathbf{v}\times(\text{curl }\mathbf{g})$ analogous to the Coriolis force which would appear (in the absence of the field) in a coordinate system rotating with angular velocity

$$\mathbf{\Omega} = \frac{c}{2}\sqrt{h}\,\text{curl }\mathbf{g}.$$

† In three-dimensional curvilinear coordinates, the unit antisymmetric tensor is defined as

$$\eta_{\alpha\beta\gamma} = \sqrt{\gamma}\,e_{\alpha\beta\gamma}, \quad \eta^{\alpha\beta\gamma} = \frac{1}{\sqrt{\gamma}}e^{\alpha\beta\gamma},$$

where $e_{123} = e^{123} = 1$, and the sign changes under transposition of indices [compare (83.13)–(83.14)]. Accordingly the vector $\mathbf{c} = \mathbf{a}\times\mathbf{b}$, defined as the vector dual to the antisymmetric tensor $c_{\beta\gamma} = a_\beta b_\gamma a_\gamma b_\beta$, has components:

$$c_\alpha = \frac{1}{2}\sqrt{\gamma}\,e_{\alpha\beta\gamma}c^{\beta\gamma} = \sqrt{\gamma}\,e_{\alpha\beta\gamma}a^\beta b^\gamma, \quad c^\alpha = \frac{1}{2\sqrt{\gamma}}e^{\alpha\beta\gamma}c_{\beta\gamma} = \frac{1}{\sqrt{\gamma}}e^{\alpha\beta\gamma}a_\beta b_\gamma.$$

Conversely,

$$c_{\alpha\beta} = \sqrt{\gamma}\,e_{\alpha\beta\gamma}c^\gamma, \quad c^{\alpha\beta} = \frac{1}{\sqrt{\gamma}}e^{\alpha\beta\gamma}c_\gamma.$$

In particular, curl **a** should be understood in this same sense as the vector dual to the tensor $a_{\beta;\,\alpha} - a_{\alpha;\,\beta} = (\partial a_\beta/\partial x^\alpha) - (\partial a_\alpha/\partial x^\beta)$, so that its contravariant components are

$$(\text{curl }\mathbf{a})^\alpha = \frac{1}{2\sqrt{\gamma}}e^{\alpha\beta\gamma}\left(\frac{\partial a_\gamma}{\partial x^\beta} - \frac{\partial a_\beta}{\partial x^\gamma}\right).$$

In this same connection we repeat that for the three-dimensional divergence of a vector [see (86.9)]:

$$\text{div }\mathbf{a} = \frac{1}{\sqrt{\gamma}}\frac{\partial}{\partial x^\alpha}(\sqrt{\gamma}\,a^\alpha).$$

To avoid misunderstandings when comparing with formulas frequently used for the three-dimensional vector operations in orthogonal curvilinear coordinates (see, for example, *Electrodynamics of Continuous Media*, appendix), we point out that in these formulas the components of the vectors are understood to be the quantities $\sqrt{g_{11}}A^1(=\sqrt{A_1A^1})$, $\sqrt{g_{22}}A^2$, $\sqrt{g_{33}}A^3$.

2. Derive Fermat's principle for the propagation of a ray in a constant gravitational field.

Solution: Fermat's principle (§ 53) states:

$$\delta \int k_\alpha dx^\alpha = 0 \, ,$$

where the integral is taken along the ray, and the integral must be expressed in terms of the frequency ω_0 (which is constant along the ray) and the coordinate differentials. Noting that $k_0 = - \partial \psi / \partial x^0 = (\omega_0/c)$, we write:

$$\frac{\omega_0}{c} = k_0 = g_{0i} k^i = g_{00} k^0 + g_{0\alpha} k^\alpha = h(k^0 - g_\alpha k^\alpha) \, .$$

Substituting this in the relation $k_i k^i = g_{ik} k^i k^k = 0$, written in the form

$$h(k^0 - g_\alpha k^\alpha)^2 - \gamma_{\alpha\beta} \, k^\alpha k^\beta = 0,$$

we obtain:

$$\frac{1}{h} \left(\frac{\omega_0}{c} \right)^2 - \gamma_{\alpha\beta} k^\alpha k^\beta = 0.$$

Noting that the vector k^α must have the direction of the vector dx^α, we then find:

$$k^\alpha = \frac{\omega_0}{c\sqrt{h}} \frac{dx^\alpha}{dl} \, ,$$

where dl (84.6) is the element of spatial distance along the ray. In order to obtain the expression for k_α, we write

$$k^\alpha = g^{\alpha i} k_i = g^{\alpha 0} k_0 + g^{\alpha\beta} k_\beta = -g^\alpha \frac{\omega_0}{c} - \gamma^{\alpha\beta} k_\beta,$$

so that

$$k_\alpha = -\gamma_{\alpha\beta} \left(k^\beta + \frac{\omega_0}{c} g^\beta \right) = -\frac{\omega_0}{c} \left(\frac{\gamma_{\alpha\beta}}{\sqrt{h}} \frac{dx^\beta}{dl} + g_\alpha \right).$$

Finally, multiplying by dx^α, we obtain Fermat's principle in the form (dropping the constant factor ω_0/c):

$$\delta \int \left(\frac{dl}{\sqrt{h}} + g_\alpha dx^\alpha \right) = 0.$$

In a static field, we have simply:

$$\delta \int \frac{dl}{\sqrt{h}} = 0.$$

We call attention to the fact that in a gravitational field the ray does not propagate along the shortest line in space, since the latter would be defined by the equation $\delta \int dl = 0$.

§ 89. Rotation

As a special case of a stationary gravitational field, let us consider a uniformly rotating reference system. To calculate the interval ds we carry out the transformation from a system at rest (inertial system) to the uniformly rotating one. In the coordinates r', ϕ', z', t of the system at rest (we use cylindrical coordinates r', ϕ', z'), the interval has the form

$$ds^2 = c^2 \, dt^2 - dr'^2 - r'^2 \, d\phi'^2 - dz'^2. \tag{89.1}$$

Let the cylindrical coordinates in the rotating system be r, ϕ, z. If the axis of rotation coincides with the axes Z and Z', then we have $r' = r$, $z' = z$, $\phi' = \phi + \Omega t$, where Ω is the angular velocity of rotation. Substituting in (89.1), we find the required expression for ds^2 in the rotating system of reference:

$$ds^2 = (c^2 - \Omega^2 r^2)\, dt^2 - 2\Omega r^2\, d\phi\, dt - dz^2 - r^2\, d\phi^2 - dr^2. \tag{89.2}$$

It is necessary to note that the rotating system of reference can be used only out to distances equal to c/Ω. In fact, from (89.2) we see that for $r > c/\Omega$, g_{00} becomes negative, which is not admissible. The inapplicability of the rotating reference system at large distances is related to the fact that there the velocity would become greater than the velocity of light, and therefore such a system cannot be made up from real bodies.

As in every stationary field, clocks on the rotating body cannot be uniquely synchronized at all points. Proceeding with the synchronization along any closed curve, we find, upon returning to the starting point, a time differing from the initial value by an amount [see (88.5)]

$$\Delta t = -\frac{1}{c}\oint \frac{g_{0\alpha}}{g_{00}}\, dx^\alpha = \frac{1}{c^2}\oint \frac{\Omega r^2\, d\phi}{1 - \dfrac{\Omega^2 r^2}{c^2}}$$

or, assuming that $\Omega r/c \ll 1$ (i.e. that the velocity of the rotation is small compared with the velocity of light),

$$\Delta t = \frac{\Omega}{c^2}\int r^2\, d\phi = \pm \frac{2\Omega}{c^2} S, \tag{89.3}$$

where S is the projected area of the contour on a plane perpendicular to the axis of rotation (the sign $+$ or $-$ holding according as we traverse the contour in, or opposite to, the direction of rotation).

Let us assume that a ray of light propagates along a certain closed contour. Let us calculate to terms of order v/c the time t that elapses between the starting out of the light ray and its return to the initial point. The velocity of light, by definition, is always equal to c, if the times are synchronized along the given closed curve and if at each point we use the proper time. Since the difference between proper and world time is of order v^2/c^2, then in calculating the required time interval t to terms of order v/c this difference can be neglected. Therefore we have

$$t = \frac{L}{c} \pm \frac{2\Omega}{c^2} S,$$

where L is the length of the contour. Corresponding to this, the velocity of light, measured as the ratio L/t, appears equal to

$$c \pm 2\Omega \frac{S}{L}. \tag{89.4}$$

This formula, like the first approximation for the Doppler effect, can also be easily derived in a purely classical manner.

PROBLEM

Calculate the element of spatial distance in a rotating coordinate system.

Solution: With the help of (84.6) and (84.7), we find

$$dl^2 = dr^2 + dz^2 + \frac{r^2 d\phi^2}{1 - \Omega^2 \dfrac{r^2}{c^2}} \, ,$$

which determines the spatial geometry in the rotating reference system. We note that the ratio of the circumference of a circle in the plane $z = $ constant (with centre on the axis of rotation) to its radius r is

$$2\pi / \sqrt{1 - \Omega^2 r^2 / c^2} \, ,$$

i.e. larger than 2π.

§ 90. The equations of electrodynamics in the presence of a gravitational field

The electromagnetic field equations of the special theory of relativity can be easily generalized so that they are applicable in an arbitrary four-dimensional curvilinear system of coordinates, i.e. in the presence of a gravitational field.

The electromagnetic field tensor in the special theory of relativity is defined as $F_{ik} = (\partial A_k / \partial x^i) - (\partial A_i / \partial x^k)$. Clearly it must now be defined correspondingly as $F_{ik} = A_{k;i} - A_{i;k}$. But because of (86.12),

$$F_{ik} = A_{k;i} - A_{i;k} = \frac{\partial A_k}{\partial x^i} - \frac{\partial A_i}{\partial x^k} \, , \tag{90.1}$$

and therefore the relation of F_{ik} to the potential A_i does not change. Consequently the first pair of Maxwell equations (26.5) also does not change its form†

$$\frac{\partial F_{ik}}{\partial x^l} + \frac{\partial F_{li}}{\partial x^k} + \frac{\partial F_{kl}}{\partial x^i} = 0 \, . \tag{90.2}$$

In order to write the second pair of Maxwell equations, we must first determine the current four-vector in curvilinear coordinates. We do this in a fashion completely analogous to that which we followed in § 28. The spatial volume element, constructed on the space coordinate elements dx^1, dx^2, and dx^3, is $\sqrt{\gamma} dV$, where γ is the determinant of the spatial metric tensor (84.7) and $dV = dx^1 \, dx^2 \, dx^3$ (see the footnote on p. 249). We introduce the charge density ϱ according to the definition $de = \varrho \sqrt{\gamma} dV$, where de is the charge located within the volume element $\sqrt{\gamma} dV$. Multiplying this equation on both sides by dx^i, we have:

$$de \, dx^i = \varrho \, dx^i \sqrt{\gamma} dx^1 dx^2 dx^3 = \frac{\varrho}{\sqrt{g_{00}}} \sqrt{-g} \, d\Omega \, \frac{dx^i}{dx^0}$$

[where we have used the formula $-g = \gamma g_{00}$. (84.10)]. The product $\sqrt{-g} \, d\Omega$ is the invariant element of four-volume, so that the current four-vector is defined by the expression

$$j^i = \frac{\varrho c}{\sqrt{g_{00}}} \frac{dx^i}{dx^0} \tag{90.3}$$

---

† It is easily seen that the equation can also be written in the form

$$f_{ik;\, l} + F_{li;\, k} + F_{kl;\, i} = 0,$$

from which its covariance is obvious.

(the quantities dx^i/dx^0 are the rates of change of the coordinates with the "time" x^0, and *do not* constitute a four-vector). The component j^0 of the current four-vector, multiplied by $\sqrt{g_{00}}/c$, is the spatial density of charge.

For point charges the density ϱ is expressed as a sum of δ-functions, as in formula (28.1). We must, however, correct the definition of these functions for the case of curvilinear coordinates. By $\delta(\mathbf{r})$ we shall again mean the product $\delta(x^1)\,\delta(x^2)\,\delta(x^3)$, regardless of the geometrical meaning of the coordinates x^1, x^2, x^3; then the integral over dV (and not over $\sqrt{\gamma}\,dV$) is unity: $\int \delta(\mathbf{r})\,dV = 1$. With this same definition of the δ-functions, the charge density is

$$\varrho = \sum_a \frac{e_a}{\sqrt{\gamma}} \delta(\mathbf{r} - \mathbf{r}_a),$$

and the current four-vector is

$$j^i = \sum_a \frac{e_a c}{\sqrt{-g}} \delta(\mathbf{r} - \mathbf{r}_a) \frac{dx^i}{dx^0}. \tag{90.4}$$

Conservation of charge is expressed by the equation of continuity, which differs from (29.4) only in replacement of the ordinary derivatives by covariant derivatives:

$$j^i{}_{;i} = \frac{1}{\sqrt{-g}} \frac{\partial}{\partial x^i}(\sqrt{-g}\,j^i) = 0 \tag{90.5}$$

[using formula (86.9)].

The second pair of Maxwell equations (30.2) is generalized similarly; replacing the ordinary derivatives by covariant derivatives, we find:

$$F^{ik}{}_{;k} = \frac{1}{\sqrt{-g}} \frac{\partial}{\partial x^k}(\sqrt{-g}\,F^{ik}) = -\frac{4\pi}{c} j^i \tag{90.6}$$

[using formula (86.10)].

Finally the equations of motion of a charged particle in gravitational and electromagnetic fields is obtained by replacing the four-acceleration du^i/ds in (23.4) by Du^i/ds:

$$mc\frac{Du^i}{ds} = mc\left(\frac{du^i}{ds} + \Gamma^i_{kl} u^k u^l\right) = \frac{e}{c} F^{ik} u_k. \tag{90.7}$$

PROBLEM

Write the Maxwell equations in a given gravitational field in three-dimensional form (in the three-dimensional space with metric $\gamma_{\alpha\beta}$), introducing the three-vectors $\mathbf{E}$, $\mathbf{D}$ and the antisymmetric three-tensors $B_{\alpha\beta}$ and $H_{\alpha\beta}$ according to the definitions:

$$E_\alpha = F_{0\alpha}, \qquad\qquad B_{\alpha\beta} = F_{\alpha\beta},$$

$$D^\alpha = -\sqrt{g_{00}}\,F^{0\alpha}, \qquad\qquad H^{\alpha\beta} = \sqrt{g_{00}}\,F^{\alpha\beta}. \tag{1}$$

Solution: The quantities introduced above are not independent. Writing out the equations

$$F_{0\alpha} = g_{0l}g_{\alpha m}F^{lm}, \qquad F^{\alpha\beta} = g^{\alpha l}g^{\beta m}F_{lm},$$

and introducing the three-dimensional metric tensor $\gamma_{\alpha\beta} = -g_{\alpha\beta} + hg_\alpha g_\beta$ [with **g** and h from (88.11)], and using formulas (84.9) and (84.12), we get:

$$D_\alpha = \frac{E_\alpha}{\sqrt{h}} + g^\beta H_{\alpha\beta}, \quad B^{\alpha\beta} = \frac{H^{\alpha\beta}}{\sqrt{h}} + g^\beta E^\alpha - g^\alpha E^\beta. \tag{2}$$

We introduce the vectors **B** and **H**, dual to the tensors $B_{\alpha\beta}$ and $H_{\alpha\beta}$, in accordance with the definition:

$$B^\alpha = -\frac{1}{2\sqrt{\gamma}} e^{\alpha\beta\gamma} B_{\beta\gamma}, \quad H_\alpha = -\frac{1}{2}\sqrt{\gamma} e_{\alpha\beta\gamma} H^{\beta\gamma} \tag{3}$$

(see the footnote on p. 270; the minus sign is introduced so that in galilean coordinates the vector: **H** and **B** coincide with the ordinary magnetic field intensity). Then (2) can be written in the forms

$$\mathbf{D} = \frac{\mathbf{E}}{\sqrt{h}} + \mathbf{H} \times \mathbf{g}, \quad \mathbf{B} = \frac{\mathbf{H}}{\sqrt{h}} + \mathbf{g} \times \mathbf{E}. \tag{4}$$

Introducing definition (1) in (90.2), we get the equations:

$$\frac{\partial B_{\alpha\beta}}{\partial x^\gamma} + \frac{\partial B_{\gamma\alpha}}{\partial x^\beta} + \frac{\partial B_{\beta\gamma}}{\partial x^\alpha} = 0,$$

$$\frac{\partial B_{\alpha\beta}}{\partial x^0} + \frac{\partial E_\alpha}{\partial x^\beta} - \frac{\partial E_\beta}{\partial x^\alpha} = 0,$$

or, changing to the dual quantities (3):

$$\text{div } \mathbf{B} = 0, \quad \text{curl } \mathbf{E} = -\frac{1}{c\sqrt{\gamma}} \frac{\partial}{\partial t}(\sqrt{\gamma}\mathbf{B}) \tag{5}$$

($x^0 = ct$; the definitions of the operations div and curl are given in the footnote on p. 272). Similarly we find from (90.6) the equations

$$\frac{1}{\sqrt{\gamma}} \frac{\partial}{\partial x^\alpha}(\sqrt{\gamma}D^\alpha) = 4\pi\varrho,$$

$$\frac{1}{\sqrt{\gamma}} \frac{\partial}{\partial x^\beta}(\sqrt{\gamma}H^{\alpha\beta}) + \frac{1}{\sqrt{\gamma}} \frac{\partial}{\partial x^0}(\sqrt{\gamma}D^\alpha) = -4\pi\varrho \frac{dx^\alpha}{dx^0},$$

or, in three-dimensional notation:

$$\text{div } \mathbf{D} = 4\pi\varrho, \quad \text{curl } \mathbf{H} = \frac{1}{c\sqrt{\gamma}} \frac{\partial}{\partial t}(\sqrt{\gamma}\mathbf{D}) + \frac{4\pi}{c}\mathbf{s}, \tag{6}$$

where **s** is the vector with components $s^\alpha = \varrho \, dx^\alpha/dt$.

We also write the continuity equation (90.5) in three-dimensional form:

$$\frac{1}{\sqrt{\gamma}} \frac{\partial}{\partial t}(\sqrt{\gamma}\varrho) + \text{div } \mathbf{s} = 0. \tag{7}$$

The reader should note the analogy (purely formal, of course) of equations (5) and (6) to the Maxwell equations for the electromagnetic field in material media. In particular, in a static gravitational field the quantity $\sqrt{\gamma}$ drop out of the terms containing time derivatives, and relation (4) reduces to $\mathbf{D} = \mathbf{E}/\sqrt{h}$, $\mathbf{B} = \mathbf{H}/\sqrt{h}$. We may say that with respect to its effect on the electromagnetic field a static gravitational field plays the role of a medium with electric and magnetic permeabilities $\varepsilon = \mu = 1/\sqrt{h}$.

CHAPTER 11

THE GRAVITATIONAL FIELD EQUATIONS

§ 91. The curvature tensor

Let us go back once more to the concept of parallel displacement of a vector. As we said in § 85, in the general case of a curved four-space, the infinitesimal parallel displacement of a vector is defined as a displacement in which the components of the vector are not changed in a system of coordinates which is galilean in the given infinitesimal volume element.

If $x^i = x^i(s)$ is the parametric equation of a certain curve (s is the arc length measured from some point), then the vector $u^i = dx^i/ds$ is a unit vector tangent to the curve. If the curve we are considering is a geodesic, then along it $Du^i = 0$. This means that if the vector u^i is subjected to a parallel displacement from a point x^i on a geodesic curve to the point $x^i + dx^i$ on the same curve, then it coincides with the vector $u^i + du^i$ tangent to the curve at the point $x^i + dx^i$. Thus when the tangent to a geodesic moves along the curve, it is displaced parallel to itself.

On the other hand, during the parallel displacement of two vectors, the "angle" between them clearly remains unchanged. Therefore we may say that during the parallel displacement of any vector along a geodesic curve, the angle between the vector and the tangent to the geodesic remains unchanged. In other words, during the parallel displacement of a vector, its component along the geodesic must be the same at all points of the path.

Now the very important result appears that in a curved space the parallel displacement of a vector from one given point to another gives different results if the displacement is carried out over different paths. In particular, it follows from this that if we displace a vector parallel to itself along some closed contour, then upon returning to the starting point, it will not coincide with its original value.

In order to make this clear, let us consider a curved two-dimensional space, i.e. any curved surface. Figure 19 shows a portion of such a surface, bounded by three geodesic curves. Let us subject the vector 1 to a parallel displacement along the contour made up of these three curves. In moving along the line AB, the vector 1, always retaining its angle with the curve

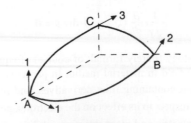

Fig. 19.

unchanged, goes over into the vector 2. In the same way, on moving along BC it goes over into 3. Finally, on moving from C to A along the curve CA, maintaining a constant angle with this curve, the vector under consideration goes over into $1'$, not coinciding with the vector 1.

We derive the general formula for the change in a vector after parallel displacement around any infinitesimal closed contour. This change ΔA_k can clearly be written in the form $\oint \delta A_k$, where the integral is taken over the given contour. Substituting in place of δA_k the expression (85.5), we have

$$\Delta A_k = \oint \Gamma^i_{kl} A_i \, dx^l \qquad (91.1)$$

(the vector A_i which appears in the integrand changes as we move along the contour).

For the further transformation of this integral, we must note the following. The values of the vector A_i at points inside the contour are not unique; they depend on the path along which we approach the particular point. However, as we shall see from the result obtained below, this non-uniqueness is related to terms of second order. We may therefore, with the first-order accuracy which is sufficient for the transformation, regard the components of the vector A_i at points inside the infinitesimal contour as being uniquely determined by their values on the contour itself by the formulas $\delta A_i = \Gamma^n_{il} A_n dx^l$, i.e. by the derivatives

$$\frac{\partial A_i}{\partial x^l} = \Gamma^n_{il} A_n. \qquad (91.2)$$

Now applying Stokes' theorem (6.19) to the integral (91.1) and considering that the area enclosed by the contour has the infinitesimal value Δf^{lm}, we get:

$$\Delta A_k = \frac{1}{2} \left[\frac{\partial (\Gamma^i_{km} A_i)}{\partial x^l} - \frac{\partial (\Gamma^i_{kl} A_i)}{\partial x^m} \right] \Delta f^{lm}$$

$$= \frac{1}{2} \left[\frac{\partial \Gamma^i_{km}}{\partial x^l} A_i - \frac{\partial \Gamma^i_{kl}}{\partial x^m} A_i + \Gamma^i_{km} \frac{\partial A_i}{\partial x^l} - \Gamma^i_{kl} \frac{\partial A_i}{\partial x^m} \right] \Delta f^{lm}.$$

Substituting the values of the derivatives (91.2), we get finally:

$$\Delta A_k = \frac{1}{2} R^i_{klm} A_i \Delta f^{lm}, \qquad (91.3)$$

where R^i_{klm} is a tensor of the fourth rank:

$$R^i_{klm} = \frac{\partial \Gamma^i_{km}}{\partial x^l} - \frac{\partial \Gamma^i_{kl}}{\partial x^m} + \Gamma^i_{nl} \Gamma^n_{km} - \Gamma^i_{nm} \Gamma^n_{kl}. \qquad (91.4)$$

That R^i_{klm} is a tensor is clear from the fact that in (91.3) the left side is a vector – the difference ΔA_k between the values of vectors at one and the same point. The tensor R^i_{klm} is called the *curvature tensor* or the *Riemann tensor*.

It is easy to obtain a similar formula for a contravariant vector A^k. To do this we note, since under parallel displacement a scalar does not change, that $\Delta(A^k B_k) = 0$, where B_k is any covariant vector. With the help of (91.3), we can then have

$$\Delta(A^k B_k) = A^k \Delta B_k + B_k \Delta A^k = \frac{1}{2} A^k B_i R^i_{klm} \Delta f^{lm} + B_k \Delta A^k =$$

$$= B_k \left(\Delta A^k + \frac{1}{2} A^i R^k_{ilm} \Delta f^{lm} \right) = 0,$$

or, in view of the arbitrariness of the vector B_k,

$$\Delta A^k = -\tfrac{1}{2} R^k{}_{ilm} A^i \Delta f^{lm}. \tag{91.5}$$

If we twice differentiate a vector A_i covariantly with respect to x^k and x^l, then the result generally depends on the order of differentiation, contrary to the situation for ordinary differentiation. It turns out that the difference $A_{i;\,k;\,l} - A_{i;\,l;\,k}$ is given by the same curvature tensor which we introduced above. Namely, one finds the formula

$$A_{i;\,k;\,l} - A_{i;\,l;\,k} = A_m R^m{}_{ikl}, \tag{91.6}$$

which is easily verified by direct calculation in the locally-geodesic coordinate system Similarly, for a contravariant vector,†

$$A^i{}_{;k;l} - A^i{}_{;l;k} = -A^m R^i{}_{mkl}. \tag{91.7}$$

Finally, it is easy to obtain similar formulas for the second derivatives of tensors [this is done most easily by considering, for example, a tensor of the form $A_i B_k$, and using formulas (91.6) and (91.7); because of the linearity, the formulas thus obtained must be valid for an arbitrary tensor A_{ik}]. Thus

$$A_{ik;l;m} - A_{ik;m;l} = A_{in} R^n{}_{klm} + A_{nk} R^n{}_{ilm}. \tag{91.8}$$

Clearly, in a flat space the curvature tensor is zero, for, in a flat space, we can choose coordinates such that over all the space all the $\Gamma^i_{kl} = 0$, and therefore also $R^i{}_{klm} = 0$. Because of the tensor character of $R^i{}_{klm}$ it is then equal to zero also in any other coordinate system. This is related to the fact that in a flat space parallel displacement is a single-valued operation, so that in making a circuit of a closed contour a vector does not change.

The converse theorem is also valid: if $R^i{}_{klm} = 0$, then the space is flat. Namely, in any space we can choose a coordinate system which is galilean over a given infinitesimal region. If $R^i{}_{klm} = 0$, then parallel displacement is a unique operation, and then by a parallel displacement of the galilean system from the given infinitesimal region to all the rest of the space, we can construct a galilean system over the whole space, which proves that the space is Euclidean.

Thus the vanishing or nonvanishing of the curvature tensor is a criterion which enables us to determine whether a space is flat or curved.

We note that although in a curved space we can also choose a coordinate system which will be locally geodesic at a given point, at the same time the curvature tensor at this same point does not go to zero (since the derivatives of the Γ^i_{kl} do not become zero along with the Γ^i_{kl}).

PROBLEMS

1. Determine the relative four-acceleration of two particles moving along infinitely close geodesic world lines.

Solution: Consider a family of geodesics differing in the value of some parameter v; in other words, the coordinates of the world point are expressed as functions $x^i = x^i(s, v)$, so that for each $v =$ const, this is the equation of a geodesic (where s is the length of interval measured along the line from its point of intersection with some given hypersurface). We introduce the four-vector

† Formula (91.7) can also be obtained directly from (91.6) by raising the index i and using the symmetry properties of the tensor R_{iklm} (§ 92).

$$\eta' = \frac{\partial x^i}{\partial v}\, \delta v \equiv v^i \delta v,$$

joining points on infinitely close geodesics (corresponding to parameter values v and $v + dv$) that have the same value of s.

From the definition of a covariant derivative and the equality $\partial u^i/\partial v = \partial v^i/\partial s$ (where $u^i = \partial x^i/\partial s$), it follows that

$$u^i_{;k} v^k = v^i_{;k} u^k. \tag{1}$$

Now consider the second derivative:

$$\frac{D^2 v^i}{ds^2} \equiv (v^i_{;k} u^k)_{;l} u^l = (u^i_{;k} v^k)_{;l} v^l = u^i_{;k;l} v^k u^l + u^i_{;k} v^k_{;l} u_l.$$

We use (1) again in the second term, and change the order of covariant differentiation in the first term using (91.7), and find:

$$\frac{D^2 v^i}{ds^2} = (u^i_{;l} u^l)_{;k} v^k + u^m R^i_{mkl} u^k v^l.$$

The first term is zero, since $u^i_{;l} u^l = 0$ along geodesics. Introducing the constant factor δv, we get the final equation:

$$\frac{D^2 \eta^i}{ds^2} = R^i_{klm} u^k u^l \eta^m, \tag{2}$$

which is called the geodesic deviation.

2. Write the Maxwell equations in the absence of charges for the 4-potential in the Lorentz gauge.

Solution: The covariant generalization of the condition (46.9) has the form:

$$A^i_{;i} = 0. \tag{1}$$

Using formula (91.7) the Maxwell equations can be written as

$$F_{ik}^{;k} = A_{k;i}^{;k} - A_{i;k}^{;k} = A_k^{;k}_{;i} + A^m R_{im} - A_{i;k}^{;k} = 0$$

with R_{ik} from (92.6). Then from Eq. (1):

$$A_{i;k}^{;k} - R_{ik} A^k = 0. \tag{2}$$

§ 92. Properties of the curvature tensor

The curvature tensor has symmetry properties which can be made completely apparent by changing from mixed components R^i_{klm} to covariant ones:

$$R_{iklm} = g_{in} R^n_{klm}.$$

By means of simple transformations it is easy to obtain the following expression:

$$R_{iklm} = \frac{1}{2}\left(\frac{\partial^2 g_{im}}{\partial x^k \partial x^l} + \frac{\partial^2 g_{kl}}{\partial x^i \partial x^m} - \frac{\partial^2 g_{il}}{\partial x^k \partial x^m} - \frac{\partial^2 g_{km}}{\partial x^i \partial x^l} \right) + g_{np}(\Gamma^n_{kl}\Gamma^p_{im} - \Gamma^n_{km}\Gamma^p_{il}). \tag{92.1}$$

From this expression one sees immediately the following symmetry properties:

$$R_{iklm} = - R_{kilm} = - R_{ikml} \tag{92.2}$$

$$R_{iklm} = R_{lmik}, \tag{92.3}$$

i.e. the tensor is antisymmetric in each of the index pairs i, k and l, m, and is symmetric under the interchange of the two pairs with one another. In particular, all components R_{iklm}, in which $i = k$ or $l = m$, are zero.

One also verifies that the cyclic sum of components of R_{iklm}, formed by permutation of any three indices, is equal to zero; e.g.,

$$R_{iklm} + R_{imkl} + R_{ilmk} = 0 .   \tag{92.4}$$

[The other relations of this type are obtained from (92.4) automatically, because of the properties (92.2)–(92.3).]

Finally, we also prove the *Bianchi identity*:

$$R^n{}_{ikl;m} + R^n{}_{imk;l} + R^n{}_{ilm;k} = 0 .   \tag{92.5}$$

It is most conveniently verified by using a locally-geodesic coordinate system. Because of its tensor character, the relation (92.5) will then be valid in any other system. Differentiating (91.4) and then substituting in it $\Gamma^l_{kl} = 0$, we find for the point under consideration

$$R^n{}_{ikl;m} = \frac{\partial R^n{}_{ikl}}{\partial x^m} = \frac{\partial^2 \Gamma^n_{il}}{\partial x^m \partial x^k} - \frac{\partial^2 \Gamma^n_{ik}}{\partial x^m \partial x^l} .$$

With the aid of this expression it is easy to verify that (92.5) actually holds.

From the curvature tensor we can, by contraction, construct a tensor of the second rank. This contraction can be carried out in only one way: contraction of the tensor R_{iklm} on the indices i and k or l and m gives zero because of the antisymmetry in these indices, while contraction on any other pair always gives the same result, except for sign. We define the tensor R_{ik} (the *Ricci tensor*) as†

$$R_{ik} = g^{lm} R_{limk} = R^l{}_{ilk} .   \tag{92.6}$$

According to (91.4), we have:

$$R_{ik} = \frac{\partial \Gamma^l_{ik}}{\partial x^l} - \frac{\partial \Gamma^l_{il}}{\partial x^k} + \Gamma^l_{ik} \Gamma^m_{lm} - \Gamma^m_{il} \Gamma^l_{km} .   \tag{92.7}$$

This tensor is clearly symmetric:

$$R_{ik} = R_{ki} .   \tag{92.8}$$

Finally, contracting R_{ik}, we obtain the invariant

$$R = g^{ik} R_{ik} = g^{il} g^{km} R_{iklm},   \tag{92.9}$$

which is called the *scalar curvature* of the space.

The components of the tensor R_{ik} satisfy a differential identity obtained by contracting the Bianchi identity (92.5) on the pairs of indices ik and ln:

$$R^l{}_{m;l} = \frac{1}{2} \frac{\partial R}{\partial x^m} .   \tag{92.10}$$

Because of the relations (92.2)–(92.4) not all the components of the curvature tensor are independent. Let us determine the number of independent components.

† In the literature one also finds another definition of the tensor R_{ik}, using contraction of R_{iklm} on the first and last indices. This definition differs in sign from the one used here.

The definition of the curvature tensor as given by the formulas written above applies to a space of an arbitrary number of dimensions. Let us first consider the case of two dimensions, i.e. an ordinary surface; in this case (to distinguish them from four-dimensional quantities) we denote the curvature tensor by P_{abcd} and the metric tensor by γ_{ab}, where the indices a, b, ... run through the values 1, 2. Since in each of the pairs ab and cd the two indices must have different values, it is obvious that all the non-vanishing components of the curvature tensor coincide or differ in sign. Thus in this case there is only one independent component, for example P_{1212}. It is easily found that the scalar curvature is

$$P = \frac{2P_{1212}}{\gamma}, \quad \gamma \equiv |\gamma_{ab}| = \gamma_{11}\gamma_{22} - (\gamma_{12})^2. \tag{92.11}$$

The quantity $P/2$ coincides with the *Gaussian curvature K* of the surface:

$$\frac{P}{2} = K = \frac{1}{\rho_1 \rho_2} \tag{92.12}$$

where the ρ_1, ρ_2 are the principal radii of curvature of the surface at the particular point (remember that ρ_1 and ρ_2 are assumed to have the same sign if the corresponding centres of curvature are on one side of the surface, and opposite signs if the centres of curvature lie on opposite sides of the surface; in the first case $K > 0$, while in the second $K < 0$.†

Next we consider the curvature tensor in three-dimensional space; we denote it by $P_{\alpha\beta\gamma\delta}$ and the metric tensor by $\gamma_{\alpha\beta}$, where the indices α, β run through values 1, 2, 3. The index pairs $\alpha\beta$ and $\gamma\delta$ run through three essentially different sets of values; 23, 31, and 12 (permutation of indices in a pair merely changes the sign of the tensor component). Since the tensor $P_{\alpha\beta\gamma\delta}$ is symmetric under interchange of these pairs, there are all together $3 \cdot 2/2$ independent components with different pairs of indices, and three components with identical pairs. The identity (92.4) adds no new restrictions. Thus, in three-dimensional space the curvature tensor has six independent components. The symmetric tensor $P_{\alpha\beta}$ has the same number. Thus, from the linear relations $P_{\alpha\beta} = g^{\gamma\delta} P_{\gamma\alpha\delta\beta}$ all the components of the tensor $P_{\alpha\beta\gamma\delta}$ can be expressed in terms of $P_{\alpha\beta}$ and the metric tensor $\gamma_{\alpha\beta}$ (see problem 1). If we choose a system of coordinates that is cartesian at the particular point, then by a suitable rotation we can bring the tensor $P_{\alpha\beta}$ to principal axes.‡ Thus the curvature tensor of a three-dimensional space at a given point is determined by three quantities.§

---

† Formula (92.12) is easy to get by writing the equation of the surface in the vicinity of the given point ($x = y = 0$) in the form $z = (x^2/2\rho_1) + (y^2/2\rho_2)$.

Then the square of the line element on it is

$$dl^2 = \left(1 + \frac{x^2}{\rho_1^2}\right)dx^2 + \left(1 + \frac{y^2}{\rho_2^2}\right)dy^2 + 2\frac{xy}{\rho_1\rho_2}dx\,dy.$$

Calculation of P_{1212} at the point $x = y = 0$ using formula (92.1) (in which only terms with second derivatives of the $\gamma_{\alpha\beta}$ are needed) leads to (92.12).

‡ For the actual determination of the principal values of the tensor $P_{\alpha\beta}$ there is no need to transform to a coordinate system that is cartesian at the given point. These values can be found by determining the roots λ of the equation $|P_{\alpha\beta} - \lambda\gamma_{\alpha\beta}| = 0$.

§ Knowledge of the tensor $P_{\alpha\beta\gamma\delta}$ enables us to determine the Gaussian curvature K of an arbitrary surface in the space. Here we note only that if the x^1, x^2, x^3 are an orthogonal coordinate system, then

$$K = \frac{P_{1212}}{\gamma_{11}\gamma_{22} - (\gamma_{12})^2}$$

is the Gaussian curvature for the "plane" perpendicular (at the given point) to the x^3 axis; by a "plane" we mean a surface formed by geodesic lines.

Finally we go to four-dimensional space. The pairs of indices ik and lm in this case run through six different sets of values: 01, 02, 03, 23, 31, 12. Thus there are six components of R_{iklm} with identical, and $6 \cdot 5/2$ with different, pairs of indices. The latter, however, are still not independent of one another; the three components for which all four indices are different are related, because of (92.4), by the identity:

$$R_{0123} + R_{0312} + R_{0231} = 0 . \tag{92.13}$$

Thus, in four-space the curvature tensor has a total of twenty independent components.

By choosing a coordinate system that is galilean at the given point and considering the transformations that rotate this system (so that the g_{ik} at the point are not changed), one can achieve the vanishing of six of the components of the curvature tensor (since there are six independent rotations of a four-dimensional coordinate system). Thus, in the general case the curvature of four-space is determined at each point by fourteen quantities.

If $R_{ik} = 0$,[†] then the curvature tensor has a total of ten independent components in an arbitrary coordinate system. By a suitable transformation we can then bring the tensor R_{iklm} (at the given point of four-space) to a "canonical" form, in which its components are expressed in general in terms of four independent quantities; in special cases this number may be even smaller.

If, however, $R_{ik} \neq 0$, then the same classification can be used for the curvature tensor after one has subtracted from it a particular part that is expressible in terms of the components R_{ik}. Namely, we construct the tensor[‡]

$$C_{iklm} = R_{iklm} - \tfrac{1}{2} R_{il} g_{km} + \tfrac{1}{2} R_{im} g_{kl} + \tfrac{1}{2} R_{kl} g_{im} - \tfrac{1}{2} R_{km} g_{il} + \tfrac{1}{6} R(g_{il} g_{km} - g_{im} g_{kl}). \tag{92.14}$$

It is easy to see that this tensor has all the symmetry properties of the tensor R_{iklm}, but vanishes when contracted on a pair of indices (il or km).

Let us show how one classifies the possible types of canonical forms of the curvature tensor when $R_{ik} = 0$. (A. Z. Petrov, 1950).

We shall assume that the metric at the given point in four-dimensional space has been brought to galilean form. We write the set of twenty independent components of the tensor R_{iklm} as a collection of three-dimensional tensors defined as follows:

$$A_{\alpha\beta} = R_{0\alpha0\beta}, \qquad C_{\alpha\beta} = \tfrac{1}{4} e_{\alpha\gamma\delta} e_{\beta\lambda\mu} R_{\gamma\delta\lambda\mu}, \qquad B_{\alpha\beta} = \tfrac{1}{2} e_{\alpha\gamma\delta} R_{0\beta\gamma\delta} \tag{92.15}$$

($e_{\alpha\beta\gamma}$ is the unit antisymmetric tensor; since the three-dimensional metric is cartesian, there is no need to deal with the difference between upper and lower indices in the summation). The tensors $A_{\alpha\beta}$ and $C_{\alpha\beta}$ are symmetric by definition; the tensor $B_{\alpha\beta}$ is asymmetric, while its trace is zero because of (92.13). According to the definitions (92.15) we have, for example,

$$B_{11} = R_{0123}, \quad B_{21} = R_{0131}, \quad B_{31} = R_{0112}, \quad C_{11} = R_{2323}, \dots$$

---

† We shall see later (§ 95) that the curvature tensor for the gravitational field in vacuum has this property.

‡ This complicated expression can be written more compactly in the form:

$$C_{iklm} = R_{iklm} - R_{l[i} g_{k]m} + R_{m[i} g_{k]i} + \tfrac{1}{3} R g_{i[i} g_{k]m},$$

where the square brackets imply antisymmetrization over the indices contained in them:

$$A_{[ik]} = \tfrac{1}{2} (A_{ik} - A_{ki}) .$$

The tensor (92.14) is called the *Weyl tensor*.

It is easy to see that the conditions $R_{km} = g^{il} R_{iklm} = 0$ are equivalent to the following relations between the components of the tensors (92.15):

$$A_{\alpha\alpha} = 0, \quad B_{\alpha\beta} = B_{\beta\alpha}, \quad A_{\alpha\beta} = -C_{\alpha\beta}. \tag{92.16}$$

We also introduce the symmetric complex tensor

$$D_{\alpha\beta} = \tfrac{1}{2}(A_{\alpha\beta} + 2iB_{\alpha\beta} - C_{\alpha\beta}) - A_{\alpha\beta} + iB_{\alpha\beta}. \tag{92.17}$$

This combining of the two real three-dimensional tensors $A_{\alpha\beta}$ and $B_{\alpha\beta}$ into one complex tensor corresponds precisely to the combination (in § 25) of the two vectors $\mathbf{E}$ and $\mathbf{H}$ into the complex vector $\mathbf{F}$, while the resulting relation between $D_{\alpha\beta}$ and the four-tensor R_{iklm} corresponds to the relation between $\mathbf{F}$ and the four-tensor F_{ik}. It then follows that four-dimensional transformations of the tensor R_{iklm} are equivalent to three-dimensional complex rotations carried out on the tensor $D_{\alpha\beta}$.

With respect to these rotations one can define eigenvalues $\lambda = \lambda' + i\lambda''$ and eigenvectors n_α (complex, in general) as solutions of the system of equations

$$D_{\alpha\beta} n_\beta = \lambda n_\alpha. \tag{92.18}$$

The quantities λ are the invariants of the curvature tensor. Since the trace $D_{\alpha\alpha} = 0$, the sum of the roots of equation (4) is zero:

$$\lambda^{(1)} + \lambda^{(2)} + \lambda^{(3)} = 0 .$$

Depending on the number of independent eigenvectors n_α, we arrive at the following classification of possible cases of reduction of the curvature tensor to the *canonical Petrov types* I–III.

(I) There are three independent eigenvectors. Then their squares $n_\alpha n^\alpha$ are different from zero and by a suitable rotation we can bring the tensor $D_{\alpha\beta}$, and with it $A_{\alpha\beta}$ and $B_{\alpha\beta}$, to diagonal form:

$$A_{\alpha\beta} = \begin{pmatrix} \lambda^{(1)'} & 0 & 0 \\ 0 & \lambda^{(2)'} & 0 \\ 0 & 0 & -\lambda^{(1)'} - \lambda^{(2)'} \end{pmatrix}, \quad B_{\alpha\beta} = \begin{pmatrix} \lambda^{(1)''} & 0 & 0 \\ 0 & \lambda^{(2)''} & 0 \\ 0 & 0 & -\lambda^{(1)''} - \lambda^{(2)''} \end{pmatrix}.$$

$$\tag{92.19}$$

In this case the curvature tensor has four independent invariants.†

The complex invariants $\lambda^{(1)}$, $\lambda^{(2)}$ are expressed algebraically in terms of the complex scalars

$$I_1 = \frac{1}{48} (R_{iklm} R^{iklm} - i R_{iklm} \overset{*}{R}{}^{iklm}),$$

$$I_2 = \frac{1}{96} (R_{iklm} R^{lmpr} R_{pr}{}^{ik} + i R_{iklm} R^{lmpr} \overset{*}{R}_{pr}{}^{ik}), \tag{92.20}$$

where the asterisk over a symbol denotes the dual tensor:

$$\overset{*}{R}_{iklm} = \tfrac{1}{2} E_{ikpr} R^{pr}{}_{lm}.$$

† The degenerate case when $\lambda^{(1)'} = \lambda^{(2)'}, = \lambda^{(1)''} = \lambda^{(2)''}$ is called type D in the literature.

Calculating I_1 and I_2 using (92.19), we obtain:

$$I_1 = \tfrac{1}{3}(\lambda^{(1)2} + \lambda^{(2)2} + \lambda^{(3)2}), \quad I_2 = \tfrac{1}{2}\lambda^{(1)}\lambda^{(2)}(\lambda^{(1)} + \lambda^{(2)}). \tag{92.21}$$

These formulas enable us to calculate $\lambda^{(1)}$, $\lambda^{(2)}$ starting from the values of R_{iklm} in any reference system.

(II) There are two independent eigenvectors. The square of one of them is then equal to zero, so that it cannot be chosen as the direction of one of the coordinate axes. One can, however, take it to lie in the x^1, x^2 plane; then $n_2 = in_1$, $n_3 = 0$. The corresponding equations (92.18) give:

$$D_{11} + i D_{12} = \lambda, \quad D_{22} - i D_{12} = \lambda,$$

so that

$$D_{11} = \lambda - i\mu, \quad D_{22} = \lambda + i\mu, \quad D_{12} = \mu.$$

The complex quantity $\lambda = \lambda' + i\lambda''$ is a scalar and cannot be changed. But the quantity μ can be given any nonzero value by a suitable complex rotation; we can therefore, without loss of generality, assume it to be real. As a result we get the following canonical type for the real tensors $A_{\alpha\beta}$ and $B_{\alpha\beta}$:

$$A_{\alpha\beta} = \begin{pmatrix} \lambda' & \mu & 0 \\ \mu & \lambda' & 0 \\ 0 & 0 & -2\lambda' \end{pmatrix}, \quad B_{\alpha\beta} = \begin{pmatrix} \lambda'' - \mu & 0 & 0 \\ 0 & \lambda'' + \mu & 0 \\ 0 & 0 & -2\lambda'' \end{pmatrix}. \tag{92.22}$$

In this case there are just two invariants λ' and λ''. Then, in accordance with (92.22), $I_1 = \lambda^2$, $I_2 = \lambda^3$, so that $I_1^3 = I_2^2$.

(III) There is just one eigenvector, and its square is zero. All the eigenvalues λ are then identical and consequently equal to zero. The solutions of equations (92.18) can be brought to the form $D_{11} = D_{22} = D_{12} = 0$, $D_{13} = \mu$, $D_{23} = i\mu$, so that

$$A_{\alpha\beta} = \begin{pmatrix} 0 & 0 & \mu \\ 0 & 0 & 0 \\ \mu & 0 & 0 \end{pmatrix}, \quad B_{\alpha\beta} = \begin{pmatrix} 0 & 0 & 0 \\ 0 & 0 & \mu \\ 0 & \mu & 0 \end{pmatrix}. \tag{92.23}$$

In this case the curvature tensor has no invariants at all and we have a peculiar situation: the four-space is curved, but there are no invariants which could be used as a measure of its curvature.†

PROBLEMS

1. Express the curvature tensor $P_{\alpha\beta\gamma\delta}$ of three-dimensional space in terms of the second-rank tensor $P_{\alpha\beta}$.

Solution: We look for $P_{\alpha\beta\gamma\delta}$ in the form

$$P_{\alpha\beta\gamma\delta} = A_{\alpha\gamma}\gamma_{\beta\delta} - A_{\alpha\delta}\gamma_{\beta\gamma} + A_{\beta\delta}\gamma_{\alpha\gamma} - A_{\beta\gamma}\gamma_{\alpha\delta},$$

† The same situation occurs in the degenerate case (II) when $\lambda' = \lambda'' = 0$; this case is called type N.

which satisfies the symmetry conditions; here $A_{\alpha\beta}$ is some symmetric tensor whose relation to $P_{\alpha\beta}$ is determined by contracting the expression we have written on the indices α and γ. We thus find:

$$P_{\alpha\beta} = A\gamma_{\alpha\beta} + A_{\alpha\beta}, \quad A_{\alpha\beta} = P_{\alpha\beta} - \tfrac{1}{4}P\gamma_{\alpha\beta},$$

and finally,

$$P_{\alpha\beta\gamma\delta} = P_{\alpha\gamma}\gamma_{\beta\delta} - P_{\alpha\delta}\gamma_{\beta\gamma} + P_{\beta\delta}\gamma_{\alpha\gamma} - P_{\beta\gamma}\gamma_{\alpha\delta} + \frac{P}{2}(\gamma_{\alpha\delta}\gamma_{\beta\gamma} - \gamma_{\alpha\gamma}\gamma_{\beta\delta}).$$

2. Calculate the components of the tensors R_{iklm} and R_{ik} for a metric in which $g_{ik} = 0$ for $i \neq k$.

Solution: We represent the nonzero components of the metric tensor in the form

$$g_{ii} = e_i e^{2F_i}, \quad e_0 = 1, \quad e_\alpha = -1.$$

The calculation according to formula (92.4) gives the following expressions for the nonzero components of the curvature tensor:

$$R_{lilk} = e_l e^{2F_i}[F_{l,k}F_{ki} + F_{i,k}F_{l,i} - F_{l,i}F_{l,k} - F_{l,i,k}], i \neq k \neq l;$$

$$R_{lili} = e_l e^{2F_i}(F_{i,i}F_{l,i} - F_{l,i}^2 - F_{l,i,i}) + e_i e^{2F_i}(F_{l,l}F_{i,l} - F_{i,l}^2 - F_{i,l,l}) -$$

$$- e_l e^{2F_i}\sum_{m \neq i, l} e_i e_m e^{2(F_i - F_m)}F_{i,m}F_{l,m}i \neq l$$

(no summation over repeated indices!). The subscripts preceded by a comma denote ordinary differentiation with respect to the corresponding coordinate.

Contracting the curvature tensor on two indices, we obtain:

$$R_{ik} = \sum_{l \neq i, k}(F_{l,k}F_{k,i} + F_{i,k}F_{l,i} - F_{l,i}F_{l,k} - F_{l,i,k}), i \neq k;$$

$$R_{ii} = \sum_{l \neq i}[F_{i,i}F_{l,i} - F_{l,i}^2 - F_{l,i,i} + e_i e_l e^{2(F_i - F_l)}(F_{l,l}F_{i,l} - F_{i,l}^2 - F_{i,l,l} - F_{i,l}\sum_{m \neq i, l} f_{m,l})].$$

§ 93. The action function for the gravitational field

To arrive at the equations determining the gravitational field, it is necessary first to determine the action S_g for this field. The required equations can then be obtained by varying the sum of the actions of field plus material particles.

Just as for the electromagnetic field, the action S_g must be expressed in terms of a scalar integral $\int G\sqrt{-g}\, d\Omega$, taken over all space and over the time coordinate x^0 between two given values. To determine this scalar we shall start from the fact that the equations of the gravitational field must contain derivatives of the "potentials" no higher than the second (just as is the case for the electromagnetic field). Since the field equations are obtained by varying the action, it is necessary that the integrand G contain derivatives of g_{ik} no higher than first order; thus G must contain only the tensor g_{ik} and the quantities Γ_{kl}^i.

However, it is impossible to construct an invariant from the quantities g_{ik} and Γ_{kl}^i alone. This is immediately clear from the fact that by a suitable choice of coordinate system we can always make all the quantities Γ_{kl}^i zero at a given point. There is, however, the scalar R (the curvature of the four-space), which though it contains in addition to the g_{ik} and its first derivatives also the second derivatives of g_{ik}, is linear in the second derivatives. Because or this linearity, the invariant integral $\int R\sqrt{-g}\, d\Omega$ can be transformed by means of Gauss' theorem to the integral of an expression not containing the second derivatives. Namely, $\int R\sqrt{-g}\,d\Omega$ can be presented in the form

$$\int R\sqrt{-g}\,d\Omega = \int G\sqrt{-g}\,d\Omega + \int \frac{\partial(\sqrt{-g}\,w^i)}{\partial x^i}\,d\Omega,$$

where G contains only the tensor g_{ik} and its first derivatives, and the integrand of the second integral has the form of a divergence of a certain quantity w^i (the detailed calculation is given at the end of this section). According to Gauss' theorem, this second integral can be transformed into an integral over a hypersurface surrounding the four-volume over which the integration is carried out in the other two integrals. When we vary the action, the variation of the second term on the right vanishes, since in the principle of least action, the variations of the field at the limits of the region of integration are zero. Consequently, we may write

$$\delta \int R\sqrt{-g}\,d\Omega = \delta \int G\sqrt{-g}\,d\Omega.$$

The left side is a scalar; therefore the expression on the right is also a scalar (the quantity G itself is, of course, not a scalar).

The quantity G satisfies the condition imposed above, since it contains only the g_{ik} and its derivatives. Thus we may write

$$\delta S_g = -\frac{c^3}{16\pi k}\,\delta \int G\sqrt{-g}\,d\Omega = -\frac{c^3}{16\pi k}\,\delta \int R\sqrt{-g}\,d\Omega, \qquad (93.1)$$

where k is a new universal constant. Just as was done for the action of the electromagnetic field in § 27, we can see that the constant k must be positive (see the end of this section).

The constant k is called the *gravitational constant*. The dimensions of k follow from (93.1). The action has dimensions gm-cm^2-sec^{-1}; all the coordinates have the dimensions cm, the g_{ik} are dimensionless, and so R has dimensions cm^{-2}. As a result, we find that k has the dimensions cm^3-gm^{-1}-sec^{-2}. Its numerical value is

$$k = 6.67 \times 10^{-8}\ \text{cm}^3\text{-gm}^{-1}\text{-sec}^{-2}. \qquad (93.2)$$

We note that we could have set k equal to unity (or any other dimensionless constant). However, this would fix the unit of mass.†

Finally, let us calculate the quantity G of (93.1). From the expression (92.10) for R_{ik}, we have

$$\sqrt{-g}\,R = \sqrt{-g}\,g^{ik}R_{ik} = \sqrt{-g}\left\{ g^{ik}\frac{\partial\Gamma_{ik}^l}{\partial x^l} - g^{ik}\frac{\partial\Gamma_{il}^l}{\partial x^k} + g^{ik}\Gamma_{ik}^l\Gamma_{lm}^m - g^{ik}\Gamma_{il}^m\Gamma_{km}^l \right\}.$$

In the first two terms on the right, we have

† If one sets $k = c^2$, the mass is measured in cm, where 1 cm $= 1.35 \times 10^{28}$ gm. Sometimes one uses in place of k the quantity

$$\varkappa = \frac{8\pi k}{c^2} = 1.86 \times 10^{-27}\ \text{cm gm}^{-1},$$

which is called the Einstein gravitational constant.

$$\sqrt{-g}\, g^{ik}\frac{\partial \Gamma_{ik}^{l}}{\partial x^{l}} = \frac{\partial}{\partial x^{l}}(\sqrt{-g}\, g^{ik}\Gamma_{ik}^{l}) - \Gamma_{ik}^{l}\frac{\partial}{\partial x^{l}}(\sqrt{-g}g^{ik}),$$

$$\sqrt{-g}\, g^{ik}\frac{\partial \Gamma_{il}^{l}}{\partial x^{k}} = \frac{\partial}{\partial x^{k}}(\sqrt{-g}\, g^{ik}\Gamma_{il}^{l}) - \Gamma_{il}^{l}\frac{\partial}{\partial x^{k}}(\sqrt{-g}g^{ik}).$$

Dropping the total derivatives, we find

$$\sqrt{-g}\, G = \Gamma_{im}^{m}\frac{\partial}{\partial x^{k}}(\sqrt{-g}g^{ik}) - \Gamma_{ik}^{l}\frac{\partial}{\partial x^{l}}(\sqrt{-g}g^{ik}) - (\Gamma_{il}^{m}\Gamma_{km}^{l} - \Gamma_{ik}^{l}\Gamma_{lm}^{m})g^{ik}\sqrt{-g}\,.$$

With the aid of formulas (86.5)–(86.8), we find that the first two terms on the right are equal to $\sqrt{-g}$ multiplied by

$$2\Gamma_{ik}^{l}\Gamma_{lm}^{i}g^{mk} - \Gamma_{im}^{m}\Gamma_{kl}^{i}g^{kl} - \Gamma_{ik}^{l}\Gamma_{lm}^{m}g^{ik} = g^{ik}(2\Gamma_{mk}^{l}\Gamma_{li}^{m} - \Gamma_{lm}^{m}\Gamma_{ik}^{l} - \Gamma_{ik}^{l}\Gamma_{lm}^{m})$$

$$= 2g^{ik}(\Gamma_{il}^{m}\Gamma_{km}^{l} - \Gamma_{ik}^{l}\Gamma_{lm}^{m})\,.$$

Finally, we have

$$G = g^{ik}(\Gamma_{il}^{m}\Gamma_{km}^{l} - \Gamma_{ik}^{l}\Gamma_{lm}^{m})\,. \tag{93.3}$$

The components of the metric tensor are the quantities which determine the gravitational field. Therefore in the principle of least action for the gravitational field it is the quantities g_{ik} which are subjected to variation. However, it is necessary here to make the following fundamental reservation. Namely, we cannot claim now that in an actually realizable field the action integral has a minimum (and not just an extremum) with respect to *all* possible variations of the g_{ik}. This is related to the fact that not every change in the g_{ik} is associated with a change in the space-time metric, i.e. with a real change in the gravitational field. The components g_{ik} also change under a simple transformation of coordinates connected merely with the shift from one system to another in one and the same space-time. Each such coordinate transformation is generally an aggregate of four independent transformations. In order to exclude such changes in g_{ik} which are not associated with a change in the metric, we can impose four auxiliary conditions and require the fulfilment of these conditions under the variation. Thus, when the principle of least action is applied to a gravitational field, we can assert only that we can impose auxiliary conditions on the g_{ik}, such that when they are fulfilled the action has a minimum with respect to variations of the g_{ik}.†

Keeping these remarks in mind, we now show that the gravitational constant must be positive. As the four auxiliary conditions mentioned, we use the vanishing of the three components $g_{0\alpha}$, and the constancy of the determinant $|g_{\alpha\beta}|$ made up from the components of $g_{\alpha\beta}$:

$$g_{0\alpha} = 0, \quad |g_{\alpha\beta}| = \text{const};$$

from the last of these conditions we have

---

† We must emphasize, however, that everything we have said has no effect on the derivation of the field equations from the principle of least action (§ 95). These equations are already obtained as a result of the requirement that the action be an extremum (i.e., vanishing of the first derivative), and not necessarily a minimum. Therefore in deriving them we can vary all of the g_{ik} independently.

$$g^{\alpha\beta}\frac{\partial g_{\alpha\beta}}{\partial x^0} = \frac{\partial}{\partial x^0} |g_{\alpha\beta}| = 0.$$

We are here interested in those terms in the integrand of the expression for the action which contain derivatives of g_{ik} with respect to x^0 (cf. p. 287). A simple calculation using (93.3) shows that these terms in G are

$$-\frac{1}{4} g^{\alpha\beta} g^{\gamma\delta} g^{00} \frac{\partial g_{\alpha\gamma}}{\partial x^0} \frac{\partial g_{\beta\delta}}{\partial x^0}.$$

It is easy to see that this quantity is essentially negative. Namely, choosing a spatial system of coordinates which is cartesian at a given point at a given moment of time (so that $g_{\alpha\beta} = g^{\alpha\beta} = -\delta_{\alpha\beta}$), we obtain:

$$-\frac{1}{4} g^{00} \left(\frac{\partial g_{\alpha\beta}}{\partial x^0} \right)^2,$$

and, since $g^{00} = 1/g_{00} > 0$, the sign of the quantity is obvious.

By a sufficiently rapid change of the components $g_{\alpha\beta}$ with the time x^0 (within the time interval between the limits of integration of x^0) the quantity G can consequently be made as large as one likes. If the constant k were negative, the action would then decrease without limit (taking on negative values of arbitrarily large absolute magnitude), that is, there could be no minimum.

§ 94. The energy-momentum tensor

In § 32 the general rule was given for calculating the energy-momentum tensor of any physical system whose action is given in the form of an integral (32.1) over four-space. In curvilinear coordinates this integral must be written in the form

$$S = \frac{1}{c} \int \Lambda \sqrt{-g}\, d\Omega \tag{94.1}$$

(in galilean coordinates $g = -1$, and S goes over into $\int \Lambda\, dV\, dt$). The integration extends over all the three-dimensional space and over the time between two given moments, i.e., over the infinite region of four-space contained between two hypersurfaces.

As already discussed in § 32, the energy-momentum tensor, calculated from the formula (32.5), is generally not symmetric, as it should be. In order to symmetrize it, we had to add to (32.5) suitable terms of the form $(\partial/\partial x^l)\psi_{ikl}$, where $\psi_{ikl} = -\psi_{ilk}$. We shall now give another method of calculating the energy-momentum tensor which has the advantage of leading at once to the correct expression.

In (94.1) we carry out a transformation from the coordinates x^i to the coordinates $x'^i = x^i + \xi^i$, where the ξ^i are small quantities. Under this transformation the g^{ik} are transformed according to the formulas:

$$g'^{ik}(x'^l) = g^{lm}(x^l)\frac{\partial x'^i \partial x'^k}{\partial x^l \partial x^m} = g^{lm}\left(\delta^i_l + \frac{\partial \xi^i}{\partial x^l}\right)\left(\delta^k_m + \frac{\partial \xi^k}{\partial x^m}\right)$$

$$\approx g^{ik}(x^l) + g^{im}\frac{\partial \xi^k}{\partial x^m} + g^{kl}\frac{\partial \xi^i}{\partial x^l}.$$

Here the tensor g'^{ik} is a function of the x'^l, while the tensor g^{ik} is a function of the original coordinates x^l. In order to represent all terms as functions of one and the same variables, we expand $g'^{ik}(x^l + \xi^l)$ in powers of ξ^l. Furthermore, if we neglect terms of higher order in ξ^l, we can in all terms containing ξ^l, replace g'^{ik} by g^{ik}. Thus we find

$$g'^{ik}(x^l) = g^{ik}(x^l) - \xi^l \frac{\partial g^{ik}}{\partial x^l} + g^{il} \frac{\partial \xi^k}{\partial x^l} + g^{kl} \frac{\partial \xi^i}{\partial x^l}.$$

It is easy to verify by direct trial that the last three terms on the right can be written as a sum $\xi^{i;k} + \xi^{k;i}$ of contravariant derivatives of the ξ^i. Thus we finally obtain the transformation of the g^{ik} in the form

$$g'^{ik} = g^{ik} + \delta g^{ik}, \quad \delta g^{ik} = \xi^{i;k} + \xi^{k;i}. \tag{94.2}$$

For the covariant components, we have:

$$g'_{ik} = g_{ik} + \delta g_{ik}, \quad \delta g_{ik} = - \xi_{i;k} - \xi_{k;i} \tag{94.3}$$

(so that, to terms of first order we satisfy the condition $g_{il} g'^{kl} = \delta_i^k$).†

Since the action S is a scalar, it does not change under a transformation of coordinates. On the other hand, the change δS in the action under a transformation of coordinates can be written in the following form. As in § 32, let q denote the quantities defining the physical system to which the action S applies. Under coordinate transformation the quantities q change by δq. In calculating δS we need not write terms containing the changes in q. All such terms must cancel each other by virtue of the "equations of motion" of the physical system, since these equations are obtained by equating to zero the variation of S with respect to the quantities q. Therefore it is sufficient to write the terms associated with changes in the g_{ik}. Using Gauss' theorem, and setting $\delta g^{ik} = 0$ at the integration limits, we find δS in the form‡

$$\delta S = \frac{1}{c} \int \left\{ \frac{\partial \sqrt{-g}\,\Lambda}{\partial g^{ik}} \delta g^{ik} + \frac{\partial \sqrt{-g}\,\Lambda}{\partial \dfrac{\partial g^{ik}}{\partial x^l}} \delta \frac{\partial g^{ik}}{\partial x^l} \right\} d\Omega$$

† We note that the equations

$$\xi^{i;k} + \xi^{k;i} = 0$$

determine the infinitesimal coordinate transformations that do not change the metric. In the literature these are often called the *Killing equations*.

‡ It is necessary to emphasize that the notation of differentiation with respect to the components of the symmetric tensor g_{ik}, which we introduce here, has in a certain sense a symbolic character. Namely, the derivative $\partial F/\partial g_{ik}$ (F is some function of the g_{ik}) actually has a meaning only as the expression of the fact that $dF = (\partial F/\partial g_{ik})dg_{ik}$. But in the sum $(\partial F/\partial g_{ik})dg_{ik}$, the terms with differentials dg_{ik}, of components with $i \neq k$, appear twice. Therefore in differentiating the actual expression for F with respect to any definite component g_{ik} with $i \neq k$, we would obtain a value which is twice as large as that which we denote by $\partial F/\partial g_{ik}$. This remark must be kept in mind if we assign definite values to the indices i, k, in formulas in which the derivatives with respect to g_{ik} appear.

$$= \frac{1}{c} \int \left\{ \frac{\partial \sqrt{-g} \Lambda}{\partial g^{ik}} - \frac{\partial}{\partial x^l} \frac{\partial \sqrt{-g} \Lambda}{\partial \frac{\partial g^{ik}}{\partial x^l}} \right\} \delta g^{ik} \, d\Omega.$$

Here we introduce the notation

$$\frac{1}{2} \sqrt{-g} \, T_{ik} = \frac{\partial \sqrt{-g} \Lambda}{\partial g^{ik}} - \frac{\partial}{\partial x^l} \frac{\partial \sqrt{-g} \Lambda}{\partial \frac{\partial g^{ik}}{\partial x^l}} \tag{94.4}$$

Then δS takes the form†

$$\delta S = \frac{1}{2c} \int T_{ik} \delta g^{ik} \sqrt{-g} \, d\Omega = -\frac{1}{2c} \int T^{ik} \delta g_{ik} \sqrt{-g} \, d\Omega \tag{94.5}$$

(note that $g^{ik} \delta g_{lk} = - g_{lk} \, \delta g^{ik}$, and therefore $T^{ik} \, \delta g_{ik} = - T_{ik} \delta g^{ik}$). Substituting for δg^{ik} the expression (94.2), we have, making use of the symmetry of the tensor T_{ik},

$$\delta S = \frac{1}{2c} \int T_{ik} \, (\xi^{i;k} + \xi^{k;i}) \sqrt{-g} \, d\Omega = \frac{1}{c} \int T_{ik} \xi^{i;k} \sqrt{-g} \, d\Omega.$$

Furthermore, we transform this expression in the following way:

$$\delta S = \frac{1}{c} \int (T_i^k \xi^i)_{;k} \sqrt{-g} \, d\Omega - \frac{1}{c} \int T_{i;k}^k \xi^i \sqrt{-g} \, d\Omega. \tag{94.6}$$

Using (86.9), the first integral can be written in the form

$$\frac{1}{c} \int \frac{\partial}{\partial x^k} (\sqrt{-g} \, T_i^k \xi^i) \, d\Omega,$$

and transformed into an integral over a hypersurface. Since the ξ^i vanish at the limits of integration, this integral drops out.

Thus, equating δS to zero, we find

$$\delta S = -\frac{1}{c} \int T_{i;k}^k \xi^i \sqrt{-g} \, d\Omega = 0.$$

Because of the arbitrariness of the ξ^i it then follows that

$$T_{i;k}^k = 0. \tag{94.7}$$

Comparing this with equation (32.4) $\partial T_{ik}/\partial x^k = 0$, valid in galilean coordinates, we see that the tensor T_{ik}, defined by formula (94.4), must be identical with the energy-momentum tensor—at least to within a constant factor. It is easy to verify, carrying out, for example, the calculation from formula (94.4) for the electromagnetic field

† In the case we are considering, the ten quantities δg_{ik} are not independent, since they are the result of a transformation of the coordinates, of which there are only four. Therefore from the vanishing of δS it does not follow that $T_{ik} = 0$!

$$\left(\Lambda = - \frac{1}{16\pi} F_{ik} F^{ik} = - \frac{1}{16\pi} F_{ik} F_{lm} g^{il} g^{km} \right).$$

that this factor is equal to unity.

Thus, formula (94.4) enables us to calculate the energy-momentum tensor by differentiating the function Λ with respect to the components of the metric tensor (and their derivatives). The tensor T_{ik} obtained in this way is symmetric. Formula (94.4) is convenient for calculating the energy-momentum tensor not only in the case of the presence of a gravitational field, but also in its absence, in which case the metric tensor has no independent significance and the transition to curvilinear coordinates occurs formally as an intermediate step in the calculation of T_{ik}.

The expression (33.1) for the energy-momentum tensor of the electromagnetic field must be written in curvilinear coordinates in the form

$$T_{ik} = \frac{1}{4\pi} \left(-F_{il} F_k^l + \frac{1}{4} F_{lm} F^{lm} g_{ik} \right). \tag{94.8}$$

For a macroscopic body the energy-momentum tensor is

$$T_{ik} = (p + \varepsilon) u_i u_k - p g_{ik}. \tag{94.9}$$

We note that the quantity T_{00} is always positive:†

$$T_{00} \geq 0. \tag{94.10}$$

(No general statement can be made about the mixed component T_0^0.)

PROBLEM

Consider the possible cases of reduction to canonical form of a symmetric tensor of second rank in a pseudo-euclidean space.

Solution: The reduction of a symmetric tensor A_{ik} to principal axes means that we find "eigenvectors" n^i for which

$$A_{ik} n^k = \lambda n_i. \tag{1}$$

The corresponding principal (or "proper") values λ are obtained from the condition for consistency of equation (1), i.e. as the roots of the fourth degree equation

$$|A_{ik} - \lambda g_{ik}| = 0, \tag{2}$$

and are invariants of the tensor. Both the quantities λ and the eigenvectors corresponding to them may be complex. (The components of the tensor A_{ik} itself are of course assumed to be real.)

From equation (1) it is easily shown in the usual fashion that two vectors $n_i^{(1)}$ and $n_i^{(2)}$ which correspond to different principal values $\lambda^{(1)}$ and $\lambda^{(2)}$ are "mutually perpendicular".

† We have $T_{00} = \varepsilon u_0^2 + p(u_0^2 - g_{00})$. The first term is always positive. In the second term we write

$$u_0 = g_{00} u^0 + g_{0\alpha} u^\alpha = \frac{g_{00} dx^0 + g_{0\alpha} dx^\alpha}{ds}$$

and obtain after a simple transformation $g_{00} p(dl/ds)^2$, where dl is the element of spatial distance (84.6); from this it is clear that the second term of T_{00} is also positive. The same result can also be shown for the tensor (94.8).

$$n_i^{(1)} n^{(2)i} = 0. \tag{3}$$

In particular, if equation (2) has complex-conjugate roots λ and λ^*, to which there correspond the complex-conjugate vectors n_i and n_i^*, then we must have

$$n_i n^{i*} = 0. \tag{4}$$

The tensor A_{ik} is expressed in terms of its principal values and the corresponding eigenvectors by the formula

$$A_{ik} = \sum \lambda \frac{n_i n_k}{n_l n^l} \tag{5}$$

(so long as none of the quantities $n_l n^l$ is equal to zero—cf. below).

Depending on the character of the roots of equation (2), the following three situations may occur.

(a) All four eigenvalues λ are real. Then the vectors n_i are also real, and since they are mutually perpendicular, three of them must have spacelike directions and one a timelike direction (and are normalized by the conditions $n_l n^l = -1$ and $n_l n^l = 1$, respectively). Choosing the directions of the coordinates along these vectors, we bring the tensor A_{ik} to the form

$$A_{ik} = \begin{pmatrix} \lambda^{(0)} & 0 & 0 & 0 \\ 0 & -\lambda^{(1)} & 0 & 0 \\ 0 & 0 & -\lambda^{(2)} & 0 \\ 0 & 0 & 0 & -\lambda^{(3)} \end{pmatrix} \tag{6}$$

(b) Equation (2) has two real roots ($\lambda^{(2)}$, $\lambda^{(3)}$) and two complex-conjugate roots ($\lambda' \pm i\lambda''$). We write the complex-conjugate vectors n_i, n_i^*, corresponding to the last two roots in the form $a_i \pm ib_i$; since they are defined only to within an arbitrary complex factor, we can normalize them by the condition $n_i n^i = n_i^* n^{i*} = 1$. Also using equation (4), we find

$$a_i a^i + b_i b^i - 0, \quad a_i b^i = 0, \quad a_i a^i - b_i b^i = 1,$$

so that

$$a_i a^i = \tfrac{1}{2}, \quad b_i b^i = -\tfrac{1}{2},$$

i.e. one of these vectors must be spacelike and the other timelike.† Choosing the coordinate axes along the vectors a^i, b^i, $n^{(2)i}$, $n^{(3)i}$, we bring the tensor to the form:

$$A_{ik} = \begin{pmatrix} \lambda' & \lambda'' & 0 & 0 \\ \lambda'' & -\lambda' & 0 & 0 \\ 0 & 0 & -\lambda^{(2)} & 0 \\ 0 & 0 & 0 & -\lambda^{(3)} \end{pmatrix} \tag{7}$$

(c) If the square of one of the vectors n^i is equal to zero ($n_l n^l = 0$), then this vector cannot be chosen as the direction of a coordinate axis. We can however choose one of the planes x^0, x^α so that the vector n^i lies in it. Suppose this is the x^0, x^1 plane; then it follows from $n_l n^l = 0$ that $n^0 = n^1$, and from equation (1) we have $A_{00} + A_{01} = \lambda$, $A_{10} + A_{11} = -\lambda$, so that $A_{11} = -\lambda + \mu$, $A_{00} = \lambda + \mu$, $A_{01} = -\mu$, where μ is a quantity which is not invariant but changes under rotations in the x^0, x^1 plane; it can always be made real by a suitable rotation. Choosing the axes x^2, x^3 along the other two (spacelike) vectors $n^{(2)i}$, $n^{(3)i}$, we bring the tensor A_{ik} to the form

† Since only one of the vectors can have a timelike direction, it then follows that equation (2) cannot have two pairs of complex-conjugate roots.

$$A_{ik} = \begin{pmatrix} \lambda + \mu & -\mu & 0 & 0 \\ -\mu & -\lambda + \mu & 0 & 0 \\ 0 & 0 & -\lambda^{(2)} & 0 \\ 0 & 0 & 0 & -\lambda^{(3)} \end{pmatrix} \tag{8}$$

This case corresponds to the situation when two of the roots $(\lambda^{(0)}, \lambda^{(1)})$ of equation (2) are equal.

We note that for the physical energy-momentum tensor T_{ik} of matter moving with velocities less than the velocity of light only case (a) can occur; this is related to the fact that there must always exist a reference system in which the flux of the energy of the matter, i.e. the components $T_{\alpha 0}$ are equal to zero. For the energy-momentum tensor of electromagnetic waves we have case (c) with $\lambda^{(2)} = \lambda^{(3)} = \lambda = 0$ (cf. p. 87); it can be shown that if this were not the case there would exist a reference frame in which the energy flux would exceed the value c times the energy density.

§ 95. The Einstein equations

We can now proceed to the derivation of the equations of the gravitational field. These equations are obtained from the principle of least action $\delta(S_m + S_g) = 0$, where S_g and S_m are the actions of the gravitational field and matter respectively. We now subject the gravitational field, that is, the quantities g_{ik}, to variation.

Calculating the variation δS_g, we have

$$\delta \int R\sqrt{-g}\,d\Omega = \delta \int g^{ik} R_{ik} \sqrt{-g}\,d\Omega$$

$$= \int \{R_{ik} \sqrt{-g}\,\delta g^{ik} + R_{ik} g^{ik} \delta\sqrt{-g} + g^{ik} \sqrt{-g}\,\delta R_{ik}\}\,d\Omega.$$

From formula (86.4), we have

$$\delta\sqrt{-g} = -\frac{1}{2\sqrt{-g}}\,\delta g = -\frac{1}{2}\sqrt{-g}\,g_{ik}\,\delta g^{ik};$$

substituting this, we find

$$\delta \int R\sqrt{-g}\,d\Omega = \int (R_{ik} - \tfrac{1}{2}g_{ik}R)\delta g^{ik} \sqrt{-g}\,d\Omega + \int g^{ik} \delta R_{ik} \sqrt{-g}\,d\Omega. \tag{95.1}$$

For the calculation of δR_{ik} we note that although the quantities Γ_{kl}^i do not constitute a tensor, their variations $\delta\Gamma_{kl}^i$ do form a tensor, for $\Gamma_{il}^k A_k dx^l$ is the change in a vector under parallel displacement [see (85.5)] from some point P to an infinitesimally separated point P'. Therefore $\delta\Gamma_{il}^k A_k dx^l$ is the difference between the two vectors, obtained as the result of two parallel displacements (one with the unvaried, the other with the varied Γ_{kl}^i) from the point P to one and the same point P'. The difference between two vectors at the same point is a vector, and therefore $\delta\Gamma_{kl}^i$ is a tensor.

Let us use a locally geodesic system of coordinates. Then at that point all the $\Gamma_{kl}^i = 0$. With the help of expression (92.10) for the R_{ik}, we have (remembering that the first derivatives of the g^{ik} are now equal to zero)

$$g^{ik}\delta R_{ik} = g^{ik}\left\{\frac{\partial}{\partial x^l}\delta\Gamma^l_{ik} - \frac{\partial}{\partial x^k}\delta\Gamma^l_{il}\right\} = g^{ik}\frac{\partial}{\partial x^l}\delta\Gamma^l_{ik} - g^{il}\frac{\partial}{\partial x^l}\delta\Gamma^k_{ik} = \frac{\partial w^l}{\partial x^l},$$

where

$$w^l = g^{ik}\delta\Gamma^l_{ik} - g^{il}\delta\Gamma^k_{ik}.$$

Since w^l is a vector, we may write the relation we have obtained, in an arbitrary coordinate system, in the form

$$g^{ik}\delta R_{ik} = \frac{1}{\sqrt{-g}}\frac{\partial}{\partial x^l}(\sqrt{-g}\,w^l)$$

[replacing $\partial\omega^l/\partial x^l$ by $w^l_{;l}$ and using (86.9)]. Consequently the second integral on the right side of (95.1) is equal to

$$\int g^{ik}\delta R_{ik}\sqrt{-g}\,d\Omega = \int \frac{\partial(\sqrt{-g}\,w^l)}{\partial x^l}\,d\Omega,$$

and by Gauss' theorem can be transformed into an integral of w^l over the hypersurface surrounding the whole four-volume. Since the variations of the field are zero at the integration limits, this term drops out. Thus, the variation δS_g is equal to†

$$\delta S_g = -\frac{c^3}{16\pi k}\int\left(R_{ik} - \frac{1}{2}g_{ik}R\right)\delta g^{ik}\sqrt{-g}\,d\Omega. \tag{95.2}$$

We note that if we had started from the expression

$$S_g = -\frac{c^3}{16\pi k}\int G\sqrt{-g}\,d\Omega$$

for the action of the field, then we would have obtained

$$\delta S_g = -\frac{c^3}{16\pi k}\int\left\{\frac{\partial(G\sqrt{-g})}{\partial g^{ik}} - \frac{\partial}{\partial x^l}\frac{\partial(G\sqrt{-g})}{\partial\dfrac{\partial g^{ik}}{\partial x^l}}\right\}\delta g^{ik}\,d\Omega.$$

Comparing this with (95.2), we find the following relation:

$$R_{ik} - \frac{1}{2}g_{ik}R = \frac{1}{\sqrt{-g}}\left\{\frac{\partial(G\sqrt{-g})}{\partial g^{ik}} - \frac{\partial}{\partial x^l}\frac{\partial(G\sqrt{-g})}{\partial\dfrac{\partial g^{ik}}{\partial x^l}}\right\}. \tag{95.3}$$

For the variation of the action of the matter we can write immediately from (94.5):

---

† We note here the following curious fact. If we calculate the variation $\delta\int R\sqrt{-g}\,d\Omega$ [with R_{ik} from (92.10)], considering the Γ^i_{kl} as independent variables and the g_{ik} as constants, and then use expression (86.3) for the Γ^i_{kl} we would obtain, as one easily verifies, identically zero. Conversely, one could determine the relation between the Γ^i_{kl} and the metric tensor by requiring that the variation we have mentioned should vanish.

$$\delta S_m = \frac{1}{2c} \int T_{ik} \delta g^{ik} \sqrt{-g}\, d\Omega, \tag{95.4}$$

where T_{ik} is the energy-momentum tensor of the matter (including the electromagnetic field). Gravitational interaction plays a role only for bodies with sufficiently large mass (because of the smallness of the gravitational constant), and therefore in studying the gravitational field we usually have to deal with macroscopic bodies. Corresponding to this we must usually write for T_{ik} the expression (94.9).

Thus, from the principle of least action $\delta S_m + \delta S_g = 0$ we find:

$$-\frac{c^3}{16\pi k} \int \left(R_{ik} - \frac{1}{2} g_{ik} R - \frac{8\pi k}{c^4} T_{ik} \right) \delta g^{ik} \sqrt{-g}\, d\Omega = 0,$$

from which, in view of the arbitrariness of the δg^{ik}:

$$R_{ik} - \frac{1}{2} g_{ik} R = \frac{8\pi k}{c^4} T_{ik}, \tag{95.5}$$

or, in mixed components,

$$R_i^k - \frac{1}{2}\, \delta_i^k R = \frac{8\pi k}{c^4} T_i^k. \tag{95.6}$$

These are the required *equations of the gravitational field*—the basic equation of the general theory of relativity. They are called the *Einstein equations.*

Contracting (95.6) on the indices i and k, we find

$$R = -\frac{8\pi k}{c^4} T; \tag{95.7}$$

$(T = T_i^i)$. Therefore the equations of the field can also be written in the form

$$R_{ik} = \frac{8\pi k}{c^4} \left(T_{ik} - \frac{1}{2} g_{ik} T \right). \tag{95.8}$$

The Einstein equations are nonlinear. Therefore for gravitational fields the principle of superposition is not valid. The principle is valid only approximately for weak fields which permit a linearization of the Einstein equations (in particular, the gravitational field in the classical Newtonian limit, cf. § 99).

In empty space $T_{ik} = 0$, and the equations of the gravitational field reduce to the equation

$$R_{ik} = 0. \tag{95.9}$$

We mention that this does not at all mean that in vacuum, spacetime is flat; for this we would need the stronger conditions $R^i_{klm} = 0$.

The energy momentum tensor of the electromagnetic field has the property that $T_i^i = 0$ [see (33.2)]. From (95.7), it follows that in the presence of an electromagnetic field without any masses the scalar curvature of spacetime is zero.

As we know, the divergence of the energy-momentum tensor is zero:

$$T_{i;k}^k = 0; \tag{95.10}$$

therefore the divergence of the left side of equation (95.6) must be zero. This is actually the case because of the identity (92.13).

Thus the equation (95.10) is essentially contained in the field equations (95.6). On the other hand, the equation (95.10), expressing the law of conservation of energy and momentum contains the equation of motion of the physical system to which the energy-momentum tensor under consideration refers (i.e., the equations of motion of the material particles or the second pair of Maxwell equations). Thus the equations of the gravitational field also contain the equations for the matter which produces this field. Therefore the distribution and motion of the matter producing the gravitational field cannot be assigned arbitrarily. On the contrary, they must be determined (by solving the field equations under given initial conditions) at the same time as we find the field produced by the matter.

We call attention to the difference in principle between the present situation and the one we had in the case of the electromagnetic field. The equations of that field (the Maxwell equations) contain only the equation of conservation of the total charge (the continuity equation), but not the equations of motion of the charges themselves. Therefore the distribution and motion of the charges can be assigned arbitrarily, so long as the total charge is constant. Assignment of this charge distribution then determines, through Maxwell's equations, the electromagnetic field produced by the charges.

We must, however, make it clear that for a complete determination of the distribution and motion of the matter in the case of the Einstein equations one must still add to them the equation of state of the matter, i.e. an equation relating the pressure and density. This equation must be given along with the field equations.†

The four coordinates x^i can be subjected to an arbitrary transformation. By means of these transformations we can arbitrarily assign four of the ten components of the tensor g_{ik}. Therefore there are only six independent quantities g_{ik}. Furthermore, the four components of the four-velocity u^i, which appear in the energy-momentum tensor of the matter, are related to one another by $u^i u_i = 1$, so that only three of them are independent. Thus we have ten field equations (95.5) for ten unknowns, namely, six components of g_{ik}, three components of u^i, and the density ε/c^2 of the matter (or its pressure p).

For the gravitational field in vacuum there remain a total of six unknown quantities (components of g_{ik}) and the number of independent field equations is reduced correspondingly: the ten equations $R_{ik} = 0$ are connected by the four identities (92.10).

We mention some peculiarities of the structure of the Einstein equations. They are a system of second-order partial differential equations. But the equations do not contain the time derivatives of all components g_{ik}. In fact it is clear from (92.1) that second derivatives with respect to the time are contained only in the components $R_{0\alpha0\beta}$ of the curvature tensor, where they enter in the form of the term $-\frac{1}{2}\ddot{g}_{\alpha\beta}$ (the dot denotes differentiation with respect to x^0); the second derivatives of the components $g_{0\alpha}$ and g_{00} do not appear at all. It is therefore clear that the tensor R_{ik}, which is obtained by contraction of the curvature tensor, and with it the equations (95.5), also contain the second derivatives with respect to the time of only the six spatial components $g_{\alpha\beta}$.

It is also easy to see that these derivatives enter only in the $^\beta_\alpha$-equation of (95.6), i.e. the equation

---

† Actually the equation of state relates to one another not two but three thermodynamic quantities, for example the pressure, density and temperature of the matter. In applications in the theory of gravitation, this point is however not important, since the approximate equations of state used here actually do not depend on the temperature (as, for example, the equation $p = 0$ for rarefied matter, the limiting extreme-relativistic equation $p = \varepsilon/3$ for highly compressed matter, etc.).

$$R_\alpha^\beta - \tfrac{1}{2}\delta_\alpha^\beta R = \frac{8\pi k}{c^4} T_\alpha^\beta .$$

(95.11)

The $\genfrac{}{}{0pt}{}{0}{0}$ and $\genfrac{}{}{0pt}{}{0}{\alpha}$ equations, i.e. the equations

$$R_0^0 - \tfrac{1}{2} R = \frac{8\pi k}{c^4} T_0^0, \quad R_\alpha^0 = \frac{8\pi k}{c^4} T_\alpha^0 ,$$

(95.12)

contain only first-order time derivatives. One can verify this by checking that in forming the quantities R_α^0 and $R_0^0 - \tfrac{1}{2}R = \tfrac{1}{2}(R_0^0 - R_\alpha^\alpha)$ from R_{iklm} by contraction, the components of the form $R_{0\alpha 0\beta}$ actually drop out. This can be seen even more simply from the identity (92.10), by writing it in the form

$$(R_i^0 - \tfrac{1}{2}\delta_i^0 R)_{;0} = - (R_i^\alpha - \tfrac{1}{2}\delta_i^\alpha R)_{;\alpha}$$

(95.13)

$(i = 0, 1, 2, 3)$. The highest time derivatives appearing on the right side of this equation are second derivatives (appearing in the quantities R_i^α, R). Since (95.13) is an identity, its left side must consequently contain no time derivatives of higher than second order. But one time differentiation already appears explicitly in it; therefore the expressions $R_i^0 - \tfrac{1}{2}\delta_i^0 R$ themselves cannot contain time derivatives of order higher than the first.

Furthermore, the left sides of equations (95.12) also do not contain the first derivatives $\dot{g}_{0\alpha}$ and $\dot{g}_{00}$ (but only the derivatives $\dot{g}_{\alpha\beta}$). In fact, of all the $\Gamma_{i,\,kl}$, only $\Gamma_{\alpha,\,00}$ and $\Gamma_{0,\,00}$ contain these quantities, but these latter in turn appear only in the components of the curvature tensor of the form $R_{0\alpha\,0\beta}$ which, as we already know, drop out when we form the left sides of equations (95.12).

If one is interested in the solution of the Einstein equations for given initial conditions (in the time), we must consider the question of the number of quantities for which the initial spatial distribution can be assigned arbitrarily.

The initial conditions for a set of equations of second order must include both the quantities to be differentiated as well as their first time derivatives. But since in the present case the equations contain second derivatives of only the six $g_{\alpha\beta}$, not all the g_{ik} and $\dot{g}_{ik}$ can be arbitrarily assigned. Thus, we may assign (in addition to the velocity and density of the matter) the initial values of the functions $g_{\alpha\beta}$ and $\dot{g}_{\alpha\beta}$, after which the four equations (95.12) determine the admissible initial values of $g_{0\alpha}$ and g_{00}; in (95.11) the initial values of $\dot{g}_{0\alpha}$ still remain arbitrary.

Among the initial conditions thus assigned there are some functions whose arbitrariness is related simply to the arbitrariness in choice of the four-dimensional coordinate system. But the only thing that has real physical meaning is the number of "physically different" arbitrary functions, which cannot be reduced by any choice of coordinate system. From physical arguments it is easy to see that this number is eight: the initial conditions must assign the distribution of the matter density and of its three velocity components, and also of four other quantities characterising the free gravitational field in the absence of matter (see later in § 107); for the free gravitational field in vacuum only the last four quantities should be fixed by the initial conditions.

PROBLEM

Write the equations for a constant gravitational field, expressing all the operations of differentiation with respect to the space coordinates as covariant derivatives in a space with the metric $\gamma_{\alpha\beta}$ (84.7).

Solution: We introduce the notation $g_{00} = h$, $g_{0\alpha} = -hg_\alpha$ (88.11) and the three-dimensional velocity v^α (88.10). In the following all operations of raising and lowering indices and of covariant differentiation are carried out in the three-dimensional space with the metric $\gamma_{\alpha\beta}$ on the three-dimensional vectors g_α, v^α and the three-dimensional scalar h.

The desired equations must be invariant with respect to the transformation

$$x^\alpha \to x^\alpha, \quad x^0 \to x^0 + f(x^\alpha), \tag{1}$$

which does not change the stationary character of the field. But under such a transformation, as is easily shown (see the footnote on p. 267), $g_\alpha \to g_\alpha - \partial f / \partial x^\alpha$, while the scalar h and the tensor $\gamma_{\alpha\beta} = -g_{\alpha\beta} + hg_\alpha g_\beta$ are unchanged. It is therefore clear that the required equations, when expressed in terms of $\gamma_{\alpha\beta}$, h and g_α, can contain g_α only in the form of combinations of derivatives that constitute a three-dimensional antisymmetric tensor:

$$f_{\alpha\beta} = g_{\beta;\alpha} - g_{\alpha;\beta} = \frac{\partial g_\beta}{\partial x^\alpha} - \frac{\partial g_\alpha}{\partial x^\beta}, \tag{2}$$

which is invariant under such transformations. Taking this fact into account, we can drastically simplify the computations by setting (after computing all the derivatives appearing in R_{ik}) $g_\alpha = 0$ and $g_{\alpha;\beta} + g_{\beta;\alpha} = 0$.†

The Christoffel symbols are:

$$\Gamma^0_{00} = \frac{1}{2} g^\alpha h_{;\alpha},$$

$$\Gamma^\alpha_{00} = \frac{1}{2} h^{;\alpha},$$

$$\Gamma^0_{\alpha 0} = \frac{1}{2h} h_{;\alpha} + \frac{h}{2} g^\beta f_{\alpha\beta} + \ldots,$$

$$\Gamma^\alpha_{0\beta} = \frac{h}{2} f_\beta{}^\alpha - \frac{1}{2} g_\beta h^{;\alpha},$$

$$\Gamma^0_{\alpha\beta} = -\frac{1}{2}\left(\frac{\partial g_\alpha}{\partial x^\beta} + \frac{\partial g_\beta}{\partial x^\alpha}\right) - \frac{1}{2h}(g_\alpha h_{;\beta} + g_\beta h_{;\alpha}) + g_\gamma \lambda^\gamma_{\alpha\beta} + \ldots,$$

$$\Gamma^\alpha_{\beta\gamma} = \lambda^\alpha_{\beta\gamma} - \frac{h}{2}(g_\beta f_\gamma{}^\alpha + g_\gamma f_\beta{}^\alpha) + \ldots.$$

The terms omitted (indicated by the dots) are quadratic in the components of g_α; these terms do drop out when we set $g_\alpha = 0$ after performing the differentiations in R_{ik} (92.10). In the calculations one uses formulas (84.9), (84.12)–(84.13); the $\lambda^\alpha_{\beta\gamma}$ are three-dimensional Christoffel symbols constructed from the metric $\gamma_{\alpha\beta}$.

The tensor T_{ik} is calculated using formula (94.9) with the u^i from (88.14) (where again we set $g_\alpha = 0$). As a result of the calculations, we obtain the following equations from (95.8):

$$\frac{1}{h} R_{00} \frac{1}{\sqrt{h}} (\sqrt{h})_{;\alpha}^{;\alpha} + \frac{h}{4} f_{\alpha\beta} f^{\alpha\beta} = \frac{8\pi k}{c^4}\left(\frac{\varepsilon + p}{1 - \frac{v^2}{c^2}} - \frac{\varepsilon - p}{2}\right), \tag{3}$$

$$\frac{1}{\sqrt{h}} R_0^\alpha = -\frac{\sqrt{h}}{2} f^{\alpha\beta}{}_{;\beta} - \frac{3}{2} f^{\alpha\beta} (\sqrt{h})_{;\beta} = \frac{8\pi k}{c^4} \frac{p + \varepsilon}{1 - \frac{v^2}{c^2}} \frac{v^\alpha}{c}, \tag{4}$$

† To avoid any misunderstanding we emphasize that this simplified method for making the computations, which gives the correct field equations, would not be applicable to the calculation of arbitrary components of the R_{ik} itself, since they are not invariant under the transformation (1). In equations (3)–(5) on the left are given those components of the Ricci tensor which are actually equal to the expressions given. These components are invariant under (1).

$$R^{\alpha\beta} = P^{\alpha\beta} + \frac{h}{2} f^{\alpha\gamma} f^{\beta}{}_{\gamma} - \frac{1}{\sqrt{h}} (\sqrt{h})^{;\,\alpha;\,\beta} = \frac{8\pi k}{c^4} \left[\frac{(p+\varepsilon) v^{\alpha} v^{\beta}}{c^2 \left(1 - \frac{v^2}{c^2}\right)} + \frac{\varepsilon - p}{2} \gamma^{\alpha\beta} \right]. \tag{5}$$

Here $P^{\alpha\beta}$ is a three-dimensional tensor constructed from the $\gamma_{\alpha\beta}$ in the same way as R^{ik} is constructed from the g_{ik}.†

§ 96. The energy-momentum pseudotensor of the gravitational field

In the absence of a gravitational field, the law of conservation of energy and momentum of the material (and electromagnetic field) is expressed by the equation $\partial T^{ik}/\partial x^k = 0$. The generalization of this equation to the case where a gravitational field is present is equation (94.7):

$$T^k_{i;\,k} = \frac{1}{\sqrt{-g}} \frac{\partial (T^k_i \sqrt{-g})}{\partial x^k} - \frac{1}{2} \frac{\partial g_{kl}}{\partial x^i} T^{kl} = 0. \tag{96.1}$$

In this form, however, this equation does not generally express any conservation law whatever.‡ This is related to the fact that in a gravitational field the four-momentum of the matter alone must not be conserved, but rather the four-momentum of matter plus gravitational field; the latter is not included in the expression for T^k_i.

To determine the conserved total four-momentum for a gravitational field plus the matter located in it, we proceed as follows (L. D. Landau and E. M. Lifshitz, 1947). We choose a system of coordinates of such form that at some particular point in spacetime all the first derivatives of the g_{ik} vanish (the g_{ik} need not, for this, necessarily have their galilean values).§ Then at this point the second term in equation (96.1) vanishes, and in the first term we can take $\sqrt{-g}$ out from under the derivative sign, so that there remains

$$\frac{\partial}{\partial x^k} T^k_i = 0,$$

---

† The Einstein equations can also be written in an analogous way for the general case of a time-dependent metric. In addition to space derivatives they will also contain time derivatives of the quantities $\gamma_{\alpha\beta}$, g_{α}, and h. See A. L. Zel'manov, *Doklady Acad. Sci., U.S.S.R.* **107**, 815 (1956).

‡ Because the integral $\int T^k_i \sqrt{-g}\, dS_k$ is conserved only if the condition

$$\frac{\partial(\sqrt{-g} T^k_i)}{\partial x^k} = 0$$

is fulfilled, and not (96.1). This is easily verified by carrying out in curvilinear coordinates all those calculations which in § 29 were done in galilean coordinates. Besides it is sufficient simply to note that these calculations have a purely formal character not connected with the tensor properties of the corresponding quantities, like the proof of Gauss' theorem, which has the same form (83.17) in curvilinear as in cartesian coordinates.

§ One might get the notion to apply to the gravitational field the formula (94.4), substituting $\Lambda = -(c^4/16\pi k)G$. We emphasize, however, that this formula applies only to physical systems described by quantities q different from the g_{ik}; therefore it cannot be applied to the gravitational field which is determined by the quantities g_{ik} themselves. Note, by the way, that upon substituting G in place of Λ in (94.4) we would obtain simply zero, as is immediately clear from the relation (95.3) and the equations of the field in vacuum.

or, in contravariant components,

$$\frac{\partial}{\partial x^k} T^{ik} = 0.$$

Quantities T^{ik}, identically satisfying this equation, can be written in the form

$$T^{ik} = \frac{\partial}{\partial x^l} \eta^{ikl},$$

where the η^{ikl}, are quantities antisymmetric in the indices k, l;

$$\eta^{ikl} = -\eta^{ilk}.$$

Actually it is not difficult to bring T^{ik} to this form. To do this we start from the field equation

$$T^{ik} = \frac{c^4}{8\pi k} \left(R^{ik} - \frac{1}{2} g^{ik} R \right),$$

and for R^{ik} we have, according to (92.1)

$$R^{ik} = \frac{1}{2} g^{im} g^{kp} g^{ln} \left\{ \frac{\partial^2 g_{lp}}{\partial x^m \partial x^n} + \frac{\partial^2 g_{mn}}{\partial x^l \partial x^p} - \frac{\partial^2 g_{ln}}{\partial x^m \partial x^p} - \frac{\partial^2 g_{mp}}{\partial x^l \partial x^n} \right\}$$

(we recall that at the point under consideration, all the $\Gamma^i_{kl} = 0$). After simple transformations the tensor T^{ik} can be put in the form

$$T^{ik} = \frac{\partial}{\partial x^l} \left\{ \frac{c^4}{16\pi k} \frac{1}{(-g)} \frac{\partial}{\partial x^m} [(-g)(g^{ik} g^{lm} - g^{il} g^{km})] \right\}.$$

The expression in the curly brackets is antisymmetric in k and l, and is the quantity which we designated above as η^{ikl}. Since the first derivatives of g_{ik} are zero at the point under consideration, the factor $1/(-g)$ can be taken out from under the sign of differentiation $\partial/\partial x^l$. We introduce the notation

$$h^{ikl} = \frac{\partial}{\partial x^m} \lambda^{iklm} \tag{96.2}$$

$$\lambda^{iklm} = \frac{c^4}{16\pi k} (-g)(g^{ik} g^{lm} - g^{il} g^{km}) \tag{96.3}$$

The quantities h^{ikl} are antisymmetric in k and l:

$$h^{ikl} = -h^{ilk}. \tag{96.4}$$

Then we can write

$$\frac{\partial h^{ikl}}{\partial x^l} = (-g) T^{ik}.$$

This relation, derived under the assumption $\partial g_{ik}/\partial x^l = 0$, is no longer valid when we go to an arbitrary system of coordinates. In the general case, the difference $\partial h^{ikl}/\partial x^l - (-g)T^{ik}$ is different from zero; we denote it by $(-g)t^{ik}$. Then we have, by definition,

$$(-g)(T^{ik} + t^{ik}) = \frac{\partial h^{ikl}}{\partial x^l}. \tag{96.5}$$

The quantities t^{ik} are symmetric in i and k:

$$t^{ik} = t^{ki}. \tag{96.6}$$

This is clear immediately from their definition, since like the tensor T^{ik}, the derivatives $\partial h^{ikl}/\partial x^l$ are symmetric quantities.† Expressing T^{ik} in terms of R^{ik}, according to the Einstein equations, we get the identity

$$(-g)\left\{ \frac{c^4}{8\pi k}(R^{ik} - \tfrac{1}{2}g^{ik}R) + t^{ik} \right\} = \frac{\partial h^{ikl}}{\partial x^l}, \tag{96.7}$$

from which, after a rather lengthy calculation, we find the following expression for t^{ik}:

$$\begin{aligned}
t^{ik} = \frac{c^4}{16\pi k} &\{ (2\Gamma^n_{lm}\Gamma^p_{np} - \Gamma^n_{lp}\Gamma^p_{mn} - \Gamma^n_{ln}\Gamma^p_{mp})(g^{il}g^{km} - g^{ik}g^{lm}) + \\
&+ g^{il}g^{mn}(\Gamma^k_{lp}\Gamma^p_{mn} + \Gamma^k_{mn}\Gamma^p_{lp} - \Gamma^k_{np}\Gamma^p_{lm} - \Gamma^k_{lm}\Gamma^p_{np}) + \\
&+ g^{kl}g^{mn}(\Gamma^k_{lp}\Gamma^p_{mn} + \Gamma^i_{mn}\Gamma^p_{lp} - \Gamma^i_{np}\Gamma^p_{lm} - \Gamma^i_{lm}\Gamma^p_{np}) + \\
&+ g^{lm}g^{np}(\Gamma^i_{ln}\Gamma^k_{mp} - \Gamma^i_{lm}\Gamma^k_{np}) \},
\end{aligned} \tag{96.8}$$

or, in terms of derivatives of the components of the metric tensor,

$$\begin{aligned}
(-g)t^{ik} = \frac{c^4}{16\pi k} &\{ \mathfrak{g}^{ik}{}_{,l}\mathfrak{g}^{lm}{}_{,m} - \mathfrak{g}^{il}{}_{,l}\mathfrak{g}^{km}{}_{,m} + \tfrac{1}{2}g^{ik}\mathfrak{g}_{lm}\mathfrak{g}^{ln}{}_{,p}\mathfrak{g}^{pm}{}_{,n} - \\
&- (g^{il}\mathfrak{g}_{mn}\mathfrak{g}^{kn}{}_{,p}\mathfrak{g}^{mp}{}_{,l} + g^{kl}\mathfrak{g}_{mn}\mathfrak{g}^{in}{}_{,p}\mathfrak{g}^{mp}{}_{,l}) + \mathfrak{g}_{lm}\mathfrak{g}^{np}\mathfrak{g}^{il}{}_{,n}\mathfrak{g}^{km}{}_{,p} + \\
&+ \tfrac{1}{8}(2g^{il}g^{km} - g^{ik}g^{lm})(2g_{np}g_{qr} - g_{pq}g_{nr})\mathfrak{g}^{nr}{}_{,l}\mathfrak{g}^{pq}{}_{,m} \}, \tag{96.9}
\end{aligned}$$

where $\mathfrak{g}^{ik} = \sqrt{-g}\,g^{ik}$, while the index $,i$ denotes a simple differentiation with respect to x^i.

An essential property of the t^{ik} is that they do not constitute a tensor; this is clear from the fact that in $\partial h^{ikl}/\partial x^l$ there appears the ordinary, and not the covariant derivative. However, t^{ik} is expressed in terms of the quantities Γ^l_{kl}, and the latter behave like a tensor with respect to linear transformations of the coordinates (see § 85), so the same applies to the t^{ik}.

From the definition (96.5) it follows that for the sum $T^{ik} + t^{ik}$ the equation

$$\frac{\partial}{\partial x^k}(-g)(T^{ik} + t^{ik}) = 0 \tag{96.10}$$

is identically satisfied. This means that there is a conservation law for the quantities

$$P^i = \frac{1}{c}\int (-g)(T^{ik} + t^{ik})\,dS_k. \tag{96.11}$$

In the absence of a gravitational field, in galilean coordinates, $t^{ik} = 0$, and the integral we

---

† For just this purpose we took $(-g)$ out from under the derivative sign in the expression for T^{ik}. If this has not been done, $\partial h^{ikl}/\partial x^l$ and therefore also t^{ik} would turn out not to be symmetric in i and k.

have written goes over into $(1/c) \int T^{ik} dS_k$, that is, into the four-momentum of the material. Therefore the quantity (96.11) must be identified with the total four-momentum of matter plus gravitational field. The set of quantities t^{ik} is called the *energy-momentum pseudo-tensor* of the gravitational field.

The integration in (96.11) can be taken over any infinite hypersurface, including all of the three-dimensional space. If we choose for this the hypersurface $x^0 = $ const, then P^i can be written in the form of a three-dimensional space integral:

$$P^i = \frac{1}{c} \int (-g)(T^{i0} + t^{i0}) dV. \tag{96.12}$$

This fact, that the total four-momentum of matter plus field is expressible as an integral of the quantity $(-g) (T^{ik} + t^{ik})$ which is symmetric in the indices i, k, is very important. It means that there is a conservation law for the angular momentum, defined as (see § 32)†

$$M^{ik} = \int (x^i dP^k - x^k dP^i) = \frac{1}{c} \int \{x^i (T^{kl} + t^{kl}) - x^k (T^{il} + t^{il})\}(-g) dS_l. \tag{96.13}$$

Thus, also in the general theory of relativity, for a closed system of gravitating bodies the total angular momentum is conserved, and, moreover, one can again define a centre of inertia which carries out a uniform motion. This latter point is related to the conservation of the components $M^{0\alpha}$ (see § 14) which is expressed by the equation

$$x^0 \int (T^{\alpha 0} + t^{\alpha 0})(-g) \, dV - \int x^\alpha (T^{00} + t^{00})(-g) dV = \text{const},$$

so that the coordinates of the centre of inertia are given by the formula

$$X^\alpha = \frac{\int x^\alpha (T^{00} + t^{00})(-g) \, dV}{\int (T^{00} + t^{00})(-g) \, dV}. \tag{96.14}$$

By choosing a coordinate system which is inertial in a given volume element, we can make all the t^{ik} vanish at any point in space-time (since then all the Γ^i_{kl} vanish). On the other hand, we can get values of the t^{ik} different from zero in flat space, i.e. in the absence of a gravitational field, if we simply use curvilinear coordinates instead of cartesian. Thus, in any case, it has no meaning to speak of a definite localization of the energy of the gravitational field in space. If the tensor T_{ik} is zero at some world point, then this is the case for any reference system, so that we may say that at this point there is no matter or electromagnetic field. On the other hand, from the vanishing of a pseudo-tensor at some point in one reference system it does not at all follow that this is so for another reference system, so that it is meaningless to talk of whether or not there is gravitational energy at a given place. This

---

† It is necessary to note that the expression obtained by us for the four-momentum of matter plus field is by no means the only possible one. On the contrary, one can, in an infinity of ways (see, for example, the problem in this section), form expressions which in the absence of a field reduce to T^{ik}, and which upon integration over dS_k, give conservation of some quantity. However, the choice made by us is the only one for which the energy-momentum pseudotensor of the field contains only first (and not higher) derivatives of g_{ik} (a condition which is completely natural from the physical point of view), and is also symmetric, so that it is possible to formulate a conservation law for the angular momentum.

corresponds completely to the fact that by a suitable choice of coordinates, we can "annihilate" the gravitational field in a given volume element, in which case, from what has been said, the pseudotensor t^{ik} also vanishes in this volume element.

The quantities P^i (the four-momentum of field plus matter) have a completely definite meaning and are independent of the choice of reference system to just the extent that is necessary on the basis of physical considerations.

Let us draw around the masses under consideration a region of space sufficiently large so that outside of it we may say that there is no gravitational field. In the course of time, this region cuts out a "channel" in four-dimensional space-time. Outside of this channel there is no field, so that four-space is flat. Because of this we must, when calculating the energy and momentum of the field, choose a four-dimensional reference system such that outside the channel it goes over into a galilean system and all the t^{ik} vanish.

By this requirement the reference system is, of course, not at all uniquely determined— it can still be chosen arbitrarily in the interior of the channel. However the P^i, in full accord with their physical meaning, turn out to be completely independent of the choice of coordinate system in the interior of the channel. Consider two coordinate systems, different in the interior of the channel, but reducing outside of it to one and the same galilean system, and compare the values of the four-momentum P^i and P'^i in these two systems at definite moments of "time" x^0 and x'^0. Let us introduce a third coordinate system, coinciding in the interior of the channel at the moment x^0 with the first system, and at the moment x'^0 with the second, while outside of the channel it is galilean. But by virtue of the law of conservation of energy and momentum the quantities P^i are constant ($dP^i/dx^0 = 0$). This is the case for the third coordinate system as well as for the first two, and from this it follows that $P^i = P'^i$.

Earlier it was mentioned that the quantities t^{ik} behave like a tensor with respect to linear transformations of the coordinates. Therefore the quantities P^i form a four-vector with respect to such transformations, in particular with respect to Lorentz transformations which, at infinity, take one galilean reference frame into another.† The four-momentum P^i can also be expressed as an integral over a distant three-dimensional surface surrounding "all space". Substituting (96.5) in (96.11), we find

$$P^i = \frac{1}{c} \int \frac{\partial h^{ikl}}{\partial x^l} \, dS_k .$$

This integral can be transformed into an integral over an ordinary surface by means of (6.17):

$$P^i = \frac{1}{2c} \oint h^{ikl} df_{kl}^* . \tag{96.15}$$

If for the surface of integration in (96.11) we choose the hypersurface $x^0 = $ const., then in (96.15) the surface of integration turns out to be a surface in ordinary space.‡

---

† Strictly speaking, in the definition (96.11) P^i is a four-vector only with respect to linear transformations with determinant equal to unity; among these are the Lorentz transformations, which alone are of physical interest. If we also admit transformations with determinant not equal to unity, then we must introduce into the definition of P^i the value of g at infinity by writing $\sqrt{-g_\infty} \, P^i$ in place of P^i on the left side of (96.11).

‡ The quantity df_{kl}^* is the "normal" to the surface element, related to the "tangential" element df^{ik} by (6.11): $df_{ik}^* = \frac{1}{2} e_{iklm} df^{lm}$. On the surface bounding the hypersurface which is perpendicular to the x^0 axis the only nonzero components of df^{lm} are those with $l, m = 1, 2, 3$, and so df_{ik}^* has only those components, in which one of i and k is 0. The components $df_{0\alpha}^*$ are just the components of the three-dimensional element of ordinary surface, which we denote by df_α.

$$P^i = \frac{1}{c} \oint h^{i0\alpha} df_\alpha . \tag{96.16}$$

To derive the analogous formula for the angular momentum, we substitute (96.5) in (96.13) and write h^{ikl} in the form (96.2). Integrating by part, we obtain:

$$M^{ik} = \frac{1}{c} \int \left(x^i \frac{\partial^2 \lambda^{klmn}}{\partial x^m \partial x^n} - x^k \frac{\partial^2 \lambda^{ilmn}}{\partial x^m \partial x^n} \right) dS_l$$

$$= \frac{1}{2c} \int \left(x^i \frac{\partial \lambda^{klmn}}{\partial x^n} - x^k \frac{\partial \lambda^{ilmn}}{\partial x^n} \right) df_{lm}^* - \frac{1}{c} \int \left(\delta_m^i \frac{\partial \lambda^{klmn}}{\partial x^n} - \delta_m^k \frac{\partial \lambda^{ilmn}}{\partial x^n} \right) dS_l$$

$$= \frac{1}{2c} \int (x^i h^{klm} - x^k h^{ilm}) df_{lm}^* - \frac{1}{c} \int \frac{\partial}{\partial x^n} (\lambda^{klin} - \lambda^{ilkn}) dS_l .$$

From the definition of the quantities λ^{iklm} it is easy to see that

$$\lambda^{ilkn} - \lambda^{klin} = \lambda^{ilnk}, \quad \lambda^{inlk} = -\lambda^{ilnk} .$$

Thus the remaining integral over dS_l is equal to

$$\frac{1}{c} \int \frac{\partial \lambda^{ilnk}}{\partial x^n} dS_l = \frac{1}{2c} \int \lambda^{ilnk} df_{ln}^* .$$

Finally, again choosing a purely spatial surface for the integration, we obtain:

$$M^{ik} = \frac{1}{c} \int (x^i h^{k0\alpha} - x^k h^{i0\alpha} + \lambda^{io\alpha k}) df_\alpha . \tag{96.17}$$

PROBLEM

Find the expression for the total four-momentum of matter plus gravitational field, using formula (32.5).

Solution: In curvilinear coordinates one has, in place of (32.1),

$$S = \int \Lambda \sqrt{-g} \, dV \, dt,$$

and therefore to obtain a quantity which is conserved we must in (32.5) write $\Lambda \sqrt{-g}$ in place of Λ, so that the four-momentum has the form

$$P_i = \frac{1}{c} \int \left\{ -\Lambda \sqrt{-g} \delta_i^k + \sum \frac{\partial q^{(l)}}{\partial x^i} \frac{\partial (\sqrt{-g} \Lambda)}{\partial \frac{\partial q^{(l)}}{\partial x^k}} \right\} dS_k .$$

In applying this formula to matter, for which the quantities $q^{(l)}$ are different from the g_{ik}, we can take $\sqrt{-g}$ out from under the sign of differentiation, and the integrand turns out to be equal to $\sqrt{-g} \, T_i^k$, where T_i^k is the energy-momentum tensor of the matter. When applying this same formula to the gravitational field, we must set $\Lambda = -(c^4/16\pi k)G$, while the quantities $q^{(l)}$ are the components g_{ik} of the metric tensor. The total four-momentum of field plus matter is thus equal to

$$P_i = \frac{1}{c} \int T_i^k \sqrt{-g} \, dS_k + \frac{c^3}{16\pi k} \int \left[G\sqrt{-g}\,\delta_i^k - \frac{\partial g^{lm}}{\partial x^i} \frac{\partial(G\sqrt{-g})}{\partial \frac{\partial g^{lm}}{\partial x^k}} \right] dS_k.$$

Using the expression (93.3) for G, we can rewrite this expression in the form:

$$P_i = \frac{1}{c} \int \left\{ T_i^k \sqrt{-g} + \frac{c^4}{16\pi k} \left[G\sqrt{-g}\,\delta_i^k + \Gamma_{lm}^k \frac{\partial(g^{lm}\sqrt{-g})}{\partial x^i} - \Gamma_{ml}^l \frac{\partial(g^{mk}\sqrt{-g})}{\partial x^i} \right] \right\} dS_k.$$

The second term in the curly brackets gives the four-momentum of the gravitational field in the absence of matter. The integrand is not symmetric in the indices i, k, so that one cannot formulate a law of conservation of angular momentum.

§ 97. The synchronous reference system

As we know from § 84, the condition for it to be possible to synchronize clocks at different points in space is that the components $g_{0\alpha}$ of the metric tensor be equal to zero. If, in addition, $g_{00} = 1$, the time coordinate $x^0 = t$ is the proper time at each point in space.† A reference system satisfying the conditions

$$g_{0\alpha} = 0, \quad g_{00} = 1 \tag{97.1}$$

is said to be *synchronous*. The interval element in such a system is given by the expression

$$ds^2 = dt^2 - \gamma_{\alpha\beta}\, dx^\alpha\, dx^\beta, \tag{97.2}$$

where the components of the spatial metric tensor are the same (except for sign) as the $g_{\alpha\beta}$:

$$\gamma_{\alpha\beta} = -g_{\alpha\beta}. \tag{97.3}$$

In the synchronous reference system the time lines are geodesics in the four-space. The four-vector $u^i = dx^i/ds$, which is tangent to the world line $x^1, x^2, x^3 = \text{const}$, has components $u^\alpha = 0$, $u^0 = 1$, and automatically satisfies the geodesic equations:

$$\frac{du^i}{ds} + \Gamma_{kl}^i u^k u^l = \Gamma_{00}^i = 0,$$

since, from the conditions (99.1), the Christoffel symbols Γ_{00}^α and Γ_{00}^0 vanish identically.

It is also easy to see that these lines are normal to the hypersurfaces $t = \text{const}$. In fact, the four-vector normal to such a hypersurface, $n_i = \partial t/\partial x^i$, has covariant components $n_0 = 1$, $n_\alpha = 0$. With the conditions (97.1), the corresponding contravariant components are also $n^0 = 1$, $n^\alpha = 0$, i.e., they coincide with the components of the four-vector u^i which is tangent to the time lines.

Conversely, these properties can be used for the geometrical construction of a synchronous reference system in any space-time. For this purpose we choose as our starting surface any spacelike hypersurface, i.e. a hypersurface whose normals at each point have a time-like direction (they lie inside the light cone with its vertex at this point); all elements of interval on such a hypersurface are spacelike. Next we construct the family of geodesic lines normal to this hypersurface. If we now choose these lines as the time coordinate lines and determine

† In this section we set $c = 1$.

the time coordinate t as the length s of the geodesic line measured from the initial hypersurface, we obtain a synchronous reference system.

It is clear that such a construction, and the selection of a synchronous reference system, is always possible in principle. Furthermore, this choice is still not unique. A metric of the form (97.2) allows any transformation of the space coordinates which does not affect the time, and also transformations corresponding to the arbitrariness in the choice of the initial hypersurface for the geometrical construction.

The transformation to the synchronous reference system can, in principle, be done analytically by using the Hamilton–Jacobi equation; the basis of this method is the fact that the trajectories of a particle in a gravitational field are just the geodesic lines.

The Hamilton–Jacobi equation for a particle (whose mass we set equal to unity) in a gravitational field is

$$g^{ik} \frac{\partial \tau}{\partial x^i} \frac{\partial \tau}{\partial x^k} = 1, \tag{97.4}$$

(where we denote the action by τ). Its complete integral has the form:

$$\tau = f(\xi^\alpha, x^i) + A(\xi^\alpha), \tag{97.5}$$

where f is a function of the four coordinates x^i and the three parameters ξ^α; the fourth constant A we treat as an arbitrary function of the three ξ^α. With such a representation for τ, the equations for the trajectory of the particle can be obtained by equating the derivatives $\partial \tau / \partial \xi^\alpha$ to zero, i.e.

$$\frac{\partial f}{\partial \xi^\alpha} = - \frac{\partial A}{\partial \xi^\alpha}. \tag{97.6}$$

For each set of assigned values of the parameters ξ^α, the right sides of equations (97.6) have definite constant values, and the world line determined by these equations is one of the possible trajectories of the particle. Choosing the quantities ξ^α, which are constant along the trajectory, as new space coordinates, and the quantity τ as the new time coordinate, we get the synchronous reference system; the transformation which takes us from the old coordinates to the new is given by equations (97.5)–(97.6). In fact, it is guaranteed that for such a transformation the time lines will be geodesics and will be normal to the hypersurfaces $\tau =$ const. The latter point is obvious from the mechanical analogy: the four-vector $-\partial \tau / \partial x^i$ which is normal to the hypersurface coincides in mechanics with the four-momentum of the particle, and therefore coincides in direction with its four-velocity u^i, i.e. with the four-vector tangent to the trajectory. Finally the condition $g_{00} = 1$ is obviously satisfied, since the derivative $-d\tau/ds$ of the action along the trajectory is the mass of the particle, which we set equal to 1; therefore $|d\tau/ds| = 1$.

We write the Einstein equations in the synchronous reference system, separating the operations of space and time differentiation in the equations.

We introduce the notation

$$\varkappa_{\alpha\beta} = \frac{\partial \gamma_{\alpha\beta}}{\partial t} \tag{97.7}$$

for the time derivatives of the three-dimensional metric tensor; these quantities also form a three-dimensional tensor. All operations of shifting indices and covariant differentiation of the three-dimensional tensor $\varkappa_{\alpha\beta}$ will be done in three-dimensional space with the metric

$\gamma_{\alpha\beta}$.[†] We note that the sum $\varkappa_\alpha^\alpha$ is the logarithmic derivative of the determinant $\gamma \equiv |\gamma_{\alpha\beta}| = -g$:

$$\varkappa_\alpha^\alpha = \gamma^{\alpha\beta} \frac{\partial \gamma_{\alpha\beta}}{\partial t} = \frac{\partial}{\partial t} \ln(\gamma). \tag{97.8}$$

For the Christoffel symbols we find the expressions:

$$\Gamma_{00}^0 = \Gamma_{00}^\alpha = \Gamma_{0\alpha}^0 = 0,$$

$$\Gamma_{\alpha\beta}^0 = \tfrac{1}{2}\varkappa_{\alpha\beta}, \quad \Gamma_{0\beta}^\alpha = \tfrac{1}{2}\varkappa_\beta^\alpha, \quad \Gamma_{\beta\gamma}^\alpha = \lambda_{\beta\gamma}^\alpha, \tag{97.9}$$

where $\lambda_{\beta\gamma}^\alpha$ are the three-dimensional Christoffel symbols formed from the tensor $\gamma_{\alpha\beta}$. A calculation using formula (92.7) gives the following expressions for the components of the tensor R_{ik}

$$R_{00} = -\frac{1}{2} \frac{\partial}{\partial t} \varkappa_\alpha^\alpha - \frac{1}{4} \varkappa_\alpha^\beta \varkappa_\beta^\alpha,$$

$$R_{0\alpha} = \frac{1}{2} (\varkappa_{\alpha;\beta}^\beta - \varkappa_{\beta;\alpha}^\beta), \tag{97.10}$$

$$R_{\alpha\beta} = \frac{1}{2} \frac{\partial}{\partial t} \varkappa_{\alpha\beta} + \frac{1}{4}(\varkappa_{\alpha\beta}\varkappa_\gamma^\gamma - 2\varkappa_\alpha^\gamma \varkappa_{\beta\gamma}) + P_{\alpha\beta}.$$

Here $P_{\alpha\beta}$ is the three-dimensional Ricci tensor which is expressed in terms of $\gamma_{\alpha\beta}$ in the same way as R_{ik} is expressed in terms of g_{ik}. All operations of raising indices and of covariant differentiation are carried out with the three-dimensional metric $\gamma_{\alpha\beta}$.

We write the Einstein equations in mixed components:

$$R_0^0 = -\frac{1}{2} \frac{\partial}{\partial t} \varkappa_\alpha^\alpha - \frac{1}{4} \varkappa_\alpha^\beta \varkappa_\beta^\alpha = 8\pi k(T_0^0 - \tfrac{1}{2}T), \tag{97.11}$$

$$R_\alpha^0 = \frac{1}{2}(\varkappa_{\alpha;\beta}^\beta - \varkappa_{\beta;\alpha}^\beta) = 8\pi k T_\alpha^0, \tag{97.12}$$

$$R_\alpha^\beta = -P_\alpha^\beta - \frac{1}{2\sqrt{\gamma}} \frac{\partial}{\partial t} (\sqrt{\gamma}\varkappa_\alpha^\beta) = 8\pi k(T_\alpha^\beta - \tfrac{1}{2}\delta_\alpha^\beta T). \tag{97.13}$$

A characteristic feature of synchronous reference systems is that they are not stationary: the gravitational field cannot be constant in such a system. In fact, in a constant field we would have $\varkappa_{\alpha\beta} = 0$. But in the presence of matter the vanishing of all the $\varkappa_{\alpha\beta}$ would contradict (97.11) (which has a right side different from zero). In empty space we would find from (97.13) that all the $P_{\alpha\beta}$, and with them all the components of the three-dimensional curvature tensor $P_{\alpha\beta\gamma\delta}$, vanish, i.e. the field vanishes entirely (in a synchronous system with a euclidean spatial metric the space-time is flat).

At the same time the matter filling the space cannot in general be at rest relative to the synchronous reference frame. This is obvious from the fact that particles of matter within

† But this does not, of course, apply to operations of shifting indices in the space components of the four-tensors R_{ik}, T_{ik} (see the footnote on p. 270). Thus T_α^β must be understood as before to be $g^{\beta\gamma}T_{\gamma\alpha} + g^{\beta 0}T_{0\alpha}$, which in the present case reduces to $g^{\beta\gamma}T_{\gamma\alpha}$ and differs in sign from $\gamma^{\beta\gamma}T_{\gamma\alpha}$.

which there are pressures generally move along lines that are not geodesics; the world line of a particle at rest is a time line, and thus is a geodesic in the synchronous reference system. An exception is the case of "dust" ($p = 0$). Here the particles interacting with one another will move along geodesic lines; consequently, in this case the condition for a synchronous reference system does not contradict the condition that it be comoving with the matter.† For other equations of state a similar situation can occur only in special cases when the pressure gradient vanishes in all or in certain directions.

From (97.11) one can show that the determinant $-g = \gamma$ of the metric tensor in a synchronous system must necessarily go to zero in a finite length of time.

To prove this we note that the expression on the right side of this equation is positive for any distribution of the matter. In fact, in a synchronous reference system we have for the energy-momentum tensor (94.9):

$$T_0^0 - \tfrac{1}{2}T = \tfrac{1}{2}(\varepsilon + 3p) + \frac{(p + \varepsilon)v^2}{1 - v^2}$$

[the components of the four-velocity are given by (88.14)]; this quantity is clearly positive. The same statement is also true for the energy-momentum tensor of the electromagnetic field ($T = 0$, T_0^0 is the positive energy density of the field). Thus we have from (97.11):

$$-R_0^0 = \frac{1}{2}\frac{\partial}{\partial t}\varkappa_\alpha^\alpha + \frac{1}{4}\varkappa_\alpha^\beta\varkappa_\beta^\alpha \le 0 \tag{97.14}$$

(where the equality sign applies in empty space).

Using the algebraic inequality‡

$$\varkappa_\beta^\alpha\varkappa_\alpha^\beta \ge \tfrac{1}{3}(\varkappa_\alpha^\alpha)^2$$

we can rewrite (97.14) in the form

$$\frac{\partial}{\partial t}\varkappa_\alpha^\alpha + \tfrac{1}{6}(\varkappa_\alpha^\alpha)^2 \le 0$$

or

$$\frac{\partial}{\partial t}\left(\frac{1}{\varkappa_\alpha^\alpha}\right) \ge \frac{1}{6}. \tag{97.15}$$

Suppose, for example, that at a certain time $\varkappa_\alpha^\alpha > 0$. Then as t decreases the quantity $1/\varkappa_\alpha^\alpha$ decreases, with a finite (nonzero) derivative, so that it must go to zero (from positive values) in the course of a finite time. In other words, $\varkappa_\alpha^\alpha$ goes to $+\infty$, but since $\varkappa_\alpha^\alpha = \partial \ln \gamma/\partial t$, this means that the determinant γ goes to zero [no faster than t^6, according to the inequality

---

† Even in this case, in order to be able to choose a "synchronously comoving" system of reference, it is still necessary that the matter move "without rotation". In the comoving system the contravariant components of the velocity are $u^0 = 1$, $u^\alpha = 0$. If the reference system is also synchronous, the covariant components must satisfy $u_0 = 1$, $u_\alpha = 0$, so that its four-dimensional curl must vanish:

$$u_{i;k} - u_{k;i} \equiv \frac{\partial u_i}{\partial x^k} - \frac{\partial u_k}{\partial x^i} = 0.$$

But this tensor equation must then also be valid in any other reference frame. Thus, in a synchronous, but not comoving system we then get the condition curl $\mathbf{v} = 0$ for the three-dimensional velocity $\mathbf{v}$.

‡ Its validity can easily be seen by bringing the tensor $\varkappa_\alpha^\beta$ to diagonal form (at a given instant of time).

(97.15)]. If on the other hand $\varkappa_\alpha^\alpha < 0$ at the initial time, we get the same result for increasing times.

This result does not, however, by any means prove that there must be a real physical singularity of the metric. A physical singularity is one that is characteristic of the space-time itself, and is not related to the character of the reference frame chosen (such a singularity should be characterized by the tending to infinity of various scalar quantities—the matter density, or the invariants of the curvature tensor). The singularity in the synchronous reference system, which we have proven to be inevitable, is in general actually fictitious, and disappears when we change to another (nonsynchronous) reference frame. Its origin is evident from simple geometrical arguments.

We saw earlier that setting up a synchronous system reduces to the construction of a family of geodesic lines orthogonal to any space-like hypersurface. But the geodesic lines of an arbitrary family will, in general, intersect one another on certain enveloping hypersurfaces—the four-dimensional analogues of the caustic surfaces of geometrical optics. We know that intersection of the coordinate lines gives rise to a singularity of the metric in the particular coordinate system. Thus there is a geometrical reason for the appearance of a singularity, associated with the specific properties of the synchronous system and therefore not physical in character. In general an arbitrary metric of four-space also permits the existence of nonintersecting families of geodesic lines. The unavoidable vanishing of the determinant in the synchronous system means that the curvature properties of a real (non-flat) space-time (which are expressed by the inequality $R_0^0 \geq 0$) that are permitted by the field equations exclude the possibility of existence of such families, so that the time lines in a synchronous reference system necessarily intersect one another.†

We mentioned earlier that for dustlike matter the synchronous reference system can also be comoving. In this case the density of the matter goes to infinity at the caustic—simply as a result of the intersection of the world trajectories of the particles, which coincide with the time lines. It is, however, clear that this singularity of the density can be eliminated by introducing an arbitrarily small nonzero pressure of the matter, and in this sense is not physical in character.

PROBLEMS

1. Find the form of the solution of the gravitational field equations in vacuum in the vicinity of a point that is not singular, but regular in the time.

Solution: Having agreed on the convention that the time under consideration is the time origin, we look for $\gamma_{\alpha\beta}$ in the form:

$$\gamma_{\alpha\beta} = a_{\alpha\beta} + t b_{\alpha\beta} + t^2 c_{\alpha\beta} + \ldots, \tag{1}$$

---

† For the analytic construction of the metric in the vicinity of a fictitious singularity in a synchronous reference system, see E. M. Lifshitz, V. V. Sudakov and I. M. Khalatnikov, *JETP* 40, 1847, 1961, (*Soviet Phys. —JETP*, 13, 1298, 1961). The general character of the metric is clear from geometrical considerations. Since the caustic hypersurface always contains timelike intervals (the line elements of the geodesic time lines at their points of tangency to the caustic), it is not spacelike. Furthermore, on the caustic one of the principal values of the metric tensor $\gamma_{\alpha\beta}$ vanishes, corresponding to the vanishing of the distance (δ) between two neighbouring geodesics that intersect one another at their point of tangency to the caustic. The quantity δ goes to zero as the first power of the distance (l) to the point of intersection. Thus the principal value of the metric tensor, and with it the determinant γ, goes to zero like l^2.

where $a_{\alpha\beta}$, $b_{\alpha\beta}$, $c_{\alpha\beta}$ are functions of the space coordinates. In this same approximation the reciprocal tensor is:

$$\gamma^{\alpha\beta} = a^{\alpha\beta} - t b^{\alpha\beta} + t^2 (b^{\alpha\gamma} b_\gamma^\beta - c^{\alpha\beta}),$$

where $a^{\alpha\beta}$ is the tensor reciprocal to $a_{\alpha\beta}$, and the raising of indices of the other tensors is done by using $a^{\alpha\beta}$. We also have:

$$\varkappa_{\alpha\beta} = b_{\alpha\beta} + 2t c_{\alpha\beta}, \quad \varkappa_\alpha^\beta = b_\alpha^\beta + t(2c_\alpha^\beta - b_{\alpha\gamma} b^{\beta\gamma}).$$

The Einstein equations (97.11)–(97.13) lead to the following relations:

$$R_0^0 = -c + \tfrac{1}{4} b_\alpha^\beta b_\beta^\alpha = 0, \tag{2}$$

$$R_\alpha^0 = \tfrac{1}{2}(b_{\alpha;\beta}^\beta - b_{;\alpha}) + t[-c_{;\alpha} + \tfrac{3}{8}(b_\gamma^\beta b_\beta^\gamma)_{;\alpha} + c_{\alpha;\beta}^\beta + \tfrac{1}{4} b_\alpha^\beta b_{;\beta} - \tfrac{1}{2}(b_\alpha^\gamma b_\gamma^\beta)_{;\beta}] = 0, \tag{3}$$

$$R_\alpha^\beta = -P_\alpha^\beta - \tfrac{1}{4} b_\alpha^\beta b + \tfrac{1}{2} b_\alpha^\gamma b_\gamma^\beta - c_\alpha^\beta = 0 \tag{4}$$

$(b \equiv b_\alpha^\alpha, c \equiv c_\alpha^\alpha)$. The operations of covariant differentiation are carried out in the three-dimensional space with metric $a_{\alpha\beta}$; the tensor $P_{\alpha\beta}$ is also defined with respect to this metric.

From (4) the coefficients $c_{\alpha\beta}$ are completely determined in terms of the coefficients $a_{\alpha\beta}$ and $b_{\alpha\beta}$. Then (2) gives the relation

$$P + \tfrac{1}{4} b^2 - \tfrac{1}{4} b_\alpha^\beta b_\beta^\alpha = 0. \tag{5}$$

From the terms of zero order in (3) we have:

$$b_{\alpha;\beta}^\beta = b_{;\alpha}. \tag{6}$$

The terms $\sim t$ in this equation vanish identically when we use (2) and (4)–(6) and the identity $P_{\alpha;\beta}^\beta = \tfrac{1}{2} P_{;\alpha}$ [see (92.10)].

Thus the twelve quantities $a_{\alpha\beta}$, $b_{\alpha\beta}$ are related to one another by the one relation (5) and the three relations (6), so that there remain eight arbitrary functions of the three space coordinates. Of these, three are related to the possibility of arbitrary transformations of the three space coordinates, and one to the arbitrariness in choosing the initial hypersurface for setting up the synchronous reference system. Therefore we are left with the correct number (see the end of § 95) of four "physically different" arbitrary functions.

2. Calculate the components of the curvature tensor R_{iklm} in the synchronous reference system.

Solution: Using the Christoffel symbols (97.9) we find from (92.1):

$$R_{\alpha\beta\gamma\delta} = -P_{\alpha\beta\gamma\delta} + \tfrac{1}{4}(\varkappa_{\alpha\delta}\varkappa_{\beta\gamma} - \varkappa_{\alpha\gamma}\varkappa_{\beta\delta}),$$

$$R_{0\alpha\beta\gamma} = \tfrac{1}{2}(\varkappa_{\alpha\gamma;\beta} - \varkappa_{\alpha\beta;\gamma}),$$

$$R_{0\alpha0\beta} = \frac{1}{2}\frac{\partial}{\partial t}\varkappa_{\alpha\beta} - \frac{1}{4}\varkappa_{\alpha\gamma}\varkappa_\beta^\gamma,$$

where $P_{\alpha\beta\gamma\delta}$ is the three-dimensional curvature tensor corresponding to the three-dimensional metric $\gamma_{\alpha\beta}$.

3. Find the general form of the infinitesimal transformation from one synchronous reference system to another.

Solution: The transformation has the form

$$t \to t + \varphi(x^1, x^2, x^3), \quad x^\alpha \to x^\alpha + \xi^\alpha(x^1, x^2, x^3, t),$$

where φ and ξ^α are small quantities. We are guaranteed that the condition $g_{00} = 1$ is satisfied by keeping φ independent of t; to maintain the condition $g_{0\alpha} = 0$, we must satisfy the equations

$$\gamma_{\alpha\beta}\frac{\partial\xi^\beta}{\partial t} = \frac{\partial\varphi}{\partial x^\alpha},$$

from which

$$\xi^\alpha = \frac{\partial \varphi}{\partial x^\beta} \int \gamma^{\alpha\beta} \, dt + f^\alpha (x^1, x^2, x^3), \tag{1}$$

where the f^α are again small quantities (forming a three-dimensional vector **f**). The spatial metric tensor $\gamma_{\alpha\beta}$ is replaced by

$$\gamma_{\alpha\beta} \, \rangle \, \gamma_{\alpha\beta} \quad \xi_{\alpha; \beta} \quad \xi_{\beta; \alpha} - \varphi \varkappa_{\alpha\beta} \tag{2}$$

[as can be easily verified using (94.3)].

Thus the transformation contains four arbitrary functions (φ, f^α) of the space coordinates.

§ 98. The tetrad representation of the Einstein equations

The determination of the components of the Ricci tensor (and the resulting formulation of the Einstein equations) for a metric of some special form generally involves quite complicated calculations. It is therefore important to consider various formulas that enable one to simplify these calculations in some cases and to represent the result in a more easily understood form. Among such formulas is the expression for the curvature tensor in tetrad form.

We introduce a set of four linearly independent coordinate four-vectors $e^i_{(a)}$ (labelled by the index a) and subject only to the condition that

$$e^i_{(a)} e_{(b)i} = \eta_{ab}, \tag{98.1}$$

where η_{ab} is a given constant symmetric matrix with signature $+ - - -$; we denote the matrix reciprocal to η_{ab} by $\eta^{ab}(\eta^{ac}\eta_{cb} = \delta^a_b)$.† Together with the tetrad of vectors $e^i_{(a)}$, we also introduce a tetrad of reciprocal vectors $e^{(a)i}$ (labeled by superscript axial indices), and defined by the conditions

$$e^{(a)}_i e^i_{(b)} = \delta^a_b, \tag{98.2}$$

i.e., each of the vectors $e^{(a)}_i$ is orthogonal to the three vectors $e^i_{(b)}$ with $b \neq a$. Multiplying (98.2) by $e^k_{(a)}$, we get $(e^k_{(a)} e^{(a)}_i) e^i_{(b)} = e^k_{(b)}$, from which we see that in addition to (98.2), the equation

$$e^{(a)}_i e^k_{(a)} = \delta^k_i \tag{98.3}$$

is also automatically satisfied.

Multiplying the equation $e^i_{(a)} e_{(c)i} = \eta_{ac}$ on both sides by η^{bc}, we get:

$$e^i_{(a)} (\eta^{bc} e_{(c)i}) = \delta^b_a;$$

comparing with (98.2), we find that

$$e^{(b)}_i = \eta^{bc} e_{(c)i}, \quad e_{(b)i} = \eta_{bc} e^{(c)}_i. \tag{98.4}$$

Thus the lowering and raising of tetrad indices is done with the matrices η_{bc} and η^{bc}. The importance of the tetrads thus introduced is that one can express the metric tensor in terms

† In this section we use early Latin letters $a, b, c, \ldots$ for indices labelling coordinate vectors; four-tensor indices are again labelled by $i, k, l \ldots$. In the literature the indices of axes are usually enclosed in brackets. To avoid too cumbersome writing of formulas we shall use brackets only when indices for axes appear along with tensor indices and omit them for quantities which by their definition involve only coordinate axis indices (for example, η_{ab} and later $\gamma_{abc}, \lambda_{abc}$). A summation is understood throughout over repeated indices of both types.

of them. In fact, according to the definition of the relation between co- and contravariant components of a four-vector, we have $e_i^{(a)} = g_{il}e^{(a)l}$: multiplying by $e_{(a)k}$ and using (98.3) and (98.4), we find:

$$g_{ik} = e_{(a)i}e_k^{(a)} = \eta_{ab}e_i^{(a)}e_k^{(b)}. \tag{98.5}$$

The square of the line element with the metric tensor (98.5) is

$$ds^2 = \eta_{ab}(e_i^{(a)}dx^i)(e_k^{(b)}dx^k). \tag{98.6}$$

As for the arbitrarily assignable matrix η_{ab}, the most natural choice is the "galilean" form (i.e. a diagonal matrix with elements $1, -1, -1, -1$); then, according to (98.1) the coordinate vectors will be orthogonal, with one of them timelike and the other three spacelike.† We emphasize, however, that this choice is by no means necessary, and that situations are possible where, for one or another reason (e.g., symmetry properties of the metric) the choice of a nonorthogonal tetrad is advantageous.‡

The tetrad components of a four-vector A^i (and similarly for a four-tensor of any rank) are defined as its "projections" on the coordinate four-vectors:

$$A_{(a)} = e_{(a)}^i A_i, \quad A^{(a)} = e_i^{(a)}A^i = \eta^{ab}A_{(b)}. \tag{98.7}$$

Conversely,

$$A_i = e_i^{(a)}A_{(a)}, \quad A^i = e_{(a)}^i A^{(a)}, \tag{98.8}$$

In this same way we define the operation of "differentiation along the a direction":

$$\phi_{,(a)} = e_{(a)}^i \frac{\partial\phi}{\partial x^i}.$$

We introduce quantities needed in the sequel:§

$$\gamma_{acb} = e_{(a)i;k}e_{(b)}^i e_{(c)}^k, \tag{98.9}$$

and their linear combinations

$$\lambda_{abc} = \gamma_{abc} - \gamma_{acb} =$$

$$= (e_{(a)i;k} - e_{(a)k;i})e_{(b)}^i e_{(c)}^k = (e_{(a)i,k} - e_{(a)k,i})e_{(b)}^i e_{(c)}^k. \tag{98.10}$$

The last equation in (98.10) follows from (86.12); we note that the quantities λ_{abc} are calculated by a simple differentiation of the frame vectors. The converse expression for the γ_{abc} in terms of the λ_{abc} is:

† Having chosen the linear forms $dx^{(a)} = e_i^{(a)}dx^i$ as the segments of the coordinate axes in the given element of four-space (and taken "galilean" η_{ab}) we bring the metric in this element to galilean form. We emphasize once again that the forms $dx^{(a)}$ are not in general exact differentials of any function of the coordinates.

‡ The advantageous choice of the tetrad may be dictated by a previous reduction of ds^2 to the form (98.6). Thus the expression for ds^2 in the form (88.13) corresponds to coordinate vectors

$$e_i^{(0)} = (\sqrt{h}, \; -\sqrt{h}\,g), \quad e_i^{(a)} = (0, e^{(a)}),$$

where the choice of the $e^{(a)}$ depends on the spatial form dl^2.

§ The quantities γ_{abc} are called the *Ricci rotation coefficients*.

$$\gamma_{abc} = \tfrac{1}{2}(\lambda_{abc} + \lambda_{bca} - \lambda_{cab}). \tag{98.11}$$

These quantities have symmetry properties:

$$\gamma_{abc} = -\gamma_{bac}, \quad \lambda_{abc} = -\lambda_{acb}. \tag{98.12}$$

Our aim is to determine the tetrad components of the curvature tensor. To do this we must start from the definition (91.6) as applied to the covariant derivatives of the frame vectors:

$$e_{(a)i;k;l} - e_{(a)i;l;k} = e^m_{(a)} R_{mikl},$$

or

$$R_{(a)(b)(c)(d)} = (e_{(a)i;k;l} - e_{(a)i;l;k}) e^i_{(b)} e^k_{(c)} e^l_{(d)}.$$

This expression is easily written in terms of the quantities γ_{abc}. We write

$$e_{(a)i;k} = \gamma_{abc} e^{(b)}_i e^{(c)}_k,$$

and after further covariant differentiation the derivatives of the frame vectors are again expressed in this same way; here the covariant derivative of the scalar quantity γ_{abc} coincides with its ordinary derivative.† As a result we get

$$R_{(a)(b)(c)(d)} = \gamma_{abc,d} - \gamma_{abd,c} + \gamma_{abf}(\gamma^f{}_{cd} - \gamma^f{}_{dc}) + \gamma_{afc}\gamma^f{}_{bd} - \gamma_{afd}\gamma^f{}_{bc}, \tag{98.13}$$

where, in accordance with the general rule, $\gamma^a{}_{bc} = \eta^{ad}\gamma_{abc}$, etc.

Contraction of this tensor on a pair of indices a, c gives the required tetrad components of the Ricci tensor; we present them as expressed in terms of the quantities λ_{abc}:

$$R_{(a)(b)} = -\tfrac{1}{2}(\lambda_{ab}{}^c{}_{,c} + \lambda_{ba}{}^c{}_{,c} + \lambda^c{}_{ca,b} + \lambda^c{}_{cb,a} +$$

$$+ \lambda^{cd}{}_b\lambda_{cda} + \lambda^{cd}{}_b\lambda_{dca} - \tfrac{1}{2}\lambda_b{}^{cd}\lambda_{acd} + \lambda^c{}_{cd}\lambda_{ab}{}^d + \lambda^c{}_{cd}\lambda_{ba}{}^d). \tag{98.14}$$

Finally, we call attention to the fact that our construction is essentially independent of the four-dimensional nature of the metric. Thus the results obtained can also be used to calculate the three-dimensional Riemann and Ricci tensors for a three-dimensional metric. Naturally, then, in place of the tetrad of four-vectors we will deal with a triad of three-vectors, while the matrix η_{ab} should have the signature + + + (we shall deal with such an application in § 116).

---

† We give for reference the transformation in similar fashion of arbitrary four-vectors and four-tensors:

$$A_{i;k} e^i_{(a)} e^k_{(b)} = A_{(a)(b)} - A^{(d)}\gamma_{dab},$$

$$A_{ik;l} e^i_{(a)} e^k_{(b)} e^l_{(c)} = A_{(a)(b)(c)} - A^{(d)}{}_{(b)}\gamma_{dac} + A_{(a)}{}^{(d)}\gamma_{dbc},$$

etc.

CHAPTER 12

THE FIELD OF GRAVITATING BODIES

§ 99. Newton's law

In the Einstein field equations we now carry out the transition to the limit of nonrelativistic mechanics. As was stated in § 87, the assumption of small velocities of all particles requires also that the gravitational field be weak.

The expression for the component g_{00} of the metric tensor (the only one which we need) was found, for the limiting case which we are considering, in § 87:

$$g_{00} = 1 + \frac{2\phi}{c^2}.$$

Further, we can use for the components of the energy-momentum tensor the expression (35.4) $T_i^k = \mu c^2 u_i u^k$, where μ is the mass density of the body (the sum of the rest masses of the particles in a unit volume; we drop the subscript 0 on μ). As for the four-velocity u^i, since the macroscopic motion is also considered to be slow, we must neglect all its space components and retain only the time component, that is, we must set $u^\alpha = 0$, $u^0 = u_0 = 1$. Of all the components T_i^k, there thus remains only

$$T_0^0 = \mu c^2. \tag{99.1}$$

The scalar $T = T_i^i$ will be equal to this same value μc^2.

We write the field equations in the form (95.8):

$$R_i^k = \frac{8\pi k}{c^4} \left(T_i^k - \frac{1}{2} \delta_i^k T \right);$$

for $i = k = 0$

$$R_0^0 = \frac{4\pi k}{c^2} \mu.$$

One easily verifies that in the approximation we are considering all the other equations vanish identically.

For the calculation of R_0^0 from the general formula (92.7), we note that terms containing products of the quantities Γ_{kl}^i are in every case quantities of the second order. Terms containing derivatives with respect to $x^0 = ct$ are small (compared with terms with derivatives with respect to the coordinates x^α) since they contain extra powers of $1/c$. As a result, there remains $R_{00} = R_0^0 = \partial \Gamma_{00}^\alpha / \partial x^\alpha$. Substituting

$$\Gamma^\alpha_{00} \simeq -\frac{1}{2} g^{\alpha\beta} \frac{\partial g_{00}}{\partial x^\beta} = \frac{1}{c^2} \frac{\partial \phi}{\partial x^\alpha}$$

we find

$$R^0_0 = \frac{1}{c^2} \frac{\partial^2 \phi}{\partial x^{u\,2}} = \frac{1}{c^2} \Delta\phi,$$

Thus the Einstein equations give

$$\Delta\phi = 4\pi k\mu. \tag{99.2}$$

This is the equation of the gravitational field in nonrelativistic mechanics. It is completely analogous to the Poisson equation (36.4) for the electric potential, where here in place of the charge density we have the mass density multiplied by $-k$. Therefore we can immediately write the general solution of equation (99.2) by analogy with (36.8) in the form

$$\phi = -k \int \frac{\mu dV}{R}. \tag{99.3}$$

This formula determines the potential of the gravitational field of an arbitrary mass distribution in the nonrelativistic approximation.

In particular, we have for the potential of the field of a single particle of mass m

$$\phi = -\frac{km}{R} \tag{99.4}$$

and, consequently, the force $F = -m'(\partial\phi/\partial R)$, acting in this field on another particle (mass m'), is equal to

$$F = -\frac{kmm'}{R^2}. \tag{99.5}$$

this is the well-known *law of attraction of Newton*.

The potential energy of a particle in a gravitational field is equal to its mass multiplied by the potential of the field, in analogy to the fact that the potential energy in an electric field is equal to the product of the charge and the potential of the field. Therefore, we may write, by analogy with (37.1), for the potential energy of an arbitrary mass distribution, the expression

$$U = \frac{1}{2} \int \mu\phi \, dV. \tag{99.6}$$

For the Newtonian potential of a constant gravitational field at large distances from the masses, producing it, we can give an expansion analogous to that obtained in §§ 40–41 for the electrostatic field. We choose the coordinate origin at the inertial centre of the masses. Then the integral $\int \mu\mathbf{r} \, dV$, which is analogous to the dipole moment of a system of charges, vanishes identically. Thus, unlike the case of the electrostatic field, in the case of the gravitational field we can always eliminate the "dipole terms". Consequently, the expansion of the potential ϕ has the form:

$$\phi = -k\left\{ \frac{M}{R_0} + \frac{1}{6} D_{\alpha\beta} \frac{\partial^2}{\partial X_\alpha \partial X_\beta} \frac{1}{R_0} + \cdots \right\} \tag{99.7}$$

where $M = \int \mu \, dV$ is the total mass of the system, and the quantity

$$D_{\alpha\beta} = \int \mu(3x_\alpha x_\beta - r^2\delta_{\alpha\beta}) \, dV \tag{99.8}$$

may be called the *mass quadrupole moment tensor*.† It is related to the usual *moment of inertia tensor*

$$J_{\alpha\beta} = \int \mu(r^2\delta_{\alpha\beta} - x_\alpha x_\beta) \, dV$$

by the obvious relation

$$D_{\alpha\beta} = J_{\gamma\gamma}\delta_{\alpha\beta} - 2J_{\alpha\beta}. \tag{99.9}$$

The determination of the Newtonian potential from a given distribution of masses is the subject of one of the branches of mathematical physics; the exposition of the various methods for this is not the subject of the present book. Here we shall for reference purposes give only the formulas for the potential of the gravitational field produced by a homogeneous ellipsoidal body.

Let the surface of the ellipsoid be given by the equation

$$\frac{x^2}{a^2} + \frac{y^2}{b^2} + \frac{z^2}{c^2} = 1, \qquad a > b > c. \tag{99.10}$$

Then the potential of the field at an arbitrary point outside the body is given by the following formula:

$$\varphi = -\pi\mu abck \int_\xi^\infty \left(1 - \frac{x^2}{a^2 + s} + \frac{y^2}{b^2 + s} - \frac{z^2}{c^2 + s}\right) \frac{ds}{R_s} \tag{99.11}$$

$$R_s = \sqrt{(a^2 + s)(b^2 + s)(c^2 + s)},$$

where ξ is the positive root of the equation

$$\frac{x^2}{a^2 + \xi} + \frac{y^2}{b^2 + \xi} + \frac{z^2}{c^2 + \xi} = 1. \tag{99.12}$$

The potential of the field in the interior of the ellipsoid is given by the formula

$$\varphi = -\pi\mu abck \int_0^\infty \left(1 - \frac{x^2}{a^2 + s} - \frac{y^2}{b^2 + s} - \frac{z^2}{c^2 + s}\right) \frac{ds}{R_s}, \tag{99.13}$$

which differs from (99.11) in having the lower limit replaced by zero; we note that this expression is a quadratic function of the coordinates x, y, z.

The gravitational energy of the body is obtained, according to (99.6), by integrating the

† We here write all indices α, β as subscripts, not distinguishing between co- and contravariant components, in accordance with the fact that all operations are carried out in ordinary Newtonian (Euclidean) space.

expression (99.13) over the volume of the ellipsoid. This integral can be done by elementary methods,† and gives:

$$U = \frac{3km^2}{8} \int_0^\infty \left[\frac{1}{5} \left(\frac{a^2}{a^2+s} + \frac{b^2}{b^2+s} + \frac{c^2}{c^2+s} \right) - 1 \right] \frac{ds}{R_s} =$$

$$= \frac{3km^2}{8} \int_0^\infty \left[\frac{2}{5} sd \left(\frac{1}{R_s} \right) - \frac{2\,ds}{5\,R_s} \right]$$

$\left(m = \dfrac{4\pi}{3} abc\,\mu \text{ is the total mass of the body} \right)$; integrating the first term by parts, we obtain finally:

$$U = -\frac{3km^2}{10} \int_0^\infty \frac{ds}{R_s}. \tag{99.14}$$

All the integrals appearing in formulas (99.11)–(99.14) can be expressed in terms of elliptic integrals of the first and second kind. For ellipsoids of rotation, these integrals are expressed in terms of elementary functions. In particular, the gravitational energy of an oblate ellipsoid of rotation ($a = b > c$) is

$$U = -\frac{3km^2}{5\sqrt{a^2-c^2}} \cos^{-1} \frac{c}{a} \tag{99.15}$$

and for a prolate ellipsoid of rotation ($a > b = c$):

$$U = -\frac{3km^2}{5\sqrt{a^2-c^2}} \cosh^{-1} \frac{a}{c}. \tag{99.16}$$

For a sphere ($a = c$) both formulas give the value $U = -3km^2/5a$, which, of course, can also be obtained by elementary methods.‡

PROBLEM

Determine the equilibrium shape of a homogeneous gravitating mass of liquid which is rotating as a whole.

Solution: The condition of equilibrium is the constancy on the surface of the body of the sum of the gravitational potential and the potential of the centrifugal forces:

† The integration of the squares x^2, y^2, z^2 is most simply done by making the substitution $x = ax'$, $y = by'$, $z = cz'$, which reduces the integral over the volume of the ellipsoid to an integral over the volume of the unit sphere.

‡ The potential of the field inside a homogeneous sphere of radius a is

$$\varphi = -2\pi k\mu \left(a^2 - \frac{r^2}{3} \right).$$

$$\phi - \frac{\Omega^2}{2}(x^2 + y^2) = \text{const.}$$

(Ω is the angular velocity; the axis of rotation is the z axis). The required shape is that of an oblate ellipsoid of rotation. To determine its parameters we substitute (99.13) in the condition of equilibrium, and eliminate z^2 by using equation (99.10); this gives:

$$(x^2 + y^2)\left[\int_0^\infty \frac{ds}{(a^2 + s)^2\sqrt{c^2 + s}} - \frac{\Omega^2}{2\pi\mu ka^2 c} - \frac{c^2}{a^2}\int_0^\infty \frac{ds}{(a^2 + s)(c^2 + s)^{3/2}}\right] = \text{const.},$$

from which it follows that the expression in the square brackets must vanish. Performing the integration, we get the equation

$$\frac{(a^2 + 2c^2)c}{(a^2 - c^2)^{3/2}}\cos^{-1}\frac{c}{a} - \frac{3c^2}{a^2 - c^2} = \frac{\Omega^2}{2\pi k\mu} = \frac{25}{6}\left(\frac{4\pi}{3}\right)^{1/3}\frac{M^2\mu^{1/3}}{m^{10/3}k}\left(\frac{c}{a}\right)^{4/3}$$

($M = \frac{2}{5}ma^2\Omega$ is the angular momentum of the body around the z axis), which determines the ratio of the semiaxes c/a for given Ω or M. The dependence of the ratio c/a on M is single-valued; c/a increases monotonically with increasing M.

However, it turns out that the symmetrical form which we have found is stable (with respect to small perturbations) only for not too large values of M.[†] The stability is lost for $M = 2.89\ k^{1/2}m^{5/3}\mu^{1/6}$ (when $c/a = 0.58$). With further increase of M, the equilibrium shape becomes a general ellipsoid with gradually decreasing values of b/a and c/a (from 1 and from 0.58, respectively). This shape in turn becomes unstable for $M = 3.84\ k^{1/2}m^{5/3}\mu^{-1/6}$ (when $a : b : c = 1 : 0.43 : 0.34$).

§ 100. The centrally symmetric gravitational field

Let us consider a gravitational field possessing central symmetry. Such a field can be produced by any centrally symmetric distribution of matter; for this, of course, not only the distribution but also the motion of the matter must be centrally symmetric, i.e. the velocity at each point must be directed along the radius.

The central symmetry of the field means that the space-time metric, that is, the expression for the interval ds, must be the same for all points located at the same distance from the centre. In euclidean space this distance is equal to the radius vector; in a non-euclidean space, such as we have in the presence of a gravitational field, there is no quantity which has all the properties of the euclidean radius vector (for example to be equal both to the distance from the centre and to the length of the circumference divided by 2π). Therefore the choice of a "radius vector" is now arbitrary.

If we use "spherical" space coordinates r, θ, ϕ, then the most general centrally symmetric expression for ds^2 is

$$ds^2 = h(r, t)\ dr^2 + k(r, t)\ (\sin^2\theta\ d\phi^2 + d\theta^2) + l(r, t)\ dt^2 + a(r, t)\ dr\ dt, \quad (100.1)$$

where $a\ h, k, l$ are certain functions of the "radius vector" r and the "time" t. But because of the arbitrariness in the choice of a reference system in the general theory of relativity, we can still subject the coordinates to any transformation which does not destroy the central symmetry of ds^2; this means that we can transform the coordinates r and t according to the formulas

$$r = f_1(r', t'), \qquad t = f_2(r', t'),$$

[†] References to the literature concerning this question can be found in the book by H. Lamb, *Hydrodynamics*, chap. XII.

where f_1, f_2 are any functions of the new coordinates r', t'.

Making use of this possibility, we choose the coordinate r and the time t in such a way that, first of all, the coefficient $a(r, t)$ of $dr\, dt$ in the expression for ds^2 vanishes and, secondly, the coefficient $k(r, t)$ becomes equal simply to $-r^2$.† The latter condition implies that the radius vector r is defined in such a way that the circumference of a circle with centre at the origin of coordinates is equal to $2\pi r$ (the element of arc of a circle in the plane $\theta = \pi/2$ is equal to $dl = r\, d\phi$). It will be convenient to write the quantities h and l in exponential form, as $-e^\lambda$ and $c^2 e^\nu$ respectively, where λ and ν are some functions of r and t. Thus we obtain the following expression for ds^2:

$$ds^2 = e^\nu c^2\, dt^2 - r^2(d\theta^2 + \sin^2\theta\, d\phi^2) - e^\lambda\, dr^2. \tag{100.2}$$

Denoting by x^0, x^1, x^2, x^3, respectively, the coordinates ct, r, θ, ϕ, we have for the nonzero components of the metric tensor the expressions

$$g_{00} = e^\nu, \qquad g_{11} = -e^\lambda, \qquad g_{22} = -r^2, \qquad g_{33} = -r^2 \sin^2\theta.$$

Clearly,

$$g^{00} = e^{-\nu}, \qquad g^{11} = -e^{-\lambda}, \qquad g^{22} = -r^{-2}, \qquad g^{33} = -r^{-2} \sin^{-2}\theta.$$

With these values it is easy to calculate the Γ^i_{kl} from formula (86.3). The calculation leads to the following expressions (the prime means differentiation with respect to r, while a dot on a symbol means differentiation with respect to ct):

$$\Gamma^1_{11} = \frac{\lambda'}{2}, \qquad \Gamma^0_{10} = \frac{\nu'}{2}, \qquad \Gamma^2_{33} = -\sin\theta\cos\theta,$$

$$\Gamma^0_{11} = \frac{\dot\lambda}{2} e^{\lambda-\nu}, \qquad \Gamma^1_{22} = -re^{-\lambda}, \qquad \Gamma^1_{00} = \frac{\nu'}{2} e^{\nu-\lambda},$$

$$\Gamma^2_{12} = \Gamma^3_{13} = \frac{1}{r}, \qquad \Gamma^3_{23} = \cot\theta, \qquad \Gamma^0_{00} = \frac{\dot\nu}{2}. \tag{100.3}$$

$$\Gamma^1_{10} = \frac{\dot\lambda}{2}, \qquad \Gamma^1_{33} = -r\sin^2\theta e^{-\lambda}.$$

All other components (except for those which differ from the ones we have written by a transposition of the indices k and l) are zero.

To get the equations of gravitation we must calculate the components of the tensor R^i_k according to formula (92.7). A simple calculation leads to the following equations:

$$\frac{8\pi k}{c^4} T^1_1 = -e^{-\lambda}\left(\frac{\nu'}{r} + \frac{1}{r^2}\right) + \frac{1}{r^2}, \tag{100.4}$$

$$\frac{8\pi k}{c^4} T^2_2 = \frac{8\pi k}{c^4} T^3_3 = -\frac{1}{2} e^{-\lambda}\left(\nu'' + \frac{\nu'^2}{2} + \frac{\nu' - \lambda'}{r} - \frac{\nu'\lambda'}{2}\right) + \frac{1}{2} e^{-\nu}\left(\ddot\lambda + \frac{\dot\lambda^2}{2} - \frac{\dot\lambda\dot\nu}{2}\right), \tag{100.5}$$

$$\frac{8\pi k}{c^4} T^0_0 = -e^{-\lambda}\left(\frac{1}{r^2} - \frac{\lambda'}{r}\right) + \frac{1}{r^2}, \tag{100.6}$$

† These conditions do not determine the choice of the time coordinate uniquely. It can still be subjected to an arbitrary transformation $t = f(t')$, not containing r.

$$\frac{8\pi k}{c^4} T_0^1 = -e^{-\lambda} \frac{\dot{\lambda}}{r}. \tag{100.7}$$

The other components of (95.6) vanish identically. Using (94.9), the components of the energy momentum tensor can be expressed in terms of the energy density ε of the matter, its pressure p, and the radial velocity v.

The equations (100.4)–(100.7) can be integrated exactly in the very important case of a centrally symmetric field in vacuum, that is, outside of the masses producing the field. Setting the energy-momentum tensor equal to zero, we get the following equations:

$$e^{-\lambda}\left(\frac{v'}{r} + \frac{1}{r^2}\right) - \frac{1}{r^2} = 0, \tag{100.8}$$

$$e^{-\lambda}\left(\frac{\lambda'}{r} - \frac{1}{r^2}\right) + \frac{1}{r^2} = 0, \tag{100.9}$$

$$\dot{\lambda} = 0 \tag{100.10}$$

[we do not write the fourth equation, that is, equation (100.5), since it follows from the other three equations].

From (100.10) we see directly that λ does not depend on the time. Further, adding equations (100.8) and (100.9), we find $\lambda' + v' = 0$, that is,

$$\lambda + v = f(t), \tag{100.11}$$

where $f(t)$ is a function only of the time. But when we chose the interval ds^2 in the form (100.2), there still remained the possibility of an arbitrary transformation of the time of the form $t = f(t')$. Such a transformation is equivalent to adding to v an arbitrary function of the time, and with its aid we can always make $f(t)$ in (100.11) vanish. And so, without any loss in generality, we can set $\lambda + v = 0$. Note that the centrally symmetric gravitational field in vacuum is automatically static.

The equation (100.9) is easily integrated and gives:

$$e^{-\lambda} = e^{v} = 1 + \frac{\text{const}}{r}. \tag{100.12}$$

Thus, at infinity ($r \to \infty$), $e^{-\lambda} = e^{v} = 1$, that is, far from the gravitating bodies the metric automatically becomes galilean. The constant is easily expressed in terms of the mass of the body by requiring that at large distances, where the field is weak, Newton's law should hold.† In other words, we should have $g_{00} = 1 + (2\phi/c^2)$, where the potential ϕ has its Newtonian value (99.4) $\phi = -(km/r)$ (m is the total mass of the bodies producing the field). From this it is clear that const $= -(2km/c^2)$. This quantity has the dimensions of length; it is called the *gravitational radius* r_g of the body:

$$r_g = \frac{2km}{c^2}. \tag{100.13}$$

† For the field in the interior of a spherical cavity in a centrally symmetric distribution, we must have const $= 0$, since otherwise the metric would have a singularity at $r = 0$. Thus the metric inside such a cavity is automatically galilean, i.e. there is no gravitational field in the interior of the cavity (just as in Newtonian theory).

Thus we finally obtain the space-time metric in the form:

$$ds^2 = \left(1 - \frac{r_g}{r}\right) c^2 dt^2 - r^2 (\sin^2 \theta \, d\phi^2 + d\theta^2) - \frac{dr^2}{1 - \dfrac{r_g}{r}} . \qquad (100.14)$$

This solution of the Einstein equations was found by K. Schwarzschild (1916). It completely determines the gravitational field in vacuum produced by any centrally-symmetric distributtion of masses. We emphasize that this solution is valid not only for masses at rest, but also when they are moving, so long as the motion has the required symmetry (for example, a centrally-symmetric pulsation). We note that the metric (100.14) depends only on the total mass of the gravitating body, just as in the analogous problem in Newtonian theory.

The spatial metric is determined by the expression for the element of spatial distance:

$$dl^2 = \frac{dr^2}{1 - \dfrac{r_g}{r}} + r^2(\sin^2 \theta \, d\phi^2 + d\theta^2). \qquad (100.15)$$

The geometrical meaning of the coordinate r is determined by the fact that in the metric (100.15) the circumference of a circle with its centre at the centre of the field is $2\pi r$. But the distance between two points r_1 and r_2 along the same radius is given by the integral

$$\int_{r_1}^{r_2} \frac{dr}{\sqrt{1 - \dfrac{r_g}{r}}} > r_2 - r_1. \qquad (100.16)$$

Furthermore, we see that $g_{00} \leq 1$. Combining with the formula (84.1) $d\tau = \sqrt{g_{00}} \, dt$, defining the proper time, it follows that

$$d\tau \leq dt. \qquad (100.17)$$

The equality sign holds only at infinity, where t coincides with the proper time. Thus at finite distances from the masses there is a "slowing down" of the time compared with the time at infinity.

Finally, we present an approximate expression for ds^2 at large distances from the origin of coordinates:

$$ds^2 = ds_0^2 - \frac{2km}{c^2 r} (dr^2 + c^2 dt^2). \qquad (100.18)$$

The second term represents a small correction to the galilean metric ds_0^2. At large distances from the masses producing it, every field appears centrally symmetric. Therefore (100.18) determines the metric at large distances from any system of bodies.

Certain general considerations can also be made concerning the behaviour of a centrally symmetric gravitational field in the interior of the gravitating masses. From equation (100.6) we see that for $r \to 0$, λ must also vanish at least like r^2; if this were not so the right side of the equation would become infinite for $r \to 0$, that is, T_0^0 would have a singular point at $r = 0$, which is physically impossible. Formally integrating (100.6) with the limiting condition $\lambda|_{r=0} = 0$, we obtain

$$\lambda = - \ln \left\{ 1 - \frac{8\pi k}{c^4 r} \int_0^r T_0^0 r^2 \, dr \right\}. \qquad (100.19)$$

Since, from (100.10), $T_0^0 = e^{-v}T_{00} \geq 0$, it is clear that $\lambda \geq 0$, that is,

$$e^\lambda \geq 1. \tag{100.20}$$

Subtracting equation (100.6) term by term from (100.4), we get;

$$\frac{e^{-\lambda}}{r}(v' + \lambda') = \frac{8\pi k}{c^4}(T_0^0 - T_1^1) = \frac{(\varepsilon + p)\left(1 + \dfrac{v^2}{c^2}\right)}{1 - \dfrac{v^2}{c^2}} \geq 0,$$

i.e. $v' + \lambda' \geq 0$. But for $r \to \infty$ (far from the masses) the metric becomes galilean, i.e. $v \to 0$, $\lambda \to 0$. Therefore, from $v' + \lambda' \geq 0$ it follows that over all space

$$v + \lambda \leq 0. \tag{100.21}$$

Since $\lambda \geq 0$, it then follows that $v \leq 0$, i.e.

$$e^v \leq 1. \tag{100.22}$$

The inequalities obtained show that the above properties (100.16) and (100.17) of the spatial metric and the behaviour of clocks in a centrally symmetric field in vacuum apply equally well to the field in the interior of the gravitating masses.

If the gravitational field is produced by a spherical body of "radius" a, then for $r > a$, we have $T_0^0 = 0$. For points with $r > a$, formula (100.19) therefore gives

$$\lambda = -\ln\left\{1 - \frac{8\pi k}{c^4 r}\int_0^a T_0^0 r^2\, dr\right\}.$$

On the other hand, we can here apply the expression (100.14) referring to vacuum, according to which

$$\lambda = -\ln\left(1 - \frac{2km}{c^2 r}\right).$$

Equating the two expressions, we get the formula

$$m = \frac{4\pi}{c^2}\int_0^a T_0^0 r^2\, dr, \tag{100.23}$$

expressing the total mass of a body in terms of its energy-momentum tensor.

In particular, for a static distribution of matter in the body we have $T_0^0 = \varepsilon$, so that

$$m = \frac{4\pi}{c^2}\int_0^a \varepsilon r^2\, dr. \tag{100.24}$$

We call attention to the fact that the integration is taken with respect to $4\pi r^2\, dr$, whereas the element of spatial volume for the metric (100.2) is $dV = 4\pi r^2 e^{\lambda/2}\, dr$, where, according to (100.20), $e^{\lambda/2} > 1$. This difference indicates the gravitational mass defect of the body.

PROBLEMS

1. Find the invariants of the curvature tensor for the Schwarzschild metric (100.14).

Solution: A calculation with (92.1) and the Γ^i_{kl} from (100.3) (or using the formulas found in problem 2 of § 92) leads to the following values for the nonzero components of the curvature tensor:

$$R_{0101} = \frac{r_g}{r^3}, \quad R_{0202} = \frac{R_{0303}}{\sin^2\theta} = -\frac{r_g(r-r_g)}{2r^2},$$

$$R_{1212} = \frac{R_{1313}}{\sin^2\theta} = \frac{r_g}{2(r-r_g)}, \quad R_{2323} = -rr_g\sin^2\theta.$$

For the invariants I_1 and I_2 of (92.20) we find:

$$I_1 = \left(\frac{r_g}{2r^3}\right)^2, \quad I_2 = -\left(\frac{r_g}{2r^3}\right)^2$$

(products including the dual tensor $\tilde{R}_{iklm}$ are identically zero). The curvature tensor belongs to type D of the Petrov classification (with real invariants $\lambda^{(1)} = \lambda^{(2)} = -r_g/2r^3$). We note that the curvature invariants have a singularity only at the point $r = 0$, and not at $r = r_g$.

2. Determine the spatial curvature for this metric.

Solution: The components of the spatial curvature tensor $P_{\alpha\beta\gamma\delta}$ can be expressed in terms of the components of the tensor $P_{\alpha\beta}$ (and the tensor $\gamma_{\alpha\beta}$) so that we need only calculate $P_{\alpha\beta}$ (see problem 1 in § 92). The tensor $P_{\alpha\beta}$ is expressed in terms of $\gamma_{\alpha\beta}$ just as R_{ik} is expressed in terms of g_{ik}. Using the values of $\gamma_{\alpha\beta}$ from (100.15), we find from the calculations:

$$P^\theta_\theta = P^\phi_\phi = \frac{r_g}{2r^3}, \quad P^r_r = -\frac{r_g}{r^3},$$

and $P^\beta_\alpha = 0$ for $\alpha \neq \beta$. We note that $P^\theta_\theta, P^\phi_\phi > 0, P^r_r < 0$, while $P \equiv P^\alpha_\alpha = 0$.

From the formula given in problem 1 of § 92, we find:

$$P_{r\theta r\theta} = (P^r_r + P^\theta_\theta)\gamma_{rr}\gamma_{\theta\theta} = -P^\phi_\phi\gamma_{rr}\gamma_{\theta\theta},$$

$$P_{r\phi r\phi} = -P^\theta_\theta\gamma_{rr}\gamma_{\phi\phi},$$

$$P_{\theta\phi\theta\phi} = -P^r_r\gamma_{\theta\theta}\gamma_{\phi\phi}.$$

It then follows (see the footnote on p. 283) that for a "plane" perpendicular to the radius, the Gaussian curvature is

$$K = \frac{P_{\theta\phi\theta\phi}}{\gamma_{\theta\theta}\gamma_{\phi\phi}} = -P^r_r > 0,$$

(which means that, in a small triangle drawn on the "plane" in the neighbourhood of its intersection with the radius perpendicular to it, the sum of the angles of the triangle is greater than π). As to the "planes" which pass through the centre, their Gaussian curvature $K < 0$; this means that the sum of the angles of a small triangle in such a "plane" is less than π (however this does not refer to the triangles embracing the centre—the sum of the angles in such a triangle is greater than π).

3. Determine the form of the surface of rotation on which the geometry would be the same as on a "plane" passing through the origin in a centrally symmetric gravitational field *in vacuo*.

Solution: The geometry on the surface of rotation $z = z(r)$ is determined (in cylindrical coordinates) by the element of length:

$$dl^2 = dr^2 + dz^2 + r^2 d\varphi^2 = dr^2(1 + z'^2) + r^2 d\varphi^2.$$

Comparing with the element of length (100.15) in the "plane" $\theta = \pi/2$

$$dl^2 = r^2 d\varphi^2 + \frac{dr^2}{1 - r_g/r},$$

we find

$$1 + z'^2 = \left(1 - \frac{r_g}{r}\right)^{-1},$$

from which

$$z = 2\sqrt{r_g(r - r_g)}.$$

For $r = r_g$ this function has a singularity—a branch point. The reason for this is that the spatial metric (100.15) in contrast to the space-time metric (100.14), actually has a singularity at $r = r_g$.

The general properties of the geometry on "planes" passing through the centre, which were mentioned in the preceding problem, can also be found by considering the curvature in the pictorial model given here.

4. Transform the interval (100.14) to such coordinates that its element of spatial distance has conformal—euclidean form, i.e. dl^2 is proportional to its euclidean expression.

Solution: Setting

$$r = \left(1 + \frac{r_g}{4\rho}\right)^2 \rho,$$

we get from (100.14)

$$ds^2 = \left(\frac{1 - \dfrac{r_g}{4\rho}}{1 + \dfrac{r_g}{4\rho}}\right)^2 c^2 \, dt^2 - \left(1 + \frac{r_g}{4\rho}\right)^4 (d\rho^2 + \rho^2 d\theta^2 + \rho^2 \sin^2\theta \, d\varphi^2).$$

The coordinates ρ, θ, φ are called *isotropic spherical coordinates;* instead of them we can also introduce isotropic cartesian coordinates x, y, z. In particular, at large distances ($\rho \gg r_g$) we have approximately:

$$ds^2 = \left(1 - \frac{r_g}{p}\right) c^2 \, dt^2 - \left(1 + \frac{r_g}{p}\right)(dx^2 + dy^2 + dz^2).$$

5. Find the equations for a centrally symmetric gravitational field in matter in the comoving reference system.

Solution: We make use of the two possible transformations of the coordinates r, t in the element of interval (100.1) in order to, first, make the coefficient $a(r, t)$ of $dr \, dt$ vanish, and, second, to make the radial velocity of the matter vanish at each point (because of the central symmetry the other components are not present). After this is done, r and t can still be subjected to an arbitrary transformation of the form $r = r(r')$ and $t = t(t')$.

We denote the radial coordinate and time selected in this way by R and τ, and the coefficients h, k, l by $-e^\lambda$, $-e^\mu$, e^ν, respectively (where λ, μ and ν are functions of R and τ). We then have for the line element:

$$ds^2 = c^2 e^\nu d\tau^2 - e^\lambda dR^2 - e^\mu(d\theta^2 + \sin^2\theta \, d\varphi^2). \tag{1}$$

In the comoving reference system the components of the energy-momentum tensor are:

$$T_0^0 = \varepsilon, \quad T_1^1 = T_2^2 = T_3^3 = -p.$$

The calculation gives the following field equations:†

$$-\frac{8\pi k}{c^4} T_1^1 = \frac{8\pi k}{c^4} p = \frac{1}{2}e^{-\lambda}\left(\frac{\mu'^2}{2} + \mu'v'\right) - e^{-\nu}(\ddot{\mu} - \frac{1}{2}\dot{\mu}\dot{\nu} + \frac{3}{4}\dot{\mu}^2) - e^{-\mu}, \tag{2}$$

$$-\frac{8\pi k}{c^4} T_2^2 = \frac{8\pi k}{c^4} p = \frac{1}{4}e^{-\lambda}(2\nu'' + \nu'^2 + 2\mu'' + \mu'^2 - \mu'\lambda' - \nu'\lambda' + \mu'\nu') +$$

$$+ \frac{1}{4}e^{-\nu}(\dot{\lambda}\dot{\nu} + \dot{\mu}\dot{\nu} - \dot{\lambda}\dot{\mu} - 2\ddot{\lambda} - \dot{\lambda}^2 - 2\ddot{\mu} - \dot{\mu}^2), \tag{3}$$

$$\frac{8\pi k}{c^4} T_0^0 = \frac{8\pi k}{c^4} \varepsilon = -e^{-\lambda}\left(\mu'' + \frac{3}{4}\mu'^2 - \frac{\mu'\lambda'}{2}\right) + \frac{1}{2}e^{-\nu}\left(\dot{\lambda}\dot{\mu} + \frac{\dot{\mu}^2}{2}\right) + e^{-\mu}, \tag{4}$$

$$\frac{8\pi k}{c^4} T_0^1 = 0 = \frac{1}{2}e^{-\lambda}(2\dot{\mu}' + \dot{\mu}\mu' - \dot{\lambda}\mu' - \nu'\dot{\mu}) \tag{5}$$

(where the prime denotes differentiation with respect to R and the dot with respect to $c\tau$).

General relations for λ, μ and ν can be found easily if we start from the equations $T_{i;k}^k = 0$ which are contained in the field equations. Using formula (86.11), we get the following two equations:

$$\dot{\lambda} + 2\dot{\mu} = -\frac{2\dot{\varepsilon}}{p + \varepsilon}, \quad \nu' = -\frac{2p'}{p + \varepsilon}. \tag{6}$$

If p is known as a function of ε, equation (6) can be integrated in the form:

$$\lambda + 2\mu = -2\int \frac{d\varepsilon}{p + \varepsilon} + f_1(R), \quad \nu = -2\int \frac{dp}{p + \varepsilon} + f_2(\tau), \tag{7}$$

where the functions $f_1(R)$ and $f_2(\tau)$ can be chosen arbitrarily in view of the possibility mentioned above of making arbitrary transformations of the form $R = R(R')$, $\tau = \tau(\tau')$.

6. Find the equations determining the static gravitational field in vacuum around an axially symmetric body at rest (H. Weyl, 1917).

Solution: The static line element in cylindrical space coordinates $x^1 = \phi$, $x^2 = \varrho$, $x^3 = z$ is assumed to have the form

$$ds^2 = e^{\nu}c^2 dt^2 - e^{\omega} d\phi^2 - e^{\mu}(d\varrho^2 + dz^2),$$

where ν, ω, μ are functions of ρ and z; such a representation fixes the choice of coordinates to within a transformation of the form $\varrho = \varrho(\varrho', z')$, $z = z(\varrho', z')$, which multiplies the quadratic form $d\varrho^2 + dz^2$ simply by a common factor.

From the equations

$$R_0^0 = \frac{1}{4}e^{-\mu}[2\nu_{,\rho\rho} + \nu_{,\rho}(\nu_{,\rho} + \omega_{,\rho}) + 2\nu_{,z,z} + \nu_{,z}(\nu_{,z} + \omega_{,z})] = 0,$$

$$R_1^1 = \frac{1}{4}e^{-\mu}[2\omega_{,\rho\rho} + \omega_{,\rho}(\nu_{,\rho} + \omega_{,\rho}) + 2\omega_{,z,z} + \omega_{,z}(\nu_{,z} + \omega_{,z})] = 0$$

(where the subscripts ,ρ and ,z denote differentiation with respect to ϱ and z), and taking their sum, we find:

$$\varrho'_{,\rho\rho} + \varrho'_{,z,z} = 0,$$

where

$$\varrho'(\varrho, z) = e^{\frac{\nu+\omega}{2}}.$$

† The components R_{ik} can be calculated directly as was done in the text or with formulas obtained in problem 2 of § 92.

Thus $\varrho'(\varrho, z)$ is a harmonic function of the variables ϱ, z. According to the well-known properties of such functions, this means that there exists a conjugate harmonic function $z'(\varrho, z)$ such that $\varrho' + iz' = f(\varrho + iz)$, where f is an analytic function of the complex variable $\varrho + iz$. If now we choose ϱ', z' as new coordinates, because of the conformality of the transformation $\varrho, z \to \varrho'z'$ we will have

$$e^\mu(d\varrho^2 + dz^2) = e^{\mu'}(d\varrho'^2 + dz'^2),$$

where $\mu'(\varrho', z')$ is some new function. At the same time $e^\omega = \varrho'^2 e^{-\nu}$; writing $\omega + \nu = \gamma$ and dropping the primes, we write ds^2 in the form

$$ds^2 = e^\nu c^2 dt^2 - \varrho^2 e^{-\nu} d\phi^2 - e^{\gamma - \nu}(d\varrho^2 + dz^2), \tag{1}$$

Forming the equations $R_0^0 = 0, R_3^3 - R_2^2 = 0, R_2^3 = 0$ for this metric, we find:

$$\frac{1}{\varrho} \frac{\partial}{\partial \varrho} \left(\varrho \frac{\partial \nu}{\partial \varrho} \right) + \frac{\partial^2 \nu}{\partial z^2} - 0, \tag{2}$$

$$\frac{\partial \gamma}{\partial z} = \varrho \frac{\partial \nu}{\partial \varrho} \frac{\partial \nu}{\partial z}, \quad \frac{\partial \gamma}{\partial \varrho} = \varrho \left[\left(\frac{\partial \nu}{\partial \varrho} \right)^2 - \left(\frac{\partial \nu}{\partial z} \right)^2 \right]. \tag{3}$$

We note that (2) has the form of the Laplace equation in cylindrical coordinates (for a function independent of ϕ). If this equation is solved, then the function $\gamma(\varrho, z)$ is completely determined by eqs. (2)–(3). At large distances from the body producing the field, the functions ν and γ should tend to zero.

§ 101. Motion in a centrally symmetric gravitational field

Let us consider the motion of a particle in a centrally symmetric gravitational field. As in every centrally symmetric field, the motion occurs in a single "plane" passing through the origin; we choose this plane as the plane $\theta = \pi/2$.

To determine the trajectory of the particle, we use the Hamilton–Jacobi equation:

$$g^{ik} \frac{\partial S}{\partial x^i} \frac{\partial S}{\partial x^k} - m^2 c^2 = 0.$$

where m is the mass of the particle (and we denote the mass of the central body by m'). Using the g^{ik} given in the expression (100.14), we find the following equation:

$$\left(1 - \frac{r_g}{r} \right)^{-1} \left(\frac{\partial S}{c \partial t} \right)^2 - \left(1 - \frac{r_g}{r} \right) \left(\frac{\partial S}{\partial r} \right)^r - \frac{1}{r^2} \left(\frac{\partial S}{\partial \phi} \right)^2 - m^2 c^2 = 0, \tag{101.1}$$

where $r_g = 2km'/c^2$ is the gravitational radius of the central body.

By the general procedure for solving the Hamilton–Jacobi equation, we look for an S in the form

$$S = - \mathscr{E}_0 t + M\phi + S_r(r), \tag{101.2}$$

with constant energy $\mathscr{E}_0$ and angular momentum M. Substituting (101.2) in (101.1), we find the derivative dS_r/dr, and thus:

$$S_r = \int \sqrt{\frac{\mathscr{E}_0^2}{c^2} \left(1 - \frac{r_g}{r} \right)^{-2} - \left(m^2 c^2 + \frac{M^2}{r^2} \right) \left(1 - \frac{r_g}{r} \right)^{-1}} \cdot dr. \tag{101.3}$$

The dependence $r = r(t)$ is given (cf. *Mechanics*, § 47) by the equation $\partial S/\partial \mathscr{E}_0 = \text{const}$, from which

$$ct = \frac{\mathscr{E}_0}{mc^2} \int \frac{dr}{\left(1 - \frac{r_g}{r}\right)\left[\left(\frac{\mathscr{E}_0}{mc^2}\right)^2 - \left(1 + \frac{M^2}{m^2c^2r^2}\right)\left(1 - \frac{r_g}{r}\right)\right]^{\frac{1}{2}}} \qquad (101.4)$$

The trajectory itself is determined by the equation $\partial S/\partial M = \text{const}$, so that

$$\phi = \int \frac{M\,dr}{r^2 \sqrt{\frac{\mathscr{E}_0^2}{c^2} - \left(m^2c^2 + \frac{M^2}{r^2}\right)\left(1 - \frac{r_g}{r}\right)}} . \qquad (101.5)$$

This integral reduces to an elliptic integral.

For the motion of a planet in the field of attraction of the Sun, the relativistic theory leads to only an insignificant correction compared to Newton's theory, since the velocities of the planets are very small compared to the velocity of light. In the integrand in the equation (101.5) for the trajectory, this corresponds to a small value for the ratio r_g/r, where r_g is the gravitational radius of the Sun.†

To calculate the relativistic corrections to the trajectory, it is convenient to start from the expression (101.3) for the radial part of the action, before differentiation with respect to M.

We make a transformation of the integration variable, writing

$$r(r - r_g) = r'^2, \quad \text{i.e.} \quad r - \frac{r_g}{r} \cong r',$$

as a result of which the term with M^2 under the square root takes the form M^2/r'^2. In the other terms we make an expansion in powers of r_g/r', and obtain to the required accuracy:

$$S_r = \int \left[\left(2\mathscr{E}'m + \frac{\mathscr{E}'^2}{c^2}\right) + \frac{1}{r}(2m^2m'k + 4\mathscr{E}'mr_g) - \frac{1}{r^2}\left(M^2 - \frac{3m^2c^2r_g^2}{2}\right)\right]^{1/2} dr, \qquad (101.6)$$

where for brevity we have dropped the prime on r' and introduced the non-relativistic energy $\mathscr{E}$ (without the rest energy).

The correction terms in the coefficients of the first two terms under the square root have only the not particularly interesting effect of changing the relation between the energy and momentum of the particle and changing the parameters of its Newtonian orbit (ellipse). But the change in the coefficient of $1/r^2$ leads to a more fundamental effect—to a systematic (secular) shift in the perihelion of the orbit.

Since the trajectory is defined by the equation $\phi + (\partial S_r/\partial M) = \text{const}$, the change of the angle ϕ after one revolution of the planet in its orbit is

$$\Delta\phi = -\frac{\partial}{\partial M}\Delta S_r,$$

where ΔS_r is the corresponding change in S_r. Expanding S_r in powers of the small correction to the coefficient of $1/r^2$, we get:

† For the Sun, $r_g = 3$ km; for Earth, $r_g = 0.9$ cm.

$$\Delta S_r = \Delta S_r^{(0)} - \frac{3m^2 c^2 r_g^2}{4M} \frac{\partial \Delta S_r^{(0)}}{\partial M},$$

where $\Delta S_r^{(0)}$ corresponds to the motion in the closed ellipse which is unshifted. Differentiating this relation with respect to M, and using the fact that

$$-\frac{\partial}{\partial M} \Delta S_r^{(0)} = \Delta \phi^{(0)} = 2\pi,$$

we find:

$$\Delta \phi = 2\pi + \frac{3\pi m^2 c^2 r_g^2}{2M^2} = 2\pi + \frac{6\pi k^2 m^2 m'^2}{c^2 M^2}.$$

The second term is the required angular displacement $\delta \phi$ of the Newtonian ellipse during one revolution, i.e. the shift in the perihelion of the orbit. Expressing it in terms of the length a of the semimajor axis and the eccentricity e of the ellipse by means of the formula

$$\frac{M^2}{km'm^2} = a(1 - e^2),$$

we obtain:†

$$\delta \phi = \frac{6\pi km'}{c'a(1 - e^2)}. \tag{101.7}$$

Next we consider the path of a light ray in a centrally symmetric gravitational field. This path is determined by the eikonal equation (87.9)

$$g^{ik} \frac{\partial \psi}{\partial x^i} \frac{\partial \psi}{\partial x^k} = 0,$$

which differs from the Hamilton–Jacobi equation only in having m set equal to zero. Therefore the trajectory of the ray can be obtained immediately from (101.5) by setting $m = 0$; at the same time, in place of the energy $\mathscr{E}_0 = -(\partial S/\partial t)$ of the particle we must write the frequency of the light, $\omega_0 = -(\partial \psi/\partial t)$. Also introducing in place of the constant M a constant ϱ defined by $\varrho = cM/\omega_0$, we get:

$$\phi = \int \frac{dr}{r^2 \sqrt{\frac{1}{\varrho^2} - \frac{1}{r^2} \left(1 - \frac{r_g}{r}\right)}}. \tag{101.8}$$

If we neglect the relativistic corrections ($r_g \to 0$), this equation gives $r = \varrho/\cos \phi$, i.e. a straight line passing at a distance ϱ from the origin. To study the relativistic corrections, we proceed in the same way as in the previous case.

For the radial part of the eikonal we have [see (101.3)]:

$$\psi_r(r) = \frac{\omega_0}{c} \int \sqrt{\frac{r^2}{(r - r_g)^2} - \frac{\varrho^2}{r(r - r_g)}} \, dr.$$

† Numerical values of the shifts determined from formula (101.7) for Mercury and Earth are equal, respectively, to 43.0″ and 3.8″ per century.

Making the same transformations as were used to go from (101.3) to (101.6), we find:

$$\psi_r(r) = \frac{\omega_0}{c} \int \sqrt{1 + \frac{2r_g}{r} - \frac{\varrho^2}{r^2}}\, dr.$$

Expanding the integrand in powers of r_g/r, we have:

$$\psi_r = \psi_r^{(0)} + \frac{r_g \omega_0}{c} \int \frac{dr}{\sqrt{r^2 - \varrho^2}} = \psi_r^{(0)} + \frac{r_g \omega_0}{c} \cosh^{-1} \frac{r}{\varrho},$$

where $\psi_r^{(0)}$ corresponds to the classical straight ray.

The total change in ψ_r during the propagation of the light from some very large distance R to the point $r = \varrho$ nearest to the centre and then back to the distance R is equal to

$$\Delta\psi_r = \Delta\psi_r^{(0)} + 2\, \frac{r_g \omega_0}{c} \cosh^{-1} \frac{R}{\varrho}.$$

The corresponding change in the polar angle ϕ along the ray is obtained by differentiation with respect to $M = \varrho_0/\omega c$:

$$\Delta\phi = -\frac{\partial \Delta\psi_r}{\partial M} = -\frac{\partial \Delta\psi_r^{(0)}}{\partial M} + \frac{2r_g R}{\varrho \sqrt{R^2 - \varrho^2}}.$$

Finally, going to the limit $R \to \infty$, and noting that the straight ray corresponds to $\Delta\phi = \pi$, we get:

$$\Delta\phi = \pi + \frac{2r_g}{\varrho}.$$

This means that under the influence of the field of attraction the light ray is bent: its trajectory is a curve which is concave toward the centre (the ray is "attracted" toward the centre), so that the angle between its two asymptotes differs from π by

$$\delta\phi = \frac{2r_g}{\varrho} = \frac{4km'}{c^2 \varrho}; \tag{101.9}$$

in other words, the ray of light, passing at a distance ϱ from the centre of the field, is deflected through an angle $\delta\phi$.†

§ 102. Gravitational collapse of a spherical body

In the Schwarzschild metric (101.14), g_{00} goes to zero and g_{11} to infinity at $r = r_g$ (on the "Schwarzschild sphere"). This could give the basis for concluding that there must be a singularity of the space-time metric and that it is therefore impossible for bodies to exist that have a "radius" (for a given mass) that is less than the gravitational radius. Actually, however, this conclusion would be wrong. This is already evident from the fact that the determinant $g = -r^4 \sin^2 \theta$ has no singularity at $r_g = r$, so that the condition $g < 0$ (82.3) is not violated. We shall see that in fact we are dealing simply with the impossibility of establishing a rigid reference system for $r < r_g$.

† For a ray just skirting the edge of the Sun, $\delta\phi = 1.75''$.

To make clear the true character of the space-time metric in this domain† we make a transformation of the coordinates of the form:

$$c\tau = \pm\, ct \pm \int \frac{f(r)dr}{1 - \dfrac{r_g}{r}}, \qquad R = ct + \int \frac{dr}{\left(1 - \dfrac{r_g}{r}\right) f(r)}. \tag{102.1}$$

Then

$$ds^2 = \frac{1 - \dfrac{r_g}{r}}{1 - f^2}(c^2 d\tau^2 - f^2 dR^2) - r^2(d\theta^2 + \sin^2\theta\, d\phi^2).$$

We eliminate the singularity at $r = r_g$ by choosing $f(r)$ so that $f(r_g) = 1$. If we set $f(r) = \sqrt{r_g/r}$, then the new coordinate system will also be synchronous ($g_{\tau\tau} = 1$). First choosing the upper sign in (102.1), we have:

$$R - c\tau = \int \frac{(1 - f^2)\, dr}{\left(1 - \dfrac{r_g}{r}\right) f} = \int \sqrt{\frac{r}{r_g}}\, dr = \frac{2}{3}\frac{r^{3/2}}{r_g^{1/2}},$$

or

$$r = \left(\frac{3}{2}(R - c\tau)\right)^{2/3} r_g^{1/3} \tag{102.2}$$

(we set the integration constant, which depends on the time origin, equal to zero). The element of interval is:

$$ds^2 = c^2 d\tau^2 - \frac{dR^2}{\left[\dfrac{3}{2r_g}(R - c\tau)\right]^{2/3}} - \left[\frac{3}{2}(R - c\tau)\right]^{4/3} r_g^{2/3}(d\theta^2 + \sin^2\theta\, d\phi^2). \tag{102.3}$$

In these coordinates the singularity on the Schwarzschild sphere [to which there corresponds the equality $\frac{3}{2}(R - c\tau) = r_g$] is absent. The coordinate R is everywhere spacelike, while τ is timelike. The metric (102.3) is nonstationary. As in every synchronous reference system, the time lines are geodesics. In other words, "test" particles at rest relative to the reference system are particles moving freely in the given field.

To given values of r there correspond world lines $R - c\tau = $ const (the sloping straight lines in Fig. 20). The world lines of particles at rest relative to the reference system are shown on this diagram as vertical lines; moving along these lines, after a finite interval of proper time the particles "fall in" to the centre of the field ($r = 0$), which is the location of the true singularity of the metric.

Let us consider the propagation of radial light signals. The equation $ds^2 = 0$ (for θ, $\phi = $ const) gives for the derivative $d\tau/dR$ along the ray:

† The physical meaning of the Schwarzschild singularity was first explained by D. Finkelstein (1958) using a different transformation. The metric (102.3) was first found by Lemaitre (1938).

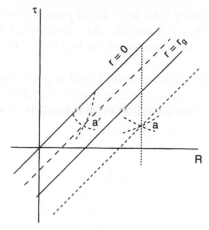

FIG. 20

$$c\frac{d\tau}{dR} = \pm \frac{1}{\left(\dfrac{3}{2r_g}(R - c\tau)\right)^{1/3}} = \pm \sqrt{\frac{r_g}{r}}, \qquad (102.4)$$

the two signs corresponding to the two boundaries of the light "cone" with its vertex at the given world point. When $r > r_g$ (point a in Fig. 20) the slope of these boundaries satisfies $|c\,d\tau/dR| < 1$, so that the straight line $r = $ const (along which $c\,d\tau/dR = 1$) falls inside the cons. But in the region $r < r_g$ (point a') we have $|c\,d\tau/dR| > 1$, so that the line $r = $ const, the world line of a particle at rest relative to the centre of the field, lies outside the cone. Both boundaries of the cone intersect the line $r = 0$ at a finite distance, approaching it along a vertical. Since no causally related events can lie on the world line outside the light cone, it follows that in the region $r < r_g$ no particles can be at rest. Here all interactions and signals propagate in the direction toward the centre, reaching it after a finite interval of time τ.

Similarly, choosing the lower signs in (102.1) we would obtain an "expanding" reference system with a metric differing from (102.3) by a change of the sign of τ. It corresponds to a space-time in which (in the region $r < r_g$) again rest is impossible, but all signals propagate outward from the centre.

The results described here can be applied to the problem of the behaviour of massive bodies in the general theory of relativity.

The investigation of the relativistic conditions for equilibrium of a spherical body shows that for a body of sufficiently large mass, states of static equilibrium cannot exist.† It is clear that such a body must contract without limit (i.e. it must undergo "*gravitational collapse*").‡

In a reference system not attached to the body and galilean at infinity [metric (100.14)], the radius of the central body cannot be less than r_g. This means that according to the clocks t of a distant observer the radius of the contracting body only approaches the gravitational radius asymptotically as $t \to \infty$. It is easy to find the limiting form of this dependence.

A particle on the surface of the contracting body is at all times in the field of attraction of a constant mass m, the total mass of the body. As $r \to r_g$ the gravitational force becomes very

† See *Statistical Physics*, § 111.

‡ The essential properties of this phenomenon were first explained by J. R. Oppenheimer and H. Snyder (1939).

large; but the density of the body (and with it, the pressure) remains finite. Neglecting the pressure forces for this reason, we reduce the determination of the time dependence $r = r(t)$ of the radius of the body to a consideration of the free fall of a test particle in the field of the mass m.

The function $r(t)$ for fall in a Schwarzschild field is given by the integral (101.4), where for purely radial motion $M = 0$. Thus, if the fall starts at a "distance" r_0 from the centre with zero velocity at some time t_0, the energy of the particle is $\mathscr{E}_0 = mc^2 \sqrt{1 - r_g/r_0}$, and for the time t for it to reach the "distance" r we have:

$$c(t - t_0) = \sqrt{1 - \frac{r_g}{r_0}} \int_r^{r_0} \frac{dr}{\left(1 - \frac{r_g}{r}\right)\sqrt{\frac{r_g}{r} - \frac{r_g}{r_0}}}. \tag{102.5}$$

This integral diverges like $r_g \ln (r - r_g)$ for $r \rightarrow r_g$. Thus we find the asymptotic formula for the approach of r to r_g:

$$r - r_g = \text{const } e^{-(ct/r_g)}. \tag{102.6}$$

Thus the final stage of approach of the collapsing body to the gravitational radius occurs according to an exponential law with a very small characteristic time $\sim r_g/c$.

Although the rate of contraction as observed from outside goes to zero asymptotically, the velocity v of fall of the particles, as measured in their proper time, increases and approaches the velocity of light. In fact, according to the definition (88.10):

$$v^2 = -\frac{g_{11}}{g_{00}} \left(\frac{dr}{dt}\right)^2.$$

Taking g_{11} and g_{00} from (100.14) and dr/dt from (102.5), we find:

$$1 - \frac{v^2}{c^2} = \frac{1 - \frac{r_g}{r}}{1 - \frac{r_g}{r_0}}. \tag{102.7}$$

The approach to the gravitational radius, which according to the clocks of the outside observer takes an infinite time, occupies only a finite interval of proper time (i.e. time in the reference system comoving with the body). This is already clear from the analysis given above, but one can also verify it directly by computing the proper time τ as the invariant integral

$$c\tau = \int ds = \int \left[c^2 g_{00} \frac{dt^2}{dr^2} + g_{11}\right]^{\frac{1}{2}} dr$$

Taking dr/dt for the falling particle from (102.5), we get for the proper time for fall from r_0 to r:

$$\tau - \tau_0 = \frac{1}{c} \int_{r_0}^{r} \left(\frac{r_g}{r} - \frac{r_g}{r_0}\right)^{-\frac{1}{2}} dr. \tag{102.8}$$

This integral converges for $r \rightarrow r_g$.

Having reached the gravitational radius (as measured by proper time), the body will continue to contract, with all of its particles arriving at the centre within a finite time; the moment of collapse of each portion of the matter into the centre is a true singularity of the space–time metric. We do not, however, observe any of this process of collapse of the body within the Schwarzschild sphere. The moment when the surface of the body crosses this sphere corresponds to $t = \infty$; we may say that the whole process of collapse within the Schwarzschild sphere occurs "after an infinite time" for the distant observer—an extreme example of the relativity of time. There are, of course, on logical contradictions in this picture. In complete accord with it is the statement above about the property of the contracting coordinate system: in this system no signals emerge from the Schwarzschild sphere. Particles or light rays may intersect this sphere (in the comoving reference system) only in one direction—toward the inside, and once crossing there, can no longer emerge. Such a "one-way valve" is called the *event horizon*.

With respect to an external observer the contraction to the gravitational radius is accompanied by a "closing up" of the body. The time for propagation of signals sent from the body tends to infinity: for a light signal $c\, dt = dr/(1 - r_g/r)$, the time for propagation from r to some $r_0 > r$ is given by the integral

$$c\Delta t = \int_r^{r_0} \frac{dr}{1 - \dfrac{r_g}{r}} = r_0 - r + r_g \ln \frac{r_0 - r_g}{r - r_g}, \tag{102.9}$$

which [like the integral [102.5]] diverges for $r \to r_g$. Intervals of proper time on the surface of the body are shortened, as compared to intervals of time t for the distant observer, in the ratio

$$\sqrt{g_{00}} = \sqrt{1 - r_g/r};$$

consequently, as $r \to r_g$ all processes on the body appear to be "frozen" with respect to the external observer.

The frequency of a spectral line emitted by the body and received by a distant observer is reduced, but this is not only an effect of the gravitational red shift, but also the effect of the Doppler shift due to the motion of the source, which is falling toward the centre along with the surface of the sphere. When the radius of the sphere is already close to r_g (so that the velocity of fall is already close to the light velocity) this effect reduces the frequency by a factor

$$\sqrt{1 - \frac{v^2}{c^2}} \bigg/ \left(1 + \frac{v}{c}\right) \approx \frac{1}{2}\sqrt{1 - \frac{v^2}{c^2}}.$$

Under the influence of both effects, the observed frequency consequently goes to zero as $r \to r_g$ according to the law

$$\omega = \text{const}\left(1 - \frac{r_g}{r}\right). \tag{102.10}$$

Thus, from the point of view of a distant observer, the gravitational collapse leads to the appearance of a "congealed" body which sends no signals into the surrounding space and interacts with the external world only through its static gravitational field. Such a structure is called a *black hole* or a *collapsar*.

In conclusion we make one further remark of a methodological nature. We have seen that for the central field in vacuum the "system of the outside observer" that is inertial at infinity is not complete: there is no place in it for the world lines of particles moving inside the Schwarzschild sphere. The metric (102.3) is still applicable inside the Schwarzschild sphere, but this system too is not complete in a certain sense. Consider, in this system, a particle carrying out a radial motion in the direction away from the centre. As $\tau \to \infty$ its world line goes out to infinity, while for $\tau \to -\infty$ it must approach asymptotically to $r = r_g$, since, in this metric, within the Schwarzschild sphere motion can occur only along the direction to the centre. On the other hand, emergence of the particle from $r = r_g$ to any given point $r > r_g$ occurs within a finite interval of proper time. In terms of proper time the particle must approach the Schwarzschild sphere from inside before it can begin to move outside it; but this part of the history of the particle is not kept by the particular reference system.†

We emphasize that this incompleteness arises only in a formal treatment of the metric of the field, where it is regarded as produced by a point mass. In a real physical problem, say the collapse of an extended body, this incompleteness does not occur: the solution obtained by matching the metric (102.3) with the solution inside of the matter, will, of course, be complete, and will describe the whole history of all possible motions of the particles. (The world lines of particles moving in the region $r > r_g$ in the direction toward the centre necessarily begin from the surface of the sphere, even before its contraction to within the Schwarzschild sphere.)

PROBLEMS

1. Find the radii of circular orbits for a particle in the field of a black hole (S.A. Kaplan, 1949).

Solution: The dependence $r = r(t)$ for a particle moving in the Schwarzschild field is given by (101.4), or, in differential form:

$$\frac{1}{1 - r_g r} \frac{dr}{cdt} = \frac{1}{\mathscr{E}_0} [\mathscr{E}_0^2 - U^2(r)]^{1/2}, \tag{1}$$

where

$$U(r) = mc^2 \left[\left(1 - \frac{r_g}{r} \right) \left(1 + \frac{M^2}{m^2 c^2 r^2} \right) \right]^{1/2}$$

(m is the mass of the particle, $r_g = 2km'/c^2$ is the gravitational radius of the central body with mass m'). The function $U(r)$ plays the role of an "effective potential energy" in the sense that the condition $\mathscr{E}_0 \geq U(r)$ determines the admissible range of the motion (in analogy to nonrelativistic theory). Figure 21 shows curves of $U(r)$ for various values of the angular momentum M of the particle.

The radii of curvature of the orbits and the corresponding values of $\mathscr{E}_0$ and M are determined by the extrema of the function $U(r)$, where the minima correspond to stable, and the maxima to unstable orbits. Simultaneous solution of the equations $U(r) = \mathscr{E}_0$, $U'(r) = 0$ gives:

$$\frac{r}{r_g} = \frac{M^2}{m^2 c^2 r_g^2} \left[1 \pm \sqrt{1 - \frac{3m^2 c^2 r_g^2}{M^2}} \right],$$

$$\mathscr{E}_0 = Mc \sqrt{\frac{2}{rr_g} \left(1 - \frac{r_g}{r} \right)},$$

† The construction of a reference system that is not incomplete in this way is considered in problem 5 of this section.

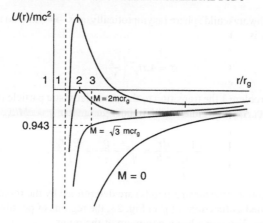

FIG. 21.

where the upper sign refers to stable, and the lower to unstable orbits. The stable orbit closest to the centre
has the parameters

$$r = 3r_g, \quad M = \sqrt{3}\, mcr_g, \quad \mathscr{E}_0 = \sqrt{\frac{8}{9}}\, mc^2.$$

The minimum radius for an unstable orbit is $3r_g/2$ and is reached in the limit $M \to \infty$, $\mathscr{E}_0 \to \infty$. Figure 22
shows the dependence of r/r_g on M/mcr_g; the upper branch gives the radius for stable, and the lower for
unstable orbits.†

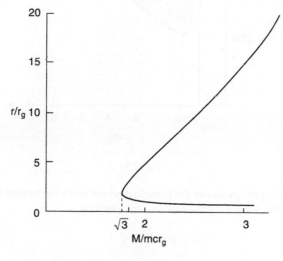

FIG. 22.

2. For motion in this same field determine the cross-section for gravitational capture of (a) nonrelativistic,
(b) ultrarelativistic, particles coming from infinity (Ya. B. Zel'dovich and I. D. Novikov, 1964).

Solution: (a) For a nonrelativistic velocity v_∞ (at infinity) the energy of the particle is $\mathscr{E}_0 \approx mc^2$. From Fig.
21 we see that the line $\mathscr{E}_0 = mc^2$ lies above all the potential curves with angular momenta $M < 2mcr_g$, i.e.
all those with impact parameters $\varrho < 2cr_g/v_\infty$. All particles with such values of ϱ undergo gravitational

† For comparison we recall that in a Newtonian field circular orbits would be possible (and stable) at any
distance from the centre (the radius being related to the angular momentum by the formula $r = M^2/km'm^2$.

capture: they reach the Schwarzschild sphere (asymptotically, as $t \to \infty$) and do not emerge again to infinity. The capture cross-section is

$$\sigma = 4\pi r_g^2 \left(\frac{c}{v_\infty} \right)^2 .$$

(b) In equation (1) of problem 1 the transition to the ultrarelativistic particle (or to a light ray) is achieved by the substitution $m \to 0$. Also introducing the impact parameter $\varrho = cM/\mathscr{E}_0$, we get:

$$\frac{1}{1 - \dfrac{r_g}{r}} \frac{dr}{c\,dt} = \sqrt{1 - \frac{\varrho^2}{r^2} + \frac{\varrho^2 r_g}{r^3}} .$$

The limits of the radial motion (the turning points) are determined by the roots of the expression under the square root. They are plotted as functions of ϱ in Fig. 23; the regions of possible motions correspond to the unshaded part of the plane. The curve has a minimum at the point

$$\varrho = \frac{3\sqrt{3}}{2} r_g , \quad r = \frac{3}{2} r_g .$$

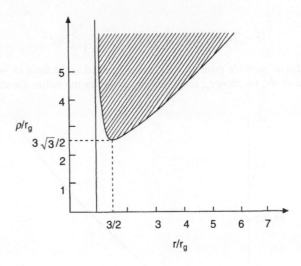

FIG. 23.

For smaller values of the impact parameter the particle does not reach the turning point, i.e. it moves to the Schwarzschild sphere. Then we get for the capture cross-section

$$\sigma = \frac{27}{4} \pi r_g^2 .$$

§ 103. Gravitational collapse of a dustlike sphere

A discussion of the course of change of the internal state of a collapsing body (including in the course of the process its compression below the Schwarzschild sphere) requires the solution of the Einstein equations for the gravitational field in a material medium. In the centrally symmetric case the field equations can be solved in general form if we neglect the pressure of the matter: $p = 0$ (R. Tolman, 1934). Although the approximation made is not usually admissible in real situations, the general solution of this problem has considerable methodological interest.

As was shown in § 97, for a dustlike medium one can choose a reference system which is both synchronous and comoving.† Denoting the time and the radial coordinate chosen in this way by τ and R, we write the spherically symmetric line element in the form‡

$$ds^2 = d\tau^2 - e^{\lambda(\tau,R)} dR^2 - r^2(\tau, R)(d\theta^2 + \sin^2 \theta \, d\phi^2). \tag{103.1}$$

The function $r(\tau, R)$ is the "radius", defined so that $2\pi r$ is the circumference of a circle (with centre at the origin). The form (103.1) fixes the choice of τ uniquely, but still permits arbitrary transformations of the radial coordinate of the form $R = R(R')$.

The calculation of the components of the Ricci tensor for this metric leads to the following system of Einstein equations:§

$$-e^{-\lambda} r'^2 + 2r\ddot{r} + \dot{r}^2 + 1 = 0, \tag{103.2}$$

$$-\frac{e^{-\lambda}}{r}(2r'' - r'\lambda') + \frac{\dot{r}\dot{\lambda}}{r} + \ddot{\lambda} + \frac{\dot{\lambda}^2}{2} + \frac{2\ddot{r}}{r} = 0, \tag{103.3}$$

$$-\frac{e^{-\lambda}}{r^2}(2rr'' + r'^2 - rr'\lambda') + \frac{1}{r^2}(rr\dot{\lambda} + \dot{r}^2 + 1) = 8\pi k\varepsilon, \tag{103.4}$$

$$2\dot{r}' - \dot{\lambda}r' = 0, \tag{103.5}$$

where the prime denotes differentiation with respect to R, and the dot with respect to τ.

Equation (103.5) is immediately integrated over the time, giving

$$e^{\lambda} = \frac{r'^2}{1 + f(R)}, \tag{103.6}$$

where $f(R)$ is an arbitrary function, subject only to the condition that $1 + f > 0$. Substituting this expression in (103.2), we get

$$2r\ddot{r} + \dot{r}^2 - f = 0$$

[substitution in (103.3) gives nothing new]. The first integral of this equation is

$$\dot{r}^2 = f(R) + \frac{F(R)}{r}, \tag{103.7}$$

where $F(R)$ is another arbitrary function. Thus

$$\tau = \pm \int \frac{dr}{\sqrt{f + \dfrac{F}{r}}}.$$

The function $r(\tau, R)$ obtained from the integration can be written in the parametric form:

---

† The matter must move "without rotation" (cf. the footnote on p. 310). In the present case this condition is surely satisfied, since spherical symmetry implies purely radial motion of the material.

‡ In this section, we set $c = 1$.

§ Compare problem 5 of § 100. Equations (103.2)–(103.5) are obtained, respectively, from eqs. (2)–(5) of the problem if we set $\nu = 0$, $e^{\mu} = r^2$, $p = 0$. We note that the second of eqs. (6) of this same problem gives $\nu' = 0$, i.e., $\nu = \nu(\tau)$, when $p = 0$; the remaining arbitrariness in the metric (1) in the choice of τ therefore allows us to make ν equal to zero, which again demonstrates the possibility of introducing a synchronous-comoving reference frame.

$$r = \frac{F}{2f} (\cosh \eta - 1), \quad \tau_0(R) - \tau = \frac{F}{2f^{\frac{3}{2}}} (\sinh \eta - \eta) \text{ for } f > 0, \tag{103.8}$$

$$r = \frac{F}{-2f} (1 - \cos \eta), \quad \tau_0(R) - \tau = \frac{F}{2(-f)^{\frac{3}{2}}} (\eta - \sinh \eta) \text{ for } f > 0, \tag{103.9}$$

where $\tau_0(R)$ is again an arbitrary function. If $f = 0$,

$$r = \left(\frac{9F}{4}\right)^{\frac{1}{3}} [\tau_0(R) - \tau]^{\frac{2}{3}} \text{ for } f = 0. \tag{103.10}$$

In all cases, substituting (103.6) in (103.4) and eliminating f by using (103.7), we get the following expression for the matter density:†

$$8\pi k \varepsilon = \frac{F'}{r' r^2}. \tag{103.11}$$

Formulas (103.6)–(103.11) determine the required general solution.‡ We remark that it depends on only two "physically different" arbitrary functions: although three functions, f, F and τ_0 appear in it, the coordinate R can still be subjected to an arbitrary transformation $R = R(R')$. This number corresponds exactly to the fact that the most general centrally symmetric distribution of matter is given by two functions (the density distribution and the radial velocity of the matter), while a free gravitational field with central symmetry does not exist.

Since the reference frame is comoving with the matter, to each particle of matter there corresponds a definite value of R; the function $r(\tau, R)$ for this value of R determines the law of motion of the particular particle, while the derivative $\dot{r}$ is its radial velocity. An important property of the solution obtained here is that the assignment of the arbitrary functions appearing in it over the interval from 0 to some R_0 completely determines the behaviour of the sphere of this radius; it does not depend on how the functions are assigned for $R > R_0$. One automatically obtains the solution of the interior problem for any finite sphere. The total mass of the sphere is given, according to (100.23), by the integral

$$m = 4\pi \int_0^{r(\tau, R_0)} \varepsilon r^2 dr = 4\pi \int_r^{R_0} \varepsilon r^2 r' dR.$$

Substituting (103.11) and noting that $F(0) = 0$ (when $R = 0$, we must have $r = 0$), we find

$$m = \frac{F(R_0)}{2k}, \quad r_g = F(R_0) \tag{103.12}$$

(r_g is the gravitational radius of the sphere).

For $F = \text{const} \neq 0$, we find $\varepsilon = 0$ from (103.11), so that the solution applies to empty space,

† The functions F, f, τ_0 have only to satisfy conditions assuring the positivity of e^λ, r and ε. In addition to the condition $1 + f > 0$ given above, it follows that $F > 0$. We shall also assume that $F' > 0$, $r' > 0$; this excludes cases that lead to crossing of spherical layers of matter during their radial motion.

‡ It does not, however, include the special case where $r = r(\tau)$, and does not depend on R, so that eq. (103.5) reduces to an identity; cf. V. A. Ruban, *JETP*, **56**, 1914 (1969); *Soviet Phys. JETP*, **29**, 1027 (1969). This case does not, however, correspond to the conditions of the problem of collapse of a finite body.

i.e. it describes the field of a point mass (located at the centre, the singular point of the metric). So, setting $F = r_g, f = 0, \tau_0 = R$, we obtain the metric (102.3).†

Formulas (103.8)–(103.10) describe both contraction and expansion of the sphere (depending on the range of values taken by the parameter η); both are equally admissible for the field equations. The important problem of behaviour of an unstable massive body corresponds to contraction—gravitational collapse. The solutions (103.8)–(103.10) are written so that contraction occurs when τ, while increasing, tends to τ_0. To the moment $\tau - \tau_0(R)$ there corresponds the arrival at the centre of the matter with a given radial coordinate R (where we must have $\tau_0' > 0$).

The limiting character of the metric inside the sphere as $\tau \to \tau_0(R)$ is the same for all three cases (103.8)–(103.10):

$$r \approx \left(\frac{9F}{4}\right)^{\frac{1}{3}} (\tau_0 - \tau)^{\frac{2}{3}}, \quad e^{\lambda/2} \approx \left(\frac{2F}{3}\right)^{\frac{1}{3}} \frac{\tau_0'}{\sqrt{1+f}} (\tau_0 - \tau)^{-\frac{1}{3}}. \tag{103.13}$$

This means that all radial distances (in the comoving reference frame used here) tend to infinity, while tangential distances go to zero (like $\tau - \tau_0$).‡ Correspondingly the matter density increases without limit:§

$$8\pi k\varepsilon \approx \frac{2F'}{3F\tau_0'(\tau_0 - \tau)}. \tag{103.14}$$

Thus, in agreement with our remarks in § 102, there is a collapse of the whole matter distribution in to the centre.¶

In the special case where the function $\tau_0(R) = $ const (i.e. all the particles reach the centre simultaneously), the metric inside the contracting sphere has a different character. In this case

$$r \approx \left(\frac{9F}{3}\right)^{\frac{1}{3}} (\tau_0 - \tau)^{\frac{1}{3}}, \quad e^{\lambda/2} \approx \left(\frac{2}{3}\right)^{\frac{1}{3}} \frac{F'}{2F^{\frac{2}{3}}\sqrt{1+f}} (\tau_0 - \tau)^{\frac{2}{3}}$$

$$8\pi k\varepsilon \approx \frac{4}{3(\tau_0 - \tau)^2}. \tag{103.15}$$

i.e. as $\tau \to \tau_0$, all distances, both tangential and radial, tend to zero according to the same law

---

† The case of $F = 0$ (where (103.7) gives $r = \sqrt{f}(\tau - \tau_0)$) corresponds to the absence of a field; by a suitable transformation of variables, the metric can be brought to Galilean form.

‡ The geometry on a "plane" passing through the centre is that which would exist on a conical surface of revolution which is stretching in the course of time along its generators and at the same time contracting along its bounding circles.

§ The fact that in this solution collapse occurs for any mass of the sphere is a natural consequence of neglecting the pressure. Clearly, as $\varepsilon \to \infty$, from the physical point of view the assumption that the matter is dustlike is never admissible, and we should use the ultrarelativistic equation of state $p = \varepsilon/3$. It appears however, that the general character of the limiting laws of compression are to a large extent independent of the equation of state of the matter (cf. E. M. Lifshitz and I. M. Khalatnikov, *JETP* **39**, 149, 1960; *Soviet Phys. JETP*, **12**, 108, 1961).

¶ The case where $\tau_0 —$ const includes, in particular, the collapse of a completely homogeneous sphere—see the problem.

$\sim (\tau_0 - \tau)^{\frac{2}{3}}$; the matter density goes to infinity like $(\tau_0 - \tau)^{-2}$ and, in the limit, its distribution becomes uniform.

We call attention to the fact that in all cases the moment of passage of the surface of the collapsing sphere through the Schwarzschild sphere $[r(\tau, R_0) = r_g]$ has no significance for its internal dynamics (described by the metric in the comoving reference frame). At each moment of time, however, a definite part of the sphere is already below its "event horizon". Just as $F(R_0)$. through (103.12), determines the gravitational radius of the sphere as a whole, so $F(R)$ for any given value of R is the gravitational radius of the part of the sphere within the spherical surface $R = $ const; thus this part of the sphere is determined at each moment of time τ by the condition $r(\tau, R) \leq F(R)$.

Finally, we show how these formulas can be used to solve the problem posed at the end of § 102: to construct the most complete reference system for the field of a point mass.†

To achieve this goal we must start from a metric in vacuum that could contain both contracting and expanding space-time regions. Equation (103.8) is such a solution, in which we must set $F = $ const $= r_g$. Also choosing

$$f = -\frac{1}{(R/r_g)^2 + 1}, \qquad \tau_0 = \frac{\pi}{2} r_g (-f)^{-\frac{3}{2}},$$

we get:

$$\frac{r}{r_g} = \frac{1}{2}\left(\frac{R^2}{r_g^2} + 1\right)(1 - \cos \eta),$$

$$\tag{103.16}$$

$$\frac{\tau}{r_g} = \frac{1}{2}\left(\frac{R^2}{r_g^2} + 1\right)^{\frac{3}{2}}(\pi - \eta + \sin \eta);$$

where the parameter η runs through values from 0 to 2π, the time τ (for a given R) decreases monotonically, while r increases from zero, goes through a maximum, and then again drops to zero.

In Fig. 24, the lines ACB and $A'C'B'$ correspond to the point $r = 0$ (parameter values $\eta = 2\pi$ and $\eta = 0$). The curves AOA' and BOB' correspond to the Schwarzschild sphere $r = r_g$. Between $A'C'B'$ and $A'OB'$ is the region of space–time in which only motion out from the centre is possible, while between ACB and AOB there is the region in which motion occurs only toward the centre.

The world line of a particle that is at rest relative to this reference system is a vertical line ($R = $ const). It starts from $r = 0$ (point a), cuts the Schwarzschild sphere at the point b,

reaches its farthest distance $\left[r = r_g\left(\dfrac{R^2}{r_g^2} + 1\right)\right]$ at time $\tau = 0$, after which the particle again

begins to fall in toward the Schwarzschild sphere, passes through it at point c, and arrives once more at $r = 0$ (point d) at the time

$$\tau = r_g\frac{\pi}{2}\left(\frac{R^2}{r_g^2} + 1\right)^{\frac{3}{2}}.$$

† Such a system was first found by M. Kruskal using other variables (M. Kruskal, *Phys. Rev.* **119**, 1743, 1960). The form of the solution given here, in which the reference system is synchronous, is due to I. D. Novikov, 1963.

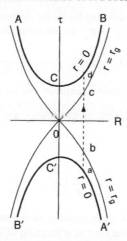

This reference system is complete: both ends of the world line of any particle moving in the field are either at the true singularity $r = 0$, or go off to infinity. The incomplete metric (102.3) covers only the region to the right of the curve AOA' (or to the left of BOB'), while the "expanding" reference system covers the region to the right of BOB' (or to the left of AOA'). The Schwarzschild reference frame with the metric (100.14) covers only the region to the right of BOA' (or to the left of AOB').

PROBLEM

Find the solution of the interior problem for the gravitational collapse of a dustlike homogeneous sphere whose material is at rest initially.

Solution: Setting

$$\tau_0 = \text{const}, \quad f = -\sin^2 R, \quad F = 2a_0 \sin^3 R,$$

we find

$$r = a_0 \sin R(1 - \cos \eta), \quad \tau - \tau_0 = a_0(\eta - \sin \eta) \tag{1}$$

(the radial coordinate R is dimensionless and runs through values from 0 to 2π). Then the density is

$$8\pi k\varepsilon = \frac{6}{a_0^2(1 - \cos \eta)^3}, \tag{2}$$

and, for a given τ, is independent of R, so the sphere is homogeneous. We can represent the metric (103.1) with r from (1) in the form

$$ds^2 = d\tau^2 - a^2(\tau)[dR^2 + \sin^2 R(d\theta^2 + \sin^2 \theta \, d\phi^2)], \tag{3}$$

$$a = a_0(1 - \cos \eta).$$

We call attention to the fact that it coincides with the Friedmann solution for the metric of a universe completely filled with uniform dustlike matter (§ 112)—a completely natural result, since a sphere cut out of a uniform distribution of matter has central symmetry.†

† The metric (3) corresponds to a space of constant positive curvature. In analogous fashion, setting $f = \sinh^2 R$, $F = 2a_0 \sinh^3 R$, we obtain the solution corresponding to a space of constant negative curvature (§ 113).

The condition originally posed can be satisfied by (1) for definite choices of the constants a_0, τ_0. Making a change of parameter for convenience ($\eta \to \pi - \eta$), we write the solution in the form

$$r = \frac{r_0 \sin R}{2 \sin R_0} (1 + \cos \eta), \quad \tau = \frac{r_0}{2 \sin R_0} (\eta + \sin \eta). \tag{4}$$

where [according to (103.12)] the gravitational radius of the sphere is $r_g = r_0 \sin^2 R_0$. At the initial time ($\tau = 0$, $\eta = 0$) the matter is at rest ($\dot{r} = 0$), and $2\pi r_0 = 2\pi r(0, R_0)$ is the initial circumference of the sphere. The collapse of this matter into the centre occurs at the time $\tau = \pi r_0 / 2 \sin R_0$.

The time t in the reference frame of a distant observer (the Schwarzschild frame) is related to the proper time τ on the sphere by the equation

$$d\tau^2 = \left(1 - \frac{r_g}{r}\right) dt^2 - \frac{dr^2}{1 - r_g/r},$$

where by r we mean the value $r(\tau, R_0)$ corresponding to the surface of the sphere. Integration of this equation leads to the following expression for t as a function of the same parameter η:

$$\frac{t}{r_g} = \ln \frac{\cot R_0 + \tan \dfrac{\eta}{2}}{\cot_t R_0 - \tan \dfrac{\eta}{2}} - \cot R_0 \left[\eta - \frac{1}{2 \sin^2 R_0} (\eta - \sin \eta)\right] \tag{5}$$

(where the time $t = 0$ corresponds to $\tau = 0$). The passage of the surface of the sphere through the Schwarzschild sphere ($r(\tau, R_0) = r_g$) corresponds to the value of the parameter η determined by the equation

$$\cos^2 \frac{\eta}{2} = \frac{r_g}{r_0} = \sin^2 R_0.$$

As we approach this value, the time t goes to infinity, in accordance with the remarks in § 102.†

§ 104. Gravitational collapse of nonspherical and rotating bodies

All the results of the preceding two sections were applicable strictly to bodies that are rigorously spherically symmetric. Simple arguments show, however, that the qualitative picture of gravitational collapse remains the same for bodies with small deviations from spherical symmetry (A. G. Doroshkevich, Ya. B. Zel'dovich and I. D. Novikov, 1965).

We shall consider, first, bodies whose deviation from central symmetry is related to the matter distribution and not to a rotation of the body as a whole.

It is obvious that if a massive centrally symmetric body is gravitationally unstable, this instability will remain for small disturbances of the symmetry, so that such a body will collapse. Treating the weak asymmetry as a small perturbation, we can follow its development (in the comoving frame) during the contraction of the body. Perturbations, generally speaking, increase with increasing density of the body. But if the perturbations are sufficiently weak at the start of the contraction, they will still be small at the time when the body reaches its gravitational radius; it was pointed out in § 103 that this moment is in no way special for the internal dynamics of the contracting body, while its density is, of course, still finite.‡

---

† The function $r(\tau, R_0)$ determined from formula (4) coincides, of course, with the function calculated from the external metric and given by the integral (102.8). The same remark applies to the function $t(r)$ defined by formulas (4) and (5); it coincides with that given by the integral (102.5).

‡ The development of perturbations in a nonstationary, unbounded homogeneous distribution of matter is treated in § 115 (where the formulas obtained apply equally to the cases of expansion and contraction). Nonuniformities of the unperturbed distribution or infinite extent of the body do not change the conclusions drawn there.

Because the internal perturbations in the body are small, the perturbations of the external centrally symmetric gravitational field produced by it also remain small. This means that the surface of the "event horizon," the Schwarzschild sphere, also remains almost unchanged, and nothing prevents the collapsing body from crossing it (in the comoving reference frame).

No information reaches the external observer concerning the further growth of perturbations inside the body since no signals emerge from beyond the event horizon; the whole process remains "infinitely delayed" for the distant observer. From this it follows, in turn, that with respect to the external reference frame the gravitational field of the collapsing body must tend to become stationary as the body approaches the gravitational radius asymptotically. The characteristic time for this approach is very small ($\sim r_g/c$), and during it we may assume that in the external space there are only those perturbations that developed earlier in the centrally symmetric field. But all disturbances must in the course of time dissipate in space as gravitational waves, going out to infinity (or passing beyond the horizon).

Time-independent static disturbances also cannot remain in the external gravitational field of a developing black hole. This conclusion can be obtained from an analysis of constant perturbations applied on the Schwarzschild sphere in vacuum. Such an analysis shows that in the static case every perturbation (which drops off at infinity) increases without limit as it approaches the Schwarzschild sphere of the unperturbed problem;† but, as already stated, in the present case there is no reason for the appearance of large perturbations of the external field.

The deviations from spherical symmetry in the density distribution of the body are described by quadrupole and higher multipole moments of the distribution; each of them gives its contribution to the external gravitational field. Our assertion means that all such perturbations of the external field damp out in the final stages (from the point of view of the external observer) of the collapse.‡ The developing gravitational field of the black hole is again the centrally symmetric Schwarzschild field, determined solely by the total mass of the body.

The question of the ultimate fate of the body in its collapse beyond the event horizon (which is not observable from the external reference frame) is not completely clear. Apparently one can assert that here, too, the collapse terminates in a true singularity of the space-time metric, but a singularity of a completely different type than in the centrally symmetric case. This question is, however, not settled completely at present.

Let us turn to the case where the weak disturbance of the spherical symmetry is associated not with the density distribution but with rotation of the body as a whole; the assumption that the perturbation is weak means that we must have a sufficiently slow rotation. All the previous remarks remain valid with one exception. It is clear from the start that, because of the conservation of the total angular momentum **M** of the body, the field of the black hole cannot depend on just the mass of the body. To this there corresponds the fact that among the stationary (but not static) time-independent perturbations of a centrally symmetric gravitational field there is one which does not increase without limit as $r \to r_g$. This perturbation is just the one that is associated with rotation of the body, and is described by adding to the

---

† Cf. T. Regge and J. A. Wheeler, *Phys. Rev.* **108**, 1063 (1957). We emphasize that we are speaking of perturbations coming from the central body itself. The condition imposed at infinity excludes cases where static perturbations arise from external sources: in such cases small perturbations only distort the Schwarzschild sphere without changing its qualitative properties and without producing a true space-time singularity on it.

‡ For this damping law, cf. R. H. Price, *Phys. Rev.* D, **5**, 2419, 2439 (1972). During collapse, initially static 2^l-pole perturbations of the external gravitational field damp like $1/t^{2i+2}$.

Schwarzschild metric tensor g_{ik} (in the coordinates $x^0 = t, x^1 = r, x^2 = \theta, x^3 = \phi$) the small off-diagonal component:[†]

$$g_{03} = \frac{2kM}{r} \sin^2 \theta \qquad (104.1)$$

(cf. the problem in § 105). This expression remains valid (in the external space) as the body approaches the gravitational radius, and thus the gravitational field of a slowly rotating black hole will (to first approximation in the small angular momentum M) be a centrally symmetric Schwarzschild field with the small correction (104.1). This field is no longer static but only stationary.

If gravitational collapse is possible for small disturbances of spherical symmetry, then collapse of the same character (with movement of the body beyond the event horizon) must also be possible in some finite range of sizeable deviations from sphericity; the conditions determining this region are not as yet established. Independent of these conditions, it appears that one can assert that, from the point of view of an external observer, the properties of the structure resulting from the collapse (a rotating collapsar) are independent of all characteristics of the initial body except for its total mass m and angular momentum M.[‡] If the body does not rotate as a whole ($M = 0$), then the external gravitational field of the collapsar is the centrally symmetric Schwarzschild field.[§]

The gravitational field of the rotating black hole is given by the following axially symmetric stationary *Kerr metric*:[¶]

$$ds^2 = \left(1 - \frac{r_g r}{\rho^2}\right) dt^2 - \frac{\rho^2}{\Delta} dr^2 - \rho^2 d\theta^2 -$$

$$- \left(r^2 + a^2 + \frac{r_g r a^2}{\rho^2} \sin^2 \theta\right) \sin^2 \theta \, d\phi^2 + \frac{2 r_g r a}{\rho^2} \sin^2 \theta \, d\phi \, dt, \qquad (104.2)$$

where we have introduced the notation

$$\Delta = r^2 - r_g r + a^2, \quad \rho^2 = r^2 + a^2 \cos^2 \theta, \qquad (104.3)$$

while r_g is again $r_g = 2mk$. This metric depends on two constant parameters, m and a, whose meaning is clear from the limiting form of the metric at large distances r. To terms of order $\sim 1/r$, we have:

$$g_{00} \approx 1 - \frac{r_g}{r}, \quad g_{03} \approx \frac{r_g a}{r} \sin^2 \theta;$$

[†] In this section, we set $c = 1$.

[‡] To avoid misunderstandings we remind the reader that we are not considering bodies that carry an uncompensated electric charge.

[§] This assertion is strongly supported by the following theorem due to Israel: among all static solutions of the Einstein equations that are galilean at infinity and have closed single-sheeted spatial surfaces $g_{00} = $ const, $t = $ const, the Schwarzschild solution is the only one that has a horizon ($g_{00} = 0$) without a singularity of the space-time metric on it (for the proof cf. W. Israel, *Phys. Rev.* **164**, 1776, 1967).

[¶] This solution of the Einstein equations was discovered by R. Kerr, 1963, in a different form, and reduced to (104.2) by R. H. Boyer and R. W. Lindquist, 1967. There is no constructive analytic derivation of the metric (104.2) in the literature that is adequate in its physical ideas, and even a direct check of this solution of the Einstein equations involves cumbersome calculations. The claim that the Kerr metric is the unique field for a rotating collapsar is supported by a theorem analogous to the theorem of Israel (cf. B. Carter, *Phys. Rev. Lett.* **26**, 331, 1971).

comparison of the first expression with (100.18) and of the second with (104.1) shows that m is the mass of the body, while the parameter a is related to the angular momentum M by

$$M = ma \tag{104.4}$$

($M = mac$ in the usual units). For $a = 0$ the Kerr metric goes over into the Schwarzschild metric in its standard form (100.14).† We also call attention to the fact that the form (104.2) exhibits explicitly the symmetry under time reversal: this transformation $(t \to - t)$ also changes the rotation direction, i.e., the sign of the angular momentum $(a \to - a)$, so that ds^2 remains unchanged.

The determinant of the metric tensor (104.2) is

$$- g = \rho^4 \sin^4 \theta. \tag{104.5}$$

We also give the contravariant components g^{ik}, by introducing them into the following expression for the square of the four-gradient operator:

$$g^{ik} \frac{\partial}{\partial x^l} \frac{\partial}{\partial x^k} = \frac{1}{\Delta} \left(r^2 + a^2 + \frac{r_g r a^2}{\rho^2} \sin^2 \theta \right) \left(\frac{\partial}{\partial t} \right)^2 - \frac{\Delta}{\rho^2} \left(\frac{\partial}{\partial r} \right)^2 -$$

$$- \frac{1}{\rho^2} \left(\frac{\partial}{\partial \theta} \right)^2 - \frac{1}{\Delta \sin^2 \theta} \left(1 - \frac{r_g r}{\rho^2} \right) \left(\frac{\partial}{\partial \phi} \right)^2 + \frac{2 r_g r a}{\rho^2 \Delta} \frac{\partial}{\partial \phi} \frac{\partial}{\partial t}. \tag{104.6}$$

When $m = 0$, in the absence of a gravitating mass, the metric (104.2) should become galilean. In fact, the expression

$$ds^2 = dt^2 - \frac{\rho^2}{r^2 + a^2} dr^2 - \rho^2 d\theta^2 - (r^2 + a^2) \sin^2 \theta \, d\phi^2 \tag{104.7}$$

is the galilean metric

$$ds^2 = dt^2 - dx^2 - dy^2 - dz^2$$

written in spatially oblate spheroidal coordinates; the transformation to cartesian coordinates is accomplished with the formulas:

$$x = \sqrt{r^2 + a^2} \sin \theta \cos \phi,$$

$$y = \sqrt{r^2 + a^2} \sin \theta \cos \phi,$$

$$z = r \cos \theta;$$

the surfaces $r = $ const are oblate ellipsoids of rotation:

$$\frac{x^2 + y^2}{r^2 + a^2} + \frac{z^2}{r^2} = 1.$$

The metric (104.2) has a fictitious singularity, just as the Schwarzschild metric (100.14)

† To terms of first order in a, the metric (104.2) for $a \ll 1$ differs from the Schwarzschild metric only in the term $(2 r_g a/r) \sin^2 \theta d\phi \, dt$, in agreement with our remarks above about the case of weak deviations from spherical symmetry.

has a fictitious singularity at $r = r_g$. But, whereas in the Schwarzschild case g_{00} goes to 0 and g_{11} to infinity simultaneously on the surface $r = r_g$, in the Kerr metric these two surfaces are separated. The equality $g_{00} = 0$ holds when $\rho^2 = r r_g$; the larger of the two roots of this quadratic equation is

$$r_0 = \frac{r_g}{2} + \sqrt{\left(\frac{r_g}{2}\right)^2 - a^2 \cos^2\theta} \qquad (g_{00} = 0). \tag{104.8}$$

The quantity g_{11} goes to infinity when $\Delta = 0$; the larger of the two roots of this equation is

$$r_{\mathrm{hor}} = \frac{r_g}{2} + \sqrt{\left(\frac{r_g}{2}\right)^2 - a^2} \qquad (g_{11} = \infty). \tag{104.9}$$

For brevity, we shall denote the surfaces $r = r_0$ and $r = r_{\mathrm{hor}}$, whose physical meaning is explained below, by S_0 and S_{hor}. The surface S_{hor} is a sphere, while S_0 is an oblate figure of rotation; S_{hor} is contained inside S_0, and the two surfaces touch at the poles ($\theta = 0$ and $\theta = \pi$).

As we see from (104.8)–(104.9), the surfaces S_0 and S_{hor} exist only when $a \leq r_g/2$. When $a > r_g/2$ the character of the metric (104.2) changes radically and begins to show physically inadmissible properties that violate the principle of causality.†

The fact that the Kerr metric is no longer meaningful for $a > r_g/2$ implies that the value

$$a_{\max} = \frac{r_g}{2}, \quad M_{\max} = \frac{m r_g}{2}, \tag{104.10}$$

gives the upper limit of possible angular momenta of a collapsar. It must be regarded as a limiting value that can be approached arbitrarily closely, while exact equality $a = a_{\max}$ is impossible. The corresponding limiting values of the radii of the surfaces S_0 and S_{hor} are

$$r_0 = \frac{r_g}{2} (1 + \sin\theta), \quad r_{\mathrm{hor}} = \frac{r_g}{2} \tag{104.11}$$

We shall show that the surface S_{hor} is the event horizon, which is passed by a moving particle or light ray in only one direction, toward the interior.

As a preliminary, we show that, from the general point of view, the property of unidirectional passage of world lines of moving particles holds for any null hypersurface (i.e. a hypersurface whose normal at every point is a null vector). Suppose the hypersurface is defined by the equation $f(x^0, x^1, x^2, x^3) = \mathrm{const}$. Its normal is directed along the four-gradient $n_i = \partial f/\partial x^i$, so that for a null hypersurface we have $n_i n^i = 0$. This means, in other words, that the direction of the normal lies in the surface itself: along the hypersurface $df = n_i dx^i = 0$, and this equation is satisfied when the directions of the four-vectors dx^i and n^i coincide. From this same property $n_i n^i = 0$, the element of length on the hypersurface in this same direction is $ds = 0$. In other words, along this direction the hypersurface is tangent, at the given point,

---

† The violations manifest themselves in the appearance of closed timelike world lines: they would make it possible to head into the past and then later develop into the future. We point out immediately that these same violations would appear if we extended the Kerr metric inside of S_{hor}, even when $a < r_g/2$, which shows the physical inapplicability of this metric inside of S_{hor} (we shall return to this point later). For this same reason, there is no physical interest in the surfaces defined by the two smaller roots of the quadratic equations $g_{00} = 0$ and $1/g_{11} = 0$, which lie inside S_{hor}; cf. B. Carter, *Phys. Rev.* **147**, 1559 (1968).

to the light cone constructed on the point. Thus, the light cones constructed at each point of a null hypersurface (say, in the direction of the future) lie entirely on one side of it, and are tangent to the hypersurface (at these points) along one of their generators. But this means precisely that (directed into the future) the world lines of particles or light rays can cross the hypersurface in only one direction.

This property of null hypersurfaces is usually physically trivial: unidirectional passage through these surfaces simply expresses the impossibility of motion with a velocity greater than the light velocity (the simplest example of this sort is the hypersurface $x = t$ in a flat space-time). A nontrivial new physical situation arises when the null hypersurface does not extend out to spatial infinity, so that its sections $t = $ const are closed spatial surfaces; these surfaces are the event horizon in the same sense as the Schwarzschild sphere was in the centrally symmetric gravitational field.

What is the surface S_{hor} in the Kerr field? The condition $n_i n^i = 0$ for the hypersurface of the form $f(r, \theta) = $ const in the Kerr field has the form

$$g^{11}\left(\frac{\partial f}{\partial r}\right)^2 + g^{22}\left(\frac{\partial f}{\partial \theta}\right)^2 = \frac{1}{\rho^2}\left[\Delta\left(\frac{\partial f}{\partial r}\right)^2 + \left(\frac{\partial f}{\partial \theta}\right)^2\right] = 0 \qquad (104.12)$$

[with g^{ik} from (104.6)]. This equation is not satisfied on S_0, but is satisfied on S_{hor} (where $\partial f/\partial\theta = 0$, $\Delta = 0$).

Continuation of the Kerr metric inside the surface of the horizon (as was done in §§ 102, 103 for the Schwarzschild metric) has no physical meaning. Such a continuation would depend on only the same two parameters (m and a) as the field outside of S_{hor}, from which it is already clear that it could have no connection with the physical question of the fate of the collapsing body after its passing under the horizon. The effects of nonsphericity are not at all damped out in the comoving reference frame, and, on the contrary, must increase with further contraction of the body, so there is no reason to expect that the field beyond the horizon should be determined solely by the mass and angular momentum of the body.†

Let us consider the properties of the surface S_0 and the space between it and the horizon (this region of the Kerr field is called the *ergosphere*).

The fundamental property of the ergosphere is that no particle in it can remain at rest relative to the frame of a distant observer: when r, θ, $\phi = $ const, we have $ds^2 < 0$, i.e. the interval is not timelike, as it should be for the world line of a particle; the variable t loses its temporal character. Thus a rigid reference frame cannot be extended from infinity into the ergosphere, and in this sense the surface S_0 might be called the limit of stationarity.

The character of the motion of a particle in the ergosphere is essentially different from what we had inside the horizon of the Schwarzschild field. In the latter case, also, particles could not be at rest relative to an external reference system, and we could not have $r = $ const; all particles had to move radially toward the centre. In the ergosphere of the Kerr field it is $\phi = $ const that is impossible for particles (the particles must necessarily rotate around the axis of symmetry of the field), while $r = $ const is possible for a particle. Furthermore, particles (and light rays) can move both with increasing and decreasing r, and can emerge from the ergosphere into the external space. Corresponding to this last point, it is possible for particles coming from the external region to reach the ergosphere: the time for such a particle (or light ray) to reach the surface S_0, as measured with the clocks t of a distant

---

† Mathematically, this situation manifests itself in the violation of the causality principle (mentioned above) when we extend the Kerr metric into the interior of S_{hor}.

observer, is finite for all of S_0 except for the poles, at which S_0 touches S_{hor}; the time for reaching these points, just as for all points of S_{hor}, is again, of course, infinite.†

Because of the inevitability of rotational motion of particles in the ergosphere, the natural form for writing the metric in this region is:

$$ds^2 = \left(g_{00} - \frac{g_{03}^2}{g_{33}} \right) dt^2 + g_{11} dr^2 + g_{22} d\theta^2 + g_{33} \left(d\phi + \frac{g_{03}}{g_{33}} dt \right)^2 \tag{104.13}$$

The coefficient of dt^2,

$$g_{00} - \frac{g_{03}^2}{g_{33}} = \frac{\Delta}{r^2 + a^2 + r_g r a^2 \sin^2 \theta / \rho^2}$$

is positive everywhere outside of S_{hor} (and does not vanish on S_0); the interval ds is timelike when $r = \text{const}$, $\theta = \text{const}$, $d\phi = -(g_{03}/g_{33})dt$. The quantity

$$-\frac{g_{03}}{g_{33}} = \frac{r_g a r}{\rho^2(r^2 + a^2) + r_g r a^2 \sin^2 \theta} \tag{104.14}$$

plays the role of a "generalized angular velocity of rotation of the ergosphere" relative to the external reference system (where this direction of rotation coincides with the direction of rotation of the central body).‡

The energy of a particle, defined as $-\partial S/\partial \tau$, the derivative of the action with respect to the proper time of the particle, synchronized along the trajectory, is always positive (cf. § 88). But, as explained in § 88, during motion of the particle in a field independent of time t, the energy $\mathcal{E}_0$ defined as $-\partial S/\partial t$, is conserved; this quantity coincides with the covariant component of four-momentum $p_0 = m u_0 = m g_{0i} dx^i$ (where m is the mass of the particle). The fact that the variable t (the time according to the clocks of the distant observer) does not have temporal character inside the ergosphere results in a peculiar situation: in this region $g_{00} < 0$, and thus the quantity

$$\mathcal{E}_0 = m(g_{00} u^0 + g_{03} u^3) = m \left(g_{00} \frac{dt}{ds} + g_{03} \frac{d\phi}{ds} \right)$$

can be negative. Since, in the external space, where t is the time, the energy $\mathcal{E}_0$ cannot be negative, a particle with $\mathcal{E}_0 < 0$ cannot fall into the ergosphere from outside. A possible source for the appearance of such a particle is the breakup of a body, entering the ergosphere, into, say, two parts, one of which is captured in a "negative energy" orbit. This part can no longer emerge from the ergosphere, and is finally captured within the horizon. The second

† The time for reaching particular points of S_0 may also turn out to be infinite in particular cases of special values of the energy and angular momentum of the particle, chosen so that the radial velocity vanishes at the particular point of S_0.

‡ We call attention to the fact that intervals of proper time for particles moving along the boundary of the ergosphere do not vanish along with g_{00}. In this sense, S_0 is not the surface of "infinite red shift"; the frequencies of light signals sent out from it by a moving source (here sources cannot be at rest) and observed by a distant observer, do not go to zero. We recall that in the centrally symmetric field there could be neither sources at rest nor moving sources on the Schwarzschild sphere (since a null surface cannot contain timelike world lines). The "infinite red shift" arose there because, as $r \to r_g$, intervals of proper time $d\tau = \sqrt{g_{00}} \, dt$ (for given dt), as measured with clocks at rest relative to the reference frame, went to zero.

part can return into the external space; since $\mathscr{E}_0$ is a conserved additive quantity, the energy of this part is greater than the energy of the initial body, so we get an extraction of energy from the rotating black hole (R. Penrose, 1969).

Finally, we note that although the surface S_0 is not singular for the space-time metric, the purely spatial metric [in the reference frame (104.2)] does have a singularity there. Outside of S_0, where the variable t has temporal character, the spatial metric tensor is calculated from (84.7), and the element of spatial distance has the form

$$dl^2 = \frac{\rho^2}{\Delta} dr^2 + \rho^2 d\theta^2 + \frac{\Delta \sin^2 \theta}{1 - rr_g/\rho^2} d\phi^2. \tag{104.15}$$

Near S_0 parallels ($\theta = \text{const}$, $r = \text{const}$) go to infinity according to the law $2\pi a \sin^2 \theta/\sqrt{g_{00}}$. The difference in readings of clocks [cf. (88.5)] also goes to infinity here when they are synchronized along this closed contour.

PROBLEMS

1. Carry out the separation of variables in the Hamilton–Jacobi equation for a particle moving in the Kerr field (B. Carter, 1968).

Solution: In the Hamilton–Jacobi equation

$$g^{ik} \frac{\partial S}{\partial x^i} \frac{\partial S}{\partial x^k} - m^2 = 0$$

(m is the mass of the particle, not to be confused with the mass of the central body) with g^{ik} from (104.6) the time t and the angle ϕ are cyclic variables; they therefore enter in the action S in the form $-\mathscr{E}_0 t + L\phi$, where $\mathscr{E}_0$ is the conserved energy and L denotes the component of the angular momentum along the axis of symmetry of the field. It turns out that the variables θ and r can also be separated. Writing S in the form

$$S = -\mathscr{E}_0 t + L\phi + S_r(r) + S_\theta(\theta), \tag{1}$$

we reduce the Hamilton–Jacobi equation to two ordinary differential equations (cf. *Mechanics*, § 48):

$$\left(\frac{dS_\theta}{d\theta}\right)^2 + \left(a\mathscr{E}_0 \sin \theta - \frac{L}{\sin \theta}\right)^2 + a^2 m^2 \cos^2 \theta = K,$$

$$\left(\frac{dS_r}{dr}\right)^2 - \frac{1}{\Delta} [(r^2 + a^2) \mathscr{E}_0 - aL]^2 + m^2 r^2 = -K, \tag{2}$$

where K (the separation parameter) is a new arbitrary constant. The functions S_θ and S_r are then determined by simple quadratures.

The four-momentum of the particle is

$$p^i = m \frac{dx^i}{ds} = g^{ik} p_k = -g^{ik} \frac{\partial S}{\partial x_k}$$

Calculating the right-hand side of this equation using (1) and (2), we get the following equations:

$$m \frac{dt}{ds} = -\frac{r_g ra}{\varrho^2 \Delta} L + \frac{\mathscr{E}_0}{\Delta} \left(r^2 + a^2 + \frac{r_g ra^2}{\varrho^2} \sin^2 \theta\right), \tag{3}$$

$$m \frac{d\phi}{ds} = \frac{L}{\Delta \sin^2 \theta} \left(1 - \frac{r_g r}{\varrho^2}\right) + \frac{r_g ra}{\varrho^2 \Delta} \mathscr{E}_0, \tag{4}$$

$$m^2 \left(\frac{dr}{ds}\right)^2 = \frac{1}{\varrho^4} \left[(r^2 + a^2)\mathscr{E}_0 - aL\right]^2 - \frac{\Delta}{\varrho^4}(K + m^2 r^2), \tag{5}$$

$$m^2 \left(\frac{d\theta}{ds}\right)^2 = \frac{1}{\varrho^4}(K - a^2 m^2 \cos^2\theta) - \frac{1}{\varrho^4}\left(a\mathscr{E}_0 \sin\theta - \frac{L}{\sin\theta}\right)^2 \tag{6}$$

These integrals are the first integrals of the equations of motion (the equations of the geodesics). The equation of the trajectory and the time dependence of the coordinates along the trajectory can be found either from (3) to (6) or directly from the equations

$$\partial S/\partial\mathscr{E}_0 = \text{const}, \quad \partial S/\partial L = \text{const}, \quad \partial S/\partial K = \text{const}.$$

For the case of light rays, we must set $m = 0$ on the right sides of equations (3)–(6) and write ω_0 in place of $\mathscr{E}_0$ (cf. § 101), while we must replace the derivatives md/ds on the left sides by the derivatives $d/d\lambda$ with respect to the parameter λ, which varies along the ray (cf. the end of § 87).

Equations (4)–(6) permit purely radial motion only along the axis of rotation of the body, as is already clear from symmetry arguments. From these same considerations it is clear that motion in a "plane" is possible only if the plane is equatorial. In that case, setting $\theta = \pi/2$ and expressing K in terms of $\mathscr{E}_0$ and L from the condition $d\theta/ds = 0$, we obtain the equations of motion in the form

$$m\frac{dt}{ds} = -\frac{r_g a}{r\Delta}L + \frac{\mathscr{E}_0}{\Delta}\left(r^2 + a^2 + \frac{r_g a^2}{r}\right), \tag{7}$$

$$m\frac{d\phi}{ds} = \frac{M}{\Delta}\left(1 - \frac{r_g}{r}\right) + \frac{r_g a}{r\Delta}\mathscr{E}_0, \tag{8}$$

$$m^2 \left(\frac{dr}{ds}\right)^2 = \frac{1}{r^4}\left[(r^2 + a^2)\mathscr{E}_0 - aL\right]^2 - \frac{\Delta}{r^4}\left[(a\mathscr{E}_0 - L)^2 + m^2 r^2\right], \tag{9}$$

2. Determine the radius of the circle, closest to the centre, that is a stable orbit for a particle moving in the equatorial plane of the limiting ($a \to r_g/2$) Kerr field (R. Ruffini and J. A. Wheeler, 1969).

Solution: Proceeding as in problem 1 of § 102, we introduce the "effective potential energy" $U(r)$, defined from

$$[(r^2 + a^2)U(r) - aL]^2 - \Delta[(aU(r) - L)^2 + r^2 m^2] = 0$$

[for $\mathscr{E}_0 = U$ the right side of eq. (9) vanishes]. The radii of stable orbits are determined by the minima of the function $U(r)$, i.e. by simultaneous solution of the equations $U(r) = \mathscr{E}_0$, $U'(r) = 0$ for $U''(r) > 0$. The orbit closest to the centre corresponds to $U''(r_{\min}) = 0$; for $r < r_{\min}$, the function $U(r)$ has no minima. As a result we obtain the following values for the parameters of the motion:

(a) When $L < 0$ (motion opposite to the direction of rotation of the collapsar)

$$\frac{r_{\min}}{r_g} = \frac{9}{2}, \quad \frac{\mathscr{E}_0}{m} = \frac{5}{3\sqrt{3}}, \quad \frac{L}{mr_g} = \frac{11}{3\sqrt{3}}.$$

(b) For $L > 0$ (motion in the direction of rotation of the collapsar) as $a \to r_g/2$ the radius $r_{\min}$ tends toward the radius of the horizon. Setting $a = (r_g/2)(1 + \delta)$, we find, for $\delta \to 0$:

$$\frac{r_{\text{hor}}}{r_g} = \frac{1}{2}(1 + \sqrt{2\delta}), \quad \frac{r_{\min}}{r_g} = \frac{1}{2}[1 + (4\delta)^{\frac{1}{3}}]$$

Then

$$\frac{\mathscr{E}_0}{m} = \frac{L}{mr_g} = \frac{1}{\sqrt{3}}[1 + (4\delta)^{\frac{1}{3}}].$$

We call attention to the fact that $r_{\min}/r_{\text{hor}}$ remains greater than 1 throughout, i.e. the orbit does not go outside the horizon. This is as it should be: the horizon is a null hypersurface, and no timelike world lines of moving particles can lie on it.

§ 105. Gravitational fields at large distances from bodies

Let us consider a stationary gravitational field at large distances from the body producing it, and determine the first terms in its expansion in powers of $1/r$.

Far from the body, the field is weak. This means that there the space-time metric is almost galilean, i.e. we can choose a reference system in which the components of the metric tensor are almost equal to their galilean values:

$$g_{00}^{(0)} = 1, \quad g_{0\alpha}^{(0)} = 0, \quad g_{\alpha\beta}^{(0)} = -\delta_{\alpha\beta}. \tag{105.1}$$

We accordingly write the g_{ik} in the form

$$g_{ik} = g_{ik}^{(0)} + h_{ik}, \tag{105.2}$$

where the h_{ik} are small corrections determined by the gravitational field.

In operating with the tensor h_{ik}, we shall agree to raise and lower its indices using the "unperturbed" metric: $h_i^k = g^{(0)kl}h_{il}$, etc. Here we must distinguish the h^{ik} from the corrections to the contravariant components of the metric tensor g^{ik}. The latter are determined by solving the equations:

$$g_{il}g^{lk} = (g_{il}^{(0)} + h_{il})g^{lk} = \delta_i^k,$$

so that, to terms of second order, we find:

$$g^{ik} = g^{ik(0)} - h^{ik} + h_l^i h^{lk}. \tag{105.3}$$

To this same accuracy, the determinant of the metric tensor is

$$g = g^{(0)}(1 + h + \tfrac{1}{2}h^2 - \tfrac{1}{2}h_k^i h_i^k), \tag{105.4}$$

where $h \equiv h_i^i$.

We emphasize immediately that the condition that the h_{ik} be small in no way fixes a unique choice of reference frame. If this condition is satisfied in any one system, it will also be satisfied after any transformation $x'^i = x^i + \xi^i$, where the ξ^i are small quantities. According to (94.3), the tensor h_{ik} then goes over into

$$h'_{ik} = h_{ik} - \frac{\partial \xi_i}{\partial x^k} - \frac{\partial \xi_k}{\partial x^i}, \tag{105.5}$$

where $\xi^i = g_{ik}^{(0)}\xi^k$ [because of the constancy of the $g_{ik}^{(0)}$ the covariant derivatives in (94.3) reduce to ordinary derivatives in the present case].†

In first approximation, to terms of order $1/r$, the small corrections to the galilean values are given by the corresponding terms in the expansion of the centrally symmetric Schwarzschild metric. Because of the indeterminacy in the choice of reference frame (galilean at infinity) mentioned above, the specific form of the h_{ik} depends on the way the radial coordinate r is defined. Thus, if the Schwarzschild metric is written in the form (100.14) the first terms in its expansion for large r are given by the expression (100.18). Changing from spherical spatial coordinates to cartesian (for which we must write $dr = n_\alpha dx^\alpha$, where $\mathbf{n}$ is the unit vector in the direction of $\mathbf{r}$), we obtain the following values:

---

† For a stationary field it is natural to admit only those transformations that preserve the time-independence of the g_{ik}, i.e. the ξ' must be functions only of the space coordinates.

$$h_{00}^{(1)} = -\frac{r_g}{r}, \quad h_{\alpha\beta}^{(1)} = -\frac{r_g}{r}\, n_\alpha n_\beta, \quad h_{0\alpha}^{(1)} = 0, \tag{105.6}$$

where $r_g = 2km/c^2$.†

Among the terms of second order proportional to $1/r^2$, there are terms having two different origins. Some of the terms result from the effect of nonlinearity of the Einstein equations on the first-order terms. Since the latter depend only on the mass (and on no other characteristics of the body), these second-order terms can depend only on the mass. It is therefore clear that these terms can be obtained by expanding the Schwarzschild metric. In these coordinates, we find

$$h_{00}^{(2)} = 0, \quad h_{\alpha\beta}^{(2)} = -\left(\frac{r_g}{r}\right)^2 n_\alpha n_\beta. \tag{105.7}$$

The remaining second-order terms arise as the corresponding solutions of the linearized field equations. Having in mind later applications, we shall carry out the linearization of the equations with the formulas written in more general form than is needed here; at the start we shall not make use of the stationarity of the field.

For small h_{ik} the quantities Γ_{kl}^i, expressed in terms of derivatives of the h_{ik}, are also small. Neglecting powers higher than the first, we can keep in the curvature tensor (92.1) only the terms in the first bracket:

$$R_{iklm} = \frac{1}{2}\left(\frac{\partial^2 h_{im}}{\partial x^k \partial x^l} + \frac{\partial^2 h_{kl}}{\partial x^i \partial x^m} - \frac{\partial^2 h_{km}}{\partial x^i \partial x^l} - \frac{\partial^2 h_{il}}{\partial x^k \partial x^m}\right). \tag{105.8}$$

For the Ricci tensor, we have to this same accuracy:

$$R_{ik} = g^{lm} R_{limk} \approx g^{lm(0)} R_{limk},$$

or

$$R_{ik} = \frac{1}{2}\left(-g^{lm(0)}\frac{\partial^2 h_{ik}}{\partial x^l \partial x^m} + \frac{\partial^2 h_i^l}{\partial x^k \partial x^l} + \frac{\partial^2 h_k^l}{\partial x^i \partial x^l} - \frac{\partial^2 h}{\partial x^i \partial x^k}\right). \tag{105.9}$$

The expression (105.9) can be simplified by making use of the remaining arbitrariness in the choice of reference frame. We can impose on the h_{ik} four (the number of arbitrary functions ξ^i) supplementary conditions

$$\frac{\partial \psi_i^k}{\partial x^k} = 0, \quad \psi_i^k = h_i^k - \delta_i^k h. \tag{105.10}$$

Then the last three terms in (105.9) cancel one another, and we are left with

---

† If, however, we start from the Schwarzschild metric in isotropic spatial coordinates (cf. problem 4 of § 100), we would get:

$$h_{00}^{(1)} = -\frac{r_g}{r}, \quad h_{\alpha\beta}^{(1)} = -\frac{r_g}{r}\, \delta_{\alpha\beta}, \quad h_{0\alpha}^{(1)} = 0. \tag{105.6a}$$

The shift from (105.6) to (105.6a) is accomplished by the transformation (105.5) with

$$\xi^0 = 0, \quad \xi^\alpha = -\frac{r_g x^\alpha}{2r}.$$

$$R_{ik} = -\tfrac{1}{2} g^{lm(0)} \frac{\partial^2 h_{ik}}{\partial x^l \partial x^m}. \tag{105.11}$$

In the stationary case that we are considering here, when the h_{ik} do not depend on the time, the expression (105.11) reduces to $R_{ik} = \tfrac{1}{2}\Delta h_{ik}$, where Δ is the Laplace operator in the three spatial coordinates. The Einstein equations for the field in vacuum thus reduce to the Laplace equation

$$\Delta h_{ik} = 0, \tag{105.12}$$

with the supplementary conditions (105.10), which take the form

$$\frac{\partial}{\partial x^\beta} (h_\alpha^\beta - \tfrac{1}{2} h \delta_\alpha^\beta) = 0, \tag{105.13}$$

$$\frac{\partial}{\partial x^\beta} h_0^\beta = 0. \tag{105.14}$$

We call attention to the fact that these conditions still do not completely fix a unique choice of reference frame. It is easy to see that if the h_{ik} satisfy eqs. (105.13)–(105.14), then the same conditions will be satisfied by the h_{ik} of (105.5) so long as the ξ^i satisfy the equations

$$\Delta \xi^i = 0. \tag{105.15}$$

The component h_{00} must be given by a scalar solution of the three-dimensional Laplace equation. We know that such a solution, proportional to $1/r^2$, has the form $\mathbf{a} \cdot \nabla(1/r)$, where $\mathbf{a}$ is a constant vector. But a term of this type in h_{00} can always be eliminated by simply shifting the coordinate origin in the term of first order in $1/r$. Thus the presence of such a term would indicate only a poor choice of the coordinate origin, and is therefore not of interest.

The components $h_{0\alpha}$ are given by a vector solution of the Laplace equation, i.e. they must have the form

$$h_{0\alpha} = \lambda_{\alpha\beta} \frac{\partial}{\partial x^\beta} \frac{1}{r},$$

where $\lambda_{\alpha\beta}$ is a constant tensor. The condition (105.14) gives

$$\lambda_{\alpha\beta} \frac{\partial^2}{\partial x^\alpha \partial x^\beta} \frac{1}{r} = 0,$$

from which it follows that the $\lambda_{\alpha\beta}$ must have the form $a_{\alpha\beta} + \lambda \delta_{\alpha\beta}$, where $a_{\alpha\beta}$ is an antisymmetric tensor. But a solution of the form $\lambda \dfrac{\partial}{\partial x^\alpha} \dfrac{1}{r}$ can be eliminated by the transformation (105.5) with $\xi^0 = \lambda/r$, $\xi^\alpha = 0$ [satisfying the condition (105.15)]. Thus the only solution that has real meaning is

$$h_{0\alpha} = a_{\alpha\beta} \frac{\partial}{\partial x^\beta} \frac{1}{r}.$$

Finally, by similar but more complicated arguments one can show that by a suitable transformation of the spatial coordinates one can always eliminate the quantities $h_{\alpha\beta}$ given by a tensor solution (symmetric in α and β) of the Laplace equation.

As for the tensor $a_{\alpha\beta}$, it is related to the total angular momentum tensor $M_{\alpha\beta}$, and the final expression for $h_{0\alpha}$ has the form

$$h_{0\alpha}^{(2)} = \frac{2k}{c^3} M_{\alpha\beta} \frac{\partial}{\partial x^\beta} \frac{1}{r} = -\frac{2k}{c^3} M_{\alpha\beta} \frac{n_\beta}{r^2} \tag{105.16}$$

We show this by calculating the integral (96.17).

The angular momentum $M_{\alpha\beta}$ is related only to the $h_{0\alpha}$, and so in calculating it we may assume that all other components of h_{ik} are absent. To terms of second order in the $h_{0\alpha}$, we have from (96.2)–(96.3) (we note that $g^{\alpha 0} = -h^{\alpha 0} = h_{\alpha 0}$, while $-g$ differs from unity only by a term of second order:

$$h^{\alpha 0 \beta} = \frac{c^4}{16\pi k} \frac{\partial}{\partial x^\gamma} (g^{\alpha 0} g^{\beta\gamma} - g^{\gamma 0} g^{\alpha\beta}) = -\frac{c^4}{16\pi k} \frac{\partial}{\partial x^\gamma} (h_{\alpha 0} \delta_{\beta\gamma} - h_{\gamma 0} \delta_{\alpha\beta}).$$

Upon substituting from (105.16), the second term under the derivative sign vanishes, while the first gives

$$h^{\alpha 0 \beta} = -\frac{c}{8\pi} M_{\alpha\gamma} \frac{\partial^2}{\partial x^\beta \partial x^\gamma} \frac{1}{r} = -\frac{c}{8\pi} M_{\alpha\gamma} \frac{3n_\beta n_\gamma - \delta_{\beta\gamma}}{r^2}.$$

Using this expression we find that, carrying out the integration in (96.16) over the surface of a sphere of radius r ($df_\gamma = n_\gamma r^2 do$):

$$\frac{1}{c} \int (x^\alpha h^{\beta 0 \gamma} - x^\beta h^{\alpha 0 \gamma}) df_\gamma =$$

$$= -\frac{1}{4\pi} \int (n_\alpha n_\gamma M_{\beta\gamma} - n_\beta n_\gamma M_{\alpha\gamma}) do = -\frac{1}{3}(\delta_{\alpha\gamma} M_{\beta\gamma} - \delta_{\beta\gamma} M_{\alpha\gamma}) = \frac{2}{3} M_{\alpha\beta}.$$

An analogous calculation gives:

$$\frac{1}{c} \int \lambda^{\alpha 0 \gamma \beta} df_\gamma = -\frac{c^3}{16\pi k} \int (h_{\alpha 0} df_\beta - h_{\beta 0} df_\alpha) = \frac{1}{3} M_{\alpha\beta}.$$

Adding these two quantities we obtain the required value of $M_{\alpha\beta}$.

We emphasize that in the general case, when the field near the body may not be weak, $M_{\alpha\beta}$ is the total angular momentum of the body together with its gravitational field. Only if the field is weak at all distances can its contribution to the angular momentum be neglected.†

Formulas (105.6)–(105.7) and (105.16) solve our problem to terms of order $1/r^2$.‡ The covariant components of the metric tensor are:

$$g_{ik} = g_{ik}^{(0)} + h_{ik}^{(1)} + h_{ik}^{(2)}. \tag{105.17}$$

According to (105.3), to this same accuracy the contravariant components are

$$g^{ik} = g^{ik(0)} - h^{ik(1)} - h^{ik(2)} + h_l^{i(1)} h^{ik(1)}. \tag{105.18}$$

† If the rotating body is spherical in shape, the direction of **M** remains the only distinguished direction for the field in all the space outside the body. If the field is weak everywhere (and not just at large distances from the body) formula (105.16) is valid over all the space outside the body. This formula remains valid over all space also for the case when the centrally symmetric part of the field is not weak everywhere, but the spherical body rotates sufficiently slowly (cf. problem 1).

‡ The transformations (105.5) with $\xi^0 = 0$, $\xi^\alpha = \xi^\alpha(x^1, x^2, x^3)$ do not change the $h_{0\alpha}$. Thus the expression (105.16) does not depend on the choice of the coordinate r.

Formula (105.16) can be rewritten in vector form as†

$$\mathbf{g} = \frac{2k}{c^3 r^2} \, \mathbf{n} \times \mathbf{M}. \tag{105.19}$$

where $\mathbf{M}$ is the total angular momentum vector of the body. It was shown in problem 1 of § 88 that in a stationary gravitational field a "Coriolis force" acts on the body which is the same as that which would act on the body in a reference frame rotating with angular velocity

$$\Omega = \frac{c}{2} \, \sqrt{g_{00}} \, \operatorname{curl} \mathbf{g}.$$

We may therefore say that in the field of a rotating body a coriolis force acts on a distant particle with strength corresponding to an angular velocity:

$$\Omega \approx \frac{c}{2} \operatorname{curl} \mathbf{g} = \frac{k}{c^2 r^3} \, [\mathbf{M} - 3\mathbf{n}(\mathbf{M} \cdot \mathbf{n})]. \tag{105.20}$$

Finally, we apply the expression (105.6) to calculate the total energy of a gravitating body using the integral (96.16). Computing the necessary components of h^{ikl} from formulas (96.2)–(96.3), we find to the required accuracy (we keep terms to order $1/r^2$):

$$h^{\alpha 0 \beta} = 0,$$

$$h^{00\alpha} = \frac{c^4}{16\pi k} \frac{\partial}{\partial x^\beta} (g^{00} g^{\alpha\beta}) = \frac{mc^2}{8\pi} \frac{\partial}{\partial x^\beta} \left(-\frac{\delta^{\alpha\beta}}{r} + \frac{x^\alpha x^\beta}{r^3} \right) = \frac{mc^2}{4\pi} \frac{n^\alpha}{r^2}.$$

Now, integrating in (96.16) over a sphere of radius r, we get, finally,

$$P^\alpha = 0, \quad P^0 = mc, \tag{105.21}$$

a result which was naturally to be expected. It is an expression of the equality of "gravitational" and "inertial" mass ("gravitational" mass is the mass that determines the gravitational field produced by the body, the same mass that appears in the metric tensor in a gravitational field, or, in particular, in Newton's law; "inertial" mass is the mas that determines the ratio of energy and momentum of the body; in particular, the rest energy of the body is equal to this mass multiplied by c^2).

In the case of a constant gravitational fleld it is possible to derive a simple expression for the total energy of matter plus filed in the form of an integral only over the space occupied by the matter. We can do this, for example, by starting from the following expression, which is valid when all quantities are independent of x^0:‡

---

† To the assumed accuracy, the vector $g_\alpha = -g_{0\alpha}/g_{00} \approx -g_{0\alpha}$. For this same reason, in the definitions of vector product and curl (cf. the footnote on p. 272) we must set $\gamma = 1$, so that they may be understood in their usual sense for cartesian vectors.

‡ From (92.7) we have

$$R_0^0 = g^{0i} R_{i0} = g^{0i} \left(\frac{\partial \Gamma_{i0}^l}{\partial x^l} - \Gamma_{i0}^l \Gamma_{lm}^m - \Gamma_{il}^m \Gamma_{0m}^l \right),$$

and, using (86.5) and (86.8), we find that this expression can be written as

$$R_0^0 = \frac{1}{\sqrt{-g}} \frac{\partial}{\partial x^l} (\sqrt{-g} \, g^{0i} \Gamma_{i0}^l) - g^{im} \Gamma_{ml}^0 \Gamma_{i0}^l;$$

using the same relation (86.8) it is easy to show that the second term on the right is identically equal to $-\frac{1}{2} \Gamma_{lm}^0 \frac{\partial g^{im}}{\partial x^0}$, and, because all quantities are independent of x^0, is equal to zero. Finally, replacing the summation over l by one over α in the first term, for this same reason, we get (105.22).

$$R_0^0 = \frac{1}{\sqrt{-g}} \frac{\partial}{\partial x^\alpha} (\sqrt{-g} g^{i0} \Gamma_{0i}^x). \tag{105.22}$$

Integrating $R_0^0 \sqrt{-g}$ over (three-dimensional) space and using the three-dimensional Gauss formula, we obtain:

$$\int R_0^0 \sqrt{-g}\, dV = \oint \sqrt{-g}\, g^{i0} \Gamma_{0i}^\alpha df_\alpha.$$

Taking a sufficiently remote surface for the integration and using the expressions (105.6) for the g_{ki}, we find, after simple calculations:

$$\int R_0^0 \sqrt{-g}\, dV = \frac{4\pi k}{c^2} m = \frac{4\pi k}{c^3} P^0.$$

Noting also that, according to the field equations,

$$R_0^0 = \frac{8\pi k}{c^4} (T_0^0 - \tfrac{1}{2}T) = \frac{4\pi k}{c^4} (T_0^0 - T_1^1 - T_2^2 - T_3^3),$$

we get the required formula:

$$P^0 = mc = \frac{1}{c} \int (T_0^0 - T_1^1 - T_2^2 - T_3^3) \sqrt{-g}\, dV. \tag{105.23}$$

This formula expresses the total energy of the matter and the constant gravitational field (i.e., the total mass of the body) in terms of the energy–momentum tensor of the matter alone (R. Tolman, 1930). We recall that in the case of central symmetry of the field we had still another expression for this quantity—formula (100.23).

PROBLEMS

1. Show that formula (105.16) remains valid for the field over all space outside of the rotating spherical body under conditions of slow rotation ($M \ll cmr_g$) but without the requirement that the centrally symmetric part of the field be small (A. G. Doroshkevich, Ya. B. Zel'dovich and I. D. Novikov, 1965; V. Gurovich, 1965).

Solution: In spherical spatial coordinates ($x^1 = r$, $x^2 = \theta$, $x^3 = \phi$) formula (105.16) is written as

$$h_{03} = \frac{2kM}{rc^2} \sin^2 \theta. \tag{1}$$

Considering this quantity to be a small correction to the Schwarzschild metric (100.14), we must verify that the equation $R_{03} = 0$, linearized in h_{03}, is satisfied (since in the other field equations the correction terms drop out identically). R_{03} can be calculated using formula (4) of the problem in § 95, where the linearization means that the three-dimensional tensor operations should be carried out with the "unperturbed" metric (100.15). As a result we get the following equation

$$\left(1 - \frac{r_g}{r}\right) \frac{\partial^2 h_{03}}{\partial r^2} + \frac{2r_g}{r^3} h_{03} + \frac{\sin\theta}{r^2} \frac{\partial}{\partial\theta} \left(\frac{1}{\sin\theta} \frac{\partial h_{03}}{\partial\theta}\right) = 0,$$

which expression (1) actually does satisfy.

2. Determine the systematic ("secular") shift of the orbit of a particle moving in the field of a central body, associated with the rotation of the latter (J. Lense, H. Thirring, 1918).

Solution: Because all the relativistic effects are small, they superpose linearly with one another, so in calculating the effects resulting from the rotation of the central body we can neglect the influence of the non-Newtonian centrally symmetric force field which we considered in § 101; in other words, we can make the computations assuming that of all the h_{ik} only the $h_{0\alpha}$ are different from zero.

The orientation of the classical orbit of the particle is determined by two conserved quantities: the angular momentum of the particle, $\mathbf{M} = \mathbf{r} \times \mathbf{p}$, and the vector

$$\mathbf{A} = \frac{\mathbf{p}}{m} \times \mathbf{M} - \frac{kmm'\mathbf{r}}{r},$$

whose conservation is peculiar to the Newtonian field $\varphi = -km'/r$ (where m' is the mass of the central body).† The vector $\mathbf{M}$ is perpendicular to the plane of the oribit, while the vector $\mathbf{A}$ is directed along the major axis of the ellipse toward the perihelion (and is equal in magnitude to $kmm'e$, where e is the eccentricity of the orbit). The required secular shift of the orbit can be described in terms of the change in direction of these vectors.

The Lagrangian for a particle moving in the field (105.19) is

$$L = -mc\frac{ds}{dt} = L_0 + \delta L, \quad \delta L = mc\mathbf{g} \cdot \mathbf{v} = \frac{2km}{c^2 r^3} \mathbf{M}' \cdot \mathbf{v} \times \mathbf{r}, \tag{1}$$

(where we denote the angular momentum of the central body by $\mathbf{M}'$ to distinguish it from the angular momentum $\mathbf{M}$ of the particle). Then the Hamiltonian is [cf. *Mechanics* (40.7)]:

$$\mathcal{H} = \mathcal{H}_0 + \delta\mathcal{H}, \quad \delta\mathcal{H} = \frac{2k}{c^2 r^3} \mathbf{M}' \cdot \mathbf{r} \times \mathbf{p}.$$

Computing the derivative $\dot{\mathbf{M}} = \dot{\mathbf{r}} \times \mathbf{p} + \mathbf{r} \times \dot{\mathbf{p}}$ using the Hamilton equations $\dot{\mathbf{r}} = \partial\mathcal{H}/\partial\mathbf{p}$, $\dot{\mathbf{p}} = -(\partial\mathcal{H}/\partial\mathbf{r})$, we get:

$$\dot{\mathbf{M}} = \frac{2k}{c^2 r^3} \mathbf{M}' \times \mathbf{M}. \tag{2}$$

Since we are interested in the secular variation of $\mathbf{M}$, we should average this expression over the period of rotation of the particle. The averaging is conveniently done using the parametric representation of the dependence of r on the time for motion in an elliptical orbit, in the form

$$r = a(1 - e\cos\xi), \quad t = \frac{T}{2\pi}(\xi - e\sin\xi)$$

(a and e are the semimajor axis and eccentricity of the ellipse; cf. *Mechanics*, § 15):

$$\overline{r^{-3}} = \frac{1}{T}\int_0^T \frac{dt}{r^3} = \frac{1}{2\pi a^3}\int_0^{2\pi} \frac{d\xi}{(1 - e\cos\xi)^2} = \frac{1}{a^3(1 - e^2)^{3/2}}.$$

Thus the secular change of $\mathbf{M}$ is given by the formula

$$\frac{d\mathbf{M}}{dt} = \frac{2k\mathbf{M}' \times \mathbf{M}}{c^2 a^3(1 - e^2)^{3/2}}, \tag{3}$$

i.e. the vector $\mathbf{M}$ rotates around the axis of rotation of the central body, remaining fixed in magnitude.

An analogous calculation for the vector $\mathbf{A}$ gives:

$$\dot{\mathbf{A}} = \frac{2k}{c^2 r^3} \mathbf{M}' \times \mathbf{A} + \frac{6k}{c^2 mr^5}(\mathbf{M} \cdot \mathbf{M}')(\mathbf{r} \times \mathbf{M}).$$

The averaging of this expression is carried out in the same way as before; from symmetry considerations it is clear beforehand that the averaged vector $\overline{\mathbf{r}/r^5}$ will be along the major axis of the ellipse, i.e. along the

† See *Mechanics*, § 15.

direction of the vector **A**. The computation leads to the following expression for the secular change of the vector **A**:

$$\frac{d\mathbf{A}}{dt} = \mathbf{\Omega} \times \mathbf{A}, \qquad \mathbf{\Omega} = \frac{2kM'}{c^2 a^3 (1 - e^2)^{3/2}} \{\mathbf{n}' - 3\mathbf{n}(\mathbf{n} \cdot \mathbf{n}')\} \tag{4}$$

(**n** and **n**' are unit vectors along the directions of **M** and **M**'), i.e. the vector **A** rotates with angular velocity **Ω**, remaining fixed in magnitude; this last point shows that the eccentricity of the orbit does not undergo any secular change.

Formula (3) can be written in the form

$$\frac{d\mathbf{M}}{dt} = \mathbf{\Omega} \times \mathbf{M},$$

with the same **Ω** as in (4); in other words, **Ω** is the angular velocity of rotation of the ellipse "as a whole". This rotation includes both the additional (compared to that considered in § 101) shift of the perihelion of the orbit, and the secular rotation of its plane about the direction of the axis of the body (where the latter effect is absent if the plane of the orbit coincides with the equatorial plane of the body).

For comparison we note that to the effect considered in § 101 there corresponds

$$\mathbf{\Omega} = \frac{6\pi k m'}{c^2 a (1 - e^2) T} \mathbf{n}.$$

§ 106. The equations of motion of a system of bodies in the second approximation

As we shall see later (§ 110), a system of moving bodies radiates gravitational waves and thus loses energy. This loss appears only in the fifth approximation in $1/c$. In the first four approximations, the energy of the system remains constant. From this it follows that a system of gravitating bodies can be described by a Lagrangian correctly to terms of order $1/c^4$ in the absence of an electromagnetic field, for which a Lagrangian exists in general only to terms of second order (§ 65). Here we shall give the derivation of the Lagrangian of a system of bodies to terms of second order. We thus find the equations of motion of the system in the next approximation after the Newtonian.

We shall neglect the dimensions and internal structure of the bodies, regarding them as "pointlike"; in other words, we shall restrict ourselves to the zero'th approximation in the expansion in powers of the ratios of the dimensions a of the bodies to their mutual separations l.

To solve our problem we must start with the determination, in this same approximation, of the weak gravitational field produced by the bodies at distances large compared to their dimensions, but at the same time small compared to the wavelength λ of the gravitational waves radiated by the system ($a \ll r \ll \lambda \sim lc/v$).

To terms of order $1/c^2$ the field far from the body is given by the expressions obtained in the preceding section and denoted there by $h_{ik}^{(1)}$; here we use these expressions in the form (105.6a). In § 105 it was tacitly assumed that the field was produced by just one body, located at the coordinate origin. But since the field $h_{ik}^{(1)}$ is a solution of the linearized Einstein equations, the principle of superposition holds for it. Thus the field far from a system of bodies is obtained by simply summing the fields of each of them; we write the field in the form

$$h_\alpha^\beta = -\frac{2}{c^2}\, \phi\, \delta_\alpha^\beta, \tag{106.1}$$

$$h_0^\alpha = 0, \qquad h_0^0 = \frac{2}{c^2}\, \phi, \tag{106.2}$$

where

$$\phi(\mathbf{r}) = -k \sum_a \frac{m_a}{|\mathbf{r} - \mathbf{r}_0|}$$

is the Newtonian gravitational potential of a system of point objects ($\mathbf{r}_a$ is the radius vector to the body with mass m_a). The expression for the line element with the metric tensor (106.1)–(106.2) is:

$$ds^2 = \left(1 + \frac{2}{c^2}\phi\right)c^2 dt^2 - \left(1 - \frac{2}{c^2}\phi\right)(dx^2 + dy^2 + dz^2). \tag{106.3}$$

We note that first order terms containing ϕ appear not only in g_{00} but also in $g_{\alpha\beta}$; in § 87 it was already stated that, in the equations of motion of the particle, the correction terms in $g_{\alpha\beta}$ give quantities of higher order than the terms coming from g_{00}; as a consequence, of this, by a comparison with the Newtonian equations of motion we can determine only g_{00}.

As will be seen from the sequel, to obtain the required equations of motion it is sufficient to know the spatial components $h_{\alpha\beta}$ to the accuracy ($\sim 1/c^2$) with which they are given in (106.1); the mixed components (which are absent in the $1/c^2$ approximation) are needed to terms of order $1/c^3$, and the time component h_{00} to terms in $1/c^4$. To calculate them we turn once again to the general equations of gravitation, and consider the terms of corresponding order in these equations.

Disregarding the fact that the bodies are macroscopic, we must write the energy-momentum tensor of the matter in the form (33.4), (33.5). In curvilinear coordinates, this expression is rewritten as

$$T^{ik} = \sum_a \frac{m_a c}{\sqrt{-g}} \frac{dx^i}{ds} \frac{dx^k}{dt} \delta(\mathbf{r} - \mathbf{r}_a) \tag{106.4}$$

[for the appearance of the factor $1/\sqrt{-g}$, see the analogous transition in (90.4)]; the summation extends over all the bodies in the system.

The component

$$T_{00} = \sum_a \frac{m_a c^3}{\sqrt{-g}} g_{00}^2 \frac{dt}{ds} \delta(\mathbf{r} - \mathbf{r}_a)$$

in first approximation (for galilean g_{ik}) is equal to $\sum_a m_a c^2 \delta(\mathbf{r} - \mathbf{r}_a)$; in the next approximation, we substitute for g_{ik} from (106.3) and find, after a simple computation:

$$T_{00} = \sum_a m_a c^2 \left(1 + \frac{5\phi_a}{c^2} + \frac{v_a^2}{2c^2}\right)\delta(\mathbf{r} - \mathbf{r}_a), \tag{106.5}$$

where $\mathbf{v}$ is the ordinary three-dimensional velocity ($v^\alpha = dx^\alpha/dt$) and ϕ_a is the potential of the field at the point $\mathbf{r}_a$. (As yet we pay no attention to the fact that ϕ_a contains an infinite part— the potential of the self-field of the particle m_a; concerning this, see below.)

As regards the components $T_{\alpha\beta}$, $T_{0\alpha}$ of the energy-momentum tensor, in this approximation it is sufficient to keep for them only the first terms in the expansion of the expression (106.4).

$$T_{\alpha\beta} = \sum_a m_a v_{a\alpha} v_{a\beta} \delta(\mathbf{r} - \mathbf{r}_a),$$

$$T_{0\alpha} = - \sum_a m_a c v_{a\alpha} \delta(\mathbf{r} - \mathbf{r}_a). \tag{106.6}$$

Next we proceed to compute the components of the tensor R_{ik}. The calculation is conveniently done using the formula $R_{ik} = g^{lm} R_{limk}$ with R_{limk} given by (92.1). Here we must remember that the quantities $h_{\alpha\beta}$ and h_{00} contain no terms of order lower than $1/c^2$, and $h_{0\alpha}$ no terms lower than $1/c^3$; differentiation with respect to $x^0 = ct$ raises the order of smallness of quantities by unity.

The main terms in R_{00} are of order $1/c^2$; in addition to them we must also keep terms of the next non-vanishing order $1/c^4$. A simple computation gives the result:

$$R_{00} = \frac{1}{c} \frac{\partial}{\partial t} \left(\frac{\partial h_0^\alpha}{\partial x^\alpha} - \frac{1}{2c} \frac{\partial h_\alpha^\alpha}{\partial t} \right) + \frac{1}{2} \Delta h_{00} + \frac{1}{2} h^{\alpha\beta} \frac{\partial^2 h_{00}}{\partial x^\alpha \partial x^\beta} - \frac{1}{4} \left(\frac{\partial h_{00}}{\partial x^\alpha} \right)^2 -$$

$$- \frac{1}{4} \frac{\partial h_{00}}{\partial x^\beta} \left(2 \frac{\partial h_\beta^\alpha}{\partial x^\alpha} - \frac{\partial h_\alpha^\alpha}{\partial x^\beta} \right).$$

In this computation we have still not used any auxiliary condition for the quantities h_{ik}. Making use of this freedom, we now impose the condition

$$\frac{\partial h_0^\alpha}{\partial x^\alpha} - \frac{1}{2c} \frac{\partial h_\alpha^\alpha}{\partial t} = 0, \tag{106.7}$$

as a result of which all the terms containing the components $h_{0\alpha}$ drop out of R_{00}. In the remaining terms we substitute

$$h_\alpha^\beta = - \frac{2}{c^2} \phi \delta_\alpha^\beta, \qquad h_{00} = \frac{2}{c^2} \phi + O\left(\frac{1}{c^4} \right),$$

and obtain, to the required accuracy,

$$R_{00} = \frac{1}{2} \Delta h_{00} + \frac{2}{c^4} \phi \Delta \phi - \frac{2}{c^4} (\nabla \phi)^2, \tag{106.8}$$

where we have gone over to three-dimensional notation. In computing the components $R_{0\alpha}$ it is sufficient to keep only the terms of the first nonvanishing order—$1/c^3$. In similar fashion, we find:

$$R_{0\alpha} = \frac{1}{2c} \frac{\partial^2 h_\alpha^\beta}{\partial t \partial x^\beta} + \frac{1}{2} \frac{\partial^2 h_0^\beta}{\partial x^\alpha \partial x^\beta} - \frac{1}{2c} \frac{\partial^2 h_\beta^\beta}{\partial t \partial x^\alpha} + \frac{1}{2} \Delta h_{0\alpha}$$

and then, using the condition (106.7):

$$R_{0\alpha} = \frac{1}{2} \Delta h_{0\alpha} + \frac{1}{2c^3} \frac{\partial^2 \phi}{\partial t \partial x^\alpha}. \tag{106.9}$$

Using the expressions (106.5)–(106.9), we now write the Einstein equations

$$R_{ik} = \frac{8\pi k}{c^4} \left(T_{ik} - \frac{1}{2} g_{ik} T \right). \tag{106.10}$$

The time component of equation (106.10) gives:

$$\Delta h_{00} + \frac{4}{c^4}\,\phi\Delta\phi - \frac{4}{c^4}\,(\nabla\phi)^2 = \frac{8\pi k}{c^4}\sum_a m_a c^2\left(1 + \frac{5\phi_a}{c^2} + \frac{3v_a^2}{2c^2}\right)\delta(\mathbf{r} - \mathbf{r}_a);$$

making use of the identity

$$4(\nabla\phi)^2 = 2\Delta(\phi^2) - 4\phi\Delta\phi$$

and the equation of the Newtonian potential

$$\Delta\phi = 4\pi k \sum_a m_a\delta(\mathbf{r} - \mathbf{r}_a), \tag{106.11}$$

we rewrite this equation in the form

$$\Delta\left(h_{00} - \frac{2}{c^4}\,\phi^2\right) = \frac{8\pi k}{c^2}\sum_a m_a\left(1 + \frac{\phi_a'}{c^2} + \frac{3v_a^3}{2c^2}\right)\delta(\mathbf{r} - \mathbf{r}_a). \tag{106.12}$$

After completing all the computations, we have replaced ϕ_a on the right side of (106.12) by

$$\phi_a' = -k\sum_b{}' \frac{m_b}{|\mathbf{r}_a - \mathbf{r}_b|},$$

i.e. by the potential at the point $\mathbf{r}_a$ of the field produced by all the bodies except for the body m_a; the exclusion of the infinite self-potential of the bodies (in the method used by us, which regards the bodies as pointlike) corresponds to a "renormalization" of their masses, as a result of which they take on their true values, which take into account the field produced by the bodies themselves.†

The solution of (106.12) can be given immediately, using the familiar relation (36.9)

$$\Delta\frac{1}{r} = -4\pi\delta(\mathbf{r}).$$

We thus find:

$$h_{00} = \frac{2\phi}{c^2} + \frac{2\phi^2}{c^4} - \frac{2k}{c^4}\sum_a \frac{m_a\phi_a'}{|\mathbf{r} - \mathbf{r}_a|} - \frac{3k}{c^4}\sum_a \frac{m_a v_a^2}{|\mathbf{r} - \mathbf{r}_a|}, \tag{106.13}$$

The mixed component of equation (106.10) gives:

$$\Delta h_{0\alpha} = -\frac{16\pi k}{c^2}\sum_a m_a v_{a\alpha}\delta(\mathbf{r} - \mathbf{r}_a) - \frac{1}{c^3}\frac{\partial^2\phi}{\partial t\partial x^\alpha}. \tag{106.14}$$

The solution of this linear equation is‡

† Actually, if there is only one body at rest, the right side of the equation will have simply $(8\pi k/c^2)m_a\delta(\mathbf{r} - \mathbf{r}_a)$, and this equation will determine correctly (in second approximation) the field produced by the body.

‡ In the stationary case, the second term on the right of equation (106.14) is absent. At large distances from the system, its solution can be written immediately by analogy with the solution (44.3) of equation (43.4)

$$h_{0\alpha} = \frac{2k}{c^3 r^2}(\mathbf{M} \times \mathbf{n})_\alpha$$

(where $\mathbf{M} = \int \mathbf{r} \times \mu\mathbf{v}dV = \sum m_a\mathbf{r}_a \times \mathbf{v}_a$ is the angular momentum of the system), in agreement with formula (105.19).

$$h_{0\alpha} = \frac{4k}{c^3} \sum_a \frac{m_a v_{a\alpha}}{|\mathbf{r} - \mathbf{r}_a|} - \frac{1}{c^3} \frac{\partial^2 f}{\partial t \partial x^\alpha},$$

where f is the solution of the auxiliary equation

$$\Delta f = \phi = - \sum_a \frac{km_a}{|\mathbf{r} - \mathbf{r}_a|}.$$

Using the relation $\Delta r = 2/r$, we find:

$$f = - \frac{k}{2} \sum_a m_a |\mathbf{r} - \mathbf{r}_a|,$$

and then, after a simple computation, we finally obtain:

$$h_{0\alpha} = \frac{k}{2c^3} \sum_a \frac{m_a}{|\mathbf{r} - \mathbf{r}_a|} [7 v_{a\alpha} \times (\mathbf{v}_a \cdot \mathbf{n}_a) n_{a\alpha}], \qquad (106.15)$$

where $\mathbf{n}_a$ is a unit vector along the direction of the vector $\mathbf{r} - \mathbf{r}_a$.

The expressions (106.1), (106.13) and (106.15) are sufficient for computing the required Lagrangian to terms of second order.

The Lagrangian for a single body, in a gravitational field produced by other bodies and assumed to be given, is

$$L_a = - m_a c \frac{ds}{dt} = - m_a c^2 \left(1 + h_{00} + 2h_{0\alpha} \frac{v_\alpha^\alpha}{c} - \frac{v_a^2}{c^2} + h_{\alpha\beta} \frac{v_\alpha^\alpha v_\alpha^\beta}{c^2} \right)^{1/2}$$

Expanding the square root and dropping the irrelevant constant $-m_a c^2$, we rewrite this expression, to the required accuracy, as

$$L_a = \frac{m_a v_a^2}{2} + \frac{m_a v_a^4}{8c^2} - m_a c^2 \left(\frac{h_{00}}{2} + h_{0\alpha} \frac{v_a^\alpha}{c} + \frac{1}{2c^2} h_{\alpha\beta} v_a^\alpha v_a^\beta - \frac{h_{00}^2}{8} + \frac{h_{00}}{4c^2} v_a^2 \right).$$

$$(106.16)$$

Here the values of all the h_{ik} are taken at the point $\mathbf{r}_a$; again we must drop terms which become infinite, which amounts to a "renormalization" of the mass m_a appearing as a coefficient in L_a.

The further course of the calculations is the following. The total Lagrangian of the system is, of course, not equal to the sum of the Lagrangians L_a for the individual bodies, but must be constructed so that it leads to the correct values of the forces $\mathbf{f}_a$ acting on each of the bodies for a given motion of the others. For this purpose we compute the forces $\mathbf{f}_a$ by differentiating the Lagrangian L_a:

$$\mathbf{f}_a = \left(\frac{\partial L_a}{\partial \mathbf{r}} \right)_{\mathbf{r}=\mathbf{r}_a}$$

the differentiation is carried out with respect to the running coordinate $\mathbf{r}$ of the "field point" in the expressions for h_{ik}). It is then easy to form the total Lagrangian L, from which all of the forces $\mathbf{f}_a$ are obtained by taking the partial derivatives $\partial L/\partial \mathbf{r}_a$.

Omitting the simple intermediate computations, we give immediately the final result for the Lagrangian:†

$$L = \sum_a \frac{m'_a v_a^2}{2} \left(1 + 3 \sum_b{}' \frac{km_b}{c^2 r_{ab}} \right) + \sum_a \frac{m_a v_a^4}{8c^2} + \sum_a \sum_b{}' \frac{km_a m_b}{2r_{ab}} -$$

$$- \sum_a \sum_b{}' \frac{km_a m_b}{4c^2 r_{ab}} [7\mathbf{v}_a \cdot \mathbf{v}_b + (\mathbf{v}_a \cdot \mathbf{n}_{ab})(\mathbf{v}_b \cdot \mathbf{n}_{ab})] - \sum_a \sum_b{}' \sum_c{}' \frac{k^2 m_a m_b m_c}{2c^2 r_{ab} r_{ac}}, \quad (106.17)$$

where $r_{ab} = |\mathbf{r}_a - \mathbf{r}_b|$, n_{ab} is a unit vector along the direction $\mathbf{r}_a - \mathbf{r}_b$, and the prime on the summation sign means that we should omit the term with $b = a$ or $c = a$.

PROBLEMS

1. Find the action function for the gravitational field in the Newtonian approximation.

Solution: Using the g_{ik} from (106.3), we find from the general formula (93.3), $G = (2/c^4)(\nabla \varphi)^2$, so that the action for the field is

$$S_g = - \frac{1}{8\pi k} \iint (\nabla \varphi)^2 \, dV \, dt \, .$$

The total action, for the field plus the masses distributed in space with density μ, is:

$$S = \iint \left[\frac{\mu v^2}{2} - \mu \varphi - \frac{1}{8\pi k} (\nabla \varphi)^2 \right] dV \, dt \, . \tag{1}$$

One easily verifies that variation of S with respect to φ gives the Poisson equation (99.2), as it should.

The energy density is found from the Lagrangian density Λ [the integrand in (1)] by using the general formula (32.5), which reduces in the present case (because of the absence of time derivatives of φ in Λ) to changing the signs of the second and third terms. Integrating the energy density over all space, where we substitute $\mu \varphi = (1/4\pi k) \varphi \Delta \varphi$ in the second term and integrate by parts, we finally obtain the total energy of field plus matter in the form

$$\int \left[\frac{\mu v^2}{2} - \frac{1}{8\pi k} (\nabla \varphi)^2 \right] dV \, .$$

Consequently the energy density of the gravitational field in the Newtonian theory is $W = -(1/8\pi k)(\nabla \varphi)^2$.‡

2. Find the coordinates of the centre of inertia of a system of gravitating bodies in the second approximation.

Solution: In view of the complete formal analogy between Newton's law for gravitational interaction and Coulomb's law for electrostatic interaction, the coordinates of the centre of inertia are given by the formula

$$\mathbf{R} = \frac{1}{\mathscr{E}} \sum_a{}' \mathbf{r}_a \left(m_a c^2 + \frac{p_a^2}{2m_a} - \frac{km_a}{2} \sum_b{}' \frac{m_b}{r_{ab}} \right),$$

$$\mathscr{E} = \sum_a \left(m_a c^2 + \frac{p_a^2}{2m_a} - \frac{km_a}{2} \sum_b{}' \frac{m_b}{r_{ab}} \right),$$

† The equations of motion corresponding to this Lagrangian were first obtained by A. Einstein, L. Infeld and B. Hoffmann (1938) and by A. Eddington and G. Clark (1938).

‡ To avoid any misunderstanding, we state that this expression is not the same as the component $(-g) t_{00}$ of the energy-momentum pseudotensor (as calculated with the g_{ik} from (106.3)); there is also a contribution to W from $(-g) T_{ik}$.

which is analogous to the formula found in Problem 1 of § 65.

3. Find the secular shift of the perihelion of the orbit of two gravitating bodies of comparable mass (H. Robertson, 1938).

Solution: The Lagrangian of the system of two bodies is

$$L = \frac{m_1 v_1^2}{2} + \frac{m_2 v_2^2}{2} + \frac{km_1 m_2}{r} + \frac{1}{8c^2}(m_1 v_1^4 + m_2 v_2^4)$$

$$+ \frac{km_1 m_2}{2c^2 r}[3v_1^2 + v_2^2) - 7\mathbf{v}_1 \cdot \mathbf{v}_2 - (\mathbf{v}_1 \cdot \mathbf{n})(\mathbf{v}_2 \cdot \mathbf{n})] - \frac{k^2 m_1 m_2(m_1 + m_2)}{2c^2 r^2}.$$

Going over to the Hamiltonian function and eliminating from it the motion of the centre of inertia (see problem 2 in § 65), we get:

$$\mathcal{H} = \frac{p^2}{2}\left(\frac{1}{m_1} + \frac{1}{m_2}\right) - \frac{km_1 m_2}{r} - \frac{p^4}{8c^2}\left(\frac{1}{m_1^3} + \frac{1}{m_2^3}\right) -$$

$$- \frac{k}{2c^2 r}\left[3p^2\left(\frac{m_2}{m_1} + \frac{m_1}{m_2}\right) + 7p^2 + (\mathbf{p} \cdot \mathbf{n})^2\right] + \frac{k^2 m_1 m_2(m_1 + m_2)}{2c^2 r^2}, \tag{1}$$

where **p** is the momentum of the relative motion.

We determine the radial component of momentum p_r as a function of the variable r and the parameters M (the angular momentum) and $\mathscr{E}$ (the energy). This function is determined from the equation $\mathcal{H} = \mathscr{E}$ (in which, in the second-order terms, we must replace p^2 by its expression from the zero'th approximation):

$$\mathscr{E} = \frac{1}{2}\left(\frac{1}{m_1} + \frac{1}{m_2}\right)\left(p_r^2 + \frac{M^2}{r^2}\right) - \frac{km_1 m_2}{r} - \frac{1}{8c^2}\left(\frac{1}{m_1^3} + \frac{1}{m_2^3}\right)\left(\frac{2m_1 m_2}{m_1 + m_2}\right)^2\left(\mathscr{E} + \frac{km_1 m_2}{r}\right)^2 -$$

$$- \frac{k}{2c^2 r}\left[3\left(\frac{m_2}{m_1} + \frac{m_1}{m_2}\right) + 7\right]\frac{2m_1 m_2}{m_1 + m_2}\left(\mathscr{E} + \frac{km_1 m_2}{r}\right) - \frac{k}{2c^2 r}p_r^2 + \frac{k^2 m_1 m_2(m_1 + m_2)}{2c^2 r^2}.$$

The further course of the computations is analogous to that used in § 98. Having determined p_r from the algebraic equation given above, we make a transformation of the variable r in the integral

$$S_r = \int p_r dr,$$

so that the term containing M^2 is brought to the form M^2/r^2. Then expanding the expression under the square root in terms of the small relativistic corrections, we obtain:

$$S_r = \int \sqrt{A + \frac{B}{r} - \left(M^2 - \frac{6k^2 m_1^2 m_2^2}{c^2}\right)\frac{1}{r^2}}\, dr$$

[see (101.6)], where A and B are constant coefficients whose explicit computation is not necessary.

As a result we find for the shift in the perihelion of the orbit of the relative motion:

$$\delta\varphi = \frac{6\pi k^2 m_1^2 m_2^2}{c^2 M^2} = \frac{6\pi k(m_1 + m_2)}{c^2 a(1 - e^2)}.$$

Comparing with (101.7) we see that for given dimensions and shape of the orbit, the shift in the perihelion will be the same as it would be for the motion of one body in the field of a fixed centre of mass $m_1 + m_2$.

4. Determine the frequency of precession of a spherical top, performing an orbital motion in the gravitational field of a central body that is rotating about its axis.

Solution: In the first approximation the effect is the sum of two independent parts, one of which is related to the non-Newtonian character of the centrally symmetric field (H. Weyl, 1923) and the other to the rotation of the central body (L. Schiff, 1960).

The first part is described by an additional term in the Lagrangian of the top, corresponding to the second term in (106.17). We write the velocities of individual elements of the top (with mass dm) in the form $\mathbf{v} = \mathbf{V} + \boldsymbol{\omega} \times \mathbf{r}$, where $\mathbf{V}$ is the velocity of the orbital motion, $\boldsymbol{\omega}$ is the angular velocity, and $\mathbf{r}$ is the radius vector of the element dm relative to the centre of the top (so that the integral over the volume of the top $\int \mathbf{r}\, dm = 0$). Dropping terms independent of $\boldsymbol{\omega}$ and also neglecting terms quadratic in $\boldsymbol{\omega}$, we have:

$$\delta^{(1)} L = \frac{3km'}{2c^2} \int 2\, \frac{\mathbf{V} \cdot \boldsymbol{\omega} \times \mathbf{r}}{R}\, dm,$$

where m' is the mass of the central body, $R = |\mathbf{R}_0 + \mathbf{r}|$ is the distance from the centre of the field to the element dm, $\mathbf{R}_0$ is the radius vector of the centre of inertia of the top. In the expansion $1/R \approx 1/R_0 - (\mathbf{n} \cdot \mathbf{r}/R_0^2)$ (where $\mathbf{n} = \mathbf{R}_0/R_0$) the integral of the first term vanishes, while integration of the second term is done using the formula

$$\int x_\alpha x_\beta\, dm = \tfrac{1}{2} I \delta_{\alpha\beta}$$

where I is the moment of inertia of the top. As a result we get:

$$\delta^{(1)} L = \frac{3km'}{2c^2 R_0^2}\, \mathbf{M} \cdot \mathbf{v} \times \mathbf{n},$$

where $\mathbf{M} = I\boldsymbol{\omega}$ is the angular momentum of the top.

The additional term in the Lagrangian, due to the rotation of the central body, can also be found from (106.17), but it is even simpler to calculate it using formula (1) of the problem in § 105:

$$\delta^{(2)} L = \frac{2k}{c^2} \int \frac{\mathbf{M}' \cdot (\boldsymbol{\omega} \times \mathbf{r}) \times \mathbf{R}}{R^3}\, dm,$$

where $\mathbf{M}'$ is the angular momentum of the central body. Expanding,

$$\frac{\mathbf{R}}{R^3} \approx \frac{\mathbf{n}}{R_0^2} + \frac{1}{R_0^3}\, (\mathbf{r} - 3\mathbf{n}(\mathbf{n} \cdot \mathbf{r}))$$

and performing the integration, we get:

$$\delta^{(2)} L = \frac{k}{c^2 R_0^3}\, \{\mathbf{M} \cdot \mathbf{M}' - 3(\mathbf{n} \cdot \mathbf{M})(\mathbf{n} \cdot \mathbf{M}')\}.$$

Thus the total correction to the Lagrangian is

$$\delta L = -\,\mathbf{M} \cdot \boldsymbol{\Omega}, \qquad \boldsymbol{\Omega} = \frac{3km'}{2c^2 R_0^2}\, \mathbf{n} \times \mathbf{v}_0 \times \frac{k}{c^2 R_0^3}\, \{3\mathbf{n}(\mathbf{n} \cdot \mathbf{M}') - \mathbf{M}'\}.$$

To this function there corresponds the equation of motion

$$\frac{d\mathbf{M}}{dt} = \boldsymbol{\Omega} \times \mathbf{M}$$

[see equation (2) of the problem in § 105]. This means that the angular momentum $\mathbf{M}$ of the top precesses with angular velocity $\boldsymbol{\Omega}$, remaining constant in magnitude.

CHAPTER 13

GRAVITATIONAL WAVES

§ 107. Weak gravitational waves

Just as in electrodynamics, in the relativistic theory of gravitation the finite velocity of propagation of interactions results in the possibility of the existence of free gravitational fields that are not linked to bodies—gravitational waves.

We consider the weak gravitational field in vacuum. As in § 105, we introduce the tensor h_{ik}, describing a weak perturbation of the galilean metric:

$$g_{ik} = g_{ik}^{(0)} + h_{ik}. \tag{107.1}$$

Then, to terms of first order in the h_{ik}, the contravariant metric tensor is:

$$g^{ik} = g^{ik(0)} \sim h^{ik}, \tag{107.2}$$

and the determinant of the tensor g_{ik}:

$$g = g^{(0)}(1 + h), \tag{107.3}$$

where $h \equiv h_i^i$; all operations of raising and lowering tensor indices are done with the unperturbed metric $g^{(0)}$.

As already pointed out in § 105, the condition that the h_{ik} be small leaves the possibility of arbitrary transformations of reference system of the form $x'^i = x^i + \xi^i$, with small ξ^i; then

$$h'_{ik} = h_{ik} - \frac{\partial \xi_i}{\partial x^k} - \frac{\partial \xi_k}{\partial x^i}. \tag{107.4}$$

Using this arbitrariness of gauge for the tensor h_{ik}, we impose on it the supplementary condition

$$\frac{\partial \psi_i^k}{\partial x^k} = 0, \quad \psi_i^k = h_i^k - \tfrac{1}{2} \delta_i^k h, \tag{107.5}$$

after which the Ricci tensor takes the simple form (105.11):

$$R_{ik} = \tfrac{1}{2} \, \Box \, h_{ik}. \tag{107.6}$$

where $\Box$ denotes the d'Alembertian operator:

$$\Box = - g^{lm(0)} \partial^2 / \partial x^l \partial x^m = \Delta - \frac{1}{c^2} \frac{\partial^2}{\partial t^2}.$$

The conditions (107.5) still do not fix a unique choice of reference frame: if certain h_{ik} satisfy these conditions, then so will the h'_{ik} of (107.4), if only the ξ^i are solutions of the equations

$$\Box \xi^i = 0. \tag{107.7}$$

Equating (107.6) to zero, we thus find the equations for the gravitational field in vacuum in the form

$$\Box h_i^k = 0. \tag{107.8}$$

This is the ordinary wave equation. Thus gravitational fields, like electromagnetic fields, propagate in vacuum with the velocity of light.

Let us consider a plane gravitational wave. In such a wave the field changes only along one direction in space; for this direction we choose the axis $x^1 = x$. Equation (107.8) then changes to

$$\left(\frac{\partial^2}{\partial x^2} - \frac{1}{c^2} \frac{\partial^2}{\partial t^2} \right) h_i^k = 0, \tag{107.9}$$

the solution of which is any function of $t \pm x/c$ (§ 47).

Consider a wave propagating in the positive direction along the x axis. Then all the quantities h_i^k are functions of $t - x/c$. The auxiliary condition (107.5) in this case gives $\dot\psi_i^1 - \dot\psi_i^0 = 0$, where the dot denotes differentiation with respect to t. This equality can be integrated by simply dropping the sign of differentiation—the integration constants can be set equal to zero since we are here interested only (as in the case of electromagnetic waves) in the varying part of the field. Thus, among the components ψ_i^k that are left, we have the relations

$$\psi_1^1 = \psi_1^0, \qquad \psi_2^1 = \psi_2^0, \qquad \psi_3^1 = \psi_3^0, \qquad \psi_0^1 = \psi_0^0. \tag{107.10}$$

As we pointed out, the conditions (107.5) still do not determine the system of reference uniquely. We can still subject the coordinates to a transformation of the form $x'^i = x^i + \xi^i(t - x/c)$. These transformations can be employed to make the four quantities $\psi_1^0, \psi_2^0, \psi_3^0, \psi_2^2, \psi_3^3$ vanish; from the equalities (107.10) it then follows that the components $\psi_1^1, \psi_1^2, \psi_1^3, \psi_0^0$ also vanish. As for the remaining quantities $\psi_2^3, \psi_2^2, -\psi_3^3$, they cannot be made to vanish by any choice of reference system since, as we see from (107.4), these components do not change under a transformation $\xi_i = \xi_i(t - x/c)$. We note that $\psi = \psi_i^i$ also vanishes, and therefore $\psi_i^k = h_i^k$.

Thus a plane gravitational wave is determined by two quantities, h_{23} and $h_{22} = -h_{33}$. In other words, gravitational waves are transverse waves whose polarization is determined by a symmetric tensor of the second rank in the yz plane, the sum of whose diagonal terms, $h_{22} + h_{33}$, is zero.

For the two independent polarizations we may choose the cases in which one of the two quantities h_{23} and $\frac{1}{2}(h_{22} - h_{33})$ differs from zero. These two polarizations are distinguished from one another by a rotation through $\pi/4$ in the yz plane.

Let us calculate the energy–momentum pseudotensor in a plane gravitational wave. The components t^{ik} are second-order quantities; we must calculate them neglecting terms of still higher order. Since, when $h = 0$, the determinant g differs from $g^{(0)} = -1$ only by terms of second order, we can, in the general formula (96.9), set $g^{ik}_{,l} \approx g^{ik}_{,l} \approx -h^{ik}_{,l}$. For a plane wave all the other nonzero terms in t^{ik} are contained in the term

$$\tfrac{1}{2} g^{il} g^{km} g_{np} g_{qr} g^{nr}{}_{,l} g^{pq}{}_{,m} = \tfrac{1}{2} h_q^{n,i} h_n^{q,k}$$

in curly brackets in (96.9) (as is easily shown by choosing one of the axes of a galilean system of reference along the direction of propagation of the wave). Thus,

$$t^{ik} = \frac{c^4}{32\pi k} h_q^{n,i} h_n^{q,k}. \tag{107.11}$$

The energy flux in the wave is given by the quantities $- c g t^{0\alpha} \approx c t^{0\alpha}$. In a plane wave, propagating along the x^1 axis, in which the nonzero quantities h_{23} and $h_{22} = - h_{33}$ depend only on the difference $t - x/c$, this flux is also along x^1 and is equal to

$$c t^{01} = \frac{c^3}{16\pi k} [h_{23}^2 + \tfrac{1}{4}(h_{22} - h_{33})^2]. \tag{107.12}$$

As initial conditions for the arbitrary field of a gravitational wave we must assign four arbitrary functions of the coordinates: because of the transversality of the field there are just two independent components of $h_{\alpha\beta}$, in addition to which we must also assign their first time derivatives. Although we have made this enumeration here by starting from the properties of a weak gravitational field, it is clear that the result, the number 4, cannot be related to this assumption and applies for any free gravitational field, i.e. for any field which is not associated with gravitating masses.

PROBLEMS

Determine the curvature tensor in a weak plane gravitational wave.

Solution: Calculating R_{iklm} from (105.8), we find the following nonzero components:

$$- R_{0202} = R_{0303} = - R_{1212} = R_{0212} = R_{0331} = R_{3131} = \sigma,$$

$$R_{0203} = - R_{1231} = - R_{0312} = R_{0231} = \mu,$$

where we use the notation

$$\sigma = - \tfrac{1}{2}\ddot{h}_{33} = \tfrac{1}{2}\ddot{h}_{22}, \qquad \mu = - \tfrac{1}{2}\ddot{h}_{23}.$$

In terms of the three-dimensional tensors $A_{\alpha\beta}$ and $B_{\alpha\beta}$ in (92.15), we have:

$$A_{\alpha\beta} = \begin{pmatrix} 0 & 0 & 0 \\ 0 & -\sigma & \mu \\ 0 & \mu & \sigma \end{pmatrix}, \quad B_{\alpha\beta} = \begin{pmatrix} 0 & 0 & 0 \\ 0 & \mu & \sigma \\ 0 & \sigma & -\mu \end{pmatrix}.$$

By a suitable rotation of the x^2, x^3 axes, we can make one of the quantities σ or μ vanish (at a given point of four-space); if we make σ vanish in this way, we reduce the curvature tensor to the degenerate Petrov type II (type N).

§ 108. Gravitational waves in curved space-time

Just as we have treated the propagation of gravitational waves "on the background" of a flat space-time, we can consider weak perturbations relative to an arbitrary (nongalilean) "unperturbed" métric $g_{ik}^{(0)}$. Also anticipating other possible applications, we shall write the necessary formulas in a more general form.

Again taking the g_{ik} in the form (107.1), we find the first order correction to the Christoffel symbols expressed in terms of the h_{ik}:

$$\Gamma^{i(1)}_{kl} = \tfrac{1}{2}(h^i_{k;l} + h^i_{l;k} - h_{kl}{}^{;i}),\tag{108.1}$$

which can be verified by direct calculation (here, as in the sequel, all tensor operations of raising and lowering of indices, and covariant differentiation, are done with the nongalilean metric $g^{(0)}_{ik}$). We find for the corrections to the curvature tensor:

$$R^{i(1)}{}_{klm} = \tfrac{1}{2}(h^i_{k;m;l} + h^i_{m;k;l} - h_{km}{}^{;i}{}_{;l} - h^i_{k;l;m} - h^i_{l;k;m} + h_{kl}{}^{;i}{}_{;m}).\tag{108.2}$$

The corrections to the Ricci tensor are then

$$R^{(1)}{}_{ik} = R^{l(1)}{}_{ilk} = \tfrac{1}{2}(h^l_{i;k;l} + h^l_{k;i;l} - h_{ik}{}^{;l}{}_{;l} - h_{;i;k}).\tag{108.3}$$

The corrections to the mixed components of the Ricci tensor are obtained from the relations

$$R^{k(0)}{}_{,i} + R^{k(1)}{}_i = (R^{(0)}{}_{il} + R^{(1)}{}_{il})(g^{kl(0)} - h^{kl}),$$

so that

$$R^{k(1)}{}_i = g^{kl(0)}R^{(1)}{}_{il} - h^{kl}R^{(0)}{}_{il}.\tag{108.4}$$

The exact metric in vacuum must satisfy the exact Einstein equations $R_{ik} = 0$. Since the unperturbed metric $g^{(0)}_{ik}$ satisfies the equations $R^{(0)}{}_{ik} = 0$, we find for the perturbation, $R^{(1)}{}_{ik} = 0$, i.e.,

$$h^l_{i;k;l} + h^l_{k;i;l} - h_{ik}{}^{;l}{}_{;l} - h_{;i;k} = 0.\tag{108.5}$$

In the general case of arbitrary gravitational waves, simplification of this equation to a form like (107.8) is not possible. This can, however, be done in the important case of waves of high frequency: when the wavelength λ and the oscillation period λ/c are small compared to the characteristic distances L and times L/c over which the "background field" changes. Each differentiation of a component h_{ik} increases the order of the quantity by a factor L/λ relative to derivatives of the unperturbed metric $g^{(0)}_{ik}$. If we limit the accuracy to terms of the two highest orders [$(L/\lambda)^2$ and (L/λ)] we can interchange the orders of differentiation; in fact, the difference

$$h^l_{i;k;l} - h^l_{i;l;k} \approx h^l_m R^{m(0)}{}_{ikl} - h^m_i R^{l(0)}{}_{mkl}$$

is of order $(L/\lambda)^0$, whereas each of the expressions $h^l_{i;k;l}$ and $h^l_{i;l;k}$ contains terms of both higher orders. Imposing on h_{ik} the supplementary conditions

$$\psi^k_{i;k} = 0\tag{108.6}$$

[analogous to (107.5)], we get the equation

$$h_{ik}{}^{;l}{}_{;l} = 0.\tag{108.7}$$

which generalizes (107.8).

For the reasons given in § 107, the condition (108.6) does not fix a unique choice of coordinates. They can still be subjected to a transformation $x'^i = x^i + \xi^i$, where the small quantities ξ^i satisfy the equation $\xi^{i;k}{}_{;k} = 0$. These transformations can be used, in particular, to impose on the h_{ik} the condition $h \equiv h^i_i = 0$. Then $\psi^k_i = h^k_i$, so that the h^k_i are subjected to conditions

$$h^k_{i;k} = 0, \quad h = 0. \tag{108.8}$$

After this the set of admissible transformations is reduced to the requirement $\xi^i_{;i} = 0$.

The pseudotensor t^{ik} contains, in addition to the unperturbed part $t^{ik(0)}$, terms of various orders in the h_{ik}. We arrive at an expression analogous to (107.11) if we consider the quantities t^{ik} averaged over regions of four-space with dimensions large compared to λ but small compared to L. Such an averaging (which we denote by the angular brackets $\langle ... \rangle$) does not affect the $g^{(0)}_{ik}$ and annihilates all quantities that are linear in the rapidly oscillating quantities h_{ik}. Of the quadratic terms, we preserve only the terms of higher (second) order in $1/\lambda$; these are the terms quadratic in the derivatives $h_{ik,l} \equiv \partial h_{ik}/\partial x^l$.

To this accuracy, all terms in t^{ik} that are expressed as four-divergences can be dropped. In fact, the integrals of such quantities over a region of four-space (the region of averaging) are transformed by Gauss' theorem, as a result of which their order of magnitude in $1/\lambda$ is reduced by unity. In addition, those terms drop out which vanish because of (108.7) and (108.8) after integration by parts. Thus, integrating by parts and dropping integrals of four-divergences, we find:

$$\langle h^{ln}_{,p} h^p_{l,n} \rangle = - \langle h^{ln} h^p_{l,p,n} \rangle = 0 \,,$$

$$\langle h^{il}_{,n} h^{k,n}_l \rangle = - \langle h^{il} h^{k,n}_{l,n} \rangle = 0 \,.$$

As a result the only second-order terms that remain are

$$\langle t^{ik(2)} \rangle = \frac{c^4}{32\pi k} \langle h^{n,i}_q h^{q,k}_n \rangle . \tag{108.9}$$

We note that to this same accuracy, $\langle t_i^{i(2)} \rangle^0 = 0$.

Since it has a definite energy, the gravitational wave is iteself the source of some additional gravitational field. Like the energy producing it, this field is a second-order effect in the h_{ik}. But in the case of high-frequency gravitational waves the effect is significantly strengthened: the fact that the pseudotensor t^{ik} is quadratic in the derivatives of the h_{ik} introduces the large factor λ^{-2}. In such a case we may say that the wave itself produces the background field on which it propagates. This field is conveniently treated by carrying out the averaging described above over regions of four-space with dimensions large compared to λ. Such an averaging smooths out the short-wave "ripple" and leves the slowly varying background metric (R.A. Isaacson, 1968).

To derive the equation determining this metric, we must, in expanding the R_{ik}, keep not only linear terms but also quadratic terms in h_{ik}: $R_{ik} = R^{(0)}_{ik} + R^{(1)}_{ik} + R^{(2)}_{ik}$. As already pointed out, the averaging does not affect the zero-order terms. Thus, the averaged field equations $\langle R_{ik} \rangle = 0$ take the form

$$R^{(0)}_{ik} = - \langle R^{(2)}_{ik} \rangle, \tag{108.10}$$

where we should keep only terms of second-order in $1/\lambda$ in $R^{(2)}_{ik}$. They are easily found from the identity (96.7). The terms quadratic in h_{ik} that arise on the right side of this identity, and have the form of a four-divergence, vanish (to the accuracy considered) when the averaging is done, and there remains

$$\langle (R^{ik} - \tfrac{1}{2} g^{ik} R)^{(2)} \rangle = - \frac{8\pi k}{c^4} \langle t^{ik(2)} \rangle,$$

or, since $\langle t^{ik(2)} \rangle = 0$, to this same accuracy:

$$\langle R_{ik}^{(2)} \rangle = - \frac{8\pi k}{c^4} \langle t_{ik}^{(2)} \rangle.$$

Finally, using (108.9), we get eq. (108.10) in the final form

$$R_{ik}^{(0)} = \tfrac{1}{4} \langle h_{q,i}^n h_{n,k}^q \rangle. \tag{108.11}$$

If the "background" is produced entirely by the waves themselves, (108.11) and (108.7) must be solved simultaneously. An estimate of the expressions on both sides of (108.11) shows that in this case the radius of curvature of the background metric, which is of order L, is related to the wavelength λ and the order of magnitude of its field h by $L^{-2} \sim h^2/\lambda^2$, i.e. $\lambda/L \sim h$.

§ 109. Strong gravitational waves

In this section we shall consider the solution of the Einstein equations which is a generalization of the weak, plane gravitational wave in a flat space-time (I. Robinson and H. Bondi, 1957).

We shall look for a solution in which, in a suitable reference frame, all the components of the metric tensor are functions of a single variable, which we call x^0 (without, however, prejudging its character). This condition still permits coordinate transformations of the form

$$x^\alpha \to x^\alpha + \phi^\alpha(x^0), \tag{109.1}$$

$$x^0 \to \phi^0(x^0), \tag{109.2}$$

where ϕ^0, ϕ^α are arbitrary functions.

The character of the solution depends essentially on whether we can make all the $g_{0\alpha}$ vanish by using the three transformations (109.1). This can be done if the determinant $| g_{\alpha\beta} | \neq 0$. In fact, under the transformation (109.1), $g_{0\alpha} \to g_{0\alpha} + g_{\alpha\beta} \dot{\phi}^\beta$ (where the dot denotes differentiation with respect to x^0); if $| g_{\alpha\beta} | \neq 0$, the system of equations

$$g_{0\alpha} + g_{\alpha\beta} \phi^\beta = 0$$

determines the $\phi^\beta(x^0)$ that accomplish the required transformation. Such a case will be treated in § 117; here we shall be interested in the solution in which

$$| g_{\alpha\beta} | = 0. \tag{109.3}$$

In this case there is no reference system in which all the $g_{0\alpha} = 0$. Instead, however, the four transformations (109.1)–(109.2) can be used to make

$$g_{01} = 1, \quad g_{00} = g_{02} = g_{03} = 0. \tag{109.4}$$

Here the variable x^0 has "lightlike" character. for $dx^\alpha = 0$, $dx^0 \neq 0$, the interval $ds = 0$; we shall denote the variable x^0 chosen in this way by $x^0 = \eta$. Under the conditions (109.4) the line element can be written in the form

$$ds^2 = 2dx^1\, d\eta + g_{ab}(dx^a + g^a dx^1)(dx^b + g^b dx^1). \tag{109.5}$$

Throughout this section, the indices a, b, c, ... take on values 2, 3; $g_{ab}(\eta)$ can be regarded as a two-dimensional tensor. Calculation of the quantities R_{ab} leads to the following field equations:

$$R_{ab} = - \tfrac{1}{2} g_{ac} \dot{g}^c g_{bd} \dot{g}^d = 0.$$

It then follows that $g_{ac}\dot{g}^c = 0$, or $\dot{g}^c = 0$, i.e. $g^c = $ const. We can, therefore, by a transformation $x^a + g^a x^1 \to x^a$, bring the metric to the form

$$ds^2 = 2dx^1\, d\eta + g_{ab}(\eta)dx^a dx^b. \tag{109.6}$$

The determinant $-g$ of this metric tensor coincides with the determinant $|g_{ab}|$, while the only nonzero Christoffel symbols are the following:

$$\Gamma^a_{b0} = \tfrac{1}{2}\kappa^a_b, \quad \Gamma^1_{ab} = -\tfrac{1}{2}\kappa_{ab},$$

where we have introduced the two-dimensional tensor $\kappa_{ab} = \dot{g}_{ab}$, $\kappa^b_a = g^{bc}\kappa_{ac}$. Of all the components of the Ricci tensor, the only one that does not vanish identically is R_{00}, so that we have the equation

$$R_{00} = -\tfrac{1}{2}\dot{\kappa}^a_b - \tfrac{1}{4}\kappa^b_a\kappa^a_b = 0. \tag{109.7}$$

Thus, the three functions $g_{22}(\eta)$, $g_{23}(\eta)$, $g_{33}(\eta)$ must satisfy just one equation. Therefore two of them can be chosen arbitrarily. It is convenient to write (109.7) in another form writing the g_{ab} in the form

$$g_{ab} = -\chi^2\gamma_{ab}, \quad |\gamma_{ab}| = 1. \tag{109.8}$$

Then the determinant $-g = |g_{ab}| = \chi^4$, and substitution in (109.7) gives, after simple transformations,

$$\ddot{\chi} + \tfrac{1}{8}(\dot{\gamma}_{ac}\gamma^{bc})(\dot{\gamma}_{bd}\gamma^{ad})\chi = 0 \tag{109.9}$$

(γ^{ab} is the two-dimensional tensor reciprocal to γ_{ab}). If we assign arbitrary functions $\gamma_{ab}(\eta)$ (related to one another through the relation $|\gamma_{ab}| = 1$) these equations determine the function $\chi(\eta)$.

We thus arrive at a solution containing two arbitrary functions. It is easy to see that it is a generalization of the case considered in § 107 of a weak plane gravitational wave propagating in one direction.† The latter is obtained if we make the transformation

$$\eta = \frac{t+x}{\sqrt{2}}, \quad x^1 = \frac{t-x}{\sqrt{2}}$$

and set $\gamma_{ab} = \delta_{ab} + h_{ab}(\eta)$ (where the h_{ab} are small quantities, subject to the condition $h_{22} + h_{33} = 0$) and $\chi = 1$; a constant value of χ satisfies (109.9) if we neglect small second-order terms.

Suppose that a weak gravitational wave of finite extent (a "wave packet") is passing some point x. At the beginning of the passage we have $h_{ab} = 0$, $\chi = 1$; at the end of the passage we again have $h_{ab} = 0$, $\partial^2\chi/\partial t^2 = 0$, but the inclusion of second order terms in (109.9) leads to the appearance of a nonzero negative value of $\partial\chi/\partial t$:

$$\partial\chi/\partial t \approx -\frac{1}{8}\int \left(\frac{\partial h_{ab}}{\partial t}\right)^2 dt < 0$$

(the integral is taken over the time of passage of the wave). Thus, after the wave has passed, $\chi = 1 - $ const $\cdot\, t$, and after a finite time interval, χ changes its sign. But vanishing of χ means

---

† A solution of similar character in a larger number of variables is given in I. Robinson and A. Trautman, *Phys. Rev. Lett.* **4**, 431 (1960); *Proc. Roy. Soc.* **A 265**, 463 (1962).

vanishing of the metric determinant g, i.e. a singularity in the metric. This singularity, however, is not physical in character; it is related only to the unsatisfactory nature of the reference frame, "spoiled" by the passing gravitational wave, and can be eliminated by a suitable transformation; after passage of the gravitational wave, the space-time does actually become flat again.

This can be shown directly. If we measure the variable η from its value corresponding to the singular point, $\chi = \eta$, so that

$$ds^2 = 2d\eta\, dx^1 - \eta^2[(dx^2)^2 + (dx^3)^2].$$

After the transformation

$$\eta x^2 = y, \quad \eta x^3 = z, \quad x^1 = \xi - \frac{y^2 + z^2}{2\eta},$$

we get

$$ds^2 = 2d\eta\, d\xi - dy^2 - dz^2,$$

and the substitution $\eta = (t + x)/\sqrt{2}$, $\xi = (t - x)/\sqrt{2}$ finally brings the metric to galilean form.

This property of the gravitational wave—the creation of a fictitious singularity, is, of course, not related to the fact that the wave is weak; it also applies to the general solution of (109.7); just as in the example considered, near the singularity $\chi \sim \eta$, i.e. $-g \sim \eta^4$.†

PROBLEM

Find the condition for a metric of the form

$$ds^2 = dt^2 - dx^2 - dy^2 - dz^2 + f(t - x, y, z)(dt - dx)^2$$

to be an exact solution of the Einstein equations for a field in vacuum (A. Peres, 1960).

Solution: The Ricci tensor is calculated most simply in the coordinates $u = (t - x)/\sqrt{2}$, $v = (t + x)/\sqrt{2}$, y, z, in which

$$ds^2 = -dy^2 - dz^2 + 2du\, dv + 2f(u, y, z)\, du^2.$$

Aside from $g_{22} = g_{33} = -1$, the only nonzero components of the metric tensor are $g_{uu} = 2f$, $g_{uv} = 1$; then $g^{vv} = -2f$, $g^{uv} = 1$, while the determinant $g = -1$. A direct calculation with (92.1) gives for the nonzero components of the curvature tensor:

$$R_{yuyu} = \frac{\partial^2 f}{\partial y^2}, \quad R_{zuzu} = -\frac{\partial^2 f}{\partial z^2}, \quad R_{yuzu} = -\frac{\partial^2 f}{\partial y \partial z}.$$

The only nonzero component of the Ricci tensor is $R_{uu} = \Delta f$, where Δ is the Laplacian in the coordinates y, z. Thus the Einstein equation is $\Delta f = 0$, i.e. the function $f(t - x, y, z)$ must be harmonic in the variables y, z.

If f is independent of y and z, or linear in them, there is no field—the space time is flat (the curvature tensor vanishes). The function $f(u, y, z) = yzf_1(u) + \frac{1}{2}(y^2 - z^2) f_2(u)$, which is quadratic in y and z, corresponds to a plane wave propagating in the positive x direction; the curvature tensor in such a field depends only on $t-x$:

$$R_{yuzu} = -f_1(u), \quad r_{yuyu} = -R_{zuzu} = -f_2(u).$$

---

† This can be shown using (109.7) in precisely the same way as in § 97 for the analogous three-dimensional equation in the synchronous reference frame. Just as there, the appearance of a fictitious singularity is related to the crossing of coordinate curves.

Corresponding to the two possible polarizations of the wave, the metric contains two arbitrary functions $f_1(u)$ and $f_2(u)$.

§ 110. Radiation of gravitational waves

Let us consider next a weak gravitational field, produced by arbitrary bodies, moving with velocities small compared with the velocity of light.

Because of the presence of matter, the equations of the gravitational field will differ from the simple wave equation of the form $\square\, h_i^k = 0$ (107.8) by having, on the right side of the equality, terms coming from the energy-momentum tensor of the matter. We write these equations in the form

$$\frac{1}{2} \square\, \psi_i^k = \frac{8\pi k}{c^4}\, \tau_i^k, \tag{110.1}$$

where we have introduced in place of the h_i^k the more convenient quantities

$$\psi_i^k = h_i^k - \tfrac{1}{2}\delta_i^k h,$$

and where τ_i^k denotes the auxiliary quantities which are obtained upon going over from the exact equations of gravitation to the case of a weak field in the approximation we are considering. It is easy to verify that the components τ_0^0 and τ_α^0 are obtained directly from the corresponding components T_i^k by taking out from them the terms of the order of magnitude in which we are interested; as for the components τ_β^α, they contain along with terms obtained from the T_β^α, also terms of second order from $R_i^k - \tfrac{1}{2}\delta_i^k R$.†

The quantities ψ_i^k satisfy the condition (107.5) $\partial\psi_i^k/\partial x^k = 0$. From (110.1) it follows that this same equation holds for the τ_i^k:

$$\frac{\partial \tau_i^k}{\partial x^k} = 0. \tag{110.2}$$

This equation here replaces the general relation $T_{i;k}^k = 0$.

Using the equations which we have obtained, let us consider the problem of the energy radiated by moving bodies in the form of gravitational waves. The solution of this problem requires the determination of the gravitational field in the "wave zone", i.e. at distances large compared with the wavelength of the radiated waves.

In principle, all the calculations are completely analogous to those which we carried out for electromagnetic waves. Equation (110.1) for a weak gravitational field coincides in form with the equation of the retarded potentials (§ 62). Therefore we can immediately write its general solution in the form

$$\psi_i^k = -\frac{4k}{c^4} \int (\tau_i^k)_{t-\frac{R}{c}}\, \frac{dV}{R}. \tag{110.3}$$

---

† From eqs. (110.1) we can again obtain the formulas (106.1)–(106.2) that were used in § 106 for the weak constant field far from bodies. In the first approximation we neglect terms with second time derivatives (containing $1/c^2$), and of all the components of τ_i^k, only $\tau_0^0 = \mu c^2$ remains. The solution of the equations $\Delta\psi_\alpha^\beta = 0$, $\Delta\psi_0^\alpha = 0$, $\Delta\psi_0^0 = 16\pi k\,\mu/c^2$ that vanishes at infinity is $\psi_\alpha^\beta = 0$, $\psi_0^\alpha = 0$, $\psi_0^0 = 4\phi/c$, where ϕ is the Newtonian gravitational potential; cf. (99.2). One then finds for the tensor $h_i^k = \psi_i^k - \tfrac{1}{2}\psi\delta_i^k$ the values (106.1)–(106.2).

Since the velocities of all the bodies in the system are small, we can write, for the field at large distances from the system (see §§ 66 and 67),

$$\psi_i^k = - \frac{4k}{c^4 R_0} \int (\tau_i^k)_{t - \frac{R_0}{c}} \, dV,$$
(110.4)

where R_0 is the distance from the origin, chosen anywhere in the interior of the system. From now on we shall, for brevity, omit the index $t - (R_0/c)$ in the integrand.

For the evaluation of these integrals we use equation (110.2). Dropping the index on the τ_i^k and separating space and time components, we write (110.2) in the form

$$\frac{\partial \tau_{\alpha\gamma}}{\partial x^\gamma} - \frac{\partial \tau_{\alpha 0}}{\partial x^0} = 0, \qquad \frac{\partial \tau_{0\gamma}}{\partial x^\gamma} - \frac{\partial \tau_{00}}{\partial x^0} = 0.$$
(110.5)

Multiplying the first equation by x^β, we integrate over all space,

$$\frac{\partial}{\partial x^0} \int \tau_{\alpha 0} x^\beta dV = \int \frac{\partial \tau_{\alpha\gamma}}{\partial x^\gamma} x^\beta \, dV = \int \frac{\partial (\tau_{\alpha\gamma} x^\beta)}{\partial x^\gamma} dV - \int \tau_{\alpha\beta} dV.$$

Since at infinity $\tau_{ik} = 0$, the first integral on the right, after transformation by Gauss' theorem, vanishes. Taking half the sum of the remaining equation and the same equation with transposed indices, we find

$$\int \tau_{\alpha\beta} \, dV = - \frac{1}{2} \frac{\partial}{\partial x^0} \int (\tau_{\alpha 0} x^\beta + \tau_{\beta 0} x^\alpha) \, dV.$$

Next, we multiply the second equation of (110.5) by $x^\alpha x^\beta$, and again integrate over all space. An analogous transformation leads to

$$\frac{\partial}{\partial x^0} \int \tau_{00} x^\alpha x^\beta \, dV = - \int (\tau_{\alpha 0} x^\beta + \tau_{\beta 0} x^\alpha) \, dV.$$

Comparing the two results, we find

$$\int \tau_{\alpha\beta} \, dV = \frac{1}{2} \frac{\partial^2}{\partial x_0^2} \int \tau_{00} x^\alpha x^\beta dV.$$
(110.6)

Thus the integrals of all the $\tau_{\alpha\beta}$ appear as expressions in terms of integrals containing only the component τ_{00}. But this component, as was shown earlier, is simply equal to the corresponding component T_{00} of the energy-momentum tensor and can be written to sufficient accuracy [see (99.1)] as:

$$\tau_{00} = \mu c^2.$$
(110.7)

Substituting this in (110.6) and introducing the time $t = x^0/c$, we find for (110.4)

$$\psi_{\alpha\beta} = - \frac{2k}{c^4 R_0} \frac{\partial^2}{\partial t^2} \int \mu x^\alpha x^\beta dV.$$
(110.8)

At large distances from the bodies, we can consider the waves as plane (over not too large regions of space). Therefore we can calculate the flux of energy radiated by the system, say along the direction of the x^1 axis, by using formula (107.12). In this formula there enter the components $h_{23} = \psi_{23}$ and $h_{22} - h_{33} = \psi_{22} - \psi_{33}$. From (110.8), we find for them the expressions[†]

$$h_{23} = - \frac{2k}{3c^4 R_0} \ddot{D}_{23}, \qquad h_{22} - h_{33} = - \frac{2k}{3c^4 R_0} (\ddot{D}_{22} - \ddot{D}_{33}) \tag{110.9}$$

(the dot denotes time differentiation), where we have introduced the mass quadrupole tensor (99.8):

$$D_{\alpha\beta} = \int \mu(3x^\alpha x^\beta - \delta_{\alpha\beta} r^2) \, dV. \tag{110.10}$$

As a result, we obtain the energy flux along the x^1 axis in the form

$$ct^{10} = \frac{k}{36\pi c^5 R_0^2} \left[\left(\frac{\dddot{D}_{22} - \dddot{D}_{33}}{2} \right)^2 + \dddot{D}_{23}^2 \right]. \tag{110.11}$$

The flux of energy into an element of solid angle in the given direction is then obtained by multiplying by $R_0^2 \, do$.

The two terms in this expression correspond to the radiation of waves of two independent polarizations. To write them in invariant form (independent of the choice of the direction of radiation) we introduce the three-dimensional unit polarization tensor $e_{\alpha\beta}$ of the plane gravitational wave, which determines the nonzero components of $h_{\alpha\beta}$ (in the gauge for the h_{ik} in which $h_{0\alpha} = h_{00} = h = 0$). The polarization tensor is symmetric and satisfies the conditions

$$e_{\alpha\alpha} = 0, \quad e_{\alpha\beta} n_\beta = 0, \quad e_{\alpha\beta} e_{\alpha\beta} = 1, \tag{110.12}$$

where $\mathbf{n}$ is a unit vector in the direction of propagation of the wave.

Using this tensor we can write the intensity of radiation of a given polarization into solid angle do in the form

$$dI = \frac{k}{72\pi c^5} (\dddot{D}_{\alpha\beta} e_{\alpha\beta})^2 \, do. \tag{110.13}$$

This expression depends implicitly on the direction of $\mathbf{n}$ through the transversality condition $e_{\alpha\beta} n_\beta = 0$. The total angular distribution for all polarizations is gotten by summing (110.13) over polarizations, or, what is equivalent, averaging over polarization and multiplying by 2 (the number of independent polarizations). The averaging is done using the formula

$$\overline{e_{\alpha\beta} e_{\gamma\delta}} = \tfrac{1}{4} \{ n_\alpha n_\beta n_\gamma n_\delta + (n_\alpha n_\beta \delta_{\gamma\delta} + n_\gamma n_\delta \delta_{\alpha\beta}) -$$
$$- (n_\alpha n_\gamma \delta_{\beta\delta} + n_\beta n_\gamma \delta_{\alpha\delta} + n_\alpha n_\delta \delta_{\beta\gamma} + n_\beta n_\delta \delta_{\alpha\gamma}) -$$
$$- \delta_{\alpha\delta} \delta_{\gamma\delta} + (\delta_{\alpha\gamma} \delta_{\beta\delta} + \delta_{\beta\gamma} \delta_{\alpha\delta}) \tag{110.14}$$

(the expression on the right is a tensor formed from the unit tensor and the components of the vector $\mathbf{n}$; it has the required symmetry in its indices, it gives unity on contraction on pairs of indices α, γ and β, δ, and vanishes after scalar multiplication with $\mathbf{n}$).

The result is

[†] The tensor (110.8) does not satisfy the conditions under which formula (107.12) was derived. However, the transformation of reference frame that brings the h_{ik} to the required gauge does not affect the values of the components of (110.9) that are used here.

$$dI = \frac{k}{36\pi c^5}\left[\frac{1}{4}(\dddot{D}_{\alpha\beta}n_\alpha n_\beta)^2 + \frac{1}{2}\dddot{D}^2_{\alpha\beta} - \dddot{D}_{\alpha\beta}\dddot{D}_{\alpha\gamma}n_\beta n_\gamma\right]do. \tag{110.15}$$

The total radiation in all directions, i.e., the energy loss of the system per unit time $(-d\mathscr{E}/dt)$, can be found by averaging dI/do over all directions and multiplying the result by 4π. The averaging is easily performed using the formulas given in the footnote on p. 205, and gives

$$-\frac{d\mathscr{E}}{dt} = \frac{k}{45c^5}\dddot{D}^2_{\alpha\beta}. \tag{110.16}$$

We note that the radiation of gravitational waves is a fifth order effect in $1/c$. This fact, together with the smallness of the gravitational constant k, makes the usual effects extremely small.

PROBLEMS

1. Two bodies, attracting each other according to Newton's law, move in circular orbits (around) their common centre of inertia). Determine the average (over a rotation period) of the intensity of radiation of gravitational waves and its distribution in polarization and direction.

Solution: Choosing the coordinate origin at the centre of inertia, we have for the radius vectors of the two bodies:

$$\mathbf{r}_1 = \frac{m_2}{m_1 + m_2}\mathbf{r}, \quad \mathbf{r}_2 = -\frac{m_1}{m_1 + m_2}\mathbf{r}, \quad \mathbf{r} = \mathbf{r}_1 - \mathbf{r}_2.$$

The components of the tensor $D_{\alpha\beta}$ are (if the xy plane coincides with the plane of motion):

$$D_{xx} = \mu r^2(3\cos^2\psi - 1), \quad D_{yy} = \mu r^2(3\sin^2\psi - 1),$$

$$D_{xy} = 3\mu r^2\cos\psi\sin\psi, \quad D_{zz} = -\mu r^2,$$

where $\mu = m_1 m_2/(m_1 + m_2)$, ψ is the polar angle of the vector $\mathbf{r}$ in the xy plane. For circular motion $r = \text{const}$, and $\dot\psi = r^{-\frac{3}{2}}\sqrt{k(m_1 + m_2)} \equiv \omega$.

We assign the direction of $\mathbf{n}$ by the polar angle θ and azimuth ϕ, with the polar axis z perpendicular to the plane of the motion. Let us consider the two polarizations for which: (1) $e_{0\phi} = 1/\sqrt{2}$; (2) $e_{\theta\theta} = -e_{\phi\phi} = 1/\sqrt{2}$. Projecting the tensor $D_{\alpha\beta}$ on the directions of the spherical unit vectors $\mathbf{e}_\theta$ and $\mathbf{e}_\phi$, calculating with formula (110.13) and averaging over the time, we find the result for these two cases and for the sum $I = I_1 + I_2$:

$$\frac{\overline{dI_1}}{do} = \frac{k\mu^2\omega^6 r^4}{2\pi c^5}4\cos^2\theta, \quad \frac{\overline{dI_2}}{do} = \frac{k\mu^2\omega^6 r^4}{2\pi c^5}(1 + \cos^2\theta)^2,$$

$$\frac{\overline{dI}}{do} = \frac{k\mu^2\omega^6 r^4}{2\pi c^5}(1 + 6\cos^2\theta + \cos^4\theta),$$

and after integrating over all directions:

$$-\frac{d\mathscr{E}}{dt} = I = \frac{32k\mu^2\omega^6 r^4}{5c^5} = \frac{32k^4 m_1^2 m_2^2(m_1 + m_2)}{5c^5 r^5}, \quad \frac{f_1}{I_2} = \frac{5}{7}.$$

[for calculating the total intensity I alone, we should, of course, have used (110.16)].

The loss of energy from the radiating system leads to a gradual (secular) approach of the two bodies. Since $\mathscr{E}\cdot = -km_1 m_2/2r$, the velocity of approach is

$$\dot{r} = \frac{2r^2}{km_1m_2}\frac{d\mathscr{E}}{dt} = -\frac{64k^3m_1m_2(m_1+m_2)}{5c^5r^3}.$$

2. Find the average (over a rotation period) of the energy radiated in the form of gravitational waves by a system of two bodies moving in elliptical orbits (P. C. Peters and J. Mathews).†

Solution: In contrast to the case of circular motion, the distance r and the angular velocity vary along the orbit according to the laws

$$\frac{a(1-e^2)}{r} = 1 + e \cos \psi, \quad \frac{d\psi}{dt} = \frac{1}{r^2}[k(m_1+m_2)\, a\, (1-e^2)]^{\frac{1}{2}},$$

where e is the eccentricity and a is the semimajor axis of the orbit (cf. *Mechanics*, § 15). A quite lengthy calculation using (110.16) gives:

$$-\frac{d\mathscr{E}}{dt} = \frac{8k^4m_1^2m_2^2(m_1+m_2)}{15a^5c^5(1-e^2)^5}(1 + e \cos \psi)^4 \, [12(1+e\cos\psi)^2 + e^2\sin^2\psi].$$

In averaging over the period of rotation, the integration over t is replaced by integration over ψ, and gives the result:

$$-\frac{\overline{d\mathscr{E}}}{dt} = \frac{32k^4m_1^2m_2^2(m_1+m_2)}{5c^5a^5}\frac{1}{(1-e^2)^{\frac{7}{2}}}\left(1 + \frac{73}{24}e^2 - \frac{37}{96}e^4\right).$$

We note the rapid increase in intensity of radiation with increasing eccentricity of the orbit.

3. Determine the time-averaged rate of loss of angular momentum from a system of bodies in stationary motion and emitting gravitational waves.

Solution: For convenience of writing formulas, we temporarily regard the body as consisting of discrete particles. We represent the average rate of loss of energy of the system as the work of the "frictional forces" **f** acting on the particles:

$$\frac{\overline{d\mathscr{E}}}{dt} = \Sigma \, \overline{\mathbf{f} \cdot \mathbf{v}} \tag{1}$$

(we omit the index labelling the particles). Then the average rate of loss of angular momentum is given by

$$\frac{\overline{dM_\alpha}}{dt} = \Sigma \, \overline{(\mathbf{r} \times \mathbf{f})_\alpha} = \Sigma \, e_{\alpha\beta\gamma} \, \overline{x_\beta f_\gamma} \tag{2}$$

(cf. the derivation of formula (75.7)). To determine **f**, we write

$$\frac{\overline{d\mathscr{E}}}{dt} = -\frac{k}{45c^5}\, \overline{\dddot{D}_{\alpha\beta}\dddot{D}_{\alpha\beta}} = -\frac{k}{45c^5}\, \overline{\dot{D}_{\alpha\beta} D^{(v)}_{\alpha\beta}}.$$

(where we have used the fact that the average values of total time derivatives vanish). Substituting $\dot{D}_{\alpha\beta} = \Sigma m(3x_\alpha v_\beta + 3x_\beta v_\alpha - 2\mathbf{r}\cdot\mathbf{v}\delta_{\alpha\beta})$ and comparing with (1), we find:

$$f_\alpha = -\frac{2k}{15c^5}\, D^{(v)}_{\alpha\beta}\, mx_\beta.$$

Substitution of this expression in (2) gives the result:

$$\frac{\overline{dM_\alpha}}{dt} = -\frac{2k}{45c^5}\, e_{\alpha\beta\gamma} \overline{D^{(v)}_{\beta\delta} D_{\gamma\delta}} = -\frac{2k}{45c^5}\, e_{\alpha\beta\gamma} \overline{\dddot{D}_{\beta\delta}\ddot{D}_{\delta\gamma}}. \tag{3}$$

† For the angular, polarization, and spectral distributions of this radiation, cf. *Phys. Rev.* **131**, 435 (1963).

4. For a system of two bodies moving in elliptical orbits, find the average loss of angular momentum per unit time.

Solution: A calculation with formula (3) of the preceding problem, analogous to that done in problem 2, gives the result:

$$-\frac{\overline{dM_z}}{dt} = \frac{32 k^{\frac{7}{2}} m_1^2 m_2^2 \sqrt{m_1 + m_2}}{5 c^5 a^{\frac{7}{2}}} \frac{1}{(1 - e^2)^2} \left(1 + \frac{7}{8} e^2\right).$$

For circular motion ($e = 0$) the values of $\mathscr{E}$ and $\dot{M}$ are, as they should be, related by $\dot{\mathscr{E}} = \dot{M}\omega$.

CHAPTER 14

RELATIVISTIC COSMOLOGY

§ 111. Isotropic space

The general theory of relativity opens new avenues of approach to the solution of problems related to the properties of the universe on a cosmic scale. The new remarkable possibilities which arise are related to the non-galilean nature of space-time (first noted by Einstein in 1917).

Before proceeding to a systematic construction of relativistic cosmological models, we make the following comment concerning the basic field equations from which we start.

The requirements formulated in § 93 as conditions for determining the action for the gravitational field will still be satisfied if we add a constant term to the scalar G, i.e. if we set

$$S_g = -\frac{c^3}{16\pi k} \int (G + 2\Lambda)\sqrt{-g}\, d\Omega,$$

where Λ is a new constant (with dimensions cm^{-2}). Such a change leads to the appearance in the Einstein equations of an auxiliary term Λg_{ik}:

$$R_{ik} - \tfrac{1}{2}R g_{ik} = \frac{8\pi k}{c^4} T_{ik} + \Lambda g_{ik}.$$

If one ascribes a small value to the "cosmological constant" Λ, the presence of this term will not significantly affect gravitational waves over not too large regions of space-time, but it will lead to the appearance of new types of "cosmological solutions" which could describe the universe as a whole.† At the present time, however, there are no cogent and convincing reasons, observational or theoretical, for such a change in the form of the fundamental equations of the theory. We emphasize that we are talking about changes that have a profound physical significance: introducing into the Lagrange density a constant term which is generally independent of the state of the field would mean that we ascribe to space-time a curvature which cannot be eliminated in principle and is not associated with either matter or gravitational waves. All of the remaining presentation in this section is therefore based on the Einstein equations in their classical form without the "cosmological constant".

We know that the stars are distributed over space in a very nonuniform manner; they are concentrated into separate star systems (galaxies). But in studying the universe on a "large

---

† In particular, one can now have stationary solutions which do not occur for $\Lambda = 0$. It was precisely for this reason that the "cosmological term" was introduced by Einstein before the discovery by Friedmann of nonstationary solutions of the field equations—cf. below.

382

scale" one should disregard "local" inhomogeneities produced by the condensation of matter into stars and star systems. Thus, by the mass density we should understand the density averaged over regions of space whose dimensions are large compared to the distances between galaxies.

The solutions of the Einstein equations considered later (in §§ 111–114), the so-called isotropic cosmological model (first discovered by A. A. Friedmann in 1922), are based on the assumption of homogeneity and isotropy of the distribution of matter in space. Existing astronomical data do not contradict such an assumption,† and at present there is every reason to believe that in general terms the isotropic model gives an adequate description not only of the present state of the universe but also of a significant part of its evolution in the past. We shall see below that the main feature of this model is its nonstationarity. There is no doubt that this property ("the expanding universe") gives a correct explanation of the phenomenon of the red shift, which is fundamental for the cosmological problem (§ 114).

At the same time it is clear that, by its very nature, the assumption of homogeneity and isotropy of the universe can have only an approximate character, since these properties surely are not valid if we go to a smaller scale. We shall turn to the question of the possible role of inhomogeneities of the universe in various aspects of the cosmological problem in §§ 115–119.

The homogeneity and isotropy of space mean that we can choose a world time so that at each moment the metric of the space is the same at all points and in all directions.

First we take up the study of the metric of the isotropic space as such, disregarding for the moment any possible time dependence. As we did previously, we denote the three-dimensional metric tensor by $\gamma_{\alpha\beta}$, i.e. we write the element of spatial distance in the form

$$dl^2 = \gamma_{\alpha\beta}\, dx^\alpha dx^\beta. \tag{111.1}$$

The curvature of the space is completely determined by its three-dimensional curvature tensor, which we shall denote by $P^\alpha_{\beta\gamma\delta}$ in distinction to the four-dimensional tensor R_{klm}. In the case of complete isotropy, the tensor $P^\mu_{\beta\gamma\delta}$ must clearly be expressible in terms of the metric tensor $\gamma_{\alpha\beta}$ alone. It is easy to see from the symmetry properties of $P^\alpha_{\beta\gamma\delta}$ that it must have the form:

$$P_{\alpha\beta\gamma\delta} = \lambda(\gamma_{\alpha\gamma}\gamma_{\beta\delta} - \gamma_{\alpha\delta}\gamma_{\gamma\beta}), \tag{111.2}$$

where λ is some constant. The Ricci tensor $P_{\alpha\beta} = P^\gamma_{\alpha\gamma\beta}$ is accordingly equal to

$$P_{\alpha\beta} = 2\lambda\gamma_{\alpha\beta} \tag{111.3}$$

and the scalar curvature

$$P = 6\lambda. \tag{111.4}$$

Thus the curvature properties of an isotropic space are determined by just one constant. Corresponding to this there are altogether three different possible cases for the spatial metric: (1) the so-called space of constant positive curvature (corresponding to a positive value of λ), (2) space of constant negative curvature (corresponding to values of $\lambda < 0$), and (3) space with zero curvature ($\lambda = 0$). Of these, the last will be a flat, i.e. euclidean, space.

To investigate the metric it is convenient to start from geometrical analogy, by considering the geometry of isotropic three-dimensional space as the geometry on a hypersurface known

---

† We have in mind data on the distribution of galaxies in space and on the isotropy of the so-called background radio radiation.

to be isotropic, in a fictitious four-dimensional space.† Such a space is a hypersphere; the three-dimensional space corresponding to this has a positive constant curvature. The equation of a hypersphere of radius a in the four-dimensional space x_1, x_2, x_3, x_4, has the form

$$x_1^2 + x_2^2 + x_3^2 + x_4^2 = a^2,$$

and the element of length on it can be expressed as

$$dl^2 = dx_1^2 + dx_2^2 + dx_3^2 + dx_4^2.$$

Considering x_1, x_2, x_3 as the three space coordinates, and eliminating the fictitious co-ordinate x_4 with the aid of the first equation, we get the element of spatial distance in the form

$$dl^2 = dx_1^2 + dx_2^2 + dx_3^2 + \frac{(x_1 dx_1 + x_2 dx_2 + x_3 dx_3)^2}{a^2 - x_1^2 - x_2^2 - x_3^2}. \tag{111.5}$$

From this expression, it is easy to calculate the constant λ in (111.2). Since we know beforehand that $P_{\alpha\beta}$ has the form (111.3) over all space, it is sufficient to calculate it only for points located near the origin, where the $\gamma_{\alpha\beta}$ are equal to

$$\gamma_{\alpha\beta} = \delta_{\alpha\beta} + \frac{x_\alpha x_\beta}{a^2}.$$

Since the first derivatives of the $\gamma_{\alpha\beta}$, and consequently the quantities $\lambda_{\beta\gamma}^\alpha$, vanish at the origin, the calculation from the general formula (92.7) turns out to be very simple and gives the result

$$\lambda = \frac{1}{a^2}. \tag{111.6}$$

We may call the quantity a the "radius of curvature" of the space. We introduce in place of the coordinates x_1, x_2, x_3, the corresponding "spherical" coordinates r, θ, ϕ. Then the line element takes the form

$$dl^2 = \frac{dr^2}{1 - \frac{r^2}{a^2}} + r^2(\sin^2\theta d\phi^2 + d\theta^2). \tag{111.7}$$

The coordinate origin can of course be chosen at any point in space. The circumference of a circle in these coordinates is equal to $2\pi r$, and the surface of a sphere to $4\pi r^2$. The "radius" of a circle (or sphere) is equal to

$$\int_0^r \frac{dr}{\sqrt{1 - r^2/a^2}} = a \sin^{-1}(r/a),$$

that is, is larger than r. Thus the ratio of circumference to radius in this space is less than 2π.

Another convenient form for the dl^2 in "four-dimensional spherical coordinates" is obtained

---

† This four-space is understood to have nothing to do with four-dimensional space-time.

by introducing in place of the coordinate r the "angle" χ according to $r = a \sin \chi$ (χ goes between the limits 0 to π).† Then

$$dl^2 = a^2[d\chi^2 + \sin^2 \chi(\sin^2 \theta \, d\phi^2 + d\theta^2)].$$ (111.8)

The coordinate χ determines the distance from the origin, given by $a\chi$. The surface of a sphere in these coordinates equals $4\pi a^2 \sin^2 \chi$. We see that as we move away from the origin, the surface of a sphere increases, reaching its maximum value $4\pi a^2$ at a distance of $\pi a/2$. After that it begins to decrease, reducing to a point at the "opposite pole" of the space, at distance πa, the largest distance which can in general exist in such a space [all this is also clear from (111.7) if we note that the coordinate r cannot take on values greater than a].

The volume of a space with positive curvature is equal to

$$V = \int_0^{2\pi} \int_0^\pi \int_0^\pi a^3 \sin^2 \chi \sin \theta \, d\chi \, d\theta \, d\phi.$$

so that

$$V = 2\pi^2 a^3.$$ (111.9)

Thus a space of positive curvature turns out to be "closed on itself". Its volume is finite though of course it has no boundaries.

It is interesting to note that in a closed space the total electric charge must be zero. Namely, every closed surface in a finite space encloses on each side of itself a finite region of space. Therefore the flux of the electric field through this surface is equal, on the one hand, to the total charge located in the interior of the surface, and on the other hand to the total charge outside of it, with opposite sign. Consequently, the sum of the charges on the two sides of the surface is zero.

Similarly, from the expression (96.16) for the four-momentum in the form of a surface integral there follows the vanishing of the total four-momentum P^i over all space. Thus the definition of the total four-momentum loses its meaning, since the corresponding conservation law degenerates into the empty identity $0 = 0$.

We now go on to consider geometry of a space having a constant negative curvature. From (111.6) we see that the constant λ is negative if a is imaginary. Therefore all the formulas for a space with negative curvature can be immediately obtained from the preceding ones by replacing a by ia. In other words, the geometry of a space with negative curvature is obtained mathematically as the geometry on a four-dimensional pseudosphere with imaginary radius.

Thus the constant λ is now

$$\lambda = -\frac{1}{a^2},$$ (111.10)

and the element of length in a space of negative curvature has, in coordinates r, θ, ϕ, the form

---

† The "cartesian" coordinates x_1, x_2, x_3, x_4 are related to the four-dimensional spherical coordinates a, θ, ϕ, χ by the relations:

$$x_1 = a \sin \chi \sin \theta \cos \phi, \quad x_2 = a \sin \chi \sin \theta \sin \phi,$$
$$x_3 = a \sin \chi \cos \theta, \quad x_4 = a \cos \chi.$$

$$dl^2 = \frac{dr^2}{1 + \dfrac{r^2}{a^2}} + r^2(\sin^2\theta\, d\phi^2 + d\theta^2), \tag{111.11}$$

where the coordinate r can go through all values from 0 to ∞. The ratio of the circumference of a circle to its radius is now greater than 2π. The expression for dl^2 corresponding to (111.8) is obtained if we introduce the coordinate χ according to $r = a \sinh\chi$ (χ here goes from 0 to ∞). Then

$$dl^2 = a^2\{d\chi^2 + \sinh^2\chi(\sin^2\theta\phi^2 + d\theta^2)\}. \tag{111.12}$$

The surface of a sphere is now equal to $4\pi a^2 \sinh^2\chi$ and as we move away from the origin (increasing χ), it increases without limit. The volume of a space of negative curvature is, clearly, infinite.

PROBLEM

Transform the element of length (111.7) to a form in which it is proportional to its euclidean expression (conformal-euclidean coordinates).

Solution: The substitution

$$r = \frac{r_1}{1 + \dfrac{r_1^2}{4a^2}}$$

leads to the result:

$$dl^2 = \left(1 + \frac{r_1^2}{4a^2}\right)^{-2}(dr_1^2 + r_1^2\, d\theta^2 + r_1^2 \sin^2\theta \cdot d\phi^2).$$

§ 112. The closed isotropic model

Going on now to the study of the space-time metric of the isotropic model, we must first of all make a choice of our reference system. The most convenient is a "co-moving" reference system, moving, at each point in space, along with the matter located at that point. In other words, the reference system is just the matter filling the space; the velocity of the matter in this system is by definition zero everywhere. It is clear that this reference system is reasonable for the isotropic model—for any other choice the direction of the velocity of the matter would lead to an apparent nonequivalence of different directions in space. The time coordinate must be chosen in the manner discussed in the preceding section, i.e. so that at each moment of time the metric is the same over all of the space.

In view of the complete equivalence of all directions, the components $g_{0\alpha}$ of the metric tensor are equal to zero in the reference system we have chosen. Namely, the three components $g_{0\alpha}$ can be considered as the components of a three-dimensional vector which, if it were different from zero, would lead to a nonequivalence of different directions. Thus ds^2 must have the form $ds^2 = g_{00}(dx^0)^2 - dl^2$. The component g_{00} is here a function only of x^0. Therefore we can always choose the time coordinate so that g_{00} reduces to 1. Denoting it by ct, we have

$$ds^2 = c^2\, dt^2 - dl^2. \tag{112.1}$$

This time t is the synchronous proper time at each point in space.

Let us begin with the consideration of a space with positive curvature; from now on we shall, for brevity, refer to the corresponding solution of the Einstein equations of gravitation as the "*closed model*". For dl we use the expression (111.8) in which the "radius of curvature" a is, in general, a function of the time. Thus we write ds^2 in the form

$$ds^2 = c^2\, dt^2 - a^2(t)\, \{d\chi^2 + \sin^2 \chi(d\theta^2 + \sin^2 \theta \cdot d\phi^2)\}. \tag{112.2}$$

The function $a(t)$ is determined by the equations of the gravitational field. For the solution of these equations it is convenient to use, in place of the time, the quantity η defined by the relation

$$c\, dt = a\, d\eta. \tag{112.3}$$

Then ds^2 can be written as

$$ds^2 = a^2(\eta)\, \{d\eta^2 - d\chi^2 - \sin^2 \chi(d\theta^2 + \sin^2 \theta \cdot d\phi^2)\}. \tag{112.4}$$

To set up the field equations we must begin with the calculation of the components of the tensor R_{ik} (the coordinates x^0, x^1, x^2, x^3 are η, χ, θ, ϕ). Using the values of the components of the metric tensor,

$$g_{00} = a^2, \quad g_{11} = -a^2, \quad g_{22} = -a^2 \sin^2 \chi, \quad g_{33} = -a^2 \sin^2 \chi \sin^2 \theta,$$

we calculate the quantities Γ^i_{kl}:

$$\Gamma^0_{00} = \frac{a'}{a}, \quad \Gamma^0_{\alpha\beta} = -\frac{a'}{a^3} g_{\alpha\beta}, \quad \Gamma^\alpha_{0\beta} = \frac{a'}{a}\delta^\alpha_\beta, \quad \Gamma^0_{\alpha 0} = \Gamma^\alpha_{00} = 0,$$

where the prime denotes differentiation with respect to η. (There is no need to compute the components $\Gamma^\alpha_{\beta\gamma}$ explicitly.) Using these values, we find from the general formula (92.7):

$$R^0_0 = \frac{3}{a^4}(a'^2 - aa'').$$

From the same symmetry arguments as we used earlier for the $g_{0\alpha}$, it is clear from the start that $R_{0\alpha} = 0$. For the calculation of the components R^β_α we note that if we separate in them the terms containing only $g_{\alpha\beta}$ (i.e. only the $\Gamma^\alpha_{\beta\gamma}$), these terms must constitute the components of a three-dimensional tensor $-P^\alpha_\beta$, whose values are already known from (111.3) and (111.6):

$$R^\beta_\alpha = -P^\beta_\alpha + \ldots = -\frac{2}{a^2}\delta^\beta_\alpha + \ldots,$$

where the dots represent terms containing g_{00} in addition to the $g_{\alpha\beta}$. From the computation of these latter terms we find:

$$R^\beta_\alpha = -\frac{1}{a^4}(2a^2 + a'^2 + aa'')\delta^\beta_\alpha,$$

so that

$$R = R^0_0 + R^\alpha_\alpha = -\frac{6}{a^3}(a + a'').$$

Since the matter is at rest in the frame of reference we are using $u^\alpha = 0$, $u^0 = 1/a$, and we have from (94.9) $T_0^0 = \varepsilon$, where ε is the energy density of the matter. Substituting these expressions in the equation

$$R_0^0 - \frac{1}{2}R = \frac{8\pi k}{c^4}T_0^0,$$

we obtain:

$$\frac{8\pi k}{c^4}\varepsilon = \frac{3}{a^4}(a^2 + a'^2). \qquad (112.5)$$

Here there enter two unknown functions ε and a; therefore we must obtain still another equation. For this is convenient to choose (in place of the spatial components of the field equations) the equation $T_{0;\,i}^i = 0$, which is one of the four equations (94.7) contained, as we know, in the equations of gravitation. This equation can also be derived directly with the help of thermodynamic relations, in the following fashion.

When in the field equations we use the expression (94.9) for the energy-momentum tensor, we are neglecting all those processes which involve energy dissipation and lead to an increase in entropy. This neglect is here completely justified, since the auxiliary terms which should be added to T_i^k in connection with such processes of energy dissipation are negligibly small compared with the energy density ε, which contains the rest energy of the material bodies.

Thus in deriving the field equations we may consider the total entropy as constant. We now use the well-known thermodynamic relation $d\mathscr{E} = T\,dS - p\,dV$, where $\mathscr{E}$, S, V, are the energy, entropy, and volume of the system, and p, T, its pressure and temperature. At constant entropy, we have simply $d\mathscr{E} = -p\,dV$. Introducing the energy density $\varepsilon = \mathscr{E}/V$ we easily find

$$d\varepsilon = -(\varepsilon + p)\frac{dV}{V}.$$

The volume V of the space is, according to (111.9), proportional to the cube of the radius of curvature a. Therefore $dV/V = 3da/a = 3d(\ln a)$, and we can write

$$-\frac{d\varepsilon}{\varepsilon + p} = 3d(\ln a),$$

or, integrating,

$$3\ln a = -\int \frac{d\varepsilon}{p + \varepsilon} + \text{const} \qquad (112.6)$$

(the lower limit in the integral is constant).

If the relation between ε and p (the "equation of state" of the matter) is known, then equation (112.6) determines ε as a function of a. Then from (112.5) we can determine η in the form

$$\eta = \pm \int \frac{da}{a\sqrt{\dfrac{8\pi k}{3c^4}\varepsilon a^2 - 1}}. \qquad (112.7)$$

Equations (112.6)–(112.7) solve, in general form, the problem of determining the metric in the closed isotropic model.

If the material is distributed in space in the form of discrete macroscopic bodies, then to calculate the gravitational field produced by it, we may treat these bodies as material particles having definite masses, and take no account at all of their internal structure. Considering the velocities of the bodies as relatively small (compared with c), we can set $\varepsilon = \mu c^2$, where μ is the sum of the masses, of the bodies contained in unit volume. For the same reason the pressure of the "gas" made up of these bodies is extremely small compared with ε, and can be neglected (from what we have said, the pressure in the interior of the bodies has nothing to do with the question under consideration). As for the radiation present in space, its amount is relatively small, and its energy and pressure can also be neglected.

Thus, to describe the present state of the Universe in terms of this model, we should use the equation of state for "dustlike" matter,

$$\varepsilon = \mu c^2, \quad p = 0.$$

The integration in (112.6) then gives $\mu a^3 = \text{const}$. This equation could have been written immediately, since it merely expresses the constancy of the sum M of the masses of the bodies in all of space, which should be so for the case of dustlike matter.† Since the volume of space in the closed model is $V = 2\pi^2 a^3$, $\text{const} = M/2\pi^2$. Thus

$$\mu a^3 = \text{const} = \frac{M}{2\pi^2}. \tag{112.8}$$

Substituting (112.8) in equation (112.7) and performing the integration, we get:

$$a = a_0 (1 - \cos \eta), \tag{112.9}$$

where the constant

$$a_0 = \frac{2kM}{3\pi c^2}.$$

Finally, for the relation between t and η we find from (112.3):

$$t = \frac{a_0}{c}(\eta - \sin \eta). \tag{112.10}$$

The equations (112.9)–(112.10) determine the function $a(t)$ in parametric form. The function $a(t)$ grows from zero at $t = 0$ ($\eta = 0$) to a maximum value of $a = 2a_0$, which is reached when $t = \pi a_0/c (\eta = \pi)$, and then decreases once more to zero when $t = 2\pi a_0/c (\eta = 2\pi)$.

For $\eta \ll 1$ we have approximately $a = a_0 \eta^2/2$, $t = a_0 \eta^3/6c$, so that

$$a \approx \left(\frac{9a_0 c^2}{2} \right)^{1/3} t^{2/3}. \tag{112.11}$$

The matter density is

$$\mu = \frac{1}{6\pi k t^2} = \frac{8 \times 10^5}{t^2} \tag{112.12}$$

(where the numerical value is given for density in gm · cm^{-3} and t in sec). We call attention

---

† To avoid misunderstandings (that might arise if one considers the remarks in § 111 that the total four-momentum of a closed universe is zero), we emphasize that M is the sum of the masses of the bodies taken one by one, without taking account of their gravitational interaction.

to the fact that in this limit the function $\mu(t)$ has a universal character in the sense that it does not depend on the parameter a_0.

When $a \to 0$ the density μ goes to infinity. But as $\mu \to \infty$ the pressure also becomes large, so that in investigating the metric in this region we must consider the opposite case of maximum possible pressure (for a given energy density ε), i.e. we must describe the matter by the equation of state

$$p = \frac{\varepsilon}{3}$$

(see the footnote on p. 93). From formula (112.6) we then get:

$$\varepsilon a^4 = \text{const} \equiv \frac{3c^4 a_1^2}{8\pi k} \tag{112.13}$$

(where a_1 is a new constant), after which equations (112.7) and (112.3) give the relations

$$a = a_1 \sin \eta, \quad t = \frac{a_1}{c}(1 - \cos \eta).$$

Since it makes sense to consider this solution only for very large values of ε (i.e. small a), we assume $\eta \ll 1$. Then $a \approx a_1 \eta$, $t \approx a_1 \eta^2/2c$, so that

$$a = \sqrt{2a_1 ct}. \tag{112.14}$$

Then

$$\frac{\varepsilon}{c^2} = \frac{3}{32\pi kt^2} = \frac{4.5 \times 10^5}{t^2} \tag{112.15}$$

(which again contains no parameters).

Thus, here too $a \to 0$ for $t \to 0$, so that the value $t = 0$ is actually a singular point of the space-time metric of the isotropic model (and the same remark applies in the closed model also to the second point at which $a = 0$). We also see from (112.14) that if the sign of t is changed, the quantity $a(t)$ would become imaginary, and its square negative. All four components g_{ik} in (112.2) would then be positive, and the determinant g would be positive. But such a metric is physically meaningless. This means that it makes no sense physically to continue the metric analytically beyond the singularity.

§ 113. The open isotropic model

The solution corresponding to an isotropic space of negative curvature ("*open model*") is obtained by a method completely analogous to the preceding. In place of (112.2), we now have

$$ds^2 = c^2 dt^2 - a^2(t)\{d\chi^2 + \sinh^2 \chi(d\theta^2 + \sin^2 \theta \, d\phi^2)\}. \tag{113.1}$$

Again we introduce in place of t the variable η, according to $c \, dt = a \, d\eta$; then we get

$$ds^2 = a^2(\eta)\{d\eta^2 - d\chi^2 - \sinh^2 \chi(d\theta^2 + \sin^2 \theta \cdot d\phi^2)\}. \tag{113.2}$$

This expression can be obtained formally from (112.4) by changing η, χ, a respectively to $i\eta$, $i\chi$, ia. Therefore the equations of the field can also be gotten directly by this same substitution in equations (112.5) and (112.6). Equation (112.6) retains its previous form:

$$3 \ln a = - \int \frac{d\varepsilon}{\varepsilon + p} + \text{const,} \tag{113.3}$$

while in place of (112.5), we have

$$\frac{8\pi k}{c^2} \varepsilon = \frac{3}{a^4} (a'^2 - a^2). \tag{113.4}$$

Corresponding to this we find, instead of (112.7)

$$\eta = \pm \int \frac{da}{a \sqrt{\frac{8\pi k}{3c^4} \varepsilon a^2 + 1}}. \tag{113.5}$$

For material in the form of dust, we find:†

$$a = a_0 (\cosh \eta - 1), \quad t = \frac{a_0}{c} (\sinh \eta - \eta), \tag{113.6}$$

$$\mu a^3 = \frac{3c^2}{4\pi k} a_0. \tag{113.7}$$

The formulas (113.6) determine the function $a(t)$ in parametric form. In contrast to the closed model, here the radius of curvature changes monotonically, increasing from zero at $t = 0$ ($\eta = 0$) to infinity for $t \to \infty$ ($\eta \to \infty$). Correspondingly, the matter density decreases monotonically from an infinite value when $t = 0$ (when $\eta \ll 1$, the monotonic decrease is given by the same approximate formula (112.12) as in the closed model).

For large densities the solution (113.6)–(113.7) is not applicable, and we must again go to the case $p = \varepsilon/3$. We again get the relation

$$\varepsilon a^4 = \text{const} \equiv \frac{3c^4 a_1^2}{8\pi k}. \tag{113.8}$$

† We note that, by the transformation

$$r = Ae^\eta \sinh \chi, \quad c\tau = Ae^\eta \cosh \chi,$$

$$Ae^\eta = \sqrt{c^2 \tau^2 - r^2}, \quad \tanh \chi = \frac{r}{c\tau},$$

the expression (113.2) is reduced to the "conformal-galilean" form

$$ds^2 = f(r, \tau)[c^2 \, d\tau^2 - dr^2 - r^2(d\theta^2 + \sin^2 \theta \, d\phi^2)].$$

Specifically, in the case of (113.6), setting A equal to $a_0/2$,

$$ds^2 = \left(1 - \frac{a_0}{2\sqrt{c^2 \tau^2 - r^2}} \right)^4 [c^2 d\tau^2 - dr^2 - r^2(d\theta^2 + \sin^2 \theta d\phi^2)]$$

(V. A. Fock, 1955). For large values of $\sqrt{c^2\tau^2 - r^2}$ (which correspond to $\eta \gg 1$), this metric tends toward a galilean form, as was to be expected since the radius of curvature tends toward infinity.

In the coordinates r, θ, ϕ, τ, the matter is not at rest and its distribution is not uniform; the distribution and motion of the matter turns out to be centrally symmetric about any point of space chosen as the origin of coordinates r, θ, ϕ.

and find for the function $a(t)$:

$$a = a_1 \sinh \eta, \qquad t = \frac{a_1}{c}(\cosh \eta - 1)$$

or, when $\eta \ll 1$,

$$a = \sqrt{2a_1 ct} \tag{113.9}$$

[with the earlier formula (112.15) for $\varepsilon(t)$]. Thus in the open model, also, the metric has a singularity (but only one, in contrast to the closed model).

Finally, in the limiting case of the solutions under consideration, corresponding to an infinite radius of curvature of the space, we have a model with a flat (euclidean) space. The interval ds^2 in the corresponding space-time can be written in the form

$$ds^2 = c^2 dt^2 - b^2(t)(dx^2 + dy^2 + dz^2) \tag{113.10}$$

(for the space coordinates we have chosen the "cartesian" coordinates x, y, z). The time-dependent factor in the element of spatial distance does not change the euclidean nature of the space metric, since for a given t this factor is a constant, and can be made unity by a simple coordinate transformation. A calculation similar to those in the previous paragraph leads to the following equations:

$$\frac{8\pi k}{c^2}\varepsilon = \frac{3}{b^2}\left(\frac{db}{dt}\right)^2, \qquad 3 \ln b = - \int \frac{d\varepsilon}{p + \varepsilon} + \text{const.}$$

For the case of low pressures, we find

$$\mu b^3 = \text{const}, \quad b = \text{const } t^{2/3}. \tag{113.11}$$

For small t we must again consider the case $p = \varepsilon/3$, for which we find

$$\varepsilon b^4 = \text{const}, \quad b = \text{const } \sqrt{t}. \tag{113.12}$$

Thus in this case also the metric has a singular point ($t = 0$)

We note that all the isotropic solutions found exist only when the matter density is different from zero; for empty space the Einstein equations have no such solutions.† We also mention that mathematically they are a special case of a more general class of solutions that contain three physically different arbitrary functions of the space coordinates (see the problem).

PROBLEM

Find the general form near the singular point for the metric in which the expansion of the space proceeds "quasihomogeneously", i.e. so that all components $\gamma_{\alpha\beta} = - g_{\alpha\beta}$ (in the synchronous reference system) tend

---

† For $\varepsilon = 0$ we would get from (113.5) $a = a_0 e^\eta = ct$ [whereas the equations (112.7) are meaningless because the roots are imaginary]. But the metric

$$ds^2 = c^2 dt^2 - c^2 t^2 \{d\chi^2 + \sinh^2 \chi(d\theta^2 + \sin^2 \theta \, d\phi^2)\}$$

can be transformed by the substitution $r = ct \sinh \chi$, $\tau = t \cosh \chi$, to the form

$$ds^2 = c^2 d\tau^2 - dr^2 - r^2(d\theta^2 + \sin^2 \theta \, d\phi^2),$$

i.e. to a galilean space-time.

to zero according to the same law. The space is filled with matter with the equation of state $p = \varepsilon/3$ (E. M. Lifshitz and I. M. Khalatnikov, 1960).

Solution: We look for a solution near the singularity ($t = 0$) in the form:

$$\gamma_{\alpha\beta} = ta_{\alpha\beta} + t^2 b_{\alpha\beta} + \ldots, \tag{1}$$

where $a_{\alpha\beta}$ and $b_{\alpha\beta}$ are functions of the (space) coordinates†; below, we shall set $c = 1$. The reciprocal tensor is

$$\gamma^{\alpha\beta} = \frac{1}{t} a^{\alpha\beta} - b^{\alpha\beta},$$

where the tensor $a^{\alpha\beta}$ is reciprocal to $a_{\alpha\beta}$, while $b^{\alpha\beta} = a^{\alpha\gamma} a^{\beta\delta} b_{\gamma\delta}$; all raising and lowering of indices and covariant differentiation is done using the time-independent metric $a_{\alpha\beta}$.

Calculating the left sides of equations (97.11) and (97.12) to the necessary order in $1/t$, we get

$$-\frac{3}{4t^2} + \frac{1}{2t} b = \frac{8\pi k}{3} \varepsilon(-4u_0^2 + 1),$$

$$\frac{1}{2}(b_{;\alpha} - b^\beta_{\alpha;\beta}) = -\frac{32\pi k}{3} \varepsilon u_\alpha u_0$$

(where $b = b^\alpha_\alpha$. Also using the identity

$$1 = u_i u^i \approx u_0^2 - \frac{1}{t} u_\alpha u_\beta a^{\alpha\beta},$$

we find:

$$8\pi k\varepsilon = \frac{3}{4t^2} - \frac{b}{2t}, \qquad u_\alpha = -\frac{t^2}{2}(b_{;\alpha} - b^\beta_{\alpha;\beta}). \tag{2}$$

The three-dimensional Christoffel symbols, and with them the tensor $P_{\alpha\beta}$, are independent of the time in the first approximation in $1/t$; the $P_{\alpha\beta}$ coincide with the expressions obtained when calculating simply with the metric $a_{\alpha\beta}$. Using this, we find that in equation (97.13) the terms of order t^{-2} cancel, while the terms $\sim 1/t$ give

$$P^\beta_\alpha + \frac{3}{4} b^\beta_\alpha + \frac{5}{12} \delta^\beta_\alpha b = 0,$$

from which

$$b^\beta_\alpha = -\frac{4}{3} P^\beta_\alpha + \frac{5}{18} \delta^\beta_\alpha P, \tag{3}$$

(where $P = a^{\beta\gamma} P_{\beta\gamma}$). In view of the identity

$$P^\beta_{\alpha;\beta} - \frac{1}{2} P_{;\alpha} = 0$$

[see (92.10)] the relation

$$b^\beta_{\alpha;\beta} - \frac{7}{9} b_{;\alpha}$$

is valid, so that the u_α can be written in the form:

$$u_\alpha = -\frac{t^2}{9} b_{;\alpha}. \tag{4}$$

† The Friedmann solution corresponds to a special choice of the functions $a_{\alpha\beta}$, corresponding to a space of constant curvature.

Thus, all six functions $a_{\alpha\beta}$ remain arbitrary, while the coefficients $b_{\alpha\beta}$ of the next term in the expansion (1) are determined in terms of them. The choice of the time in the metric (1) is completely determined by the condition $t = 0$ at the singularity; the space coordinates still permit arbitrary transformations that do not involve the time (which can be used, for example, to bring $a_{\alpha\beta}$ to diagonal form). Thus the solution contains all together three "physically different" arbitrary functions.

We note that in this solution the spatial metric is inhomogeneous and anisotropic, while the density of the matter tends to become homogeneous as $t \to 0$. In the approximation (4) the three-dimensional velocity $\mathbf{v}$ has zero curl, while its magnitude tends to zero according to the law

$$v^2 = v_\alpha v_\beta \gamma^{\alpha\beta} \sim t^3.$$

§ 114. The red shift

The main feature characteristic of the solutions we have considered is the nonstationary metric; the radius of curvature of the space is a function of the time. A change in the radius of curvature leads to a change in all distances between bodies in the space, as is already seen from the fact that the element of spatial distance dl is proportional to a. Thus as a increases the bodies in such a space "run away" from one another (in the open model, increasing a corresponds to $\eta > 0$, and in the closed model, to $0 < \eta < \pi$).

From the point of view of an observer located on one of the bodies, it will appear as if all the other bodies were moving in radial directions away from the observer. The speed of this "running away" at a given time t is proportional to the separation of the bodies.

This prediction of the theory must be compared with a fundamental astronomical fact— the red shift of lines in the spectra of galaxies. If we regard this as a Doppler shift, we arrive at the conclusion that the galaxies are receding, i.e. at the present time the Universe is expanding.†

Let us consider the propagation of a light ray in an isotropic space. For this purpose it is simplest to use the fact that along the world line of the propagation of a light signal the interval $ds = 0$. We choose the point from which the light emerges as the origin of coordinates χ, θ, ϕ. From symmetry considerations it is clear that the light ray will propagate "radially", i.e. along a line $\theta = $ const, $\phi = $ const. In accordance with this, we set $d\theta = d\phi = 0$ in (112.4) or (113.2) and obtain $ds^2 = a^2(d\eta^2 - d\chi^2)$. Setting this equal to zero we find $d\eta = \pm d\chi$ or, integrating,

$$\chi = \pm \eta + \text{const.} \tag{114.1}$$

The plus sign applies to a ray going out from the coordinate origin, and the minus sign to a ray approaching the origin. In this form, equation (114.1) applies to the open as well as to the closed model. With the help of the formulas of the preceding section, we can from this express the distance traversed by the beam as a function of the time.

In the open model, a ray of light, starting from some point, in the course of its propagation recedes farther and farther from it. In the closed model, a ray of light, starting out from the initial point, can finally arrive at the "conjugate pole" of the space (this corresponds to a change in χ from 0 to π); during the subsequent propagation, the ray begins to approach the

---

† The conclusion that the bodies are running away with increasing $a(t)$ can only be made if the energy of interaction of the matter is small compared to the kinetic energy of its motion in the recession; this condition is always satified for sufficiently distant galaxies. In the opposite case the mutual separations of the bodies is determined mainly by their interactions; therefore, for example, the effect considered here should have practically no influence on the dimensions of the nebulae themselves, and even less so on the dimensions of stars.

initial point. A circuit of the ray "around the space", and return to the initial point, would correspond to a change of χ from 0 to 2π. From (114.1) we see that then η would also have to change by 2π, which is, however, impossible (except for the one case when the light starts at a moment corresponding to $\eta = 0$). Thus a ray of light cannot return to the starting point after a circuit "around the space".

To a ray of light approaching the point of observation (the origin of coordinates), there corresponds the negative sign on η in equation (114.1). If the moment of arrival of the ray at this point is $t(\eta_0)$, then for $\eta = \eta_0$ we must have $\chi = 0$, so that the equation of propagation of such rays is

$$\chi = \eta_0 - \eta . \tag{114.2}$$

From this it is clear that for an observer located at the point $\chi = 0$, only those rays of light can reach him at the time $\tau(\eta_0)$, which started from points located at "distances" not exceeding $\chi = \eta_0$.

This result, which applies to the open as well as to the closed model, is very essential. We see that at each given moment of time $t(\eta)$, at a given point in space, there is accessible to physical observation not all of space, but only that part of it which corresponds to $\chi \leq \eta$. Mathematically speaking, the "visible region" of the space is the section of the four-dimensional space by the light cone. This section turns out to be finite for the open as well as the closed model (the quantity which is infinite for the open model is its section by the hypersurface $t = $ const, corresponding to the space where all points are observed at one and the same time t). In this sense, the difference between the open and closed models turns out to be much less drastic than one might have thought at first glance.

The farther the region observed by the observer at a given moment of time recedes from him, the earlier the moment of time to which it corresponds. Let us look at the spherical surface which is the geometrical locus of the points from which light started out at the time $t(\eta - \chi)$ and is observed at the origin at the time $t(\eta)$. The area of this surface is $4\pi a^2(\eta - \chi)\sin^2\chi$ (in the closed model), or $4\pi a^2(\eta - \chi)\sinh^2\chi$ (in the open model). As it recedes from the observer, the area of the "visible sphere" at first increases from zero (for $\chi = 0$) and then reaches a maximum, after which it decreases once more, dropping back to zero for $\chi = \eta$ (where $a(\eta - \chi) = a(0) = 0$). This means that the section through the light cone is not only finite but also closed. It is as if it closed at the point "conjugate" to the observer; it can be seen by observing along any direction in space. At this point $\varepsilon \to \infty$, so that matter in all stages of its evolution is, in principle, accessible to observation.

The total amount of *observed* matter is equal in the open model to

$$M_{\text{obs}} = 4\pi \int_0^\eta \mu a^3 \sinh^2 \chi \cdot d\chi .$$

Substituting μa^3 from (113.7), we get

$$M_{\text{obs}} = \frac{3c^2 a_0}{2k}(\sinh \eta \cosh \eta - \eta). \tag{114.3}$$

This quantity increases without limit as $\eta \to \infty$. In the closed model, the increase of M_{obs} is limited by the total mass M; in similar fashion, we find for this case:

$$M_{\text{obs}} = \frac{M}{\pi}(\eta - \sin \eta \cos \eta). \tag{114.4}$$

As η increases from 0 to π, this quantity increases from 0 to M; the further increase of M_{obs} according to this formula is fictitious, and corresponds simply to the fact that in a "contracting" universe distant bodies would be observed twice (by means of the light "circling the space" in the two directions).

Let us now consider the change in the frequency of light during its propagation in an isotropic space. For this we first point out the following fact. Let there occur at a certain point in space two events, separated by a time interval $dt = (1/c)\, a(\eta)\, d\eta$. If at the moments of these events light signals are sent out, which are observed at another point in space, then between the moments of their observation there elapses a time interval corresponding to the same change $d\eta$ in the quantity η as for the starting point. This follows immediately from equation (114.1), according to which the change in the quantity η during the time of propagation of a light ray from one point to another depends only on the difference in the coordinates χ for these points. But since during the time of propagation the radius of curvature a changes, the time interval t between the moments of sending out of the two signals and the moments of their observation are different; the ratio of these intervals is equal to the ratio of the corresponding values of a.

From this it follows, in particular, that the periods of light vibrations, measured in terms of the world time t, also change along the ray, proportionally to a. Thus, during the propagation of a light ray, along its path,

$$\omega a = \text{const.} \tag{114.5}$$

Let us suppose that at the time $t(\eta)$ we observe light emitted by a source located at a distance corresponding to a definite value of the coordinate χ. According to (114.1), the moment of emission of this light is $t(\eta - \chi)$. If ω_0 is the frequency of the light at the time of emission, then from (114.5), the frequency ω observed by us is

$$\omega = \omega_0 \frac{a(\eta - \chi)}{a(\eta)}. \tag{114.6}$$

Because of the monotonic increase of the function $a(\eta)$, we have $\omega < \omega_0$, that is, a decrease in the light frequency occurs. This means that when we observe the spectrum of light coming toward us, all of its lines must appear to be shifted toward the red compared with the spectrum of the same matter observed under ordinary conditions. The "red shift" phenomenon is essentially the Doppler effect of the galaxies "running away" from each other.

The magnitude of the red shift measured, for example, as the ratio ω/ω_0 of the displaced to the undisplaced frequency, depends (for a given time of observation) on the distance at which the observed source is located [in relation (114.6) there enters the coordinate χ of the light source]. For not too large distances, we can expand $a(\eta - \chi)$ in a power series in χ, limiting ourselves to the first two terms:

$$\frac{\omega}{\omega_0} = 1 - \chi \frac{a'(\eta)}{a(\eta)}$$

(the prime denotes differentiation with respect to η). Further, we note that the product $\chi a(\eta)$ is here just the distance l from the observed source. Namely, the "radial" line element is equal to $dl = a\, d\chi$; in integrating this relation the question arises of how the distance is to be determined by physical observation. In determining this distance we must take the values of a at different points along the path of integration at different moments of time (integration for $\eta = \text{const}$ would correspond to simultaneous observation of all the points along the path, which is physically not feasible). But for "small" distances we can neglect the change in a along the path of integration and write simply $l = a\chi$, with the value of a taken for the moment of observation.

As a result, we find for the percentage change z in the frequency the following formula:

$$z = \frac{\omega - \omega_0}{\omega_0} = -\frac{H}{c} l, \tag{114.7}$$

where we have introduced the notation

$$H = c \frac{a'(\eta)}{a^2(\eta)} = \frac{1}{a}\frac{da}{dt} \tag{114.8}$$

for the so-called "*Hubble constant*". For a given instant of observation, this quantity is independent of l. Thus the relative shift in spectral lines must be proportional to the distance to the observed light source.

Considering the red shift as a result of a Doppler effect, one can determine the corresponding velocity v of recession of the galaxy from the observer. Writing $z = v/c$, and comparing with (114.7), we have

$$v = Hl \tag{114.9}$$

(this formula can also be obtained directly by calculating the derivative $v = d(a\chi)/dt$.

Astronomical data confirm the law (114.7), but the determination of the value of the Hubble constant is hampered by the uncertainty in the establishment of a scale of cosmic distances suitable for distant galaxies. The latest determinations give the value

$$H \cong 0.8 \times 10^{-10} \ yr^{-1} = 0.25 \times 10^{-17} \ sec^{-1}, \tag{114.10}$$

$$1/H \approx 4 \times 10^{17} \ sec = 1.3 \times 10^{10} \ yr.$$

It corresponds to an increase in the "velocity of recession" by 75 km/sec for each megaparsec distance.†

Substituting in equation (113.4), $\varepsilon = \mu c^2$ and $H = ca'/a^2$, we get for the open model the following relation:

$$\frac{c^2}{a^2} = H^2 - \frac{8\pi k}{3}\mu. \tag{114.11}$$

Combining this equation with the equality

$$H = \frac{c \sinh \eta}{a_0(\cosh \eta - 1)^2} = \frac{c}{a} \coth \frac{\eta}{2},$$

we obtain

$$\cosh \frac{\eta}{2} = H\sqrt{\frac{3}{8\pi k\mu}} \tag{114.12}$$

For the closed model we would get:

$$\frac{c^2}{a^2} = \frac{8\pi k}{3}\mu - H^2. \tag{114.13}$$

† There also exist estimates leading to a smaller value of H, corresponding to an increase of the velocity of recession by 55 km/sec in each megaparsec; then $1/H \approx 18 \times 10^9$ yr.

$$\cos \frac{\eta}{2} = H \sqrt{\frac{3}{8\pi k \mu}}. \tag{114.14}$$

Comparing (114.11) and (114.13), we see that the curvature of the space is negative or positive according as the difference $(8\pi k/3)\mu - h^2$ is negative or positive. This difference goes to zero for $\mu = \mu_k$, where

$$\mu_k = \frac{3H^2}{8\pi k}. \tag{114.15}$$

With the value (114.10), we get $\mu_k \cong 1 \times 10^{-29}$ g/cm^3. In the present state of astronomical knowledge, the value of the average density of matter in space can be estimated only with very low accuracy. For an estimate, based on the number of galaxies and their average mass, one now takes a value of about 3×10^{-31} g/cm^3. This value is 30 times less than μ_k and thus would speak in favour of the open model. But even if we forget about the doubtful reliability of this number, we should keep in mind that it does not take into account the possible existence of a metagalactic dark gas, which could greatly increase the average matter density.

Let us note here a certain inequality which one can obtain for a given value of the quantity H. For the open model we have $H = \dfrac{c \sinh \eta}{a_0 (\cosh \eta - 1)^2}$,

and therefore

$$t = \frac{a_0}{c} (\sinh \eta - \eta) = \frac{\sinh \eta (\sinh \eta - \eta)}{H (\cosh \eta - 1)^2}.$$

Since $0 < \eta < \infty$, we must have

$$\frac{2}{3H} < t < \frac{1}{H}. \tag{114.16}$$

Similarly, for the closed model we obtain

$$t = \frac{\sin \eta (\eta - \sin \eta)}{H (1 - \cos \eta)^2}.$$

To the increase of $a(\eta)$ there corresponds the interval $0 < \eta < \pi$; therefore we get

$$0 < t < \frac{2}{3H}. \tag{114.17}$$

Next we determine the intensity I of the light arriving at the observer from a source located at a distance corresponding to a definite value of the coordinate χ. The flux density of light energy at the point of observation is inversely proportional to the surface of the sphere, drawn through the point under consideration with centre at the location of the source; in a space of negative curvature the area of the surface of the sphere equals $4\pi a^2 \sinh^2 \chi$. Light emitted by the source during the interval $dt = (1/c) a(\eta - \chi) d\eta$ will reach the point of observation during a time interval

$$dt \frac{a(\eta)}{a(\eta - \chi)} = \frac{1}{c} a(\eta) \, d\eta.$$

Since the intensity is defined as the flux of light energy per unit time, there appears in I a factor $a(\eta - \chi)/a(\eta)$. Finally, the energy of a wave packet is proportional to its frequency [see (53.9)]; since the frequency changes during propagation of the light according to the law (114.5), this results in the factor $a(\eta - \chi)/a(\eta)$ appearing in I once more. As a result, we finally obtain the intensity in the form

$$I = \text{const} \frac{a^2(\eta - \chi)}{a^4(\eta) \sinh^2 \chi}.$$ (114.18)

For the closed model we would similarly obtain

$$I = \text{const} \frac{a^2(\eta - \chi)}{a^4(\eta) \sin^2 \chi}.$$ (114.19)

These formulas determine the dependence of the apparent brightness of an observed object on its distance (for a given absolute brightness). For small χ we can set $a(\eta - \chi) \simeq a(\eta)$, and then $I \sim 1/a^2(\eta)\chi^2 = 1/l^2$, that is, we have the usual law of decrease of intensity inversely as the square of the distance.

Finally, let us consider the question of the so-called proper motions of bodies. In speaking of the density and motion of matter, we have always understood this to be the average density and average motion; in particular, in the system of reference which we have always used, the velocity of the average motion is zero. The actual velocities of the bodies will undergo a certain fluctuation around this average value. In the course of time, the velocities of proper motion of the bodies change. To determine the law of this change, let us consider a freely moving body and choose the origin of coordinates at any point along its trajectory. Then the trajectory will be a radial line, $\theta = \text{const}$, $\phi = \text{const}$. The Hamilton–Jacobi equation (87.6), after substitution of the values of g^{ik}, takes the form

$$\left(\frac{\partial S}{\partial \chi}\right)^2 - \left(\frac{\partial S}{\partial \eta}\right)^2 + m^2 c^2 a^2(\eta) = 0.$$ (114.20)

Since χ does not enter into the coefficients in this equation (i.e., χ is a cyclic coordinate), the conservation law $\partial S/\partial \chi = \text{const}$ is valid. The momentum p of a moving body is equal, by definition, to $p = \partial S/\partial l = \partial S/a \, \partial \chi$. Thus for a moving body the product pa is constant:

$$pa = \text{const}.$$ (114.21)

Introducing the velocity v of proper motion of the body according to

$$p = \frac{mv}{\sqrt{1 - \dfrac{v^2}{c^2}}},$$

we obtain

$$\frac{va}{\sqrt{1 - \dfrac{v^2}{c^2}}} = \text{const}.$$ (114.22)

The law of change of velocity with time is determined by these relations. With increasing a, the velocity v decreases monotonically.

PROBLEMS

1. Find the first two terms in the expansion of the apparent brightness of a galaxy as a function of its red shift; the absolute brightness of a galaxy varies with time according to an exponential law, $I_{abs} = \text{const} \cdot e^{\alpha t}$ (H. Robertson, 1955).

Solution: The dependence on distance χ of the apparent brightness of a galaxy at the "instant" η, is given (for the closed model) by the formula

$$I = \text{const} \cdot e^{\alpha[t(\eta - \chi) - t(\eta)]} \frac{a^2(\eta - \chi)}{a^4(\eta) \sin^2 \chi}.$$

We define the red shift as the relative change in wavelength:

$$z = \frac{\lambda - \lambda_0}{\lambda_0} = \frac{\omega_0 - \omega}{\omega} = \frac{a(\eta) - a(\eta - \chi)}{a(\eta - \chi)}.$$

Expanding I and z in powers of χ [using the functions $a(\eta)$ and $t(\eta)$ from (112.9) and (112.10)] and then eliminating χ from the resulting equations, we find the result:

$$I = \text{const} \frac{1}{z^2} \left[1 - \left(1 - \frac{q}{2} + \frac{\alpha}{H} \right) z \right],$$

where we have introduced the notation

$$q = \frac{2}{1 + \cos \eta} = \frac{\mu}{\mu_k} < 1.$$

For the open model, we get the same formula with

$$q = \frac{2}{1 + \cosh \eta} = \frac{\mu}{\mu_k} < 1.$$

2. Find the leading terms in the expansion of the number of galaxies contained inside a "sphere" of given radius, as a function of the red shift at the boundary of the sphere (where the spatial distribution of galaxies is assumed to be uniform).

Solution: The number N of galaxies at "distances" $\leq \chi$ is (in the closed model)

$$N = \text{const} \cdot \int_0^\chi \sin^2 \chi \, d\chi \cong \text{const} \cdot \chi^3.$$

Substituting the first two terms in the expansion of the function $\chi(z)$, we obtain:

$$N = \text{const} \cdot z^3 \left[1 - \frac{3}{4} (2 + q) z \right].$$

In this form the formula also holds for the open model.

§ 115. Gravitational stability of an isotropic universe

Let us consider the question of the behaviour of small perturbations in the isotropic model, i.e. the question of its gravitational stability (E. M. Lifshitz, 1946). We shall restrict our treatment to perturbations over relatively small regions of space—regions whose linear dimensions are small compared to the radius a.†

† A more detailed presentation of this question, including the investigation of perturbations over regions whose size is comparable to a, is given in *Adv. in Physics* **12**, 208 (1963).

In every such region the spatial metric can be assumed to be euclidean in the first approximation, i.e. the metric (111.8) or (111.12) is replaced by the metric

$$dl^2 = a^2(\eta)(dx^2 + dy^2 + dz^2), \tag{115.1}$$

where x, y, z are cartesian coordinates, measured in units of the radius a. We again use the parameter η as time coordinate.

Without loss of generality we shall again describe the perturbed field in the synchronous reference system, i.e. we impose on the variations δg_{ik} of the metric tensor the conditions $\delta g_{00} = \delta g_{0\alpha} = 0$. Varying the identity $g_{ik}u^i u^k = 1$ under these conditions (and remembering that the unperturbed values of the components of the four-velocity of the matter are $u^0 = 1/a$, $u^\alpha = 0$,† we get $g_{00}u^0\delta u^0 = 0$, so that $\delta u^0 = 0$. The perturbations δu^α are in general different from zero, so that the reference system is no longer co-moving.

We denote the perturbations of the spatial metric tensor by $h_{\alpha\beta} \equiv \delta\gamma_{\alpha\beta} = -\delta g_{\alpha\beta}$. Then $\delta\gamma^{\alpha\beta} = -h^{\alpha\beta}$, where the raising of indices on $h_{\alpha\beta}$ is done by using the unperturbed metric $\gamma_{\alpha\beta}$.

In the linear approximation, the small perturbations of the gravitational field satisfy the equations

$$\delta R_i^k - \tfrac{1}{2}\delta_i^k \delta R = \frac{8\pi k}{c^4}\,\delta T_i^k. \tag{115.2}$$

In the synchronous reference system the variations of the components of the energy-momentum tensor (94.9) are:

$$\delta T_\alpha^\beta = -\delta_\alpha^\beta \delta p, \quad \delta T_0^\alpha = a(p + \varepsilon)\delta u^\alpha, \quad \delta T_0^0 = \delta\varepsilon. \tag{115.3}$$

Because of the smallness of $\delta\varepsilon$ and δp, we can write $\delta p = (dp/d\varepsilon)\delta\varepsilon$, and we obtain the relations:

$$\delta T_\alpha^\beta = -\delta_\alpha^\beta \frac{dp}{d\varepsilon}\,\delta T_0^0. \tag{115.4}$$

Formulas for δR_i^k can be gotten by varying the expression (97.10). Since the unperturbed metric tensor $\gamma_{\alpha\beta} = a^2\delta_{\alpha\beta}$, the unperturbed values are

$$\varkappa_{\alpha\beta} = \frac{2\dot{a}}{a}\,\gamma_{\alpha\beta} = \frac{2a'}{a^2}\,\gamma_{\alpha\beta}, \quad \varkappa_\alpha^\beta = \frac{2a'}{a^2}\,\delta_\alpha^\beta,$$

where the dot denotes differentiation with respect to ct, and the prime with respect to η. The perturbations of $\varkappa_{\alpha\beta}$ and $\varkappa_\alpha^\beta = \varkappa_{\alpha\gamma}\gamma^{\gamma\beta}$ are:

$$\delta\varkappa_{\alpha\beta} = \dot{h}_{\alpha\beta} = \frac{1}{a}h'_{\alpha\beta}, \quad \delta\varkappa_\alpha^\beta = -h^{\beta\gamma}\varkappa_{\alpha\gamma} + \gamma^{\beta\gamma}\dot{h}_{\alpha\gamma} = \dot{h}_\alpha^\beta = \frac{1}{a}h_\alpha^{\beta'},$$

where $h_\alpha^\beta = \gamma^{\beta\gamma}h_{\alpha\gamma}$. For the euclidean metric (115.1) the unperturbed values of the three-dimensional P_α^β are zero. The variations δP_α^β are calculated from formulas (108.3)–(108.4): it is obvious that δP_α^β is expressed in terms of the $\delta\gamma_{\alpha\beta}$ just as the four-tensor δR_{ik} is expressed in terms of the δg_{ik}, all tensor operations being done in the three-dimensional space with the metric (115.1); because this metric is euclidean, all the covariant differentiations reduce to simple differentiations with respect to the coordinates x^α (for the contravariant

---

† In this section we denote unperturbed values of quantities by letters without the auxiliary superscript[(0)].

derivatives we must still divide by a^2). Taking all this into account (and changing from derivatives with respect to t to derivatives with respect to η), we get, after some simple calculations:

$$\delta R_\alpha^\beta = -\frac{1}{2a^2}(h_{\alpha,\ \gamma}^{\gamma,\ \beta} + h_{\gamma,\ \alpha}^{\beta,\ \gamma} - h_{\alpha,\ \gamma}^{\beta,\ \gamma} - h_{,\alpha}^{,\beta}) - \frac{1}{2a^2}h_\alpha^{\beta''} - \frac{a'}{a^3}h_\alpha^{\beta'} - \frac{a'}{2a^3}h'\delta_\alpha^\beta,$$

$$\delta R_0^0 = -\frac{1}{2a^2}h'' - \frac{a'}{2a^3}h',$$

$$\delta R_0^\alpha = \frac{1}{2a^2}(h^{,\alpha} - h_\beta^{\alpha,\beta})',$$

(115.5)

$(h \equiv h_\alpha^\alpha)$. Here both the upper and lower indices following the comma denote simple differentiations with respect to the x^α (we continue to write indices above and below only to retain uniformity of the notation).

We obtain the final equations for the perturbations h_α^β by substituting in (115.4) the components δT_i^k, expressed in terms of the δT_i^k according to (115.2). For these equations it is convenient to choose the equations obtained from (115.4) for $\alpha \neq \beta$, and those obtained by contracting on α, β. They are:

$$(h_{\alpha,\ \gamma}^{\gamma,\ \beta} + h_{\gamma,\ \alpha}^{\beta,\ \gamma} - h_{;,\alpha}^{,\beta} - h_{\alpha,\ \gamma}^{\beta,\ \gamma}) + h_\alpha^{\beta''} + \frac{2a'}{a}h_\alpha^{\beta'} = 0, \quad \alpha \neq \beta,$$

$$\frac{1}{2}(h_{\gamma,\ \delta}^{\delta,\ \gamma} - h_{;\ \gamma}^{,\ \gamma})\left(1 + 3\frac{dp}{d\varepsilon}\right) + h'' + h'\frac{a'}{a}\left(2 + 3\frac{dp}{d\varepsilon}\right) = 0.$$

(115.6)

The perturbations of the density and matter velocity can be determined from the known h_α^β using formulas (115.2)–(115.3). Thus we have for the relative change of the density:

$$\frac{\delta\varepsilon}{\varepsilon} = \frac{c^4}{8\pi k\varepsilon}\left(\delta R_0^0 - \frac{1}{2}\delta R\right) = \frac{c^4}{16\pi k\varepsilon a^2}\left(h_{a,\ \beta}^{\beta,\ \alpha} - h_{;\ \alpha}^{,\ \alpha} + \frac{2a'}{a}h'\right).$$

(115.7)

Among the solutions of equations (115.6) there are some that can be eliminated by a simple transformation of the reference system (without destroying the condition of synchronism), and so do not represent a real physical change of the metric. The form of such solutions can be established by using formulas (1) and (2) in problem 3 of § 97. Substituting the unperturbed values $\gamma_{\alpha\beta} = a^2\delta_{\alpha\beta}$, we get from them the following expressions for fictitious perturbations of the metric:

$$h_\alpha^\beta = f_{0,\alpha}^{\ \ \beta}\int\frac{d\eta}{a} + \frac{a'}{a^2}f_0\delta_\alpha^\beta + (f_{\alpha}^{,\beta} + f^{\beta}_{,\alpha}),$$

(115.8)

where the f_0, f_α are arbitrary (small) functions of the coordinates x, y, z.

Since the metric in the small regions of space we are considering is assumed to be euclidean, an arbitrary perturbation in such a region can be expanded in plane waves. Using x, y, z for cartesian coordinates measured in units of a, we can write the periodic space factor for the plane waves in the form $e^{i\mathbf{n}\cdot\mathbf{r}}$, where $\mathbf{n}$ is a dimensionless vector, which represents the wave vector measured in units of $1/a$ (the wave vector is $\mathbf{k} = \mathbf{n}/a$). If we have a perturbation over a portion of space of dimensions $\sim l$, the expansion will involve waves of length $\lambda = 2\pi a/n \sim l$. If we restrict the perturbations to regions of size $l \ll a$, we automatically assume the number n to be quite large ($n \gg 2\pi$).

Gravitational perturbations can be divided into three types. This classification reduces to a determination of the possible types of plane waves in terms of which the symmetric tensor $h_{\alpha\beta}$ can be represented. We thus obtain the following classification:

1. Using the scalar function

$$Q = e^{i\mathbf{n}\cdot\mathbf{r}},\tag{115.9}$$

we can form the vector $\mathbf{P} = \mathbf{n}Q$ and the tensors†

$$Q_\alpha^\beta = \frac{1}{3}\delta_\alpha^\beta Q, \quad P_\alpha^\beta = \left(\frac{1}{3}\delta_\alpha^\beta - \frac{n_\alpha n^\beta}{n^2}\right)Q.\tag{115.10}$$

These plane waves correspond to perturbations in which, in addition to the gravitational field, there are changes in the velocity and density of the matter, i.e. we are dealing with perturbations accompanied by condensations or rarefactions of the matter. The perturbation of h_α^β is expressed in terms of the tensors Q_α^β and P_α^β, the perturbation of the velocity is expressed in terms of the vector $\mathbf{P}$, and the perturbation of the density, in terms of the scalar Q.

2. Using the transverse vector wave

$$\mathbf{S} = \mathbf{s}\, e^{i\mathbf{n}\cdot\mathbf{r}}, \quad \mathbf{s}\cdot\mathbf{n} = 0,\tag{115.11}$$

we can form the tensor $(n^\beta S_\alpha + n_\alpha S^\beta)$; the scalar corresponding to this does not exist, since $\mathbf{n}\cdot\mathbf{S} = 0$. These waves correspond to perturbations in which, in addition to the gravitational field, we have a change in velocity but no change of the density of the matter; they may be called rotational perturbations.

3. The transverse tensor wave

$$G_\alpha^\beta = g_\alpha^\beta e^{i\mathbf{n}\cdot\mathbf{r}}, \quad g_\alpha^\beta n_\beta = 0.\tag{115.12}$$

We can construct neither a vector nor a scalar by using it. These waves correspond to perturbations of the gravitational field in which the matter remains at rest and uniformly distributed throughout space. In other words, these are gravitational waves in an isotropic universe.

The perturbations of the first type are of principal interest. We set

$$h_\alpha^\beta = \lambda(\eta)P_\alpha^\beta + \mu(\eta)Q_\alpha^\beta, \quad h = \mu Q.\tag{115.13}$$

From (115.7) we find for the relative change of the density

$$\frac{\delta\varepsilon}{\varepsilon} = \frac{c^4}{24\pi k\varepsilon a^2}\left[n^2(\lambda + \mu) + \frac{3a'}{a}\mu'\right]Q.\tag{115.14}$$

The equations for determining λ and μ are gotten by substituting (115.13) in (115.6):

$$\lambda'' + \frac{2a'}{a}\lambda' - \frac{n^2}{3}(\lambda + \mu) = 0,$$

$$\mu'' + \mu'\frac{a'}{a}\left(2 + 3\frac{dp}{d\varepsilon}\right) + \frac{n^2}{3}(\lambda + \mu)\left(1 + 3\frac{dp}{d\varepsilon}\right) = 0.\tag{115.15}$$

---

† We write upper and lower indices on the components of ordinary cartesian tensors only to preserve uniformity of notation.

These equations have the following two partial integrals, corresponding to those fictitious changes of metric that can be eliminated by transforming the reference system:

$$\lambda = -\mu = \text{const},$$ (115.16)

$$\lambda = -n^2 \int \frac{d\eta}{a}, \quad \mu = n^2 \int \frac{d\eta}{a} - \frac{3a'}{a^2}$$ (115.17)

[the first of these is gotten from (115.8) by choosing $f_0 = 0$, $f_\alpha = P_\alpha$; the second by choosing $f_0 = Q$, $f_\alpha = 0$].

In the early stages of expansion of the universe, when the matter is described by the equation of state $p = \varepsilon/3$, we have $a \approx a_1\eta$, $\eta \ll 1$ (in both the open and closed models). Equations (115.15) take the form:

$$\lambda'' + \frac{2}{\eta}\lambda' - \frac{n^2}{3}(\lambda + \mu) = 0, \quad \mu'' + \frac{3}{\eta}\mu' + \frac{2n^2}{3}(\lambda + \mu) = 0.$$ (115.18)

These equations are conveniently investigated separately for the two limiting cases depending on the ratio of the large quantities n and $1/\eta$.

Let us assume first that n is not too large (or that η is sufficiently small), so that $\eta\eta \ll 1$. To the order of accuracy for which the equations (115.18) are valid, we find from them for this case:

$$\lambda = \frac{3C_1}{\eta} + C_2\left(1 + \frac{n^2}{9}\eta^2\right), \quad \mu = -\frac{2n^2}{3}C_1\eta + C_2\left(1 - \frac{n^2}{6}\eta^2\right),$$

where C_1, C_2 are constants; solutions of the form (115.16) and (115.17) are excluded (in the present case these are the solution with $\lambda = -\mu = \text{const}$ and the one with $\lambda + \mu \sim 1/\eta^2$). Calculating $\delta\varepsilon/\varepsilon$ from (115.14) and (112.15), we get the following expressions for the perturbations of the metric and the density:

$$h_\alpha^\beta = \frac{3C_1}{\eta}P_\alpha^\beta + C_2(Q_\alpha^\beta + P_\alpha^\beta),$$

$$\frac{\delta\varepsilon}{\varepsilon} = \frac{n^2}{9}(C_1\eta + C_2\eta^2)Q \quad \text{for} \quad p = \frac{\varepsilon}{3}, \quad \eta \ll \frac{1}{n}.$$ (115.19)

The constants C_1 and C_2 must satisfy conditions expressing the smallness of the perturbation at the time η_0 of its start: we must have $h_\alpha^\beta \ll 1$ (so that $\lambda \ll 1$ and $\mu \ll 1$) and $\delta\varepsilon/\varepsilon \ll 1$. As applied to (115.19) these conditions give the inequalities $C_1 \ll \eta_0$, $C_2 \ll 1$.

In (115.19) there are various terms that increase in the expanding universe like different powers of the radius $a = a_1\eta$. But this growth does not cause the perturbation to become large: if we apply formula (115.19) for an order of magnitude to $\eta \sim 1/n$, we see that (because of the inequalities found above for C_1 and C_2) the perturbations remain small even at the upper limit of application of these formulas.

Now suppose that n is so large that $\eta\eta \gg 1$. Solving (115.18) for this condition, we find that the leading terms in λ and μ are:†

---

† The factor $1/\eta^2$ in front of the exponential is the first term in the expansion in powers of $1/n\eta$. To find it we must consider the first two terms in the expansion simultaneously [which is justified within the limits of accuracy of (115.18)].

$$\lambda = -\frac{\mu}{2} = \text{const} \cdot \frac{1}{\eta^2} e^{in/\eta\sqrt{3}}.$$

We then find for the perturbations of the metric and the density:

$$h_\alpha^\beta = \frac{C}{n^2\eta^2}(P_\alpha^\beta - 2Q_\alpha^\beta)e^{in\eta/\sqrt{3}}, \quad \frac{\delta\varepsilon}{\varepsilon} = -\frac{C}{9}Qe^{in\eta/\sqrt{3}}$$

$$\text{for} \quad p = \frac{\varepsilon}{3}, \quad \frac{1}{n} \ll \eta \ll 1, \tag{115.20}$$

where C is a complex constant satisfying the condition $|C| \ll 1$. The presence of a periodic factor in these expressions is entirely natural. For large n we are dealing with a perturbation whose spatial periodicity is determined by the large wave vector $k = n/a$. Such perturbations must propagate like sound waves with velocity

$$u = \sqrt{\frac{dp}{d(\varepsilon/c^2)}} = \frac{c}{\sqrt{3}}.$$

Correspondingly the time part of the phase is determined, as in geometrical acoustics, by the large integral $\int ku\,dt = n\eta/\sqrt{3}$. As we see, the amplitude of the relative change of density remains constant, while the amplitude of the perturbations of the metric itself decreases like a^{-2} in the expanding universe.†

Now we consider later stages of the expansion, when the matter is already so rarefied that we can neglect its pressure ($p = 0$). We shall limit ourselves to the case of small η, corresponding to that stage of the expansion when the radius a was still small compared to its present value, but the matter was already quite rarefied.

For $p = 0$ and $\eta \ll 1$, we have $a \approx a_0\eta^2/2$, and (115.15) takes the form:

$$\lambda'' + \frac{4}{\eta}\lambda' - \frac{n^2}{3}(\lambda + \mu) = 0,$$

$$\mu'' + \frac{4}{\eta}\mu' + \frac{n^2}{3}(\lambda + \mu) = 0.$$

The solution of these equations is

$$\lambda + \mu = 2C_1, \quad \lambda - \mu = 2n^2\left(\frac{C_1\eta^2}{15} + \frac{2C_2}{\eta^3}\right).$$

Also calculating $\delta\varepsilon/\varepsilon$ by using (115.14) and (112.12), we find:

$$h_\alpha^\beta = C_1(P_\alpha^\beta + Q_\alpha^\beta) + \frac{2n^2C_2}{\eta^3}(P_\alpha^\beta - Q_\alpha^\beta) \quad \text{for} \quad \eta \ll \frac{1}{n},$$

---

† It is easy to verify that (for $p = \varepsilon/3$) $n\eta \sim L/\lambda$, where $L \sim u/\sqrt{k\varepsilon/c^2}$. It is natural that the characteristic length L, which determines the behaviour of perturbations with wavelength $\lambda \ll a$, contains only hydrodynamic quantities—the matter density ε/c^2 and the sound velocity u (and the gravitational constant k). We note that there is a growth of the perturbations when $\lambda \gg L$ [in (115.19)].

$$h_\alpha^\beta = \frac{C_1}{15} n^2 \eta^2 (P_\alpha^\beta - Q_\alpha^\beta) + \frac{2n^2 C_2}{\eta^3} (P_\alpha^\beta - Q_\alpha^\beta) \quad \text{for} \quad \frac{1}{n} \ll \eta \ll 1, \qquad (115.21)$$

$$\frac{\delta\varepsilon}{\varepsilon} = \left(\frac{C_1 n^2 \eta^2}{30} + \frac{C_2 n^2}{\eta^3} \right) Q.$$

We see that $\delta\varepsilon/\varepsilon$ contains terms that increase proportionally with a.† But if $n\eta \ll 1$, then $\delta\varepsilon/\varepsilon$ does not become large even for $\eta \sim 1/n$ because of the condition $C_1 \ll 1$. If, however, $\eta n \gg 1$, then for $\eta \sim 1$ the relative change of density becomes of order $C_1 n^2$, while the smallness of the initial perturbation requires only that $C_1 n^2 \eta_0^2 \ll 1$. Thus, although the growth of the perturbation occurs slowly, nevertheless its total growth may be considerable, so that it becomes quite large.

One can similarly treat perturbations of the second and third types listed above. But the laws for the damping of these perturbations can also be found without detailed calculations by starting from the following simple arguments.

If over a small region of the matter (with linear dimensions l) there is a rotational perturbation with velocity δv, the angular momentum of this region is $\sim (\varepsilon/c^2)l^3 \cdot l \cdot v$. During the expansion of the universe l increases proportionally with a, while ε decreases like a^{-3} (in the case of $p = 0$) or like a^{-4} (for $p = \varepsilon/3$). From the conservation of angular momentum, we have

$$\delta v = \text{const} \quad \text{for} \quad p = \varepsilon/3, \ \delta v \sim \frac{1}{a} \text{ for } p = 0. \qquad (115.22)$$

Finally, the energy density of gravitational waves must decrease during the expansion of the universe like a^{-4}. On the other hand, this density is expressed in terms of the perturbation of the metric by $\sim k^2 (h_\alpha^\beta)^2$, where $k = n/a$ is the wave vector of the perturbation. It then follows that the amplitude of perturbations of the type of gravitational waves decreases with time like $1/a$.

§ 116. Homogeneous spaces

The assumption of homogeneity and isotropy of space determines the metric completely (leaving free only the sign of the curvature). Considerably more freedom is left if one assumes only homogeneity of space, with no additional symmetry. Let us see what metric properties a homogeneous space can have.

We shall be discussing the metric of a space at a given instant of time t. We assume that the space-time reference system is chosen to be synchronous, so that t is the same synchronized time for the whole space.

Homogeneity implies identical metric properties at all points of the space. An exact definition of this concept involves considering sets of coordinate transformations that transform the space into itself, i.e. leave its metric unchanged: if the line element before transformation is

$$dl^2 = \gamma_{\alpha\beta}(x^1, x^2, x^3) \, dx^\alpha \, dx^\beta,$$

then after transformation the same line element is

† A more detailed analysis taking into account the small pressure $p(\varepsilon)$ shows that the possibility of neglecting the pressure requires that one satisfy the condition $u\eta n/c \ll 1$ (where $u = c\sqrt{dp/d\varepsilon}$ is the small sound velocity); it is easy to show that in this case also it coincides with the condition $\lambda/L \gg 1$. Thus, growth of the perturbation always occurs if $\lambda/L \gg 1$.

$$dl^2 = \gamma_{\alpha\beta}(x'^1, x'^2, x'^3)\, dx'^\alpha\, dx'^\beta,$$

with the same functional dependence of the $\gamma_{\alpha\beta}$ on the new coordinates. A space is homogeneous if it admits a set of transformations (a *group of motions*) that enables us to bring any given point to the position of any other point. Since space is three-dimensional the different transformations of the group are labelled by three independent parameters.

Thus, in euclidean space the homogeneity of space is expressed by the invariance of the metric under parallel displacements (translations) of the cartesian coordinate system. Each translation is determined by three parameters—the components of the displacement vector of the coordinate origin. All these transformations leave invariant the three independent differentials (dx, dy, dz) from which the line element is constructed.

In the general case of a noneuclidean homogeneous space, the transformations of its group of motions again leave invariant three independent linear differential forms, which do not, however, reduce to total differentials of any coordinate functions. We write these forms as

$$e_\alpha^{(a)}\, dx^\alpha, \tag{116.1}$$

where the Latin index (a) labels three independent vectors (coordinate functions); we call these vectors a frame.

Using the forms (116.1) we construct a spatial metric invariant under the given group of motions:

$$dl^2 = \eta_{ab}(e_\alpha^{(a)} dx^\alpha)(e_\beta^{(b)} dx^\beta), \tag{116.2}$$

i.e. the metric tensor is

$$\gamma_{\alpha\beta} = \eta_{ab} e_\alpha^{(a)} e_\beta^{(b)}. \tag{116.3}$$

where the coefficients η_{ab}, which are symmetric in the indices a and b, are functions of the time.

Thus we arrive at a "triad" representation of the spatial metric using a triple of coordinate vectors; all the formulas obtained in § 98 are applicable to this representation. The choice of basis vectors is dictated by the symmetry properties of the space and, in general, these basis vectors are not orthogonal (so that the matrix η_{ab} is not diagonal).

As in § 98, along with the triple of vectors $e_\alpha^{(a)}$ we introduce the reciprocal triple of vectors $e_{(a)}^\alpha$, for which

$$e_{(a)}^\alpha e_\alpha^{(b)} = \delta_a^b; \quad e_{(a)}^\alpha e_\beta^{(a)} = \delta_\beta^\alpha. \tag{116.4}$$

In the three-dimensional case, the relation between the two vector triples can be written explicitly:

$$e_{(1)} = \frac{1}{V}\, \mathbf{e}^{(2)} \times \mathbf{e}^{(3)}, \quad \mathbf{e}_{(2)} = \frac{1}{V} \mathbf{e}^{(3)} \times \mathbf{e}^{(1)}, \quad \mathbf{e}_{(3)} = \frac{1}{V}\, \mathbf{e}^{(1)} \times \mathbf{e}^{(2)}, \tag{116.5}$$

where

$$v = |e_\alpha^{(a)}| = \mathbf{e}^{(1)} \cdot \mathbf{e}^{(2)} \times \mathbf{e}^{(3)},$$

and $\mathbf{e}_{(a)}$ and $\mathbf{e}^{(a)}$ should be regarded as cartesian vectors with components $e_{(a)}^\alpha$ and $e_\alpha^{(a)}$ respectively. The determinant of the metric tensor (116.3) is

$$\gamma = \eta v^2, \tag{116.6}$$

where η is the determinant of the matrix η_{ab}.

The invariance of the differential forms (116.1) means that

$$e_\alpha^{(a)}(x)dx^\alpha = e_\alpha^{(a)}(x')dx'^\alpha,$$ (116.7)

where the $e_\alpha^{(a)}$ on the two sides of the equation are the same functions of the old and new

coordinates, respectively. Multiplying this equation by $e_{(a)}^\beta(x')$, setting $dx'^\beta = \dfrac{\partial x'^\beta}{\partial x^\alpha}dx^\alpha$,

and comparing coefficients of the same differentials dx^α, we find

$$\frac{\partial x'^\beta}{\partial x^\alpha} = e_{(a)}^{(\beta)}(x')e_\alpha^{(a)}(x).$$ (116.8)

These equations are a system of differential equations that determine the functions $x'^\beta(x)$ for a given frame.† In order to be integrable, the equations (116.8) must satisfy identically the conditions

$$\frac{\partial^2 x'^\beta}{\partial x^\alpha \partial x^\gamma} = \frac{\partial^2 x'^\beta}{\partial x^\gamma \partial x^\alpha}.$$

Calculating the derivatives, we find

$$\left[\frac{\partial e_{(a)}^\beta(x')}{\partial x'^\delta} e_{(b)}^\delta(x') - \frac{\partial e_{(b)}^\beta(x')}{\partial x'^\delta} e_{(a)}^\delta(x') \right] e_\gamma^{(b)}(x)e_\alpha^{(a)}(x) = e_{(a)}^\beta(x') \left(\frac{\partial e_\gamma^{(a)}(x)}{\partial x^\alpha} - \frac{\partial e_\alpha^{(a)}(x)}{\partial x^\gamma} \right).$$

Multiplying both sides of the equations by $e_{(d)}^\alpha(x)e_{(c)}^\gamma(x)e_\beta^{(J)}(x')$ and shifting the differentiation from one factor to the other by using (116.4), we get for the left side:

$$e_\beta^{(f)}(x') \left[\frac{\partial e_{(d)}^\beta(x')}{\partial x'^\delta} e_{(c)}^\delta(x') - \frac{\partial e_{(c)}^\beta(x')}{\partial x'^\delta} e_{(d)}^\delta(x') \right] = e_{(c)}^\beta(x')e_{(d)}^\delta(x') \left[\frac{\partial e_\beta^{(f)}(x')}{\partial x'^\delta} - \frac{\partial e_\delta^{(f)}(x')}{\partial x'^\beta} \right]$$

and for the right, the same expression in the variable x. Since x and x' are arbitrary, these expressions must reduce to constants:

$$\left(\frac{\partial e_\alpha^{(c)}}{\partial x^\beta} - \frac{\partial e_\beta^{(c)}}{\partial x^\alpha} \right) e_{(a)}^\alpha e_{(b)}^\beta = C^c_{ab}.$$ (116.9)

The constants C^c_{ab} are called the *structure constants* of the group. Multiplying by $e_{(c)}^\gamma$ we can rewrite (116.9) in the form

† For a transformation of the form $x'^\beta = x^\beta + \xi^\beta$, where the ξ^β are small quantities, we obtain from (116.8) the equations

$$\frac{\partial \xi^\beta}{\partial x^\alpha} = \xi^\gamma \frac{\partial e_{(a)}^\beta}{\partial x^\gamma} e_\alpha^{(a)}.$$ (116.8a)

The three linearly independent solutions of these equations, $\xi_b^\beta (b = 1, 2, 3)$, determine the infinitesimal transformations of the group of motions of the space. The vectors $\xi_{(b)}^\beta$ are called the *Killing vectors*. (Cf. the footnote on p. 291).

$$e^\alpha_{(a)}\frac{\partial e^\gamma_{(b)}}{\partial x^\alpha} - e^\beta_{(b)}\frac{\partial e^\gamma_{(a)}}{\partial x^\beta} = C^c{}_{ab}e^\gamma_{(c)}. \tag{116.10}$$

These are the required conditions for homogeneity of the space. The expression on the left side of (116.9) conicides with the definition of the quantities $\lambda^c{}_{ab}$ (98.10), which are therefore constants.

As we see from their definition, the structure constants are antisymmetric in their lower indices:

$$C^c{}_{ab} = - C^c{}_{ba}. \tag{116.11}$$

We can obtain still another condition on them by noting that (116.10) can be written in the form of commutation relations

$$[X_a, X_b] \equiv X_a X_b - X_b X_a = C^c{}_{ab} X_c \tag{116.12}$$

for the linear differential operators†

$$X_a = e^\alpha_{(a)}\frac{\partial}{\partial x^\alpha}. \tag{116.13}$$

Then the condition mentioned above follows from the identity

$$[[X_a, X_b], X_c] + [[X_b, X_c], X_a] + [[X_c, X_a], X_b] = 0$$

(the *Jacobi identity*), and has the form:

$$C^e{}_{ab}C^d{}_{ec} + C^e{}_{bc}C^d{}_{ea} + C^e{}_{ca}C^d{}_{eb} = 0. \tag{116.14}$$

It is a definite advantage to use, in place of the three-index constants C^c_{ab} a set of two-index quantities, obtained by the dual transformation

$$C^c{}_{ab} = e_{abd}C^{dc}, \tag{116.15}$$

where $e_{abc} = e^{abc}$ is the unit antisymmetric symbol (with $e_{123} = +1$). With these constants the commutation relations (116.12) are written as

$$e^{abc}X_b X_c = C^{ad}X_d. \tag{116.16}$$

The property (116.11) is already taken into account in the definition (116.15), while property (116.14) takes the form

$$e_{bcd}C^{cd}C^{ba} = 0. \tag{116.17}$$

We also mention that the definition (116.9) for the quantities C^{ab} can be written in vector form:

$$C^{ab} = - \frac{1}{V}\mathbf{e}^{(a)}\cdot\boldsymbol\nabla \times \mathbf{e}^{(b)}, \tag{116.18}$$

---

† In the mathematical theory of continuous groups (*Lie groups*) the operators X_a satisfying conditions of the form (116.12) are called the *generators* of the group. We mention, however (to avoid confusion when comparing with other presentations), that the systematic theory usually starts from operators defined using the Killing vectors:

$$X_a = \xi^\alpha_{(a)}\frac{\partial}{\partial x^\alpha}.$$

where the vector operations are again carried out as if the coordinates x^α were cartesian.

The choice of the three frame vectors in the differential forms (116.1) (and with them the operators X_a) is, of course, not unique. They can be subjected to any linear transformation with constant coefficients:

$$\mathbf{e}_{(a)} = A_a^b \, \mathbf{e}_{(b)}. \tag{116.19}$$

The quantities η_{ab} and C^{ab} behave like tensors with respect to such transformations.

The conditions (116.17) are the only ones that the structure constants must satisfy. But among the constants admissible by these conditions, there are equivalent sets, in the sense that their difference is related to a transformation of the type (116.19). The question of the classification of homogeneous spaces reduces to determining all nonequivalent sets of structure constants. This can be done, using the "tensor" properties of the quantities C^{ab}, by the following simple method (C. G. Behr, 1962).

The unsymmetric "tensor" C^{ab} can be resolved into a symmetric and an antisymmetric part. We denote the first by n^{ab}, and we express the second in terms of its "dual vector" a_c:

$$C^{ab} = n^{ab} + e^{abc} a_c. \tag{116.20}$$

Substitution of this expression in (116.17) leads to the condition

$$n^{ab} a_b = 0. \tag{116.21}$$

By means of the transformations (116.19) the symmetric "tensor" n^{ab} can be brought to diagonal form: let n_1, n_2, n_3 be its eigenvalues. Equation (116.21) shows that the "vector" a_b (if it exists) lies along one of the principal directions of the "tensor" n^{ab}, the one corresponding to the eigenvalue zero. Without loss of generality we can therefore set $a_b = (a, 0, 0)$. Then (116.21) reduces to $an_1 = 0$, i.e. one of the quantities a or n_1 must be zero. The commutation relations (116.12) take the form:

$$[X_1, X_2] = -aX_2 + n_3 X_3,$$
$$[X_2, X_3] = n_1 X_1, \tag{116.22}$$
$$[X_3, X_1] = n_2 X_2 + aX_3.$$

The only remaining freedom is a change of sign of the operators X_a and arbitrary scale transformations of them (multiplication by constants). This permits us simultaneously to change the sign of all the n_a and also to make the quantity a positive (if it is different from zero). We can also make all the structure constants equal to ± 1, if at least one of the quantities a, n_2, n_3 vanishes. But if all three of these quantities differ from zero, the scale transformations leave invariant the ratio $a^2/n_2 n_3$.†

Thus we arrive at the following list of possible types of homogeneous spaces; in the first column of the table we give the roman numeral by which the type is labelled according to the Bianchi classification (L. Bianchi, 1918):‡

---

† Strictly speaking, to conform to the "tensor" properties of the C^{ab} we should introduce a factor $\sqrt{\eta}$ in the definition (116.15) (cf. the remarks in § 83, which describe how the antisymmetric unit tensor must be defined with respect to arbitrary transformations of coordinates). We shall not, however, enter on these details here: for our purpose we can extract the law of transformation of the structure constants directly from eqs. (116.22).

‡ The parameter a runs through all positive values. The corresponding types are actually one-parameter families of different groups. Their assignment to types VI and VII is a matter of convention.

Type	a	$n^{(1)}$	$n^{(2)}$	$n^{(3)}$
I	0	0	0	0
II	0	1	0	0
VII_0	0	1	1	0
VI_0	0	1	−1	0
IX	0	1	1	1
VIII	0	1	1	−1
V	1	0	0	0
IV	1	0	0	1
VII_a	a	0	1	1
$\left.\begin{array}{l}\text{III } (a=1)\\[2pt]\text{VI}_a(a\neq1)\end{array}\right\}$	a	0	1	−1

Type I is euclidean space; all components of the spatial curvature tensor vanish (cf. formula (116.24) below). In addition to the trivial case of a galilean metric, this also includes the time-dependent metric that will be considered in the next section.

Type IX contains, as a special case, the space of constant positive curvature. It is obtained if, in the line element (116.2), we set $\eta_{ab} = \delta_{ab}/4\lambda$, where λ is a positive constant. In fact, calculating from (116.24) with $C^{11} = C^{22} = C^{33} = 1$ (the structure constants for type IX) gives $P_{(a)(b)} = \tfrac{1}{2}\delta_{ab}$ and so

$$P_{\alpha\beta} = P_{(a)(b)}e_\alpha^{(a)}e_\beta^{(b)} = 2\lambda\gamma_{\alpha\beta},$$

which just corresponds to such a space [cf. (111.3)].

In analogous fashion the space of constant negative curvature is contained as a special case in type V. Setting $\eta_{ab} = \delta_{ab}/\lambda$ and calculating $P_{(a)(b)}$ from (116.24) with $C^{23} = -C^{32} = 1$, we get

$$P_{(a)(b)} = -2\delta_{ab}, \qquad P_{\alpha\beta} = -2\lambda\gamma_{\alpha\beta},$$

which corresponds to a constant negative curvature.

Finally, we show how the Einstein equations for a universe with a homogeneous space reduce to a system of ordinary differential equations containing only functions of the time. To do this we must resolve the spatial components of four-vectors and four-tensors along the triad of basis vectors of the space:

$$R_{(a)(b)} = R_{\alpha\beta}e_{(a)}^\alpha e_{(b)}^\beta, \qquad R_{0(a)} = R_{0\alpha}e_{(a)}^\alpha, \qquad u^{(a)} = u^\alpha e_\alpha^{(a)},$$

where all these quantities are now functions of t alone; the scalar quantities, the energy density ε and the pressure of the matter p, are also functions of the time.

According to (97.11)–(97.13), the Einstein equations in the synchronous frame are expressed in terms of three-dimensional tensors $\varkappa_{\alpha\beta}$ and $P_{\alpha\beta}$. For the first we have simply

$$\varkappa_{(a)(b)} = \dot{\eta}_{ab}, \qquad \varkappa_{(a)}^{(b)} = \dot{\eta}_{ac}\eta^{cb} \qquad (116.23)$$

(the dot denotes differentiation with respect to t). The components of $P_{(a)(b)}$ can be expressed in terms of the quantities η_{ab} and the structure constants of the group by using (98.14). After replacing the three-index symbols $\lambda^a_{bc} = C^a_{bc}$ by two-index symbols C^{ab} and various transformations† we get:

$$P^{(b)}_{(a)} = \frac{1}{2\eta} \{2C^{bd}C_{ad} + C^{db}C_{ad} + C^{bd}C_{da} - C^d{}_d(C^b{}_a + C_a{}^b) +$$

$$+ \delta^b_a[(C^d{}_d)^2 - 2C^{df}C_{df}]\}. \tag{116.24}$$

Here, in accordance with the general rule,

$$C_a{}^b = \eta_{ac}C^{cb}, \quad C_{ab} = \eta_{ac}\eta_{bd}C^{cd}.$$

We also note that the Bianchi identities for the three-dimensional tensor $P_{\alpha\beta}$ in the homogeneous space take the form

$$P^c{}_bC^b{}_{ca} + P^c{}_aC^b{}_{cb} = 0. \tag{116.25}$$

The final expressions for the triad components of the Ricci four-tensor are:‡

$$R^0_0 = -\tfrac{1}{2}\dot{\varkappa}^{(a)}_{(a)} - \tfrac{1}{4}\varkappa^{(b)}_{(a)}\varkappa^{(a)}_{(b)},$$

$$R^0_{(a)} = -\tfrac{1}{2}\varkappa^{(c)}_{(b)}(C^b{}_{ca} - \delta^b_a C^d{}_{dc}),$$

$$R^{(b)}_{(a)} = -\frac{1}{2\sqrt{\eta}}(\sqrt{\eta}\,\varkappa^{(b)}_{(a)} - P^{(b)}_{(a)}). \tag{116.26}$$

We emphasize that in setting up the Einstein equations there is thus no need to use explicit expressions for the basis vectors as functions of the coordinates.

§ 117. The flat anisotropic model

The adequacy of the isotropic model for the description of the later stages of evolution of the universe is in itself no reason for expecting that it will be equally suited for the description of early stages of the evolution, near the time singularity. This question will be discussed in detail in § 119, but in this and the following sections we shall, as a preliminary, consider solutions of the Einstein equations that also have a time singularity, but of an essentially different type from the Friedmann singularity.

We shall look for solutions in which, for a suitable choice of reference frame, all components of the metric tensor are functions of a single variable, the time $x^0 = t$.§ Such a question was already considered in § 109, where, however, we treated only the case when the determinant $|g_{\alpha\beta}| = 0$. As was shown in § 109, in such a case, we can, without loss of generality, set all the $g_{0\alpha} = 0$.

By a transformation of the variable t according to $\sqrt{g_{00}}\,dt \to dt$, we can then make g_{00} equal to unity, so that we obtain a synchronous reference system, in which

$$g_{00} = 1, \quad g_{0\alpha} = 0, \quad g_{\alpha\beta} = -\gamma_{\alpha\beta}(t). \tag{117.1}$$

† In which we use the formulas

$$\eta_{ad}\eta_{bc}\eta_{cf}e^{def} = \eta e_{abc}, \quad e_{abf}e^{cdf} = \delta^c_a\delta^d_b - \delta^d_a\delta^c_b.$$

‡ The covariant derivatives $\varkappa^\beta_{\alpha;\gamma}$ which enter in R^0_α, are transformed using the formulas derived in the footnote on p. 315.

§ To simplify writing of formulas, we set $c = 1$ in §§ 117–118.

We can now use the Einstein equations of gravitation in the form (97.11)–(97.13). Since the quantities $\gamma_{\alpha\beta}$, and with them the components of the three-dimensional tensor $\varkappa_{\alpha\beta} = \dot{\gamma}_{\alpha\beta}$, do not depend on the coordinates x^{α}, $R_{0\alpha} \equiv 0$. For the same reason, $P_{\alpha\beta} \equiv 0$, and as a result the equations of the gravitational field in vacuum reduce to the following system:

$$\dot{\varkappa}_{\alpha}^{\alpha} + \tfrac{1}{2}\varkappa_{\alpha}^{\beta}\varkappa_{\beta}^{\alpha} = 0, \tag{117.2}$$

$$\frac{1}{\sqrt{\gamma}}\,(\sqrt{\gamma}\,\varkappa_{\alpha}^{\beta})^{\cdot} = 0 \tag{117.3}$$

From equation (117.3) it follows that

$$\sqrt{\gamma}\,\varkappa_{\alpha}^{\beta} = 2\lambda_{\alpha}^{\beta}, \tag{117.4}$$

where the λ_{α}^{β} are constants. Contracting on the indices α and β, we then obtain

$$\varkappa_{\alpha}^{\alpha} = \frac{\dot{\gamma}}{\gamma} = \frac{2}{\sqrt{\gamma}}\,\lambda_{\alpha}^{\alpha},$$

from which we see that $\gamma = \text{const} \cdot t^2$. Without loss of generality we may set the constant equal to unity (simply by a scale change of the coordinates x^{α}); then $\lambda_{\alpha}^{\alpha} = 1$. Substitution of (117.4) into equation (117.2) now gives the relation

$$\lambda_{\alpha}^{\beta}\lambda_{\beta}^{\alpha} = 1 \tag{117.5}$$

which relates the constants λ_{α}^{β}.

Next we lower the index β in equations (117.4) and rewrite them as a system of ordinary differential equations:

$$\dot{\gamma}_{\alpha\beta} = \frac{2}{t}\,\lambda_{\alpha}^{\gamma}\gamma_{\gamma\beta}. \tag{117.6}$$

The set of coefficients $\lambda_{\alpha}^{\gamma}$ may be regarded as the matrix of some linear substitution. By a suitable linear transformation of the coordinates x^1, x^2, x^3 (or, what is equivalent, of $g_{1\beta}, g_{2\beta}, g_{3\beta}$), we can in general bring this matrix to diagonal form. We shall denote its principal values (roots of the characteristic equation) by p_1, p_2, p_3, and assume that they are all real and distinct (concerning other cases, cf. below); the unit vectors along the corresponding principal axes are $\mathbf{n}^{(1)}$, $\mathbf{n}^{(2)}$ and $\mathbf{n}^{(3)}$. Then the solution of equations (117.6) can be written in the form

$$\gamma_{\alpha\beta} = t^{2p_1} n_{\alpha}^{(1)} n_{\beta}^{(1)} + t^{2p_2} n_{\alpha}^{(2)} n_{\beta}^{(2)} + t^{2p_3} n_{\alpha}^{(3)} n_{\beta}^{(3)} \tag{117.7}$$

(where the coefficients of the powers of t have been made equal to unity by a suitable scale change of the coordinates). Finally, choosing the directions of the vectors $\mathbf{n}^{(1)}$, $\mathbf{n}^{(2)}$, $\mathbf{n}^{(3)}$ as the directions of our axes (we call them x, y, z), we bring the metric to the final form (E. Kasner, 1922):

$$ds^2 = dt^2 - t^{2p_1}\,dx^2 - t^{2p_2}\,dy^2 - t^{2p_3}\,dz^2. \tag{117.8}$$

Here p_1, p_2 and p_3 are any three numbers satisfying the two relations

$$p_1 + p_2 + p_3 = 1, \qquad p_1^2 + p_2^2 + p_3^2 = 1 \tag{117.9}$$

[the first of these follows form $-g = t^2$, and the second—from (117.5)].

The three numbers p_1, p_2 and p_3 obviously cannot all have the same value. The case where two of them are equal occurs for the triples 0, 0, 1 and $-1/3, 2/3, 2/3$. In all other cases the numbers p_1, p_2 and p_3 are all different, one of them being negative and the other two positive. If we arrange them in the order $p_1 < p_2 < p_3$, their values will lie in the intervals

$$-\tfrac{1}{3} \le p_1 \le 0, \quad 0 \le p_2 \le \tfrac{2}{3}, \quad \tfrac{2}{3} \le p_3 \le 1. \tag{117.10}$$

Thus the metric (117.8) corresponds to a flat homogeneous but anisotropic space whose total volume increases (with increasing t) proportionally to t; the linear distances along two of the axes (y and z) increase, while they decrease along the third axis (x). The moment $t = 0$ is a singular point of the solution; at this point the metric has a singularity which cannot be eliminated by any transformation of the reference system where the invariants of the four-dimensional curvature tensor vanish at infinity.†

The metric (117.8) is an exact solution of the Einstein equations for empty space. But near the singular point, for small t, it remains an approximate solution (to terms of highest order in $1/t$) even for the case of matter uniformly distributed in space. How and how fast the matter density changes is then determined simply by its equations of motion in the given gravitational field, while the influence of the matter back on the field is negligible. As $t \to \infty$, the matter density tends to infinity—in accordance with the physical character of the singularity (cf. problem 3).

PROBLEMS

1. Find the solution of equations (117.6) corresponding to the case where the characteristic equation of the matrix λ_α^β has one real (p_3) and two complex ($p_{1,2} = p' \pm ip''$) roots.

Solution: In this case the parameter x^0, on which all the quantities depend, must have spacelike character; we denote it by x. Correspondingly, we must now have $g_{00} = -1$ in (117.1). Equations (117.2)–(117.3) are not changed.

The vectors $\mathbf{n}^{(1)}, \mathbf{n}^{(2)}$ in (117.7) become complex: $\mathbf{n}^{(1,2)} = (\mathbf{n}' \pm i\mathbf{n}'')/\sqrt{2}$, where $\mathbf{n}', \mathbf{n}''$ are unit vectors. Choosing the axes x^1, x^2, x^3 along the directions $\mathbf{n}', \mathbf{n}'', \mathbf{n}^{(3)}$, we obtain the solution in the form

$$-g_{11} = g_{22} = x^{2p'} \cos\left(2p'' \ln \frac{x}{a}\right), \qquad g_{12} = -x^{2p'} \sin\left(2p'' \ln \frac{x}{a}\right),$$

$$g_{33} = -x^{3p_3}, \qquad -g = -g_{00}|g_{\alpha\beta}| = x^2,$$

† The only exception is the case where $p_1 = p_2 = 0, p_3 = 1$. For these values we simply have a flat spacetime; by the transformation $t \sinh z = \zeta$, $t \cosh z = \tau$ we can bring the metric (117.8) to galilean form. We also make reference to a paper that gives a variety of exact solutions of various types for the Einstein equations in vacuum, depending on a larger number of variables. B. K. Harrison, *Phys. Rev.* **116**, 1285 (1959).

A solution of the type of (117.8) also exists in the case where the parameter is spacelike; we need only make the appropriate changes of sign, for example,

$$ds^2 = x^{2p_1} dt^2 - dx^2 - x^{2p_2} dy^2 - x^{2p_3} dz^2.$$

However, in this case there also exist solutions of another type, which occur when the characteristic equation of the matrix λ_α^β in equations (117.6) has complex or coincident roots (cf. problems 1 and 2). For the case of a timelike parameter t, these solutions are not possible, since the determinant g in them would not satisfy the necessary condition $g < 0$.

where a is a constant (which can no longer be eliminated by a scale change along the x axis, without changing other coefficients in the expressions given). The numbers, p_1, p_2, p_3 again satisfy the relations (117.9), where the real number p_3 is either $> -1/3$ or > 1.

2. Do the same for the case where two of the roots coincide ($\bar{p_2} = p_3$).

Solution: We know from the general theory of linear differential equations that in this case the system (117.6) can be brought to the following canonical form:

$$\dot{g}_{11} = \frac{2p_1}{x}\, g_{11}, \quad \dot{g}_{2\alpha} = \frac{2p_2}{x}\, g_{2\alpha}, \quad \dot{g}_{3\alpha} = \frac{2p_3}{x}\, g_{3\alpha} + \frac{\lambda}{x}\, g_{2\alpha}, \quad a = 2, 3,$$

where λ is a constant. If $\lambda = 0$, we return to (117.8). If $\lambda \neq 0$, we can put $\lambda = 1$; then

$$g_{11} = -x^{2p_1}, \quad g_{2\alpha} = a_\alpha x^{2p_2}, \quad g_{3\alpha} = b_\alpha x^{2p_2} + a_\alpha x^{2p_2} \ln x.$$

From the condition $g_{23} = g_{32}$, we find that $a_2 = 0$, $a_3 = b_2$. By an appropriate choice of scale along the x^2 and x^3 axes, we finally bring the metric to the following form:

$$ds^2 = -dx^2 - x^{2p_1}(dx^1)^2 \pm 2x^{2p_2}\, dx^2\, dx^3 + x^{2p_2} \ln x/a\, (dx^3)^2.$$

The numbers p_1, p_2 can have the values 1, 0 or $-1/3$, 2/3.

3. In the neighbourhood of the singular point $t = 0$ find the law of variation with time of the density of matter distributed uniformly in the space with metric (117.8).

Solution: Neglecting the back influence of the matter on the field, we start from the hydrodynamic equations of motion

$$\frac{1}{\sqrt{-g}}\frac{\partial}{\partial x^i}(\sqrt{-g}\,\sigma u^i) = 0,$$

$$(p + \varepsilon)u^k\left(\frac{\partial u_i}{\partial x^k} - \frac{1}{2}u^i\frac{\partial g_{ki}}{\partial x^i}\right) = -\frac{\partial p}{\partial x^i} - u_i u^k\frac{\partial p}{\partial x^k}, \tag{1}$$

contained in the equations $T^k_{i;k} = 0$ (cf. *Fluid Mechanics*, § 125). Here σ is the entropy density; near the singularity we must use the ultrarelativistic equation of state, $p = \varepsilon/3$, and then $\sigma \sim \varepsilon^{3/4}$.

We denote the time factors in (117.8) by $a = t^{p_1}$, $b = t^{p_2}$, $c = t^{p_3}$. Since all quantities depend only on the time, and $\sqrt{-g} = abc$, eqs. (1) give

$$\frac{d}{dt}(abcu_0\varepsilon^{3/4}) = 0, \quad 4\varepsilon\frac{du_\alpha}{dt} + u_\alpha\frac{d\varepsilon}{dt} = 0.$$

Then

$$abcu_0\varepsilon^{3/4} = \text{const}, \tag{2}$$

$$u_\alpha\varepsilon^{1/4} = \text{const}. \tag{3}$$

According to (3) all the covariant components u_α are of the same order of magnitude. Of the contravariant components, the largest (for $t \to 0$) is $u^3 = u_3/c^2$. Keeping only the largest terms in the identity $u_i u^i = 1$, we therefore get $u_0^2 \approx u_3 u^3 = (u_3)^2/c^2$, and then, from (2) and (3):

$$\varepsilon \sim \frac{1}{a^2 b^2}, \quad u_\alpha \sim \sqrt{ab},$$

or

$$\varepsilon \sim t^{-2(p_1+p_2)} = t^{-2(1-p_3)}, \quad u_\alpha \sim t^{(1-p_3)/2}. \tag{4}$$

As it should, ε tends to infinity as $t \to 0$ for all values of p_3 except $p_3 = 1$, in accordance with the fact that a singularity in the metric with exponents (0, 0, 1) is not physical.

The validity of the approximations used is verified by estimating the components T_i^k that were dropped on the right sides of Eqs. (117.2)–(117.3). Their main terms are:

$$T_0^0 \sim \varepsilon u_0^2 \sim t^{-(1+p_3)}, \quad T_1^1 \sim \varepsilon \sim t^{-2(1-p_3)},$$

$$T_2^2 \sim \varepsilon u_2 u^2 \sim t^{-(1+2p_2-p_3)}, \quad T_3^3 \sim \varepsilon u_3 u^3 \sim t^{-(1+p_3)}.$$

They all actually increase more slowly as $t \to 0$ than the left sides of the equations, which grow like t^{-2}.

§ 118. Oscillating regime of approach to a singular point

Using the model of a universe with a homogeneous space of type IX we shall study the time singularity of the metric which has oscillatory character (V. A. Belinskii, E. M. Lifshitz, I. M. Khalatnikov, 1968). We shall see in the next section that such a stituation has a very general significance.

We shall be interested in the behaviour of the model near the singularity (which we choose as the time origin $t = 0$). As in the Kasner solution treated in § 117, the presence of matter does not affect the qualitative properties of this behaviour. For simplicity we shall therefore assume at first that the space is empty.

We take the quantities $\eta_{ab}(t)$ in (116.3) to be diagonal, denoting the diagonal elements by a^2, b^2, c^2; we here denote the three frame vectors $\mathbf{e}^1, \mathbf{e}^2, \mathbf{e}^3$ by $\mathbf{l}, \mathbf{m}, \mathbf{n}$. Then the spatial metric is written as:

$$\gamma_{\alpha\beta} = a^2 l_\alpha l_\beta + b^2 m_\alpha m_\beta + c^2 n_\alpha n_\beta. \tag{118.1}$$

For a space of type IX the structure constants are:†

$$C^{11} = C^{22} = C^{33} = 1 \tag{118.2}$$

$$(\text{and } C^1_{23} = C^2_{31} = C^3_{12} = 1).$$

From (116.26) it can be seen that for these constants and a diagonal matrix η_{ab}, the components $R^0_{(a)}$ of the Ricci tensor vanish identically in the synchronous reference system. According to (116.26), the nondiagonal components $P_{(a)(b)}$ also vanish. The remaining components of the Einstein equations give the following system of equations for the functions a, b, c:

$$\frac{(\dot{abc})\cdot}{abc} = \frac{1}{2a^2 b^2 c^2} [(b^2 - c^2)^2 - a^4],$$

$$\frac{(\dot{abc})\cdot}{abc} = \frac{1}{2a^2 b^2 c^2} [(a^2 - c^2)^2 - b^4], \tag{118.3}$$

† The frame vectors corresponding to these constants are:

$$\mathbf{l} = (\sin x^3, -\cos x^3 \sin x^1, 0), \quad \mathbf{m} = (\cos x^3, \sin x^3 \sin x^1, 0), \quad \mathbf{n} = (0, \cos x^1, 1).$$

The coordinates run through values in the ranges $0 \le x^1 \le \pi, 0 \le x^2 \le 2\pi, 0 \le x^3 \le 4\pi$. The space is closed, and its volume

$$V = \int \sqrt{\gamma} \, dx^1 dx^2 dx^3 = abc \int \sin x^1 dx^1 dx^2 dx^3 = 16\pi^2 abc.$$

When $a = b = c$ it goes over into a space of constant positive curvature with radius of curvature $2a$.

$$\frac{(ab\dot{c})\cdot}{abc} = \frac{1}{2a^2b^2c^2}[(a^2 - b^2)^2 - c^4],$$

$$\frac{\ddot{a}}{a} + \frac{\ddot{b}}{b} + \frac{\ddot{c}}{c} = 0. \tag{118.4}$$

[Equations (118.3) are the equation set $R^1_{(1)} = R^2_{(2)} = R^3_{(3)} = 0$; equation (118.4) is the equation $R^0_0 = 0$.

The time derivatives in the system (118.3)–(118.4) take on a simpler form if we introduce in place of the functions a, b, c, their logarithms α, β, γ:

$$a = e^\alpha, \qquad b = e^\beta, \qquad c = e^\gamma, \tag{118.5}$$

and in place of t, the variable τ:

$$dt = abc \, d\tau. \tag{118.6}$$

Then:

$$2\alpha_{,\tau\tau} = (b^2 - c^2)^2 - a^4,$$

$$2\beta_{,\tau\tau} = (a^2 - c^2)^2 - b^4, \tag{118.7}$$

$$2\gamma_{,\tau\tau} = (a^2 - b^2)^2 - c^4;$$

$$\tfrac{1}{2}(\alpha + \beta + \gamma)_{,\tau\tau} = \alpha_{,\tau}\beta_{,\tau} + \alpha_{,\tau}\gamma_{,\tau} + \beta_{,\tau}\gamma_{,\tau}, \tag{118.8}$$

where the subscript, τ denotes differentiation with respect to τ. Adding equations (118.7) and replacing the sum of second derivatives on the left by (118.8). we obtain:

$$\alpha_{,\tau}\beta_{,\tau} + \alpha_{,\tau}\gamma_{,\tau} + \beta_{,\tau}\gamma_{,\tau} = \tfrac{1}{4}(a^4 + b^4 + c^4 - 2a^2b^2 - 2a^2c^2 - 2b^2c^2). \tag{118.9}$$

This relation contains only first derivatives, and is a first integral of the equations (118.7).

Equations (118.3)–(118.4) cannot be solved exactly in analytic form, but permit a detailed qualitative study in the neighbourhood of the singular point.

We note first that if the right sides of equations (118.3) or (118.7) were absent, the system would have an exact solution, in which

$$a \sim t^{p_l}, \qquad b \sim t^{p_m}, \qquad c \sim t^{p_n}, \tag{118.10}$$

where p_l, p_m, and p_n are numbers connected by the relations

$$p_l + p_m + p_n = p_l^2 + p_m^2 + p_n^2 = 1 \tag{118.11}$$

[the analog of the Kasner solution (117.8) for a homogeneous flat space]. We have denoted the exponents by p_l, p_m, p_n, without assuming any order of their size; we shall retain the notation p_1, p_2, p_3 of § 117 for the triple of numbers arranged in the order $p_1 < p_2 < p_3$ and taking on values in the intervals (117.10) respectively. These numbers can be written in parametric form as

$$p_1(u) = \frac{-u}{1 + u + u^2}, \qquad p_2(u) = \frac{1 + u}{1 + u + u^2}, \qquad p_3(u) = \frac{u(1 + u)}{1 + u + u^2}. \tag{118.12}$$

All the different values of the p_1, p_2, p_3 (preserving the assumed order) are obtained if the parameter u runs through values in the range $u \geq 1$. The values $u < 1$ are reduced to this same region as follows:

$$p_1\left(\frac{1}{u}\right) = p_1(u), \qquad p_2\left(\frac{1}{u}\right) = p_3(u), \qquad p_3\left(\frac{1}{u}\right) = p_2(u). \qquad (118.13)$$

Figure 25 shows graphs of p_1, p_2 and p_3 as functions of $1/u$.

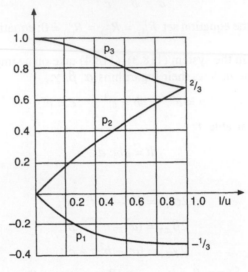

FIG. 25.

We assume that the right sides of (118.7) are small over some time interval, so that they can be neglected and we have a "Kasner-like" regime (118.10). Such a situation cannot continue indefinitely as $t \to 0$, since some of the terms must be increasing. Thus, if the negative exponent belongs to the function $a(t)$ (so that $p_l = p_1 < 0$), then the disruption of the Kasner regime arises from the terms a^4; the other terms will drop off as t decreases.

Keeping only these terms on the right of (118.7), we get the system of equations

$$\alpha_{,\tau\tau} = -\tfrac{1}{2}e^{4\alpha}, \quad \beta_{,\tau\tau} = \gamma_{,\tau\tau} = \tfrac{1}{2}e^{4\alpha}. \qquad (118.14)$$

The solution of these equations should describe the evolution of the metric from the "initial" state,† in which it is described by (118.10) with a definite choice of exponents (with $p_l < 0$); let $p_l = p_1$, $p_m = p_2$, $p_n = p_3$, so that

$$a = t^{p_1}, \quad b = t^{p_2}, \quad c = t^{p_3}$$

(the proportionality coefficients in these expressions can be set equal to unity without any loss of generality in the result obtained below). Since $abc = t$, $\tau = \ln t + \text{const}$, the initial conditions for (118.14) are formulated in the form

$$\alpha_{,\tau} = p_1, \quad \beta_{,\tau} = p_2, \quad \gamma_{,\tau} = p_3.$$

The first of equations (118.14) has the form of the equation of one-dimensional motion of a particle in the field of an exponential potential wall, where α plays the role of the coordinate. In this analogy, to the initial Kasner regime there corresponds a free motion with constant velocity $\alpha_{,\tau} = p_l$. After reflection from the wall, the particle will again move freely with the

† We recall that we are considering the evolution of the metric as $t \to 0$, so that the initial conditions correspond to a later, and not to an earlier time.

opposite sign of the velocity: $\alpha_{,\tau} = -p_l$. We also note that from equations (118.14), $\alpha_{,\tau} + \beta_{,\tau}$ = const, and $\alpha_{,\tau} + \gamma_{,\tau}$ = const, hence we find that $\beta_{,\tau}$ and $\gamma_{,\tau}$ take the values

$$\beta_{,\tau} = p_2 + 2p_1, \quad \gamma_{,\tau} = p_3 = 2p_1.$$

Now determining α, β, γ, and then t, using (118.6), we find

$$e^\alpha \sim e^{-p_1\tau}, \quad e^\beta \sim e^{(p_2+2p_1)\tau}, \quad e^\gamma \sim e^{(p_3+2p_1)\tau}, \quad t \sim e^{(1+2p_1)\tau},$$

i.e.

$$a \sim t^{p'_l}, \quad b \sim t^{p'_m}, \quad c \sim t^{p'_n},$$

where

$$p'_l = \frac{|p_1|}{1 - 2|p_1|}, \quad p'_m = \frac{p_2 - 2|p_1|}{1 - 2|p_1|}, \quad p'_n = \frac{p_3 - 2|p_1|}{1 - 2|p_1|}. \tag{118.15}$$

Thus the action of the perturbation results in the replacement of one Kasner regime by another, with the negative power of t shifting from one direction $\mathbf{l}$ to another $\mathbf{m}$: if we had $p_l < 0$, then now $p'_m < 0$. In the course of the shift the function $a(t)$ goes through a maximum, and the function $b(t)$ through a minimum: the quantity $b(t)$ which previously decreased, begins to increase, the rising function $a(t)$ starts to drop, while $c(t)$ continues to fall off. The perturbation itself [the terms a^4 in (118.7)] which were rising, begin to drop and are damped out. Further evolution of the perturbation leads in an analogous way to increase of the perturbation due to the terms b^4 in (118.7), another shift of Kasner regime, etc.

The rule for the shift of exponents (118.15) is conveniently represented by the parametrization (118.12): if

$$p_l = p_1(u), \quad p_m = p_2(u), \quad p_n = p_3(u),$$

then

$$p'_l = p_2(u - 1), \quad p'_m = p_1(u - 1), \quad p'_n = p_3(u - 1). \tag{118.16}$$

The larger of the two positive exponents remains positive.

This process of shifting of Kasner regimes is the key to understanding the character of the evolution of the metric as we approach the singularity.

The successive shifts (118.16) with bouncing of the negative exponent between the directions $\mathbf{l}$ and $\mathbf{m}$ continues so long as the integral part of the initial value of u is not exhausted and u becomes less than unity. The values $u < 1$ is transformed into $u > 1$, according to (118.13); at that moment, either p_l or p_m is negative, while p_n becomes the smaller of the two positive numbers ($p_n = p_2$). The following series of shifts will now bounce the negative exponent between the directions $\mathbf{n}$ and $\mathbf{l}$ or between $\mathbf{n}$ and $\mathbf{m}$. For an arbitrary (irrational) initial value of u, the process of shifting continues without end.

In an exact solution the exponents p_1, p_2 and p_3 will, of course, lose their literal meaning. We note that the smearing that this produces in the definition of these numbers (and with them, the parameter u) although small, makes it meaningless to consider any special (e.g. rational) values of u. This is why it is only meaningful to consider those regularities that are typical of the general case with irrational u.

Thus the process of evolution of the model toward the singular point is made up of successive series of oscillations, during which the distances along two of the space axes oscillate, while they fall off monotonically along the third; the volume drops off according to a law that is close to $\sim t$. In going from one series to the next, the direction along which there is a monotonic dropoff of distances shifts from one direction to another. Asymptotically the order of these shifts takes on the character of a random process. The order of succession of lengths of successive series of oscillations (i.e. the number of Kasner epochs in each series)† also takes on stochastic character.

The successive series of oscillations crowd together as we approach the singularity. An infinite number of oscillations are contained between any finite world time t and the moment $t = 0$. The natural variable for describing the time behaviour of this evolution is not the time t, but its logarithm, $\ln t$, in terms of which the whole process of approach to the singular point is spread out to $-\infty$.

In the solution described here we simplified the problem from the very start, assuming that the matrix $\eta_{ab}(t)$ in (116.3) is diagonal. The inclusion of nondiagonal components of η_{ab} does not change the oscillatory character of the evolution of the metric or the law (118.16) for the shift of the exponents p_l, p_m, p_n of successive Kasner epochs. It leads, however, to the appearance of an additional property: the shifting of the exponents is also accompanied by a change in the directions of the axes to which the exponents apply.‡

§ 119. The time singularity in the general cosmological solution of the Einstein equations

It has already been pointed out that the adequacy of the Friedmann model for describing the present state of the Universe is no basis for expecting that it is equally suitable for describing its early stages of evolution. One may even ask to what extent the existence of a time singularity is a necessary general property of cosmological models, or whether it is really caused by the specific simplifying assumptions on which the models are based (in particular, symmetry assumptions). We emphasize that when we speak of a singularity, we have in mind a physical singularity, a place where the density of matter and the invariants of the curvature tensor become infinite.

If the presence of the singularity were independent of these assumptions, it would mean that it is inherent not only to special solutions, but also to the general solution of the Einstein equations. The criterion of generality of the solution is the number of "physically arbitrary" functions contained in it. In the general solution the number of such functions must be sufficient for arbitrary assignment of initial conditions at any chosen time [4 for empty space, 8 for space filled with matter (cf. § 95)].§

---

† If the "initial" value of the parameter u is $u_0 = k_0 + x_0$ (where k_0 is an integer and $x_0 < 1$) the length of the first series of oscillations will be k_0, while the initial value for the next series will be $u_1 = 1/x_0 \equiv k_1 + x_1$, etc. From this it is easy to conclude that the lengths of successive series will be given by the elements $k_0, k_1, k_2 \ldots$ of the expansion of u_0 in an infinite (for irrational u_0) continued fraction

$$u_0 + k_0 + \cfrac{1}{k_1 + \cfrac{1}{k_2 + \cfrac{1}{k_3 + \ldots}}}$$

The succession of values of further elements in such an expansion is distributed according to statistical laws.

‡ Concerning this and other details of the behaviour of homogeneous cosmological models of this type, cf. V. A. Belinskii, E. M. Lifshitz and I. M. Khalatnikov, *Adv. in Physics* **19**, 525 (1970); *Adv. in Physics*, **31**, 639 (1982).

§ We emphasize that for a system of nonlinear equations, such as the Einstein equations, the notion of a general solution is not unambiguous. In principle more than one general integral may exist, each of the integrals covering not the entire manifold of conceivable initial conditions, but only some finite part of it. The existence of a general solution possessing a singularity does not therefore preclude the existence of other general solutions that do not have a singularity. For example, there is no reason to doubt the existence of a general solution, without singularities, that describes a stable, isolated body with not too large mass.

Finding such a solution in exact form, for all space and over all time, is clearly impossible. But to solve our problem it is sufficient to study the form of the solution only near the singularity.

The singularity of the Friedmann solution is characterized by the fact that the vanishing of spatial distances occurs according to the same law in all directions. This type of singularity is not sufficiently general: it is typical of a class of solutions that contain only three physically arbitrary coordinate functions (cf. the problem in § 113). We also note that these solutions exist only for a space filled with matter.

The singularity of oscillatory type, considered in the preceding section, is general in character—there exists a solution of the Einstein equations with such a singularity and containing the required set of arbitrary functions. We shall briefly describe the way to construct such a solution, without going into the details of the calculation.†

As in the homogeneous model (§ 118) the regime of approach to the singularity consists of successive series of Kasner epochs replacing one another. In the course of each epoch, the main terms (in $1/t$) in the spatial metric tensor (in the synchronous reference frame) have the form (118.1) with functions of the time a, b, c, from (118.10), but the vectors $\mathbf{l}$, $\mathbf{m}$, $\mathbf{n}$ are now arbitrary functions of the space coordinates (and not definite functions, as in the homogeneous model). Now the p_l, p_m, p_n are also functions (and not simply numbers), again related by the formulas (118.11). The metric constructed in this way satisfies the equations $R_0^0 = 0$ and $R_\alpha^\beta = 0$ for the vacuum field (at least, their leading terms) over a certain finite time interval. The equations $R_\alpha^0 = 0$ give rise to three relations (not containing the time) that must be imposed on the arbitrary functions of the space coordinates that are contained in the $\gamma_{\alpha\beta}$. These relations connect ten different functions: three components of each of the three vectors $\mathbf{l}$, $\mathbf{m}$, $\mathbf{n}$ and one function of the time exponents (since the functions p_l, p_m, p_n are connected by two conditions (118.11). In determining the number of physically arbitrary functions we must also consider that the synchronous reference frame still admits transformations of the three spatial coordinates that do not affect the time. Thus the metric contains altogether $10 - 3 - 3 = 4$ arbitrary functions—just the number one should have in the general solution for the field in vacuum.

The replacement of one Kasner regime by another occurs (just as in the homogeneous model) because of the presence in three of the six equations $R_\alpha^\beta = 0$ of terms which, with decreasing t, increase more rapidly than the others, thus playing the role of a perturbation destroying the Kasner regime. These equations, in the general case, have a form differing from (118.14) only in a factor depending on the space coordinates $(\mathbf{l} \cdot \text{curl } \mathbf{l}/\mathbf{l} \cdot \mathbf{m} \times \mathbf{n})$ on their right sides (where it is understood that of the three exponents p_l, p_m, p_n, p_l is negative).‡ Since, however, eqs. (118.14) are a system of ordinary differential equations with respect to the time, this difference in no way affects their solution or the law that follows from this solution concerning the shift of Kasner exponents (118.16), or any of the further conclusions presented in § 118.§

---

† They can be found in V.A. Belinskii, E. M. Lifshitz and I. M. Khalatnikov, *Adv. in Physics.* **31**, 639 (1982).

‡ For the homogeneous model, this factor coincides with the square of the structure constant C^{11}, and is constant by definition.

§ If we impose on the arbitrary functions in the solution the supplementary condition $\mathbf{l} \cdot \text{curl } \mathbf{l} = 0$, the oscillations disappear, and the Kasner regime will continue right up to the point $t = 0$. Such a solution, however, contains one function fewer than is required in the general case.

The degree of generality of the solution is not reduced if matter is introduced: the matter "inscribes" itself on the metric through all of the four new coordinate functions that are needed for assigning the initial distribution of its density and its three velocity components. The energy-momentum tensor T_i^k of the matter introduces terms in the field equations that are of higher order in $1/t$ than the leading terms (in precise analogy to what was shown in problem 3 of § 117 for the plane homogeneous model).

Thus the existence of a singular point in the time is an extremely general property of solutions of the Einstein equations, where the process of approach to the singular point in the general case is oscillatory in character.† We emphasize that this character is not related to the presence of matter (and therefore not related to its equation of state), and is already typical of the empty space-time itself. The singularity of the monotonic, isotropic type, typical of the Friedmann solution, and dependent on the presence of matter, has only special significance. When we speak of singularities in the cosmological sense, we have in mind a singular point that is reached over all the space and not just over some limited part, as in the gravitational collapse of a finite body. But the generality of the oscillatory solution gives one a basis for assuming that the singularity reached by a finite body in its collapse below the event horizon in the comoving reference frame has this same character.

We have continually spoken of the direction of approach to the singular point as the direction of decreasing time; but in view of the symmetry of the Einstein equations under time reversal, we could with equal justification speak of approach to the singularity in the direction of increasing time. Actually, however, in view of the physical nonequivalence of future and past, there is an essential difference between these two cases in the very formulation of the question. A singularity in the future can have physical meaning only if it is reachable for arbitrary initial conditions assigned at some preceding moment of time. Clearly there is no reason why the distribution of matter and field that was reached at some moment in the process of evolution of the universe should have corresponded to the precise conditions required for the appearance of some particular solution of the Einstein equations.

Concerning the question of the type of singularity in the past, an investigation based solely on some equations of gravitation can hardly give a general answer. It is natural to think that the choice of solution corresponding to the real world is related to some profound physical requirements, whose establishment on the basis of the existing theory of gravitation alone is impossible, and whose explanation will come only as the result of a further synthesis of physical theories. In this sense it could in principle turn out that this choice corresponds to some particular (e.g. isotropic) type of singularity.

Finally, it is necessary still to make the following remark. The domain of applicability of the Einstein equations in themselves is in no way limited in the region of small distances or large densities of matter in the sense that the equations in this limit do not lead to any internal contradictions (in contrast, for example, to the classical equations of electrodynamics). In this sense, the investigation of the singularity of the space-time metric on the basis of the Einstein equations is entirely correct. There is no doubt, however, that in this limit quantum phenomena become important, and we can say nothing about them in the present state of the theory. Only in a furture synthesis of the theory of gravitation and quantum theory will it become clear which of the results of classical theory remain meaningful. At the same time

---

† The fact that a singular point exists in the general solution of the Einstein equations was first shown by R. Penrose, 1965, by topological methods, which, however, do not enable one to establish the specific analytic character of the singularity. A presentation of these methods and the theorems obtained using them is given in R. Penrose, *Structure of Space-Time*, W. A. Benjamin, N.Y., 1968.

there is no doubt that the very fact of the appearance of a singularity in the solutions of the Einstein equations (both in their cosmological aspect and for the collapse of finite bodies) has a profound physical meaning. One must not forget that the achievement during gravitational collapse of densities so enormous and, yet, still undoubtedly correctly described by the classical theory of gravitation, is enough for us to speak of a physically "singular" phenomenon.

INDEX

Printed and bound by CPI Group (UK) Ltd, Croydon, CR0 4YY

03/10/2024

01040331-0006